Digital Crossroads

Digital Crossroads

Telecommunications Law and Policy in the Internet Age

second edition

Jonathan E. Nuechterlein and Philip J. Weiser

The MIT Press
Cambridge, Massachusetts
London, England

MIT Press books may be purchased at special quantity discounts for business or sales promotional use. For information, please email special_sales@mitpress.mit.edu or write to Special Sales Department, The MIT Press, 55 Hayward Street, Cambridge, MA 02142.

This book was set in Sabon by Toppan Best-set Premedia Limited, Hong Kong. Printed and bound in the United States of America.

Library of Congress Cataloging-in-Publication Data

Nuechterlein, Jonathan E.
Digital crossroads: telecommunications law and policy in the Internet age / Jonathan E. Nuechterlein and Philip J. Weiser.—Second edition.
p. cm
Includes bibliographical references and index.
ISBN 978-0-262-51960-1 (pbk.: alk. paper)
1. Telecommunication policy—United States. 2. Telecommunication—Deregulation—United States. 3. Internet. 4. United States. Telecommunications Act of 1996. I. Weiser, Philip J. II. Title.
HE7781.N84 2013
384.0973—dc23
2012051747

10 9 8 7 6 5 4 3 2 1

To Stephanie and Heidi and to our own next generation: Zoe, Kate, Aviva, and now Sammy, too

Contents

Acknowledgments

Many friends and colleagues in the telecommunications policy field played essential roles in both editions of this book. One of our goals has been to integrate the distinct perspectives of the various communities of academics and practitioners involved in shaping telecommunications policy. To that end, we enlisted the help of reviewers from each such community, and they responded with very constructive comments on one or (usually) more chapters of the manuscript. For this second edition, we greatly benefited from suggestions and insights provided by Coleman Bazelon, Lynn Charytan, Norton Cutler, Paul de Sa, Pierre de Vries, Ellen Goodman, Jack Goodman, Russ Hanser, Dale Hatfield, Lisa Hone, Samir Jain, Paul Margie, Preston Padden, John Ryan, Doug Sicker, Bryan Tramont, Joe Waz, Steve Williams, and Heather Zachary. For the first edition, we remain indebted to Brad Bernthal, Brad Berry, Marc Blitz, Dan Brenner, Craig Brown, Paul Campos, Lynn Charytan, Nestor Davidson, Alison Eid, Gerry Faulhaber, John Flynn, Jon Frankel, Ray Gifford, Paul Glist, Ellen Goodman, Melissa Hart, John Harwood, Dale Hatfield, Roy Hoffinger, Clare Huntington, Samir Jain, Alfred Kahn, Marty Katz, Sarah Krakoff, Bill Lake, Jeff Lanning, Mark Loewenstein, Tom Lookabaugh, Marsha MacBride, Richard Metzger, Melissa Newman, Tom Olson, Adam Peters, Dorothy Raymond, Bill Richardson, Joel Rosenbloom, Patrick Ryan, Scott Savage, Doug Sicker, Jim Speta, Jane Thompson, Nan Thompson, Molly van Houweling, Kathy Wallman, Kevin Werbach, Steve Williams, and Chris Yoo. Of course, none of these reviewers will agree with every proposition in the finished book. Also, as is always the case, we bear responsibility for any remaining errors.

We are also grateful to Elizabeth Murry and John Covell, our editors at the MIT Press for the first and second editions, respectively. Also at MIT, Virginia Crossman and Anne Barva provided very helpful editorial support. We wish to convey our profuse thanks to Kim Morris, who cite-checked the entire second edition, and Madelaine Maior, who coordinated and contributed to the outside review process.

We are also greatly indebted to our wonderful team of proofreaders and indexers: Katherine Brownlow, Adele Wilson Burrous, Tyler Cox, Michelle Hersh, Amanda Levin, Laura Littman, Maggie Macdonald, Katherine Nelson, Courtney Marin Shephard, Kimberly Skaggs, Ryan Tharp, and Anna Uhls.

We also wish to express deep gratitude to our respective institutions—Wilmer Cutler Pickering Hale & Dorr LLP and the University of Colorado—for supporting us in this project. WilmerHale once again proved its storied commitment to public interest projects by enthusiastically encouraging our undertaking from conception to time-consuming completion. CU similarly provided an intellectually supportive environment, a wealth of research assistants, and a platform—the Silicon Flatirons Center for Law, Technology, and Entrepreneurship—for the exchange of ideas between academics and practitioners throughout the field.

Last but certainly not least, both of us owe an enormous debt to our families. Our parents gave us the curiosity, drive, and discipline needed to undertake this project. And our wives, Stephanie Marcus and Heidi Wald, showed remarkably good humor in bearing with us while we secluded ourselves on countless nights and weekends to write and revise the manuscript. We promise to let at least a few years go by before we contemplate a third edition.

Preface to the Second Edition

Over the past few years, readers have given us many helpful suggestions on the now dated first edition of this book. Three have stood out: the book should be (1) more concise, (2) less wireline-centric, and (3) more current. We wrote this new edition to meet all three challenges. We have relentlessly streamlined our prose and have eliminated more than 200 pages in the process. To reflect the increasing centrality of mobile broadband services, we have moved up our discussion of spectrum reform and the wireless industry closer to the beginning of the book (chapters 3 and 4). Last but not least, we have updated every section of every chapter and refocused our analysis on the developments and policy trends of the past eight years.

The result falls somewhere between a new edition and a new book, and every chapter extensively analyzes new policy debates that had not even arisen—either in the same form or at all—when we put pens down on the first edition in late 2004. For example, this edition contains entirely new discussions relating to

- industry consolidation and the SBC–AT&T and Verizon–MCI mergers (chapter 2);
- the FCC's complex and sometimes conflicted approach to the deregulation of legacy wireline services, including special access (chapter 2);
- the FCC's initiatives to address the "spectrum crunch" for mobile broadband services, Congress's complementary 2012 incentive-auction legislation, and proposals for spectrum-sharing arrangements between carriers and public agencies (chapter 3);
- the LightSquared/GPS controversy and the need for a modernized spectrum policy for addressing interference issues (chapter 3);
- the FCC's data-roaming and 700 MHz handset interoperability initiatives (chapter 4);
- the AT&T/T-Mobile and Verizon Wireless–SpectrumCo transactions (chapter 4);

• the rise of content delivery networks and the disintermediation of traditional Internet backbone providers (chapter 5);
• the Comcast–BitTorrent controversy, the FCC's *Open Internet Order*, and proposals to reclassify broadband Internet access under Title II of the Communications Act (chapter 6);
• the Comcast–Level 3 peering dispute, the Comcast–Netflix streaming-video dispute, and the fault lines each reveals in the FCC's net neutrality regime (chapters 6 and 7);
• the FCC's 2011 overhaul of intercarrier-compensation rules and its embrace of bill-and-keep as the guiding principle for public switched telephone network (PSTN) traffic exchanges (chapter 7);
• the fast-emerging debate about whether prescriptive interconnection rules, for voice traffic or any other kind, will be necessary once the PSTN sunsets and all communications converge onto unified IP networks (chapter 7);
• the *National Broadband Plan* and the FCC's 2011 decision to refocus the federal universal-service program from rural telephony to rural broadband deployment (chapter 8);
• the Comcast–NBCU merger proceeding and the extensive remedial conditions pursued by the FCC and the Department of Justice (chapter 9);
• the cable à la carte controversy and the FCC's dueling staff reports (chapter 9); and
• the rise of online video services and their ambiguous legal status under copyright and telecommunications law (chapter 9).

Despite all these changes, our overall ambitions for this book remain exactly what they were in 2004: to help nonspecialists climb the field's formidable learning curve as efficiently as possible and to make substantive contributions to the major policy debates within the field. We also repeat our original disclaimer and offer. First, the disclaimer: one of us (Jon Nuechterlein) remains a practicing lawyer; his clients include broadband providers; and he has advised those clients in many of the proceedings described in this book. This book reflects our views, not those of any client, and we have taken pains to keep the discussion as evenhanded as possible. Now the offer: as before, our readers should still feel free to contact either of us with substantive reactions to the text. We can be reached, respectively, at jon.nuechterlein@wilmerhale.com and phil.weiser@colorado.edu.

J.N. and P.W.
Washington, D.C., and Boulder, Colorado
December 2012

Preface to the First Edition

This book is about the regulation of competition in the telecommunications industry. Our purpose is twofold. First, we aim to help nonspecialists climb this field's formidable learning curve as efficiently as possible. Second, we seek to make substantive contributions to the major policy debates within the field. We have given equal priority to these two quite distinct objectives, and we believe that telecommunications policy veterans as well as newcomers to the field will benefit from our analysis.

Each of us knows from firsthand experience about this discipline's intellectual barriers to entry. When we first met more than eight years ago in the Justice Department, we were generalist lawyers who knew very little about the nuts and bolts of telecommunications regulation. But we needed to become specialists quickly because our respective jobs—in the Solicitor General's office and the Antitrust Division—required us to explain and help formulate federal telecommunications policy in the wake of the Telecommunications Act of 1996. After learning the field the hard way—through years of intensive firsthand immersion—we resolved to shorten the process for others by writing a book that clearly explains telecommunications competition policy in the Internet era. This book is the result.

We offer a few points of clarification up front about the nature of our project. First, this book addresses competition policy issues in the United States, with particular emphasis on the regulatory dimensions of (i) competition in wireline and wireless telephone service, (ii) competition among rival platforms for broadband Internet access and video-programming distribution, and (iii) the Internet's transformation of every corner of the telecommunications industry, particularly through the emergence of voice over Internet protocol (VoIP). Except where relevant to our discussion of competition policy, we do not address issues concerning, for example, consumer privacy, government regulation of broadcasting content, or international matters.

Second, while lawyers and law students may find this book particularly useful, it is not a typical "law book" designed exclusively for a legal audience. We examine legal issues and court decisions insofar as they have significantly altered the shape of the telecommunications industry. Our analysis of the industry's deep structure, including its peculiar economic characteristics and rapidly changing technology, drives our analysis of legal developments, not vice versa. We have ordered the discussion this way precisely because we expect that many of our readers will be lawyers, whose understanding of this field is often distorted by too much exposure to legal details and too little exposure to the economics and technology of the industry.

Third, we have worked hard to explain in clear, accessible prose the many complexities of telecommunications regulation. To balance the needs of a general readership with the needs of readers with more specialized interests, we have included detailed endnotes for each chapter.

Fourth, we hope to earn the trust of our readers by remaining objective and strictly nonpartisan throughout our analysis. This is no small challenge. When we searched for a reliable explanatory book in the early years after passage of the 1996 Act, we were told that no such book *could* exist because the only people who truly understood the nuts and bolts of telecommunications competition policy were already beholden to one industry faction or another. One of our central ambitions in writing this book is to disprove that proposition. For the sake of full disclosure, one of us, Jon Nuechterlein, is a practicing lawyer in this field, and his clients currently include incumbent local exchange carriers. From late 1995 through early 2001, however, he represented the FCC itself, often against the interests of these incumbent carriers. Phil Weiser is a law professor who does not generally represent private telecommunications clients, although, most recently, he consulted with the consumer plaintiffs (against the same incumbent carriers) in the *Trinko* Supreme Court case discussed in the final chapter. No opinions expressed in this book should be attributed to any of these clients, past or present; these views are ours alone. In all events, our analysis focuses on how, as a threshold matter, policymakers should conceptualize the basic trade-offs presented in current policy debates. With a few exceptions, we steer clear of advocating any precise outcome for such debates.

Fifth, technological and marketplace developments in the telecommunications industry move very quickly, and there is of course no way to keep any discussion of this industry fully current once the manuscript has been sent to press. For this reason and others, readers should not view this book as a source of specific legal or investment advice. We have nonetheless sought to guard against premature obsolescence by focusing as much on the first principles of telecommunications policy in the Internet age as on the fleeting controversies of the moment. And we plan future editions that will take full account of the changing face of the industry.

Finally, readers should feel free to contact either of us with substantive reactions to the text. Those reactions will prove helpful in revising the text for future editions. We can be reached, respectively, at jon.nuechterlein@wilmerhale.com and phil.weiser@colorado.edu.

J.N. and P.W.
Washington, D.C., and Boulder, Colorado
November 2004

1

The Big Picture

The word *telecommunications*, a twentieth-century amalgam of Greek and Latin roots, literally means the art of conveying information "from a distance." For millennia, people had to rely on messengers to perform this task, which was as costly per message sent as it was time-consuming. When the Greeks repelled the Persians at Marathon in 490 BC, the legendary messenger Pheidippides could not shout the good news back to Athens, for it was 26.2 miles away, nor could he call anyone up, for there were no telephones; instead, he had to run. Several hours later Pheidippides arrived in Athens, gasped out the news, and died of exhaustion. There had to be a better way—but for the next 2,300 years or so sending a flesh-and-blood messenger on a trip was the normal method of delivering information from one place to another.

One dramatic break from that convention appeared in postrevolutionary France. In the early 1790s, Claude Chappe invented a system of relaying *visual* messages hundreds of miles across the French countryside over a network of towers spaced about 20 miles apart. For example, someone in Paris would manipulate the mechanical arms at the top of one of these towers to spell out a coded message; his counterpart in another tower 20 miles away would read the message and duplicate it for the benefit of the person manning the next tower down the line, and so on. Weather permitting, this system could be used to transmit a message from Paris to the border of Germany within ten minutes. Other societies had used visual communications techniques, such as ship-to-ship semaphore signals and such land-based mechanisms as smoke signals and torches. But the French, quickly joined by several other European countries, improved greatly on the idea by developing a nationwide communications *network*. By the Napoleonic era of the early 1800s, the French had developed a sprawling tower system radiating from Paris to such far-flung destinations as Cherbourg, Boulogne, Strasbourg, Marseille, Toulouse, and Bayonne.[1]

Before long, these networks, which could be used only in daylight and good weather, confronted the first revolutionary technology in telecommunications: the telegraph. Developed by Samuel Morse in the 1830s, the telegraph sent encoded messages down copper wires by rapidly opening and closing electrical circuits. The telegraph dominated telecommunications until it too was gradually replaced by the next revolutionary technology: the telephone system, invented by Alexander Graham Bell in 1876 and widely deployed throughout much of the United States within a generation. In the 1890s, Guglielmo Marconi exploited the discovery that the airwaves, like copper wires, could propagate electromagnetic signals, and so "radio" technology was born.

Today, although precise definitions differ, the term *telecommunications* is broadly defined as the transmission of information by means of electromagnetic signals: over copper wires, coaxial cable, fiber-optic strands, or the airwaves. This technology—which underpins radio and television, wireless and wireline telephone service, the World Wide Web, email, instant messaging, streaming video, and every other Internet application—is the sine qua non of contemporary global culture.

Telecommunications is also a uniquely volatile field economically, technologically, and politically. The disputes that arise within and among the different sectors of the telecommunications industry, often in response to these rapidly changing conditions, have triggered some of the fiercest public policy wars ever waged. In the United States, the very structure of the industry turns on the decisions of various governmental agencies, most notably the Federal Communications Commission (FCC). The policy questions answered at the FCC and elsewhere influence not just how we communicate with one another and what video programming we watch (and how we watch it), but the fate of an industry that in the United States alone accounts for hundreds of billions of dollars in annual revenues and more than a million employees.

Nonspecialists, however, confront a vexing conundrum in trying to learn this field: to comprehend the whole of telecommunications policy, one must first understand its parts; but to understand the parts, one must first comprehend the whole. This chapter aims to overcome these difficulties by covering the major themes of telecommunications competition policy at a high enough level, and with as little jargon as possible, to help nonspecialists understand how each of the policy issues discussed in subsequent chapters fits into the big picture. To this end, the first part of this chapter introduces the peculiar economic characteristics of the telecommunications industry that drive most forms of regulation in the United States. The second part then introduces the market-transforming phenomenon of *convergence*—the competitive offering of familiar communications services through nontraditional technologies, such as the provision of video programming over telephone lines and telephone service over cable TV wires. As we discuss, the key to ultimate convergence is the greatest invention of the late twentieth century: the Internet. You can access the Internet through any number of dissimilar fixed-line and mobile technologies; and, once there, you can exploit virtually every form of communication ever created, including written correspondence (email and instant messaging), phone calls (voice over Internet protocol, or VoIP), newspapers and books (the Web), radio broadcasts (streaming audio), and television entertainment (streaming video).

I. Economic Principles

Why does competition in the telecommunications world—unlike, say, competition in the world of home appliance manufacturing—present public policy issues of such importance and complexity? Answering that question requires a familiarity with the basic economic phenomena that regulators have long cited to justify regulatory intervention in telecommunications markets. At the risk of some oversimplification, we sum up the most important of these phenomena in three concepts: *network effects*, *economies of scale and density*, and *monopoly leveraging*. We address each of these concepts in turn.

A. Network effects and interconnection

Flash back to the infancy of the U.S. telephone industry at the turn of the twentieth century. Different telephone companies often refused to interconnect with one another, and each had its own set of subscribers. Few consumers, of course, wanted to buy several telephones and pay subscription charges to several telephone companies simply to make sure they could reach anyone else they wished to call. Unfortunately, this was the choice many consumers faced.

Such arrangements are quite wasteful in that they misallocate society's scarce resources away from their most productive uses. To be sure, the prospect of extra profits from the successful deployment of a closed (non-interconnected) telephone network may well have encouraged some entrepreneurs to build a better product and reach customers more quickly than they otherwise would.[2] Apart from those incentive effects, however, consumers typically received little added value from multiple subscriptions that they would not have received from one subscription to a single carrier if the various networks were interconnected and exchanged traffic at reasonable rates. For the most part, consumers simply paid more money for the same thing, which meant that they had less money to spend on purchasing things of value in other markets.

In the absence of any interconnection obligation, virtually every telephone market in early-twentieth-century America reached a tipping point in which the largest network—the one with the most subscribers—became perceived as the single network that everyone had to join, and the rest withered away. The potential for certain industries to slide into monopoly in this manner illustrates an economic phenomenon known as *network effects*. In many markets, individual consumers care very little how many other consumers purchase the same products that they buy. For example, the bottle of shampoo you just bought does not become significantly more or less valuable to you as the number of other purchasers of the brand increases or falls. The telecommunications industry, like several other "network industries," is different: the value of the network to *each* user generally increases or decreases, respectively, with every addition or subtraction of *other* users to the network.

Suppose, for example, that you lived in a midwestern American city in 1900, and two noninterconnecting telephone companies were offering you service. You would have been much more inclined (all else being equal) to select the company operating 80% of the lines rather than the one operating 20% because the odds would have been much greater that the people you wished to call would be on the larger network. The absence of interconnection arrangements among rival networks thus created a cutthroat race to build the largest customer base in the shortest time frame—and then put all rivals out of business by pointing out the limited value of their smaller networks. Economies of scale—a carrier's ability to reduce its per customer costs by increasing its total number of customers—further accelerated this process by permitting larger carriers to undersell smaller ones.

By the early twentieth century, the U.S. telephone market had tipped. In most population centers, the victor was the mammoth Bell System: a corporate family of very large "operating companies" that provided local exchange (i.e., telephone) service and were eventually bound together by a long-distance network known as "Long Lines." All of the far-flung operations of the Bell System were owned by American Telephone &

Telegraph (AT&T), which maintained its own equipment-manufacturing arm (Western Electric) and for a time also held the rights to patented technologies developed by the Bell System's namesake: Alexander Graham Bell.

In the areas AT&T did not control, which typically were the less populous ones, the so-called "independent" local telephone companies vied for market share. In many cases, AT&T sought to coerce these independent companies into joining the Bell System by refusing to interconnect them to AT&T Long Lines, which was then the only long-distance network in the United States. The independent companies were in no position to build a rival long-distance network. Even if they could have cooperated to construct the needed transcontinental facilities (and done so without infringing any remaining AT&T patents), they still could not have used that shared network to send calls through to the increasing majority of Americans who were served by the noninterconnecting Bell System. As a result, without interconnection rights, these independent companies could not provide their customers with satisfactory telephone service—that is, service extending beyond the local serving area—unless they could somehow duplicate the nationwide physical infrastructure the Bell System had built up over several decades of sharp dealing and self-reinforcing good fortune. That was an economic impossibility.

AT&T's coercion of the independent companies ultimately aroused the attention of the antitrust authorities at the U.S. Department of Justice. In the Kingsbury Commitment of 1913, AT&T resolved the ensuing dispute by agreeing to interconnect its Long Lines division with these independent local companies and to curb its practice of buying up independent rivals.[3] In exchange, the government placed its effective imprimatur on AT&T's monopoly control over all U.S. telecommunications markets in which it was already dominant. This episode provides an instructive contrast to the anticompetitive conduct that ultimately led to the breakup of the Bell System 70 years later into its local and long-distance components. In 1913, AT&T used its control of the *long-distance* market to suppress other *local* carriers. As explained later in this chapter, AT&T would subsequently leverage its control of most *local* markets to suppress the *long-distance* competition that technological advances had made possible by the 1960s.

The network effects phenomenon presents different competitive questions in different industries, and reasonable people can disagree about when the government should require a firm to share access to its customer base. But when such intervention is deemed necessary, the usual solution is an *interconnection* requirement. Suppose you own a telephone network, and one of your subscribers wants to place a call to someone who subscribes to Provider X's network. If Provider X's network is larger than yours, it may have the incentives just described to refuse to interconnect, in which event your subscriber learns that the call has failed and so considers defecting to Provider X. But if the government forces Provider X to take the call onto its network and route it to the intended recipient, your customer remains satisfied, and you stay in business. Interconnection obligations work the other way as well: Provider X cannot preclude its subscribers from reaching yours.

For more than one hundred years, much of telecommunications policy has focused on the rules governing interconnection among the many conventional circuit-switched telephone companies that together make up the *public switched telephone network* (PSTN). Those rules govern, for example, what (if anything) one telephone company must pay another for access to the called party on the receiving end of a call. And they often govern the physical details of the interconnection arrangements between different telephone companies.

One critical set of interconnection arrangements, however, remains unregulated: those between the mostly private data networks that in the aggregate constitute the Internet. For example, to stream movies to consumers, Netflix contracts with a variety of Internet intermediaries such as Level 3 to transmit and store its data content across the globe. In turn, those intermediaries contract with consumer broadband providers such as Comcast and Verizon to receive the data content and deliver it to Netflix's millions of subscribers. No regulator prescribes the terms of such private contracts, known as *peering* and *transit* agreements, or requires any Internet-based network to interconnect with any other. To date, policymakers have expressed confidence that market forces (the mutual self-interest of the Internet's many constituent networks) and other pressures (including social norms and the remote threat of regulatory intervention) will generally produce efficient ways for each user of the Internet to reach every other. Whether regulators *should* play a greater role in supervising these peering and transit agreements is a key subject of chapter 7. Significantly, these Internet interconnection issues are distinct from but closely related to the debate about *net neutrality* rules, which govern whether and when consumer broadband providers may block or discriminate among particular types of Internet content. In chapter 6, we explore the complex relationship between these two regulatory topics, a relationship that the FCC largely overlooked when adopting net neutrality rules in 2010.

Although we have focused so far on physical transmission providers, network effects are endemic to information industries in general. For example, Facebook now has more subscribers by far than other social-networking sites, and it is valuable to its users precisely because so many of their friends and acquaintances are on it. And network effects can sometimes appear in more subtle forms that at first blush do not appear to implicate "networks" at all. For example, the historical dominance of Microsoft Windows in the market for desktop computer operating systems arose from network effects and specifically from what antitrust courts have called the *applications barrier to entry*.[4] At some point in the 1980s, software designers realized that more users were choosing Microsoft's operating system than the alternatives. In response, more and more applications developers created programs only for Windows, leaving would-be rivals (such as IBM) to sell operating systems that did not have as many programs designed for them and were therefore less popular. As a result, Microsoft won an increasing share of the operating system market, which in turn reinforced the software designers' predictions about the dominance of Windows and their desire to produce applications for it, often to the exclusion of applications for rival operating systems.[5]

In a variety of contexts, policymakers have disagreed about whether network effects create any problems for which the government should offer a solution. The proponents of government intervention argue that monopolization is virtually always an evil to be avoided, reasoning that monopolization of any industry necessarily produces higher consumer prices, less product variety, lower quality, and potentially less innovation.[6] Opponents of government intervention, by contrast, point to a theory of competition, first developed by economist Joseph Schumpeter, that focuses on the "creative destruction" of old incumbents by new insurgents, who are rewarded with monopolies of their own until knocked off their perch by the next round of insurgents.[7] Under this theory, the most significant competition takes place not *within* a market—in the form of price wars or incremental increases in quality—but *for the market itself:* that is, in establishing the next great invention that will displace the old monopoly with a new one. According to modern-day Schumpeterians, monopolies in the digital ecosystem are both temporary and desirable in that the prospect of even short-lived monopoly profits will encourage entrepreneurs to innovate.[8] The Schumpeterians therefore argue for strong intellectual property protection and freedom from both competition-oriented regulation and aggressive antitrust enforcement.

Of course, some monopolies in the digital ecosystem are more enduring and harmful than others because the barriers to competitive entry are much higher in some markets than others. And policymakers have traditionally concluded that those barriers are particularly high—and the rationale for prophylactic regulation therefore strongest—in the case of physical telecommunications networks that provide last-mile transmission to individual homes and businesses, as telephone and cable television networks do. The reason relates not only to network effects, but also to the topic of our next subsection: *economies of scale and density*.

B. Economies of scale and density

By themselves, interconnection obligations significantly lower the entry barriers posed by the combination of network effects and scale economies because, as discussed, they exempt a new entrant from the need to build a ubiquitous network before competing for the dominant carrier's customers. But interconnection obligations do not eliminate those entry barriers altogether. Although they reduce any advantage that incumbents derive from network effects, they do not ensure that new entrants will benefit from the enormous scale economies enjoyed by a provider with a large, established customer base.*

What are these scale economies? Any telecommunications provider contemplating the construction of a new network faces immense initial costs. For wireline carriers, these include the costs of digging trenches and laying thousands of miles of wires to reach different customer locations. For wireless carriers, the start-up costs can include the price of acquiring spectrum rights, deploying a network of cell towers, and installing (or leasing) wires to connect those cell towers to network

* Although our discussion treats network effects and scale economies as two separate phenomena, they are in fact closely related. Each describes a characteristic of markets in which, all else held constant, increasing the scale of a firm's operations improves the ratio of (1) the value of the firm's services to each customer and thus the revenues the firm can obtain from that customer to (2) the per customer cost to the firm of providing those services. Network effects improve this ratio by increasing the value of the service to each customer, whereas scale economies improve it by decreasing the per customer cost of providing that service. In the absence of regulation, each result can help favor larger-scale telecommunications firms over their smaller rivals.

switches. These costs are *fixed* in that the carrier must incur them up front before it can provide any volume of service. In many cases, these costs are also *sunk* in that the investment, once made, cannot be put to some other use—a fact that makes the investment particularly risky.[9] In contrast, once the network is up and running, the *marginal* cost of providing service to each additional customer is often tiny by comparison, particularly for wireline networks. Given these enormous fixed costs and negligible marginal costs, it is often cheaper *per customer* for a carrier to provide service to one million customers than to one thousand customers.

Closely related to such economies of scale are economies of density. The latter are best explained by way of example. Imagine a 1,000-unit beach condominium complex that is both distant from any telephone company switching station and, because of zoning restrictions, isolated from other buildings. If the fixed costs of laying a cable from the nearest switch to that complex were $100,000, a single telephone provider serving the entire complex could spread the recovery of those costs among all 1,000 subscribers for a cost of $100 per subscriber. But if ten providers divided up that customer base equally after laying their own cables to the same complex—each digging up the streets at different times and incurring the same fixed $100,000 cost—the average cost of that tenfold effort would rise to $1,000 per customer, for each provider could spread its $100,000 in costs only over 100 customers rather than 1,000.*

Similar economic considerations explain why fixed-line broadband service is much more costly to provide in rural than urban areas. Suppose your company runs a telecommunications network on a rigidly fixed budget. Assuming the same level of revenues, would you rather build one line to each of 1,000 customers living on widely dispersed farms or 1,000 lines to one apartment building with 1,000 units? Even if the average line length were the same in each example (say, because the apartment building is farther away from your switching station than half of the farms), you would still much rather serve the apartment building because you would only have to dig up the ground once to lay the lines needed to serve those 1,000 units. If you picked the farms option, you would

* Economies of density can be roughly conceptualized as scale economies within a particular geographic area, such as the condominium complex in our example. For ease of exposition, we use the term *scale economies* broadly to include these economies of density.

need to dig up the ground many more times to lay 1,000 different cables, and you would have to pay far more to obtain the rights of way. This radical difference in per customer costs explains why the broadband revolution has come more slowly or not at all to much of rural America—and why policymakers have focused so heavily in recent years on how to encourage greater broadband deployment in rural areas (see chapter 8).

Of course, high fixed costs and low marginal costs lead to large scale economies in many industries, from auto manufacturing to applications software production, and most such industries have never been subject to pervasive schemes of prescriptive economic regulation. The difference is one of degree. In most settings, scale economies do not increase "over the entire extent of the market"[10] because at some point average costs stop declining with each incremental unit. In other settings, however, scale economies keep increasing until a provider is serving all customers in the market. In that context, because a single firm can serve the whole market (however defined) with lower overall costs per customer than can multiple firms, the market is said to be a *natural monopoly*.[11]

The government has traditionally addressed such a market by awarding a monopoly to a single firm and heavily regulating its rates, on the theory that rate regulation is the best way to keep consumer prices low. Although our discussion to this point has focused on the local telephone market, the government applied the same natural-monopoly premise to the cable TV market, too. As Judge Richard Posner once explained,

> You can start with a competitive free-for-all—different cable television systems frantically building out their grids and signing up subscribers in an effort to bring down their average costs faster than their rivals—but eventually there will be only a single company, because until a company serves the whole market it will have an incentive to keep expanding in order to lower its average costs. In the interim there may be wasteful duplication of facilities. This duplication may lead not only to higher prices to cable television subscribers, at least in the short run, but also to higher costs to other users of the public ways, who must compete with the cable television companies for access to them. An alternative procedure is to pick the most efficient competitor at the outset, give him a monopoly, and extract from him in exchange a commitment to provide reasonable service at reasonable rates.[12]

These considerations led regulators for many years to conclude—somewhat controversially in hindsight—that telephone and cable television markets were each a natural monopoly in this sense and that the "alternative procedure" Posner described would be the optimal means of ensuring dependable service at low rates in any geographic area.

This natural-monopoly premise provided a convenient solution to the problem of network effects as well. Because (the thinking went) there was no reason to allow a second or third provider into the same geographic market to begin with (as that would only dilute the incumbent's economies of scale), there was no need to worry about forcing the incumbent to interconnect with competitors. The principal exception, illustrated by the Kingsbury Commitment, seems almost trivial in this light: different geographic regions would be served by different monopoly providers of local service, and the government would ensure simply that neighboring monopolists interconnected with each other for the exchange of calls between their respective regions and that the national monopoly provider of long-distance service (AT&T) allowed all of these monopolies access to the rest of the country.

Relying on this natural-monopoly premise, many regulators not only refused to order interconnection among potential rivals in the same geographic market, but straightforwardly prohibited new market entry by granting exclusive franchises to the monopolists. In part, policymakers resisted competition not just because they believed in the economics of natural-monopoly theory, but also because they relied on regulated monopolies to advance various social policies, most notably *universal service*—the promotion of affordable telecommunications for all communities. For example, regulators deliberately kept prices for business customers high (compared to the underlying cost of serving them) as a means of cross-subsidizing affordable rates for other users, such as residential customers in rural areas where economies of scale and density are low and per line costs are therefore high.[13] As we discuss later, this scheme can work over the long term only to the extent that rival providers are barred from competing for the customers who pay the above-cost rates that subsidize low rates for others.

For many years, regulators acquiesced when AT&T's Bell System invoked universal service concerns to justify suppressing competition in all telephone-related markets, including equipment manufacturing as well as local and long-distance services. AT&T's long-lived regulatory success in this respect provides a classic case study in *public choice theory*—the economic analysis of relations between market participants and the government officials they seek to influence.[14] Public choice theory holds that private economic actors will exploit regulatory schemes to obtain or protect "rents"—that is, special benefits that arise from political influence rather than economically valuable contributions to social welfare. Successful rent seeking need not and usually does not take the form of outright bribery. Instead, private actors look for ways to match their own pecuniary interests with regulators' political goals. In the case of telephone regulation, the suppression of competition in the name of "universal service" gave both AT&T what it wanted—formally protected monopoly status—and the regulators what they wanted—an opaque scheme for underwriting low residential rates that avoided all the political costs presented by a more explicit and taxlike system.[15] The victims of such Faustian bargains are consumers, who in the long run are generally better off, at least in the aggregate, when regulators make the hard political choices necessary to remove barriers to competition.

Starting in the 1970s, policymakers began questioning the natural-monopoly assumptions that had been conventional wisdom almost since the inception of the industry.[16] This process followed a predictable pattern, as we discuss in chapter 2. After the FCC adopted rules allowing competition in the provision of telecommunications equipment, the markets that next fell prey to competition were the ones in which overall call volumes were so huge and the incumbent's retail prices were so far above economic cost that a competitor could efficiently build a rival network and earn large profits even though it had only a small share of the total customer base. The first such market was for business-oriented long-distance services between major cities, a market that MCI and other firms entered in the 1970s and 1980s with the help of both microwave technology and the courts. The second was the market for so-called *access* services: the high-speed links between local networks and long-distance networks. In each case, the companies that owned the core natural-monopoly assets—the local exchanges, with their "last-mile" connections to every home and business in a given calling area—tried to thwart this nascent competition by (among other things) refusing to interconnect with the upstarts or by making interconnection unnecessarily burdensome. In each case, the U.S. government stepped in and mandated nondiscriminatory interconnection.

Finally, in the Telecommunications Act of 1996, also known as the "1996 Act," Congress seemed to dispense with the natural-monopoly premise altogether. It abolished all exclusive franchises, ordered all telephone companies to interconnect with any requesting carrier, and declared all "local exchange" markets—in addition to the long-distance and "access" markets—open for competition. But Congress could not repeal the laws of economics. In many settings, it remained commercially infeasible for new wireline competitors to build brand-new telephone networks bridging the last mile to all of their subscribers' buildings. The main exception to this rule lay in some local exchange markets—such as densely populated, downtown business districts—where high volumes of voice and data traffic enabled new entrants to exploit fiber-optic technology by building telecommunications networks all the way to their customers. In less densely populated areas, however, such as many suburbs and most rural areas, call volumes could not support the efficient construction and operation of wholly duplicative telephone networks replete with thousands upon thousands of fixed connections to all homes and businesses.

As discussed in chapter 2, Congress attempted to address that concern in part by granting new entrants qualified rights to *lease capacity* on the facilities owned by the incumbent telephone company at regulated rates, thus enabling them to "participate" in the incumbent's economies of scale by availing themselves of the same low per unit costs. For the ensuing decade, the telecommunications industry was consumed with bitter arguments about how best to implement those leasing rules.

Those debates now seem almost antiquated because they have been largely overshadowed by the rise of technological alternatives to traditional wireline telephone service. Few residential consumers today buy telephone service from competitive providers that lease last-mile lines from incumbent telephone companies. As discussed later in this chapter, however, roughly three in ten American households have "cut the cord" by relying entirely on their mobile phones for voice communications. Tens of millions more subscribe to VoIP services offered by cable companies and Internet-based providers such as Vonage and Skype. Meanwhile, satellite providers such as DISH and DirecTV and telephone companies such as Verizon and AT&T now provide subscription video services in competition with traditional cable television companies.

In short, the natural-monopoly rationale underlying traditional regulation appears to have succumbed to commercial and technological developments, at least in retail consumer markets within most population centers. That does not mean, however, that these natural-monopoly concerns have disappeared altogether. For example, as discussed in chapter 2, regulators still view incumbent telephone companies (telcos), in some contexts, as monopolists in their provision of certain business-line services known as "special access." And as discussed in chapters 5 and 6, the FCC now tends to view most regional markets for fixed-line broadband Internet access as duopolies, dominated in a typical area by the legacy telco and cable incumbents. From a competitive perspective, a duopoly is better than a monopoly, but it is hardly optimal.[17]

Some industry observers fear that these consumer broadband markets may become even more concentrated over time, ultimately veering in many areas toward a new cable monopoly for broadband pipes into the home.[18] They predict that as consumers come to expect ever-faster speeds over their broadband connections, cable incumbents, with their higher-capacity pipes, will begin pulling away from telcos for market share. And they fear that the telcos, if they can hope at best for a fraction of the customers in any given neighborhood, will be unable to justify the multi-billion-dollar network investments needed to supply a viable competitive alternative to cable. These observers cite the example of Verizon, the only major U.S. telco to deploy on a wide-scale basis fiber-optic lines all the way to individual homes. Dampening hopes for greater cable TV competition, Verizon announced in 2010 that it would begin winding down new deployments of that immensely high-capacity but costly infrastructure, known as Verizon FiOS.[19] A year later Verizon's majority-owned wireless affiliate entered into a cross-marketing arrangement with cable companies, which the Justice Department and the FCC approved in 2012 with various conditions.[20]

Critics viewed these developments as evidence that, at least in many areas, telcos will put up only tepid resistance to growing cable dominance for fixed-line broadband connections.[21] But that conclusion is subject to sharp debate. It is also by no means clear that telcos actually need to deploy fiber to the home in order to meet consumer bandwidth demands, at least in the near-to-intermediate term.[22] And, of course, consumers consider a range of service features when choosing a broadband provider, including not only speed but also price. In short, predictions about the future intensity of residential broadband competition remain highly speculative.

C. Information platforms, monopoly leveraging, and net neutrality

So far, we have addressed the regulation of *horizontal* relationships within the telecommunications industry: the relationships between competing providers of substitutable services. Now we introduce the equally complex set of issues presented by *vertical* relationships between providers of communications-related goods or services in complementary markets. Vertical relationships arise across the economy: for example, between wheat farms and bakeries or between bakeries and grocery stores. Vertical integration by a firm across adjacent markets is often desirable because it can produce significant *economies of scope*: cost efficiencies obtained by producing several products at once. In most industries, moreover, competition in each of the adjacent markets liberates these vertical relationships from the need for heavy governmental oversight. To the extent the government gets involved, it is typically through ad hoc enforcement of the antitrust laws.

The government has long treated the communications marketplace differently. To govern vertical relationships in the telecommunications industry, policymakers have relied not only on after-the-fact antitrust enforcement, but also on prescriptive regulation. The rationale for that policy choice has its roots in the same natural-monopoly premise discussed above. If there is only one provider of a given communications service in a particular locality—one local telephone company for local voice service, one cable operator for multichannel video service—then, the thinking went, the provider would have strong incentives to harm competing providers of complementary services, such as independent long-distance companies (in the case of local phone companies) or unaffiliated cable channel programmers (in the case of cable operators), in order to help its own business or that of its vertically related affiliate.

Two developments now challenge that traditional premise. First, as the natural-monopoly premise has yielded to technological breakthroughs, competition among rival providers has reduced (but not eliminated) concerns that any given provider will harm consumers by denying them efficient access to complementary applications. Consider, for example, a cable operator that offers consumers both broadband Internet access and video-programming services. Depending on complex economic factors noted below, that cable operator might have incentives to impede its customers' access to unaffiliated Internet-based video services (such as Netflix streaming video or peer-to-peer file sharing) if it fears that those services will reduce consumer demand for its own lucrative video-programming services, such as video on demand. But it will be less likely to *act* on those incentives if it faces sufficient competition from a rival broadband/video provider (such as Verizon FiOS) because frustrated customers could respond by switching to the rival. As discussed in chapter 6, a key element of the *net neutrality* debate concerns whether competition between rival broadband providers is (and will remain) sufficient to keep each broadband provider from engaging in harmful discrimination against unaffiliated Internet-based applications providers that rely on unimpeded access to broadband platforms in order to reach consumers.

Second, quite apart from the degree of competition in the broadband market, many economists today, influenced by the Chicago School of antitrust economics, take a more skeptical view of "vertical leveraging" claims than did policymakers throughout much of the twentieth century.[23] That skepticism has roots in the *one-monopoly-profit* principle first developed by nineteenth-century economist Antoine Cournot. As Cournot observed, the total profits a monopolist can earn if it seeks to leverage its monopoly in one market by monopolizing an adjacent market are often no greater than the extra profits it can earn anyway simply by charging more for the monopoly product itself. As a result, even a monopoly provider of platform services may lack strong incentives to harm unaffiliated applications providers because doing so would create no extra profits but would deprive the platform of value-enhancing applications and thus weaken the platform monopoly.

There are, however, important exceptions to this general principle. One arises where the platform provider fears that unaffiliated applications providers pose a threat to the underlying platform monopoly. The turn-of-the-millennium Microsoft antitrust case illustrates this exception. Although the Microsoft Windows platform dominated the market for desktop operating systems, Microsoft had generally cooperated with unaffiliated developers of complementary applications even when those applications competed with Microsoft's own applications. But federal antitrust authorities successfully argued that Microsoft had tried to crush Netscape in the 1990s—not because Netscape had designed an ordinary Internet browser that could run on top of Windows (and thus enhance its value), but because Microsoft feared that the Netscape browser might develop into a rival platform in its own right and thus devalue the underlying Windows monopoly.[24]

Another exception to the one-monopoly-profit phenomenon arises where the platform service is subject to price regulation. If so, the provider may well have incentives to discriminate against firms in adjacent markets because it will be unable to recoup all otherwise available monopoly profits from the sale of the platform service itself and will need to extract them instead from those other markets. This exception is sometimes called *Baxter's Law* in honor of William Baxter, the Justice Department official who cited it in the early 1980s as a reason for breaking up AT&T's Bell System. As Baxter understood, AT&T had a strong incentive to leverage its (price-regulated) monopoly in local markets to suppress competition in the adjacent long-distance market (see chapter 2). Unlike traditional telephone service, however, broadband Internet access is not subject to price regulation, and Baxter's Law is therefore generally inapplicable in that context. We return to these complex economic issues in chapter 6, which explores the legal and economic underpinnings of the net neutrality debate.

II. Technological Convergence and Statutory Obsolescence

Until recently, most forms of electronic communications fell neatly into one of two general categories: *point-to-point* communications and *broadcasting*. The first category describes the traditional world of telephone companies: the transmission of content from a person or machine to a discrete recipient. Examples include ordinary telephone calls and fax transmissions. Broadcasting involves the transmission of content to a broad community or at least anyone who cares to watch or listen. Examples include television and radio.

For most of the twentieth century, people closely identified each of these categories of service with a particular medium of transmission. In particular, they assumed that commercial point-to-point voice services (telephony) would be conveyed over the copper wires of the telephone system and that radio and television broadcasting services would be provided over the airwaves. The Communications Act of 1934 was originally written with this assumption in mind. Congress designed Title II of the Act to govern wireline "common carriers"—that is, the companies that provided telephone service indiscriminately to the public at large. And it designed Title III to govern "radio communications," a category that grew to encompass both radio and television broadcasting. Under Title III, the FCC licensed radio and television stations to broadcast programming "in the public interest" over the airwaves.

And so the world remained until the 1960s, when something peculiar happened: companies increasingly began to transmit television signals not over the airwaves, but over wires. For a long time, such "cable television" service provided no new programming; it was designed only to transmit stronger signals of conventional broadcast programming to people whose homes were too far away from a transmission tower to receive clear pictures (or any pictures). Even so, the seeming anomaly of wires being used for broadcasting threw the regulatory world into tumult, for it raised questions about how the FCC could legally follow through on its expressed intent to regulate this new creature and preempt contrary state and local regulation. After all, Title II addressed common carriage, not broadcasting, and Title III addressed use of the airwaves, not wires.

The FCC ultimately invoked the general enabling language in Title I of the Act to assert what it called *ancillary authority* to regulate this and other new technologies that substantially affect the explicit subjects of its regulatory authority. In 1984, long after the Supreme Court upheld the FCC's initial rules exercising this strikingly open-ended regulatory authority (in 1968),[25] Congress stepped in and added a new Title VI to the Communications Act to govern federal, state, and local regulation of cable television services.

A similar need for statutory reform arose in the 1980s, when "cellular" wireless technology gave consumers an altogether new means of placing telephone calls. This technology used the radio spectrum—long the province of specialized broadcasts by taxi dispatchers and police officers in addition to television and radio stations—for regular communications among members of the public at large. That development produced another anomaly unanticipated in the structure of the 1934 Act: the use of the *airwaves* to provide point-to-point common carrier services. Congress eventually patched this legal hole by adding provisions to Title III to govern the regulation of this new service.[26]

The use of radio signals to carry telephone calls and of wires to carry broadcast programming are examples of *technological convergence*: the use of different technologies to provide similar services. But the examples of convergence just discussed are tame in comparison to the upheavals triggered by the Internet. By placing a "call" over your broadband Internet connection to a distant website, you can listen, along with the citizens of Prague, to the broadcast of a Czech radio station. Or you can log onto Pandora, an Internet-based service that keeps track of the music you like and sends you—and you alone—a personalized stream of songs. Or you can use software to chat with a friend across the world through instant messaging or VoIP. Or you can cancel your cable or satellite TV service and watch video entertainment over the Internet from streaming-video services such as YouTube, Netflix, and Hulu.

Here is the critical point: you can do all these things no matter what type of broadband connection you use to reach the Internet—whether you use your telephone line, your television cable, or a mobile broadband device such as a smartphone, tablet, or wireless-enabled laptop. The Internet supports every type of communications service ever invented. And as a general matter, you can run each of those services over any high-performance fixed or mobile broadband platform, irrespective of the platform's underlying technology.

The Internet owes its birth and explosive development to digital technology, which came of age commercially in the 1980s and 1990s. As chapter 5 discusses in depth, digital technology provides concise mathematical representations of the world in the form of 1s and 0s. The software inside your computer or mobile device decodes and converts those 1s and 0s into everything from voice conversations to photographs to documents to Prague radio broadcasts. The Internet, in turn, is a conceptual aggregation of many individual networks, most of them privately owned, that use a common protocol and addressing scheme—the *Internet protocol* (IP)—for transporting packets of 1s and 0s among computers and other smart devices. The computers on each end of a data session do not "care" what physical conduits link them together so long as the packets are delivered quickly enough for the relevant software programs to run properly. And, for the most part, the Internet's physical infrastructure has not traditionally "cared" what software programs those packets are associated with; it just delivers the packets and lets the computers do the rest. In part because "a bit is just a bit" in this sense, the Internet severs any strong logical or practical link between communications *services* and the physical *media* over which they are transmitted to consumers.

Technological convergence has uprooted some of the most basic premises of telecommunications policy. Take, for example, the case of telephony, the transmission of ordinary voice conversations between two or more people. For many decades, this was the exclusive province of heavily regulated wireline telephone companies operating on the PSTN, the interconnected universe of telephone networks that rely on legacy circuit-switched technologies. Because of that historical pedigree, wireline telephone companies remain subject, in a gradually declining number of areas, to monopoly-era *dominant-carrier* regulations that prescribe the rates they can charge consumers for conventional telephone service. And as part of their universal service requirements, such companies, unlike their competitors, are often compelled to provide telephone service to rural areas where per line costs far exceed regulated revenues.

These PSTN-centric regulatory obligations are increasingly anachronistic because conventional circuit-switched wireline networks are becoming steadily less relevant to modern communications. The percentage of American households that have cut the cord to their local phone company and rely exclusively on their cellphones for voice service has surged from about 6% in 2004, when this book's first edition was written, to 31.6% in 2011.[27] And even one-third of the nation's *fixed-line* residential voice connections—28.2 million out of 89.8 million total as of mid-2010—were provided by cable companies and other nonincumbent telephony providers.[28]

These alternative voice providers have already integrated, or within a few years will have integrated, all of their services onto unified IP platforms. And some incumbent telephone companies have followed suit, offering their own triple-play bundles of voice, video, and broadband Internet access over converged IP platforms as well. Data traffic used to constitute a tiny percentage of the signals flowing over a voice-centric telephone network; soon, however, voice traffic—VoIP—will become just a small percentage of the bits flowing over a data-centric network.

If the Internet ecosystem is to function properly, this comprehensive technological convergence must now be matched by equally comprehensive reforms to a regulatory regime still marked by the preconvergence assumptions of the original Communications Act of 1934. First, except where lingering natural-monopoly conditions make one provider dominant in a particular market, like services should generally be regulated alike, no matter what physical medium is used to provide them. For example, it is an increasingly questionable policy to single out wireline telephone companies for dominant-carrier regulation in their provision of retail voice services in areas where they now may have only a minority share of voice subscribers. As cross-platform competition increases, a regulatory regime that still treats substitutable platforms differently will distort the marketplace by, among other things, creating artificial regulatory advantages for one set of competitors over another.

Second, policymakers should aggressively follow through on plans to replace the traditional PSTN-based universal service system, which focuses on extending narrowband telephone service, with a competitively neutral program for supporting affordable *broadband Internet access* to otherwise unserved communities. The need for this transition is the central topic of the FCC's influential *National Broadband Plan*, issued in 2010, and of a sweeping 2011 FCC order that makes broadband deployment the new focus of federal subsidies for rural and other "high-cost" areas.[29] As we discuss in chapter 8, however, that reorientation of universal service goals is easier to embrace in principle than in practice, and it poses a host of legal, political, and practical challenges.

Third, Congress and the FCC need to work together to devise a unified and stable legal framework for addressing the policy issues raised by the new IP-centric communications ecosystem. As discussed, the Communications Act is divided up into "titles" corresponding to different categories of communications services: Title II for telephone companies and other providers of "telecommunications services" (i.e., common carrier services); Title III for broadcasters; and Title VI for cable television operators. When, in 1996, Congress last enacted major revisions to the Act, it did not clearly foresee the rise of broadband Internet access services, let alone their eventual centrality to all forms of electronic communications. And it therefore did not specify how such services should be classified within the Act's existing structure or how, if at all, they should be regulated.

Given this lack of statutory direction, the telecommunications policy world has become mired in repeated controversy since 1996 about whether broadband Internet access services are properly classified as "telecommunications services" subject to common carrier regulation under Title II or, as the FCC determined in 2002, "information services" subject only to minimal regulation under the Commission's Title I ancillary authority.[30] That debate assumed new prominence when in early 2010 a federal appellate court invalidated the FCC's initial rationale for asserting Title I authority to police net neutrality violations.[31] In response, the FCC floated—and then, in the face of political opposition, all but abandoned—a proposal to repudiate its 2002 decision and reclassify broadband Internet access as a Title II service, a reversal that itself would have been subject to legal challenge (see chapter 6). Broadband Internet access service thus remains what it was in 2002: a Title I information service. But because the contours of the FCC's Title I ancillary powers are highly uncertain, so is the Commission's authority to adopt broadband-related policy, from the net neutrality rules it adopted in late 2010 (under a new Title I theory that remains pending on appeal at press time) to the broadband subsidy regime the Commission created in 2011. That uncertainty is unlikely to be eliminated anytime soon unless Congress steps back into the fray.

Finally, the rise of convergence and broadband competition creates a new long-term aspiration for telecommunications policy: creating the market conditions needed to phase out most forms of public-utility-style regulation while supporting collaborative industry-led initiatives to develop basic norms of good Internet citizenship. As the *National Broadband Plan* observes, the U.S. broadband market structure is "relatively unique in that people in most parts of the country have been able to choose" among different fixed-line broadband platforms, whereas most foreign consumers have not.[32] The *Plan* notes that this "competition appears to have induced broadband providers to invest in network upgrades" and that "[c]onsumers are benefiting from these investments" in the form of steadily improving broadband performance.[33] Nevertheless, although competitive conditions vary, the typical American has a choice of only two fixed-line broadband providers: the local telephone company and the local cable TV company.

This is where mobile broadband services and spectrum policy come in. Americans are turning increasingly to their wireless devices for broadband Internet access, from smartphones to tablets to laptop cards. Indeed, wireless broadband services are so explosively popular that bandwidth demands, fueled by streaming video and audio, are creating what the FCC has labeled a "spectrum crunch." The federal government oversees who can use particular frequency bands and for what purposes, and there is broad consensus that far too little of it is allocated to meet surging mobile broadband needs. As FCC Chairman Julius Genachowski explained in 2011, "If we do nothing in the face of the looming spectrum crunch, many consumers will face higher prices—as the market is forced to respond to supply and demand—and frustrating service—connections that drop, apps that run unreliably or too slowly. The result will be downward pressure on consumer use of wireless service, and a slowing down of innovation and investment in the space."[34]

Mobile and fixed-line broadband services may never be full substitutes: fixed-line services will never be mobile, and mobile services may never support the same high-bandwidth usage as fixed-line services at the same price points. That said, freeing up additional spectrum for mobile broadband services not only will make them faster and better in their own right but also might well impose additional competitive discipline on fixed-line broadband services (in the form of a "good enough" broadband service).[35] Like universal service reform, however, spectrum reform is easier to propose than to accomplish, and it confronts enormous political obstacles erected by incumbent spectrum holders, ranging from local broadcasters to the military. As we discuss in chapter 3, navigating those obstacles may be the single most important ambition of telecommunications policymakers in the early twenty-first century, and meeting that policy challenge is a key to resolving many others.

2

Competition Policy in Wireline Telecommunications

As discussed in the previous chapter, wireline telephone companies were long deemed natural-monopoly providers of "last-mile" transmission for point-to-point voice and data services. Today, cross-platform competition—primarily from wireless and cable broadband services—has drawn into question many of the regulations designed to curb wireline telephone companies' now-fading monopoly power. Understanding the debate about whether such regulations remain necessary requires some familiarity with the basic technology and regulatory history of traditional telephone networks. Part I of this chapter addresses the technology; part II addresses the basics of retail rate regulation of telephone service; and part III discusses the advent of competition in telephone markets from the 1970s to the eve of the Telecommunications Act of 1996. Part IV then discusses the Act and its implementation, and part V concludes by examining present-day controversies about the pace and extent of deregulation.

One caveat is necessary at the outset. This chapter addresses the regulation of incumbent telephone companies' *networks* and their traditional *common carrier services*, ranging from ordinary residential telephone service to high-capacity "special-access" services sold to large businesses and other carriers. This chapter does not, however, address the quite different regulatory treatment of telco-provided broadband Internet-access services, addressed in chapter 6, or subscription video services, addressed in chapter 9.

I. A Primer on Wireline Technology

Placing a call from a conventional telephone is so routine that it is easy to forget how astonishingly complex the process is. By picking up the receiver and punching some numbers, you can reach anyone with access to a telephone anywhere in the world. This is an uncanny feat: the global telephone system must locate that person among a billion other telephone subscribers—and then establish a connection with her that twists and turns through aerial wires and underground cables and a succession of computerized switches, all within an instant or two. And even this description oversimplifies the matter because the call may well be handled not by one network but by several, and the companies that own those networks must arrange the multiple transfer of your voice signals like a baton in a very, very fast relay race.

But this brief sketch is merely the beginning, for the technological accomplishments just described take us only up to around 1980. Then came the rise of the Internet, which relies on many of the same facilities as the telephone network but uses the power and versatility of digital technology to convert a telecommunications infrastructure originally designed for voice calls into a worldwide network of computer networks. With a few keystrokes, you can use your personal computer, tablet, or smartphone to reach one of millions of other computers across the globe. As the 1s and 0s generated by a single webpage flit unpredictably along different paths across the Internet, the sights and sounds of foreign countries come streaming into your device at the speed of light.

Chapter 5 describes this digital overlay we call the Internet. Until then, keep the following in mind. The Internet is not some mysterious set of wires unrelated to the telephone networks over which we place ordinary voice calls. The local telephone lines that tens of millions of Americans use to connect to the Internet (via dial-up, DSL, or fiber-optic technologies) are the same telephone lines they may use to call up their friends. And although there are dedicated lines used solely for data traffic, long-distance telephone networks often use the same fiber-optic routes to transmit voice calls across the country that they use to transmit webpages, streaming video, and Facebook posts.

When you conceptualize the differences between the *public switched telephone network* (PSTN) and the Internet, you should think not so much about any differences in the underlying physical facilities (although there are some important ones), but about the differences in how those facilities are used. Although the analogy is imprecise, Internet services and conventional voice services "ride on top of" wireline network facilities in much the same way that cars and bicycles ride on the same paved roads, albeit often in different lanes. Indeed, as we discuss, these services are increasingly riding in the *same* lanes, as voice becomes one application among many transmitted over the Internet.

The conventional telephone network and the Internet are examples of switched networks: from any given point on the network, you can direct calls or data to any other single point on the network. At the most fundamental physical level, such networks consist of *transmission pipes*, including copper wires and fiber-optic cables, and the *switches* that route calls from one such pipe to another. The pipes are further subdivided into the *loops* that connect customers to switches and the high-capacity *transport links* that connect switches to other switches. Although the distinctions can blur at the edges, these three elements—loops, transport, and switches—are the fundamental building blocks of any point-to-point telecommunications network.*

A. Transmission pipes: loops and transport

Loops are so-called *last-mile* facilities: the wires or cables a wireline provider uses to connect its customers to the nearest switch and from there to the rest of the world. Strung through the air or laid underground, loops constitute by far the costliest portion of most telecommunications networks.

As deployed by conventional telephone companies, most residential loops have traditionally consisted of a twisted pair of copper wires used to establish an electrical connection between the telephone equipment in a customer's house and the telephone company's switch. When a customer lifts the receiver off the hook to place a call, the switch sends her a dial tone to confirm that the circuit has been established and is available to carry her call. When someone else calls her, the switch sends another electrical current down the line, this time to trigger the ringing of her telephone. The electrical currents come from giant batteries at the telephone company's switching station, known as a *central office*—usually identifiable as a bland-looking building with few windows and the telephone company's logo outside.

Suppose that you place a telephone call to someone whose loop is connected not to your local switch, but to a neighboring switch on the same network or to a switch operated by a different telecommunications provider altogether. To get from your switch to hers, your voice signals must travel along a high-capacity transport link. Most modern transport facilities use *optical fiber* technology. In effect, a fiber-optic cable is an extremely thin glass tube that transmits light over long distances through

* When reviewing the following discussion, readers may wish to glance forward to figure 2.1, which appears near the conclusion of this section.

various forms of internal reflection. Laser-originated light waves, carrying signals, bounce from one end to the other. (One common though quite rough analogy compares fiber-optic technology with shining a flashlight down the interior of a long tube with a mirrored surface on the inside; although the tube may bend and twist, the light shines out of the other end.) Attached to each fiber strand are expensive electronic devices that aggregate all of the signals from different customers' individual lines onto the same physical strand of optical fiber. This technique is called *multiplexing*. Copper wires often carry multiplexed signals, too, but not with the phenomenal capacity of fiber.

In the traditional wireline telephone world, the most common form of multiplexing, *time-division multiplexing* (TDM), "samples" the signal for a given call many times a second and transmits those samples along with the corresponding samples taken of other calls. A "sample" is a kind of digital snapshot of the signals in a call at any given moment, much as a frame in a movie is a snapshot of the action in progress. Each call is preassigned time slots in the multicall transmission; at the other end, this aggregated signal is "demultiplexed" back into individual signals. By means of this multiplexing technology, a single strand of fiber thinner than a human hair can carry thousands of simultaneous voice conversations. This aspect of telecommunications technology often comes as a surprise to people outside the industry: the calls you place to your friends across the country typically coexist on the same fiber strand with many other calls taking place between people you don't know.

The use of optical fiber is routine on any transport route handling large volumes of telecommunications traffic, such as between cities, between central offices, or between two interconnected telecommunications networks. Fiber-optic technology is also prevalent in the loop facilities used by businesses in major cities, where fiber "rings" pass beneath the streets in downtown business districts to collect the enormous telecommunications traffic volumes coming from large office buildings.

The prevalence of fiber in the loops that connect local switches to *residential* customers is less predictable because the economic calculus varies so much from place to place. Replacing longstanding copper loops with new fiber-optic loops requires immense up-front costs, including the expense of necessary electronic equipment and the cost of hiring skilled labor to dig up the streets and lay the cable. A phone company may nonetheless choose to incur those enormous fixed costs in order to keep up with cable companies and other competitors in the provision of high-bandwidth *triple-play* services: voice, video (i.e., cable TV), and high-end broadband Internet access services.

In particular, fiber has exceptionally high bandwidth—that is, data-carrying capacity—that does not vary significantly with the distance between the telephone company's central office and a customer's home. The bandwidth of copper wires is much more limited and varies inversely and dramatically with length. The broadband technology offered over copper loops is known as *digital subscriber line* (DSL). The fastest variants of DSL—those capable of supporting subscription video as well as Internet access—are available only to customers for whom the copper portion of the loop is very short. As a result, phone companies that wish to provide triple-play services over their own facilities must generally deploy fiber-optic cable deep into individual neighborhoods to cover much of the distance between the central office and each customer.

For example, some companies often use fiber cables in the aggregated multiloop *feeder* cables closest to the central office but leave in place the existing copper wires within the more diffuse *distribution* portion of the loops closest to customers. (Think of feeder cables as the main branches of a tree and of distribution cables as the smaller branches and twigs.) This type of network architecture is commonly known as *fiber to the node* (FTTN), and recent advances have enabled telcos to offer considerable bandwidth (theoretically up to 100 megabits per second (Mbps)) using this hybrid fiber–copper infrastructure.[1] A few companies have adopted a more aggressive *fiber to the home* (FTTH) strategy, laying fiber-optic cables all the way to millions of individual houses, incurring the greatest upfront costs while ensuring the greatest bandwidth for the long term. As noted in the previous chapter, Verizon has been the only large telco to build FTTH on a large scale, and it began winding down new fiber deployments in 2010 amid concerns about whether the returns could justify the immense capital costs.

The loop facilities discussed in this chapter are those used by traditional telephone companies. Of course, cable TV companies and mobile wireless providers must also arrange for last-mile transmission to their customers. Mobile providers rely on licensed electromagnetic spectrum to connect their customers to their cell towers. And cable operators typically use a combination of optical fiber and coaxial cable to connect customers to the cable *headend*, the rough equivalent of a telco central office. As discussed in chapter 5, cable operators have generally had higher-capacity last-mile facilities than telephone companies because they traditionally specialized in transmitting bandwidth-intensive video signals rather than narrowband voice signals. Until the turn of the millennium, however, they lacked the advanced network technologies needed to provide point-to-point voice and data services. Today, these wireless and cable networks can be used to provide many of the same voice and data services once dominated by wireline telephone companies. Keep that point in mind as you read the story of wireline telephone regulation, which originates from the now obsolescent premise that telephone company loops are the only feasible means of providing such point-to-point services.

B. Switches

Imagine trying to connect every home or business in the United States to every other home or business without the use of a switch. The number of required lines and thus the cost would be astronomical. In fact, if we estimated the number of wireline telephone connections in the United States as roughly equivalent to the number of Americans, the tangled mess of lines criss-crossing the country to connect each telephone to every other would amount to more than 40 *quadrillion* lines.[2]

Switches are built to solve this problem in the most economical way. They direct a voice or data call from one transmission pipe (a loop or transport link) to another en route to the call's destination. Although the distinction blurs at the margins, there are two basic kinds of switching technologies—*circuit switching* and *packet switching*—that are used, respectively, in conventional voice networks and more advanced data networks, including the Internet. As mentioned, the physical infrastructure of wireline telephone networks overlaps significantly with that of the Internet. The major exception to this rule lies in switching technology, for reasons we discuss later in this chapter and in chapter 5.

Circuit switching

Circuit switches range from the early hand-operated switchboard to its modern-day automated equivalents. A circuit switch sets up a dedicated transmission path from the calling party to the recipient for the duration of a call. At any point during the call, a particular increment of capacity is reserved for that call on the loop, switch, and transport pipe, even if no one is talking and no information is being sent. That is true even of calls subject to the TDM (time-division multiplexing) technologies discussed earlier: even though the physical line itself is host to a stream of signals, that line reserves capacity, in the form of dedicated time slots, for any given call for its entire duration, whether anyone is talking or not.

To save money, a telecommunications carrier does not build enough capacity on its switches and interswitch transport links to carry calls from all customers at once. Instead, like a bank, it keeps just enough in reserve to cover the greatest reasonably expected demand. The size and cost of switches and transport links are thus determined by the network's expected capacity needs at peak calling hours. This is one reason why many callers in the nation's capital received "all circuits busy" signals when calling home on September 11, 2001: telephone engineers had not built in enough network capacity to serve this unexpectedly high call volume.

Today's circuit switches are essentially very large computers that, in addition to establishing circuits for given calls, perform a variety of other "intelligent" functions, including call forwarding, caller identification, call waiting, and billing. A modern circuit-switched network is usually shadowed by a parallel, packet-switched *signaling network*, which tells the circuit switches how to route particular calls to avoid network congestion and how to implement specific customer requests, such as where 800-number calls should be directed and how calling card calls should be handled. The "brains" of a circuit-switched network are said to reside in the switch and the parallel signaling network, not at the "edge" of the network in an end user's computer, and they are centrally owned and controlled by the telecommunications company. As a result, it is difficult, if not impossible, to introduce new intelligent features to the circuit-switched network without first obtaining the permission of the telephone company that owns the switch.

Finally, circuit switches are often arranged "hierarchically" to minimize the number of switches and transport links needed to keep a circuit open during the duration of a call from one place to another (see figure 2.1). At the low end of the hierarchy, established long ago by AT&T's Bell System, are the so-called *local* (or "end office") switches, to which most loops are connected. The next level up consists of the *tandem switches*, which, among other things, route large call volumes from one local switch to another. In the years before the breakup of the Bell System in 1984, the switches higher in the hierarchy were associated with the routing of long-distance calls. After 1984, AT&T kept those switches, and the newly independent Bell Operating Companies ("BOCs" or "Bell companies") kept the local and tandem switches.

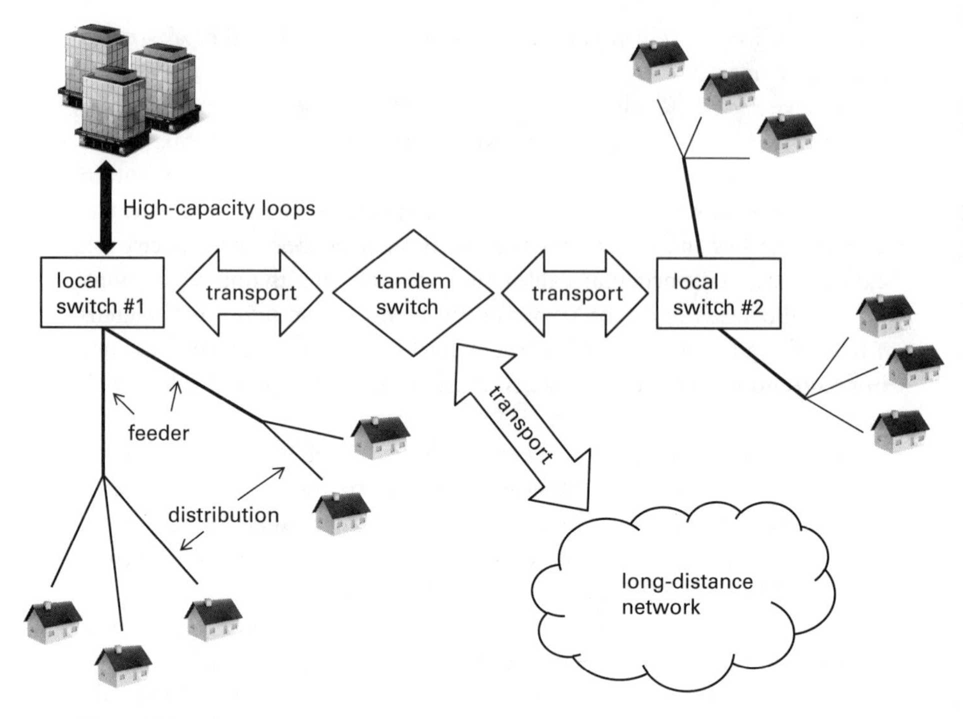

Figure 2.1
Traditional wireline telephone architecture

Packet switching

Unlike circuit-switched networks, "connectionless" packet-switching technologies do not set up a dedicated circuit for the duration of a communication.[3] Instead, the transmitted information is converted into discrete digital packets, and the packet switch sends each of them separately from the others, potentially along different transmission paths. As noted more fully in chapter 5, these packets are encoded strings of 1s and 0s that can contain information of any kind, ranging from streaming video to a photograph embedded in an email to the sound of a human voice. Each packet may be held in queue at the packet switch while yet other packets from unrelated "calls" (or "sessions") pass through. All the packets in a given transmission ideally end up at the same place because each of them has an "address header" that tells the network who is supposed to receive it. At the receiving end, the recipient's computer or other smart device unjumbles the packets and displays the message (e.g., a webpage). The difference between circuit-switched networks and connectionless packet-switched networks has been roughly compared to the difference between, on the one hand, sending an envelope down a chute intact and, on the other, ripping the envelope into pieces, sending the pieces through different chutes, and reassembling them at the other end.

Why would anyone ever prefer the second option? For most (but not all) data communications, a packet-switched network is more efficient than a circuit-switched network because it need not dedicate capacity for the duration of a particular call or session, including the many portions of the call that include the transmission of no information. In concrete terms, the circuit that remains dedicated to you during even short pauses in an ordinary circuit-switched voice call is a waste of network capacity because the facilities carrying the call are standing idle instead of carrying someone else's transmissions. So the real question is why any provider would ever prefer a more wasteful circuit-switched network to a more efficient packet-switched network, even for voice calls.

The main reason that circuit switches have persisted into the digital age is that connectionless packet-switching technology did not until recently lend itself easily to managing voice calls and certain other real-time applications. The flip side of that technology's efficiency can be perceptible delays, known as *latency*, and disruptive packet-to-packet variability in delay, known as *jitter*. Latency and jitter arise from the unpredictable and sometimes slow transmission of jumbled packets through the network and their final reorganization into a coherent message that can be understood and translated by the recipient's applications software. One-second delays are barely noticeable when someone is downloading a webpage, but they are quite distracting in an ordinary telephone conversation. These challenges are surmountable, however, as demonstrated by the rise of over-the-top VoIP services (see chapter 5).

We began this chapter by noting the significant overlap between the physical infrastructure underlying the Internet, on the one hand, and local and long-distance telephone networks, on the other. Switches, as noted, mark the major exception to this rule. You can still gain access to the Internet through a conventional circuit switch by placing an ordinary "local" call to an old-fashioned dial-up *Internet service provider* (ISP), which will in turn convert the signals into packets and route them over the packet-switched Internet. But the circuit switch at the threshold of that transmission is more a hindrance than a help even though it connects you to the dial-up ISP. That switch was designed with the almost single-minded purpose of carrying high-quality voice traffic as efficiently as possible. Although the switch ensures real-time dependability by holding a dedicated circuit open for the duration of each call, it makes up for all the capacity "wasted" by that *temporal* reservation of bandwidth by limiting the *amount* of bandwidth allocated to each circuit held open. Specifically, it filters out all but the core frequencies associated with the typical human voice; as a result, the person on the other end of the line sounds more or less like herself, but the music you hear while being placed on hold sounds tinny. Economizing on bandwidth this way keeps network costs down, but it makes circuit switches inefficient processors of the many data transmissions for which high bandwidth is essential. Just as a conventional circuit switch squelches the acoustic texture of tubas and piccolos, it also excludes the myriad frequencies on which computer data can be exchanged.

Depending on what loops are made of, they have limited bandwidth, too, but not nearly as limited as that of a conventional circuit switch. The key to providing broadband Internet access over wireline telephone facilities is thus to *bypass the circuit switch* by connecting the loop directly to a packet-switched network linked to the Internet. Later in this chapter and in chapter 5, we address how network engineers perform that feat.

II. Traditional Telephone Rate Regulation

We continue our survey of the wireline telephone industry with a brief overview of the rate-regulation framework that has traditionally governed the commercial relationships between traditional telephone companies and retail customers. As we shall see, that framework—which originated in the early twentieth century—can work as intended only if there is very little competition in telecommunications markets, and it is collapsing now that such competition has arisen. Critically, the retail rate-regulation regime discussed in this chapter applies only to *wireline telephone incumbents* on the PSTN. It does not apply to other providers of voice service, such as cable operators or mobile wireless providers.

A. The basics of price regulation

In ordinary, nonmonopolized markets, companies compete against one another for customers, and this competition theoretically keeps the price of goods and services at reasonably efficient levels. By definition, however, there is no competition in a market that regulators treat as a natural monopoly, which is how the market for local telephony services was treated for most of the twentieth century. And if left to their own devices, monopoly providers of any product, including wireline telephony services, can be expected to maximize their profits by raising their retail prices to inefficiently high levels. These high prices include monopoly profits that enrich the monopolist at the expense of consumer welfare, artificially reducing the demand for telecommunications services and keeping total output (i.e., sales volume) below optimal levels. In markets without formidable barriers to entry, monopoly profits tend to be a temporary phenomenon, for upstart firms will enter the market and undercut the incumbent's prices. Federal and state policymakers have long treated local telephone markets differently, however, because of the traditionally high entry barriers discussed in the previous chapter. In particular, they have relied on rate regulation—not Adam Smith's invisible hand—to protect consumers from monopoly pricing.

The regulatory compact, rate-of-return regulation, and price caps

Some foreign governments originally addressed the problem of monopoly pricing by owning the telephone system outright, an approach that many of them have now phased out.[4] In the United States, by contrast, private companies have owned virtually every telephone network, but they have traditionally been regulated as *public utilities*—that is, as commercial enterprises charged with providing an essential public service and subject to pervasive regulation to protect the public interest.[5] The Supreme Court explains: "At the dawn of modern utility regulation, in order to offset monopoly power and ensure affordable, stable public access to a utility's goods or services, legislatures enacted rate schedules to fix the prices a utility could charge. As this job became more complicated, legislatures established specialized administrative agencies, first local or state, then federal, to set and regulate rates."[6]

Like other public utilities, each incumbent local telephone company in any given geographic area made a kind of regulatory compact with the government. In exchange for agreeing to serve all consumers at affordable rates as the *carrier of last resort*, the company was given an opportunity to earn a "reasonable return" on its overall regulated investment. Regulators would traditionally set a dominant local telephone company's retail rates by means of *rate-of-return* regulation. Under that approach, regulators give local telephone incumbents an opportunity to charge retail rates sufficient in the aggregate to cover their anticipated expenses plus a reasonable return on their net investment. This approach involves, among other things, calculating a company's "historical costs"—the costs it has actually incurred and lists on its books—and making various judgments about them. Such judgments include (1) the extent to which these costs were "prudently incurred"; (2) how quickly the telephone company should be able to recover the total costs of given facilities (through its monthly retail rates)—a judgment that in turn depends on estimates of how long those facilities will be in service (depreciation lives); and (3) how much of a return on investment (cost of capital) the company needs to continue attracting capital to finance new investment.

If dissatisfied with the outcome of this analysis, the telephone company can file suit for "just compensation"—in the form of increases to its retail rates—if it believes that its current rates are so low that they leave the company "insufficient operating capital" or "imped[e] [its] ability to raise future capital" in violation of the Takings Clause of the U.S. Constitution.[7] Although this option is an important backstop in theory, courts are unlikely in practice to find that a public utility has suffered such a "regulatory taking," as the Supreme Court showed once more in 2002.[8]

Traditional rate-of-return regulation tends to give any public utility perverse incentives to gold-plate its assets—that is, incentives to spend more than is efficient or necessary simply to increase the rate base on which it earns its profits. Rate-of-return regulation also can make it easier for a firm to harm competition by overassigning joint and common costs to its monopoly markets and thereby to *cross-subsidize* its operations in competitive markets. In the 1980s and 1990s, federal and most state regulators sought to address these problems by adopting a *price-cap* scheme for retail rate regulation of the largest local telephone companies.

A price-cap analysis starts with the retail rates produced in a given year under traditional rate-of-return regulation. In succeeding years, however, retail rates will be determined on the basis not of new rate-of-return proceedings, but of mathematical adjustments designed principally to reflect (1) expected industry-wide increases in efficiency due to technological and other innovation (known as the *X-factor*) and (2) fluctuations in inflation and other macroeconomic variables.[9] A price-cap approach, unlike a traditional rate-of-return regime, rewards the incumbents for their efficiency over time by entitling them to keep much of the extra profit they generate as the result of cutting unnecessary costs.[10] As a consequence, price-cap regulation can create the opposite of a gold-plating incentive: incumbent providers might seek additional profit by "starving the network." To address this concern, many state regulatory authorities have adopted "quality of service" requirements and have even sanctioned providers that unduly sacrificed service quality by underinvesting in their networks. Today, the largest incumbent telephone companies, such as Verizon, AT&T, and CenturyLink, are generally subject to price-cap rules, whereas smaller rural incumbents are generally still subject to rate-of-return regulation.

Dual jurisdiction and access charges

Because every aspect of telecommunications can be characterized as an instrumentality of interstate commerce, Congress could have preempted all state regulation in this area under the Commerce Clause of the U.S. Constitution and placed the entire industry within the exclusive province of a federal regulator. When it enacted the Communications Act of 1934, however, it chose a model of *dual jurisdiction*, which gave the newly created FCC plenary jurisdiction over interstate services and, under section 2(b) of the Act,[11] precluded the FCC from intruding on state regulation of purely intrastate services.

The federal government and the states divide their traditional retail rate-regulation responsibilities roughly as follows. The FCC and the states first allocate a percentage of a telephone company's total costs into "interstate" and "intrastate" categories. The interstate costs are recovered by rates for interstate services regulated by the FCC, and the intrastate costs are recovered by rates for intrastate services regulated by the states. The process of dividing up costs this way is called the *jurisdictional separations* process and dates back to the Supreme Court's 1930 decision in *Smith v. Illinois Bell.*[12] The criteria used can be quite arbitrary because many facilities, such as the loop and switch, are typically used for both interstate and intrastate calls.[13]

Each set of regulators—federal and state—enables a telephone company to recover costs through a combination of (1) flat rates and (2) usage-sensitive rates. Subscribers pay two flat monthly rates—one overseen by the states, the other by the FCC—to receive basic local telephone service. The federal rate, called the *subscriber line charge*, is similar to the monthly "local service" rate set by state regulators, but it is designed to permit the telephone company to recover the portion of loop costs allocated to the interstate side of the cost ledger.[14]

As relevant here, "usage-sensitive" rates take two general forms. First, a telephone company may collect per minute *toll charges* from its own subscribers for certain types of calls (mainly long-distance calls). Increasingly, however, telephone companies simply charge their customers a flat monthly rate for unlimited all-distance calling within the United States. Second, a telephone company collects *intercarrier compensation*, including the per minute *access charges* that it assesses on other carriers that hand off long-distance calls bound for its subscribers.* Those charges are regulated by federal or state regulators depending on whether the call is deemed to have crossed state lines.

Access charges were first created in the 1970s when upstarts such as MCI began competing with then-monopolist AT&T in the long-distance market.[15] Access-charge rules required these independent long-distance carriers to pay a share of the local network costs incurred by AT&T's local telephone affiliates (the Bell companies) in building and maintaining local facilities that, as we have seen, are used for both local and long-distance calls. And the independent long-distance companies in turn passed these charges through to their retail subscribers in the form of higher long-distance rates.

As we explain in chapter 7, access charges—although still enormous in the aggregate—are increasingly anachronistic, and the FCC announced in 2011 that it would begin phasing them out over time. Because of their historical pedigree and key FCC policy decisions, access charges generally apply only to "long-distance" (not "local") voice calls routed to the called party through circuit switches (rather than through Internet routers) of local wireline (not wireless) telephone companies. These distinctions have very little to do with the underlying costs that access charges are said to "recover." And communications providers have found ingenious and increasingly successful ways to manipulate these distinctions in order to avoid paying access charges.

To this point, we have discussed *switched access charges*—the per minute rates that a local telephone company charges interconnecting carriers to complete individual long-distance calls through the local company's circuit switches. Switched access should not be confused with *special access*. The latter term denotes the high-capacity lines that telephone companies lease to other telecommunications providers (or to large business customers) in order to connect two discrete points. Examples include the high-capacity lines connecting a cell tower with a wire-

* These access charges are so-called *terminating* access charges. Local telephone companies may also impose *originating* access charges when they hand off calls to their customers' chosen long-distance carriers, but such charges have assumed less significance now that most people who subscribe to wireline telephone service generally purchase both local and long-distance service from the same company (or from close corporate affiliates).

less carrier's switching station or connecting a large office building with a carrier that lacks last-mile facilities of its own. Special-access rates, which are partially deregulated and overseen mostly by the FCC rather than by the states,[16] are highly controversial in their own right, and we discuss them later in this chapter.

Tariffs

Incumbent wireline telephone companies are not generally permitted to enter into wholly private contractual arrangements with ordinary consumers for the provision of circuit-switched telephone services. Instead, the basic premise of the incumbents' common carriage commitment is that each customer should have the same opportunity as any other similarly situated customer to buy the same services on the same terms. Thus, when an incumbent wishes to introduce a new wireline telephone service, it traditionally must file a *tariff* with the relevant regulatory authority, spelling out the terms and conditions of its services and offering them for sale to the public at large. Such tariffs are often permitted to take only temporary effect while regulators conduct an inquiry into the reasonableness of their terms.[17]

Once approved, the terms of these tariffs govern the retail relationships between carriers and their customers, even if a carrier offered different terms in a sales call or even agreed to other prices in a contract. In theory, this *filed rate doctrine*, which allows companies to charge only tariffed rates and (to their great benefit) shields them from litigation concerning the legitimacy of those rates,[18] protects consumers by ensuring that they receive the prices, terms, and conditions approved by the regulators. In practice, however, requiring tariff filings creates significant social costs by slowing down the rough and tumble of free market competition (where it exists) and facilitating collusion between rivals by enabling them to see one another's prices before they go into effect.[19]

The FCC has freed "nondominant" carriers from tariffing obligations in many markets and has sometimes, over their opposition, affirmatively forbidden them to file tariffs.[20] As discussed at the end of this chapter, the FCC has further freed wireline incumbents from tariffing obligations in specific markets that it has deemed competitive, such as packet-switched or very high-capacity special-access services. Finally, as discussed in chapters 4 and 6, no provider is subject to tariffs in the provision of mobile wireless services (because they are deemed sufficiently competitive) or broadband Internet access (because it is not considered a common carrier service).

B. Introduction to universal service policies

Recall that wireline telephone incumbents have traditionally been subject to carrier-of-last-resort obligations, which require them to build out facilities and provide conventional telephone service even to remote areas where per line costs are immense. Regulators have nonetheless set rates for service to such areas below the high costs of serving them. To cover the difference, regulators have historically allowed the regulated wireline monopolist to set the rates for other services, such as those provided to business customers, above the cost of providing them. Theoretically, the monopolist receiving these rates in the aggregate should be indifferent to such *implicit cross-subsidies* because—so long as it faces no competition for the customers paying the above-cost rates—its books come out even in the end.

The term *universal service* is used somewhat overbroadly to describe the regulatory manipulation that produces low residential rates, even though such manipulation may be completely unnecessary to ensure that the beneficiaries actually remain hooked up to the network. Chapter 8 discusses the changing face of universal service in some detail, but we cover the basics here.

The telecommunications industry has traditionally been riddled with different types of implicit cross-subsidies justified as necessary for "universal service." If you live in a highly populated urban or suburban neighborhood, you may well be paying the same basic rate for telephone service as someone living deep in the countryside an hour down the state highway. This itself is a form of implicit cross-subsidy. Because of economies of density, it costs much less to provide service to you than to your rural counterpart. This practice of setting the same rate for all residential customers in a large geographic region such as a state—known as *geographic rate averaging*—means that you are in effect paying a hidden surcharge on your telephone bill to enable the rural customer to receive service at a rate well below the cost incurred in providing it to him.

On the other hand, you should not complain too much if you are an urban residential customer, for the owner of the grocery store down the street from your apartment has even greater cause for dissatisfaction: he may have to pay up to twice as much for a telephone line as either you or the rural inhabitant. That may seem odd because it may well cost the telephone company no more to provide that service to the grocer than to you. But this is another way in which regulators have traditionally kept residential rates low: by authorizing the telephone company to maintain artificially high rates for "business lines." In effect, when the grocer orders telephone service, he may be signing a tacit agreement to pay more than a 100% tax on that service to underwrite low rates for rural subscribers.

Long-distance calls have been another traditional source of implicit cross-subsidies. The reason is somewhat complex. Before the 1980s, when there was very little competition in the long-distance market, regulators set per minute rates for such calls far above cost. Those rates helped underwrite low monthly rates for local telephone service. Throughout the 1980s and 1990s, as robust competition developed in the long-distance market, long-distance rates dropped significantly, but they often remained higher than "cost."

The reason lies in access charges, discussed earlier. If you, sitting in New York circa 2013, want to call your friend on his landline in Iowa, your phone company (wireline or wireless) will have to pay your friend's phone company access charges to receive the call and route it to your friend. Indeed, it has no choice but to pay them because it is generally obligated by federal interconnection requirements to hand the call off to the friend's local carrier and pay the tariffed access charges. Although the FCC is slowly phasing out access charges, they often have been set above any rigorous measure of "cost," in part to offset the losses such companies are said to incur when they are forced to provide basic local service to residential customers at low rates. Because local companies impose access charges on long-distance carriers, which in turn pass them on to their own subscribers, above-cost access charges ultimately force consumers to pay above-cost retail rates for long-distance services. Heavy long-distance callers have therefore traditionally subsidized below-cost rates for "basic" local service.

All these implicit subsidies—geographic rate averaging, above-cost business rates, above-cost access charges, among others—are politically convenient. They all are economically equivalent to a special tax imposed on some customers or services to subsidize below-cost rates for other customers or services. And precisely because they are "implicit" rather than "explicit," they come without the political baggage of an explicit tax or universal service fee.

But these charges are "taxes" with a special drawback: customers can avoid paying them altogether if they use a provider *other than* the regulated local telephone company to perform the same services at a lower price—that is, without the implicit tax inherent in the regulated rate. Alternative providers know this, and they make it their first priority to cherry-pick the very customers who are paying the largest implicit taxes.

This "cherry-picking" is a form of *arbitrage*—a low-risk profit opportunity arising from arbitrary regulatory distinctions. The new entrants that exploit such opportunities inexorably undermine the whole scheme of implicit cross-subsidies, but they are doing nothing wrong. They are merely delivering the message that this traditional scheme, designed for monopoly market conditions, is unsustainable in a competitive era. In chapter 8, we discuss how policymakers have responded by gradually converting implicit subsidies into explicit funds underwritten by competitively neutral taxlike fees and why they are now shifting the focus of universal service from circuit-switched voice telephony to broadband Internet access.

III. Wireline Competition Policy Before 1996

To review, a market is said to have "natural-monopoly" characteristics if, because of high fixed costs and large scale economies, a firm's unit costs always decline "over the entire extent of the market"[21] with any increase in output. In that context, it will always be more efficient for one firm to supply all demand in a market than for multiple firms to compete for market share. Throughout most of the twentieth century, the entire wireline telecommunications industry in the United States was treated as a natural monopoly. The principal beneficiary of that policy was AT&T's Bell System. AT&T and its subsidiaries were permitted to monopolize the market for telephones and telephone equipment (Western Electric), the market for long-distance services (AT&T Long Lines), and most major local exchange markets (the Bell companies). And those calling areas not served by the Bell System's local exchange operations were nonetheless served by some other state-sanctioned monopoly, such as GTE.

In chapter 1, we briefly introduced the major story in wireline communications since 1970: the slow but steady peeling back of this natural-monopoly premise from one market to the next until only fragments of the local exchange market were left. Over time, often as the result of technological advances and the efforts of reform-minded regulators, policymakers repeatedly recognized that some segments of the telecommunications industry do not exhibit natural-monopoly characteristics and should not be left as the exclusive preserve of the local telephone company. In each instance, the government—either telecommunications regulators or antitrust authorities—adopted one means or another of ensuring that local telephone monopolies could not leverage their ownership of bottleneck facilities to preclude efficient competition in adjacent markets for wireline telecommunications-related services. The result was a hodge-podge of antitrust and regulatory responses to different monopoly leveraging concerns, which persisted until partially superseded by the Telecommunications Act of 1996.

During this period, three basic opportunities tended to invite upstart carriers into particular telecommunications markets. First, competitors were attracted to markets in which call volumes were great enough that they, like the incumbent, could enjoy significant scale economies while serving only a fraction of the total customer base. Second, competitors were also attracted to any market, such as those that traditionally generated implicit cross-subsidies, in which the incumbent could be expected to hold a *price umbrella* over new entrants by charging its own retail customers rates higher than cost. This price umbrella enabled the entrants to undersell the incumbent while nonetheless earning substantial profit margins of their own. Finally, competitors were eager to fill niche markets—particularly for the provision of sophisticated data services—that AT&T's Bell System had largely disregarded through its century-long focus on ordinary voice telephony and its desire to avoid cannibalizing its existing revenues.

Between 1970 and 1996, competition arose in the markets that most clearly met one or more of these conditions: the "long-distance" market for intercity transport, the "access" market for connecting large business customers directly to a long-distance carrier's network, and the emerging markets for computer-to-computer data-transmission services. In each case, what the upstarts needed most from the government were robust interconnection guarantees, in the form of "equal access" requirements, to keep incumbents from leveraging their dominance in the local exchange markets to exclude rivals from adjacent markets. In the pages that follow, we summarize the history of competition in the long-distance and access markets and the relevant regulatory measures. We then turn briefly to the first, pre-1996 regulatory initiatives for bringing competition to local exchange markets. We begin, however, with the industry segment that confronted competition before any service market did: the market for telecommunications equipment. The debate about competition in that market centered not so much on natural-monopoly theory, for no one seriously suggested that equipment manufacturing itself had natural-monopoly characteristics, but on the technical consequences of letting customers attach "foreign" (i.e., non-AT&T) devices to the telephone network.

A. Telecommunications equipment manufacturing

In the telecommunications world, the market for *customer premises equipment* includes not just ordinary telephones themselves, but also the *private branch exchanges* used in large office buildings, the modems that enable computers to communicate with each other over telephone lines, and a variety of other devices. For decades, AT&T had cited dubious technical concerns as a pretext for prohibiting its customers from attaching to its network so-called "foreign" devices manufactured by companies other than Western Electric, AT&T's equipment-manufacturing unit. As AT&T's chairman warned as late as 1973, "If consumers can plug anything they want into the network—any old piece of junk made who knows where—the system will break down. A faulty telephone in one house could conceivably disrupt service to an entire city."[22] As may seem obvious in retrospect, such concerns did not support AT&T's argument for granting it an exclusive franchise to manufacture all telephone equipment. Instead, they supported, at most, the adoption of industry-wide standards that multiple manufacturers could follow in ensuring that use of their products would not harm the telephone network. But it took regulators decades to recognize this fact, to overcome AT&T's lobbying prowess, and to write the rules needed for robust competition in the equipment-manufacturing market.

AT&T's motives for resisting competition warrant a brief recap. As discussed in chapter 1, an unregulated platform monopolist ordinarily recovers all supracompetitive profits from the sale of the platform itself (here, telephone service) and thus often welcomes competition in a complementary market (equipment manufacturing). This is because the more attractive and less expensive the complementary products are, the more valuable the underlying platform monopoly becomes. One wrinkle in telecommunications markets is that retail rate regulation has long limited the price the monopolist may charge consumers for use of the platform. Under Baxter's Law, discussed in the previous chapter, such price regulation gives the monopolist strong incentives to leverage its platform monopoly to obtain supracompetitive profits in adjacent, less price-constrained markets. The market for customer premises equipment fell into this category. As with long-distance and business services, regulators had permitted the price of telephone equipment—which often took the form of monthly lease rates—to remain well above its underlying cost in order to subsidize inexpensive local service for residential customers.[23]

AT&T's monopolistic hold on the equipment market showed its first signs of erosion in the 1950s and 1960s. During the multiyear Hush-A-

Phone controversy, AT&T prohibited its customers from attaching an independently manufactured cuplike device to a telephone receiver for the modest purpose of limiting background noise. The FCC absurdly agreed with AT&T's submission that the use of such "foreign devices" threatened the integrity of the telephone system, even though the practical effect of the device was equivalent to covering the receiver with one's hand.[24] A bemused court of appeals reversed the FCC's decision in 1956 on the ground that it made no sense.[25] The FCC eventually learned its lesson: in 1968, after much hand wringing, it rejected AT&T's efforts to bar the use of the Carterfone, a device that connected a telephone line to a two-way radio so that people using the radio could gain access to the telephone network and those on the network could communicate with those using the radio.[26]

Throughout the 1970s, the FCC built on these precedents in two basic respects. First, in 1975 it created the *Part 68 rules*, a set of technical standards that, once met, entitle any equipment manufacturer to sell its wares to the public and demand cooperation from the telephone companies.[27] The Part 68 rules supplanted AT&T's last-gasp efforts to discriminate against equipment-manufacturing rivals by forcing them to purchase various "protective coupling devices" from AT&T. The ostensible purpose of those devices was to protect the integrity of the telephone network, but their actual effect was to raise rivals' costs and hamstring competition in violation of the antitrust laws, as the courts later found.[28]

Second, in connection with a set of orders in the 1970s and 1980s known as the *Computer Inquiries*, which we discuss more fully in chapter 6, the FCC required telephone companies for a time to sell equipment through structurally separated subsidiaries and to "unbundle" such sales from their telephone service offerings. This "separate subsidiary" requirement was designed (1) to keep telephone monopolists from anticompetitively linking their products in these two markets and (2) to help regulators detect any effort by these monopolists to cross-subsidize their equipment operations, to the detriment of competition in the equipment market, by allocating excessive joint and common costs to their monopoly telephone services. (We discuss cross-subsidization concerns and the nature of joint and common costs in greater detail later in this chapter.) By facilitating competition in the equipment market, the FCC's new rules triggered not just an enormous decline in prices for telephones and other equipment, but also an explosive growth in the variety of end-user devices. These included computer modems, whose proliferation helped launch the Internet into public life.

Finally, when the antitrust court broke up AT&T's Bell System in 1984 under the consent decree discussed later in this chapter, it prohibited the seven newly independent regional Bell companies (the "Baby Bells" such as Bell Atlantic, Ameritech, and BellSouth) from manufacturing telecommunications equipment. This line-of-business restriction included not just customer premises equipment, but also the very core of the telephone network, such as central-office switches. In 1996, Congress replaced that antitrust prohibition with similar, statutory line-of-business restrictions, but it enabled the Bell companies to escape many of those restrictions once they satisfied certain conditions for opening their local markets to competition, as they all did by 2003.[29]

B. Long-distance competition and the AT&T consent decree

For readers too young to remember, a typical residential telephone subscriber in the previous century subscribed to two distinct service plans: one for "local" service and one for "long distance"; distance-agnostic plans were the rare exception rather than, as today, the rule. Much of telecommunications policy in the final quarter of the twentieth century involved efforts by the Justice Department's Antitrust Division and a federal district court to keep AT&T's Bell System and, after divestiture, its Bell company progeny from leveraging their control over local markets to dominate the long-distance market.

Throughout most of the twentieth century, this was no issue at all because AT&T owned the only significant city-to-city transport facilities (AT&T Long Lines). For quite some time, these facilities were also considered part of the vast natural monopoly of telecommunications, and the FCC permitted AT&T to charge above-cost rates for long-distance service as one mechanism among many for subsidizing low residential rates for basic service.

The assumption that the long-distance market was a natural monopoly changed when in the late 1960s and early 1970s a small upstart called Microwave Communications, Inc., later famous as MCI, first offered business customers city-to-city services using a new technology, microwave relay towers, to bypass AT&T's Long Lines. It is no surprise that the first major competition in the telecommunications industry came in the long-distance market. As discussed, long-distance rates were priced far above cost, and call volumes between cities were typically great enough to provide large economies of scale to more than one carrier. A third factor was at work as well. The birth of modern computing had made businesses across the economy hungry for new data services involving communication between distant computers over telephone lines. For example, a given business might need ready access to airline flight schedules, financial market data, or the databases of Lexis-Nexis. As a monopoly, the Bell System had been slow to adapt to the demand for these innovative data services, which at the time occupied only a niche market, and had continued to focus on its bread and butter: providing ordinary voice service. The entry of specialized data carriers such as Datran, Telenet, and Tymnet, which designed specialized digital networks to carry computer traffic, came as a welcome contrast to the Bell System's continued reliance on outdated analog technology.

The vanguard of competition, however, was MCI, and its first ambitions were modest. It began by offering "private lines"—that is, closed, point-to-point circuits—connecting the branch offices of large businesses in different cities. In granting MCI's application to offer service as a "specialized common carrier," the FCC noted the unmet demand for these private lines, including for use in data communications. Over time, MCI persuaded the FCC to let it go one step further and provide so-called FX ("foreign exchange") lines to its business customers.[30] These were private lines with a twist: one end connected to the Bell System's local exchange. Thus, MCI's FX line might connect two offices of the same company—say, one in New York and one in Chicago—but in New York the line was assigned a local telephone number, and calls to that number appeared, from the perspective of the Bell company switch, to be ordinary local calls. Equipped with an access code, the company's employees could dial that number from anywhere in New York and, for the cost of a local call, be connected to the company's office in Chicago. AT&T's Long Lines division would receive no toll revenues, and MCI would charge only a flat rate for the private line.

The final step in MCI's development, which marked its full emergence into the long-distance (and not just private-line) market, came in the form of a controversial new service called Execunet, first offered in 1975 and judicially validated in 1978.[31] Execunet involved, among other things, the use of private lines that were "open"—connected to a Bell System local exchange—on *both* ends rather than just one. To continue with the example in the previous paragraph, this meant that anyone authorized to gain access to the private line could call from anywhere in Chicago to anywhere in New York and vice versa without paying toll charges to AT&T. Before long, MCI used such lines to make general-purpose long-distance services available to the public at large: that is, not just to the employees of large subscribing businesses, but also to individual consumers who signed up with MCI. MCI also sought and received regulatory enforcement of the right to purchase AT&T's long-distance services in bulk at the standard volume discounts available to AT&T's large business customers and then resell those services to customers of its own.[32]

AT&T's effective cooperation in providing nondiscriminatory access to its local exchanges was essential to MCI's prospects in the long-distance market. Because MCI could not possibly duplicate the Bell System's local facilities, its own long-distance network would be of little use if, because of Bell System recalcitrance, it had no feasible way to connect its network to its customers *and* to the parties those customers wished to call.

AT&T understood that the denial of effective interconnection was a powerful anticompetitive tool, just as it had understood the same fact 60 years earlier during the events leading up to the Kingsbury Commitment (see chapter 1). As a consequence, AT&T fought tooth and nail to deprive MCI of effective access to its network. At one point, it even unplugged from its local networks the supposedly "open" ends of the FX lines MCI had sold to its customers.[33] AT&T sought to justify its anticompetitive conduct on several grounds. Most fundamentally, it argued that permitting MCI to cherry-pick AT&T's highest-margin customers in the long-distance market—those who had been paying supracompetitive rates for many years—would undermine the commitment to "universal service" and would require substantial increases in local service rates in order to support the maintenance of the nation's telephone network.

In several different contexts, courts and regulators rejected AT&T's policy justifications and awarded its competitors a series of incremental victories until in 1982 AT&T and the Justice Department entered into an antitrust consent decree (consummated in 1984) that required the complete divestiture of the Bell System's local exchange facilities from AT&T.[34] The decree also prohibited the newly independent Bell companies from providing long-distance services themselves until they had satisfied the antitrust court that doing so posed no threat of anticompetitive behavior.

There were two basic rationales for splitting up AT&T and for quarantining the Bell companies, for the most part, to local telecommunications markets. The first was a concern about operational *discrimination*. AT&T had already agreed to interconnect with MCI's long-distance network—that is, to allow the use of the Bell companies' local exchange facilities to originate calls bound for MCI's network and to complete calls on the other end. But there are many subtle ways in which a dominant carrier can create interconnection problems for new entrants. For example, it can reserve insufficient capacity on its interconnection trunks to meet customer demands for access to the rival long-distance company's network during peak calling periods. AT&T Long Lines had given customers good reasons to perceive that because of AT&T's affiliation with the Bell companies, they would get more dependable service from AT&T than from MCI. The fear was that if the newly independent Bell companies were permitted to enter the long-distance market, they—like the integrated AT&T before them—would find ways to discriminate in favor of their own long-distance operations.

For good measure, the antitrust decree further subjected the Bell companies not just to this line-of-business restriction, but to affirmative *equal-access* obligations as well. These requirements directed the Bell companies to upgrade their equipment to give AT&T's long-distance rivals the same access as AT&T itself to the Bells' local networks. For example, in the 1980s many subscribers who wished to avail themselves of MCI's low rates had to dial an access code first to reach MCI's network, wait for a second dial tone, and only then enter the digits of the party they wished to call. These extra dialing steps, which AT&T's long-distance customers never had to take, inconvenienced MCI's customers and thus disadvantaged MCI in the long-distance market. *Dialing parity* requirements, a type of equal-access obligation, fixed that problem by directing each local telephone company to reconfigure the connections between its switches and the networks of various long-distance companies so that an end user could "presubscribe" to the long-distance provider of her choice and needed to dial only "1" plus the called party's number to have her long-distance calls carried by that provider.

The other basic concern underlying the Bell companies' exclusion from the long-distance market related to predatory *cross-subsidization*, to which we referred in the equipment-manufacturing context. Under traditional rate-of-return regulation, the rates that the predivestiture AT&T could charge its local customers for particular services were set in part on the basis of the "costs" that appeared in its accounting books in connection with those services, plus a reasonable profit. But AT&T had a great many costs and a great many services, and matching particular costs to the particular services that "caused" them was an exercise in extreme subjectivity. Costs that were "joint and common" to a number of services—such as loop costs—were especially subject to manipulation because they did not truly belong to any one service category. Thus, whenever competition arose, AT&T enjoyed considerable discretion to assign costs away from competitive markets—thereby lowering prices and underselling rivals—and to attribute those costs instead to its operations in uncontested markets, where its captive customers would be forced to pay marginally higher rates. If successful, these surgical strikes on new entrants would enable AT&T to retain its monopoly position in most markets. Here again, the fear was that the Bell companies, if permitted to enter the long-distance market after their separation from AT&T, would exploit the same anticompetitive opportunities as AT&T's integrated Bell System.

In the years following entry of the consent decree, these twin concerns—operational discrimination and predatory cross-subsidization—gradually became less compelling as a justification for the Bells' line-of-business restrictions. First, price-cap regulation, both on the federal level and in many states, significantly alleviated the risk of predatory cross-subsidization. In particular, it prevented the Bell companies from obtaining near-automatic recovery of their book costs and thereby reduced their incentive to allocate all joint and common costs to their regulated operations while slashing prices to undercut competition in contested markets.[35] Indeed, the many local telephone companies throughout the country that were *not* the offspring of AT&T's Bell System *were* permitted to offer long-distance service during this period, and the FCC had developed accounting and other safeguards to protect unaffiliated long-distance carriers from predatory conduct.[36] As for concerns about operational discrimination, years of successful administration of the equal-access rules—in both Bell and non-Bell territories—had raised questions about whether there was still a need for full-blown line-of-business restrictions to protect competition in the long-distance market.[37]

Citing these developments, the Bell companies argued to the antitrust court in the years between divestiture and 1996 that the decree's outright restriction on their entry into the long-distance market was no longer warranted. In 1996, Congress largely rejected these arguments by perpetuating these line-of-business restrictions in statutory form, but it did give the Bell companies a statutory mechanism for overcoming those restrictions, as we discuss later in this chapter.

C. The rise of competitive access providers

Telecommunications competition first arose in the long-distance market largely because the enormous call volumes *between* population centers, as opposed to *within* them, were adequate to support profitable entry by more than one carrier. Just as it is much cheaper (all else being equal) to deploy a single cable to a 100-unit apartment building than 100 different cables to 100 farms, it is also much cheaper to run a single multicircuit transport pipe from one city to another than to disperse an equivalent number of circuits among many smaller pipes into hundreds of neighborhoods. In the 1980s, the next site of facilities-based competition predictably appeared in the market for special-access links. This was not just because the Bell incumbents were sometimes providing technologically inferior access services and not just because they were often providing access services at rates well above cost, but also because high traffic volumes in the access market enabled competing carriers to enjoy large scale economies even if they served only part of the customer base.

As noted earlier in this chapter, special-access circuits are leased high-capacity lines that connect two different geographic points. For example, such a line might connect an office building that has large call volumes directly to a global voice and data network, bypassing the incumbent telephone company's local switch. Until the 1980s, these access services were provided overwhelmingly by the incumbent local telephone companies themselves and were priced well above cost. In the 1980s, *competitive access providers* began enabling long-distance carriers to bypass increasingly large segments of local Bell company networks en route to the long-distance carriers' business customers. Beneath the streets of America's major cities, these competitive access providers laid extensive fiber "rings" that offered not just enormous bandwidth, but also critical "self-healing" properties that ensured network reliability: because signals could move in either direction around a given "ring" en route to a network node, customers remained connected even if a line was cut; the signals would simply move in the opposite direction. In short, if a long-distance company wished to provide its customers with a range of voice and data services, it could bypass the Bell networks in whole or in part by contracting with a competitive access provider. In so doing, the long-distance company would avoid Bell's above-cost access charges (at least to some extent),[38] pay lower rates to these competitive access providers, receive more responsive service, and enjoy higher-quality performance from the state-of-the-art digital technology that the Bells themselves were slow to provide.

In the years immediately preceding passage of the 1996 Act, the predominant "local competition" disputes in the telecommunications industry concerned the terms on which these competitive access providers could demand interconnection at an incumbent's central office when scale economies did not permit them to lay their own cables all the way to each individual end user. In the early 1990s, the FCC issued its *Expanded Interconnection Orders*, in which, among other things, it entitled these new providers to "collocate" (i.e., "co-locate") their own equipment for this purpose in specially designated areas within a central office.[39] As discussed later, that initiative sparked a round of legal wrangling that extended through 1996, when Congress codified the collocation rights of competing carriers, and even then the disputes persisted for another six years until in 2002 the courts finally upheld FCC regulations defining exactly how far those rights should go.[40]

D. The first steps toward local exchange competition

By the mid-1990s, regulators had taken their first tentative steps toward promoting competition not just for *access* services (direct connectivity with independent long-distance networks), but also for *local exchange* services: the provision of local telephone service over competitive networks that were (1) capable of local switching, (2) interconnected with the incumbent telephone company, and (3) owned by companies unaffiliated with that incumbent. Where local exchange competition arose, it gave consumers a choice not only among long-distance providers, but also among local telephone companies. As we shall see, however, much of this new local exchange competition was "non-facilities-based," in whole or in part: the new entrants were often leasing lines and sometimes switching capacity from the local telephone incumbents.

Regulatory interest in promoting local exchange competition first arose in three principal and complementary contexts. First, a number of states, such as New York and California, began experimenting with creative schemes under which new entrants could interconnect with the incumbent's network and lease capacity on its facilities at low wholesale rates to provide competing local exchange services. Such leasing arrangements were indispensable to local exchange competition in all but the most densely populated business districts because, given scale economies, there was generally little economic justification for a new entrant to build its own lines all the way out to each customer location. Among the first carriers to take advantage of these opportunities were the competitive access providers, which had already built transport networks (i.e., fiber rings) throughout many of the major downtown business districts.

Second, as part of the *Computer Inquiries,* the FCC devoted several years in the late 1980s and early 1990s to the development of complex regulatory schemes—known by the names *comparably efficient intercon-*

nection and *open network architecture*—designed to give data carriers and other information service providers "unbundled" access to an incumbent's local network services on the same terms enjoyed by the incumbent's own information services affiliate. These schemes are briefly noted in chapter 6. Third, in 1995 the Justice Department and Ameritech (a midwestern Bell company) entered into an agreement, quickly aborted after the 1996 Act became law, under which the Department promised to seek relaxation of the consent decree's line-of-business restrictions if Ameritech implemented various interconnection and leasing requirements designed to open local markets to competition.[41]

The immediate practical consequences of all this regulatory experimentation were meager. By early 1996, the incumbent telephone companies still provided nearly all local exchange services throughout the United States. But the legacy of these early initiatives was nonetheless profound. When the bill that later became the 1996 Act was being written, the drafters' immediate influences were these three regulatory experiments: by the states, by the FCC, and by the Justice Department.

IV. Wireline Competition Under the 1996 Act

Congress enacted the Telecommunications Act of 1996[42] with much fanfare, and many observers hailed it as the most important regulatory legislation of any kind since the New Deal. In characteristically turgid prose, the House–Senate Conference Committee called the Act a "pro-competitive, de-regulatory national policy framework designed to accelerate rapidly private sector deployment of advanced telecommunications and information technologies and services to all Americans by opening all telecommunications markets to competition."[43] In retrospect, such sentiments seem almost quaint. As we shall see, the 1996 Act has become increasingly anachronistic because its drafters did not fully anticipate, among other developments, the rise of the broadband Internet and its radical reordering of the telecommunications industry. The Act remains on the books, however, and interpretive disputes about its more cryptic provisions still have profound influence over today's policy debates, ranging from net neutrality to federal funding for broadband deployment.

A. The 1996 Act's main objectives

We begin with what Congress hoped to achieve through this legislation. As relevant to wireline telecommunications policy, the Act's objectives can be broken down into three main categories: eliminating economic barriers to entry in local telecommunications markets, eliminating regulatory barriers to entry in all telecommunications markets, and universal service reform.

First, Congress added sections 251 and 252 to the Communications Act in order to jump-start competition in local exchange markets.* These so-called local competition provisions mandate interconnection between rival carriers and give new competitors vaguely defined rights to lease the incumbents' network facilities or capacity on those facilities as a means of providing competing local services. The 1996 Act immediately spawned protracted litigation about the precise scope of those leasing rights—litigation pitting the Bell companies and other *incumbent local exchange carriers*, or ILECs (EYE-lecks), against their new local exchange rivals, *competitive local exchange carriers*, or CLECs (SEE-lecks). We discuss these leasing controversies in greater detail later in this chapter.

Second, Congress sought to eliminate regulatory, not just economic, barriers to competitive entry. In section 253(a), it provided that "[n]o State or local statute or regulation, or other State or local legal requirement, may prohibit or have the effect of prohibiting the ability of any entity to provide any interstate or intrastate telecommunications service." In that one sentence, Congress drove the last nail into the coffin of exclusive franchise arrangements, which in a number of areas still protected telephone monopolies against competitive entry.[44] A separate provision relaxed federal restrictions on the provision of video programming over a telephone company's lines.[45] Although Congress took limited steps to ensure competitive parity between telephone and cable companies in the provision of voice and video services, the real competition between wireline and cable platforms arose in a market that hardly existed in 1996: the market for broadband Internet access. Because Congress did not foresee that market, it provided no clear legal framework for it. Chapter 6 discusses the long-term regulatory uncertainty that this omission has created.

* The 1996 Act formally "amends" the Communications Act of 1934. For the most part, the section numbers we cite to denote provisions added by the 1996 Act are technically sections of the amended Communications Act itself, not of the 1996 Act. Most, but not all, of the major sections of the Communications Act correspond to the sections of Title 47 of the United States Code. Readers may assume that correspondence when we discuss a particular section number without providing a formal U.S. Code citation.

The 1996 Act further eliminated the quasi-regulatory wall that the AT&T consent decree had erected between "local" and "long-distance" providers of wireline telecommunications services. It struck a grand political compromise between the Bell companies and the major long-distance carriers, including the still-independent AT&T and MCI. In particular, it nullified the consent decree and provided, in section 271, that a Bell company could offer long-distance service to its wireline customers in a given state once it persuaded the FCC, among other things, that it had opened its local exchange markets in that state to competition.[46]

Because AT&T and MCI were expected to rank among the primary providers of local exchange competition, the 1996 Act thus triggered a race between the Bells and the long-distance companies to see who would be the first to provide a complete bundle of local and long-distance services to consumers. The Bells eventually won that race. They obtained section 271 authorization from the FCC for all fifty states by late 2003, but the long-distance carriers, undermined by judicial invalidation of liberal FCC leasing rules, could find no cost-effective way to provide local exchange services to their quickly shrinking lists of residential long-distance customers. As discussed later in this chapter, AT&T and MCI, left without a sustainable business plan, ultimately agreed in 2005 to be acquired by their Bell adversaries.

In another provision designed to eliminate regulatory barriers to competition, Congress directed the FCC to "forbear from applying" federal statutory or regulatory requirements that in the FCC's view are no longer necessary to preserve "just and reasonable" terms of service, protect consumers, or serve the public interest.[47] This *forbearance authority* is extraordinary: in effect, it authorizes the FCC to veto acts of Congress that it deems superseded by technological or competitive developments. The FCC has relied on this forbearance authority to achieve a variety of deregulatory objectives, ranging from the detariffing of consumer long-distance services to the elimination of certain facilities-leasing obligations.[48]

Third, in the universal service provisions of section 254, Congress took steps to maintain low-priced telephone service for residential customers in sparsely populated areas. This third objective can be challenging to reconcile with the first two goals. As explained earlier, new entrants in local telecommunications markets can often cherry-pick the customers who would otherwise have paid the incumbent the above-cost rates that the incumbent needs to collect in order to support "affordable" (below-cost) rates for households in rural areas with unusually high per line costs. In the face of such cherry-picking, the incumbents will gradually forfeit these lucrative customers altogether or will be forced to lower their rates in order to keep them. In either event, they will eventually lose the source of the implicit cross-subsidies that allowed them to serve other customers at a loss and still balance their books.

Section 254 seeks to address this challenge by ordering the FCC to set up a competitively neutral "universal service fund" that operates like a specialized taxation system. Under this system as the FCC has implemented it, carriers contribute money into the fund on the basis of their retail "interstate" revenues, and a federal administrator then doles out the money in the form of explicit subsidies to ensure "affordable" and "reasonably comparable" rates throughout the country. As we discuss in chapter 8, section 254 gave the FCC considerable leeway (working in tandem with the states) to manage the transition toward competitively neutral subsidy mechanisms, but this transition moved far more slowly than policymakers initially expected.

B. The nuts and bolts of the 1996 Act

Most of the policy disputes arising under the 1996 Act have concerned the "local competition provisions": sections 251 and 252. Understanding those disputes requires some familiarity with the Act's obscure terminology and regulatory classifications. As discussed in chapter 6, this terminology is relevant not only to the local competition provisions themselves, but also to the FCC's authority over broadband Internet access services.

A taxonomy of carriers and services

Sections 251 and 252, like the rest of Title II of the Communications Act (into which they were inserted), address the relationships among "telecommunications carriers." These are the carriers that provide "telecommunications services," defined as basic transmission services offered "for a fee directly to the public, or to such classes of users as to be effectively available directly to the public, regardless of the facilities used."[49] The term *telecommunications service* is essentially synonymous with the more traditional term *common carriage service*.[50] Under the usual test, a "common carrier" is a provider of transmission services that (1) holds itself out to serve all customers interested in buying any services the carrier offers and (2) allows customers to transmit whatever content they wish by means of its facilities.[51] Stated simply, common carriers—as opposed to "private carriers"—do "not make individualized decisions, in particular cases, whether and on what terms to deal."[52] Moreover, many carriers—such as conventional telephone companies—are not generally permitted to act as private carriers in their traditional service markets but face regulatory obligations to act as common carriers whether they would like to do so in a particular context or not.[53]

"Telecommunications services" run the gamut from conventional wireline and wireless voice services to high-capacity special-access services. But not every service that involves telecommunications is a "telecommunications service" or "common carriage service." For example, when the telephone company sells you broadband access to the Internet, it is providing that service *by means of* "telecommunications." But under the FCC's interpretation of the relevant statutory definitions, formalized in 2002 and upheld by the Supreme Court in 2005,[54] the company is selling you an "information service";[55] it is *not* selling you a "telecommunications service" and is therefore not acting as a "telecommunications carrier." Keep these esoteric details in the back of your mind as you read through the following chapters; they will reappear with a vengeance in chapter 6.

"Local exchange carriers" (LECs) are a species of telecommunications carriers and are defined as companies that provide either local exchange service or "access" services or both.[56] Not all telecommunications carriers are LECs; for example, pure long-distance companies on the PSTN are telecommunications carriers, but not LECs. By special statutory fiat, mobile wireless carriers are also not LECs unless and until the FCC says otherwise, but they are telecommunications carriers.[57]

The world of LECs, as noted earlier, is subdivided into ILECs and all other LECs (CLECs).[58] The largest incumbents by far are the regional Bell companies, the local exchange progeny of the pre-divestiture Bell System. There were seven regional Bell companies in 1996, but now, as the result of consolidation, there are only three (discussed here at the holding company level).[59] The largest of them, AT&T, is the product of several waves of consolidation since 1996. Three original Bell companies—Southwestern Bell, Pacific Telesis, and Chicago-based Ameritech (plus the non-BOC Southern New England Telephone)—combined in successive mergers in the late 1990s. The ensuing company (SBC Communications) then acquired its former parent, the long-distance carrier (and CLEC) AT&T Corp., in 2005 and took its better-recognized brand name. And that company then acquired BellSouth, an original BOC, in

2006. The company's wireless affiliate—AT&T Mobility—is itself the product of many mergers, as discussed in chapter 4.

The second-largest Bell company is Verizon, a combination of original BOCs Bell Atlantic and NYNEX (plus the non-BOC GTE), which acquired the long-distance carrier (and CLEC) MCI in 2005. Verizon owns a controlling interest in Verizon Wireless, but U.K.-based Vodafone holds a large minority share. The final Bell company in 1996 was US WEST, which covered a fourteen-state region in the western half of the United States. In 2000, it was acquired by Qwest, which in turn was acquired by Louisiana-based ILEC CenturyLink in 2011.

There are hundreds of smaller ILECs throughout the country, particularly in rural areas, that were never part of the Bell System and remain "independent carriers." Many of these independents are small enough to qualify as "rural" carriers, a characterization that entitles them to larger universal service subsidies than the Bell companies and, under section 251(f), special statutory exemptions from the most intrusive requirements that the 1996 Act imposes on ILECs generally.

CLECs are defined by their customers, which, at the highest level of generality, fall into two different categories. The most important category today consists of *enterprise customers:* large businesses, often with multiple branch offices, that generate massive voice and data traffic on a daily basis (think Merrill Lynch or General Electric). CLECs such as tw telecom and XO Communications typically connect an enterprise customer directly to their networks over high-capacity lines. In the most densely populated urban areas, they often own these lines outright, but particularly as one moves from urban cores to more suburban areas, they often need to lease capacity on the facilities of other telecom companies, including ILECs (usually in the form of special-access services).

The second category of CLEC customers consists of smaller business and residential subscribers, known as *mass-market customers*. For example, in the first years of the new millennium, the major long-distance carriers—including premerger AT&T Corp. and MCI—served millions of mass-market customers by exploiting FCC regulations entitling them to lease, directly from the ILECs at regulated rates, capacity on all ILEC facilities needed to provide a complete bundle of local and long-distance services. As discussed later in this chapter, a 2004 court decision invalidated those regulations and deprived competitive carriers of these virtually unlimited leasing rights. Today, the most important fixed-line competitors in the mass market are not companies that rely heavily on leasing rights under the 1996 Act, but cable companies such as Comcast and Time Warner Cable that do not need such rights because they are completely facilities based.

The basic distinction between CLECs and ILECs, statutory definitions aside, is that the latter have traditionally possessed market power in local telephone markets. ILECs are therefore regulated much more heavily than CLECs, and ILEC-specific obligations are the focus of our discussion. Nonetheless, in sections 251(a) and (b) Congress imposed a handful of obligations on all LECs, including all CLECs, with potential exceptions for only certain rural carriers.[60] For example, section 251(b)(2) directs CLECs and ILECs alike to cooperate with regulatory efforts to make telephone numbers "portable" for customers that change carriers. Section 251(b)(4) requires all LECs to share their "poles, ducts, conduits, and rights-of-way" with other telecommunications carriers.[61] All LECs must further allow other carriers to purchase their services for resale to the public (section 251(b)(1)); interconnect with other carriers (section 251(a)); and strike "reciprocal compensation" deals to ensure mutual recovery of the costs of calls involving the networks of more than one carrier (section 251(b)(5)).

Apart from questions about the "reciprocal compensation" mandate (see chapter 7), the lion's share of controversy about the 1996 Act involves the provisions, all set out in section 251(c), that specify obligations imposed only on ILECs. The rules we are about to describe are *default rules* only. In theory, ILECs and CLECs are free to negotiate whatever arrangements they like so long as they do not discriminate against third parties. But given the inability of ILECs and CLECs to agree on very much, in part because no company has any incentive to agree to outcomes less favorable than what it can receive from regulators, these default rules end up governing the most important aspects of local competition.

Interconnection and collocation rights

Although section 251(a) requires all "telecommunications carriers" to interconnect directly or indirectly on *some* terms, ILECs alone are subject to a highly specific set of rigorous interconnection obligations. Under sections 251(c)(2) and (c)(6), CLECs may demand interconnection with the ILEC's network at "any technically feasible point," not just at a location of the ILEC's choosing, and may assert rights to collocate their facilities on the ILEC's property at tightly regulated rates based on "cost."

In plain English, this provision means, among other things, that any CLEC may rent space in an ILEC's central office (the building that houses the ILEC's switch); place its equipment there to interconnect with the ILEC's network; and purchase various related services, such as power and air conditioning, from the ILEC. A CLEC may also place its equipment in the ILEC's central office if it has leased some of the ILEC's loops for purposes of serving the customers at the other end and needs to link all these loops, via a high-capacity transport pipe, to its own switch.

For many years, the FCC and the courts engaged in a game of legal ping-pong about the limits on the Commission's authority to impose liberal collocation rules at the expense of an ILEC's property interests. The ultimate result, reached in 2002, is that CLECs may demand room to place necessary equipment in an ILEC's central office so long as (1) the primary function of the collocated equipment is to allow interconnection or access to an ILEC's transmission facilities, (2) any additional functions are logically related to that primary function, and (3) the additional functions do not increase the relative burden on the ILEC's property interests.[62]

The rights described here are those of a CLEC to interconnect with an ILEC by placing its equipment on the ILEC's property (or, in an arrangement known as "virtual collocation," by directing the ILEC to dedicate some of its own equipment to the same task). Under section 252(d)(1), regulators normally limit the rates a competitor must pay for such interconnection to the ILEC's costs of hosting the relevant equipment in its central office. These collocation rights are distinct from *intercarrier compensation* rules, which, for PSTN calls traversing multiple networks, govern how much the calling party's carrier owes the called party's network for delivering the call from the point of hand-off to the called party. We discuss the latter rules in chapter 7.

Network elements and leasing rights

The interconnection rights just described went a long way toward reducing any anticompetitive advantage that network effects might otherwise have given ILECs in the local exchange market. As discussed in chapter 1, however, interconnection rights do very little by themselves to help new entrants match the scale economies of incumbent providers with their ubiquitous local networks. Section 251(c)(3), which entitles new entrants to lease capacity on certain ILEC facilities at regulated prices, is designed to enable the entrants to share in those scale economies up to a point.

Suppose that you manage an upstart telecommunications carrier and wish to provide both local and long-distance service to residential customers in a suburban neighborhood. Precisely because your potential customer base is small and your scale economies are low, it might well be financially infeasible for you to incur the enormous fixed costs needed to build a complete new wireline network out to your prospective new subscribers. One way out of that Catch-22 is to lease existing lines from the ILEC so long as the lease rates are low enough. Over time, if your customer base grows, increased scale economies might then justify the construction of your own network.

In sections 251(c)(3) and 252(d)(1), Congress facilitated this process by giving new entrants a right not just to interconnect with incumbents, but to obtain "access to [the ILECs'] network elements on an unbundled basis"—that is, to lease capacity on the incumbents' network facilities—at regulated "cost"-based rates. These facilities are called *unbundled network elements*, or UNEs (pronounced YOO-neez). In this context, to say that network elements are available "on an unbundled basis" is simply to say that the competitor may, if it wishes, lease them individually at separate rates or in combinations of its choosing.[63]

Although we often use the phrase "leasing the incumbent's facilities" as an intuitive shorthand for the statutory concept of "obtaining access to network elements," the latter phrase is sometimes technically more accurate. What a competitor receives when it invokes rights under section 251(c)(3) is not always a discrete physical "facility" as such—although it can be, as in the case of a copper loop. Often, the competitor receives only *capacity* on such a facility, along with its "features, functions, and capabilities."[64] For example, when a competitor leases "dedicated transport" as a network element from an incumbent, it does not normally lease an entire fiber-optic cable; instead, it leases a fixed increment of capacity on that cable. Indeed, the FCC went one step further and, until 2005, permitted competitors to lease, as "network elements," not just *fixed* increments of capacity, but *variable* (per minute) increments as well—as in the case of access to an incumbent's switch (as discussed later in this chapter).

A related provision, section 251(d)(2), directs the Commission to limit the network elements subject to unbundling under section 251(c)(3) by "consider[ing], at a minimum, whether . . . the failure to provide access to such network elements would impair the ability of the telecommunications carrier seeking access to provide the services that it seeks to offer."[65] This statutory mandate is known as the *impairment standard*. In essence, this provision tells the FCC to identify, at some level of generality, the network elements that competitors truly need to lease from the ILEC in order to compete and to limit the unbundling obligation to those elements alone. The FCC construes the phrase "at a minimum" as giving it some discretion to impose or withhold unbundling obligations in the service of larger statutory goals, even when doing so may be in tension with the formal outcome of the "impairment" inquiry.[66]

Which network elements are subject to leasing rights constitutes only half the regulatory battle; the other half involves *how much CLECs must pay* to exercise those rights. Section 252(d)(1) provides that leasing rates, like the rates for collocation arrangements, "shall be based on the cost . . . of providing" the relevant facilities. In 1996, just after the Telecommunications Act was passed, the FCC construed this standard to require states to base network element rates on *forward-looking cost* rather than on "historical" or "embedded" cost. This means that when a competitor leases an incumbent's network assets to provide services of its own, the rates it pays the incumbent are calculated on the basis of what it would cost today to obtain those assets or their functional equivalent, not what it actually cost the incumbent to obtain the particular facilities at issue, as recorded on its books.

Since 1996, the FCC has required the use of a particular forward-looking methodology it has dubbed *total element long-run incremental cost* (TELRIC) for all network elements subject to leasing rights under sections 251(c)(3) and (d)(2). TELRIC controversially bases forward-looking costs not on any given ILEC's network design and technology mix, but on what it would cost a hypothetical "most efficient" carrier to build an entirely new network from scratch today, taking as given only the locations of the ILEC's existing switches.[67] This approach sometimes produces "cost" estimates for network elements—and thus prescribes rates at which CLECs may lease those elements—far below the ILECs' historical costs. In its 2002 decision in *Verizon Communications Inc. v. FCC*, after six years of litigation and uncertainty, the Supreme Court upheld TELRIC against claims by ILECs that this methodology necessarily produces network element rates so low that they unlawfully "strand" (deny recovery on) past investments and undermine both ILECs' and CLECs' incentives to invest in new facilities.[68]

Procedures for implementing the local competition provisions

The text of the 1996 Act is notoriously unclear on the FCC's and states' respective roles in giving practical effect to the local competition provisions. In October 1996, the Eighth Circuit, sitting in St. Louis, stayed many of the FCC's implementing rules, including the TELRIC pricing methodology, on the ground that the Commission had no statutory authority to adopt such rules.* The court believed that in designing the 1996 Act, Congress had largely meant to follow the traditional "dual-jurisdiction" framework of the Communications Act of 1934. As discussed earlier, that framework, where applicable, divides the subject matter of telecommunications law into separate interstate and intrastate spheres and under section 2(b) fences off the FCC from the latter. The Eighth Circuit concluded that matters relating to local exchange competition are essentially intrastate in character and that Congress meant to give individual state public utility commissions, not the FCC, authority to resolve most of those issues. In the Eighth Circuit's words, section 2(b) "is hog tight, horse high, and bull strong, preventing the FCC from intruding on the states' intrastate turf."[69]

This barnyard metaphor remained the law until January 1999, when the Supreme Court reversed the Eighth Circuit in *AT&T Corp. v. Iowa Utilities Board*.[70] By a five–three margin, the Court ruled that Congress had implicitly given the FCC general jurisdiction to adopt preemptive regulations fleshing out the local competition provisions. The Court relied for this conclusion on section 201(b) of the original Communications Act, which provides that "[t]he Commission may prescribe such rules and regulations as may be necessary in the public interest to carry out the provisions of this [Act]." Writing for the majority, Justice Antonin Scalia held that this provision "means what it says: The FCC has rule-making authority to carry out the 'provisions of this Act,' which include §§ 251 and 252, added by the Telecommunications Act of 1996."[71] In the sphere of local competition, the Court thus replaced the dual-jurisdiction framework with a new model of cooperative federalism, which erases the distinction between "interstate" and "intrastate" matters and directs the FCC and state commissions to work together in complementary capacities to implement the local competition provisions.[72] In particular, the FCC establishes the basic rules governing local competition matters, and state public utility commissions apply those rules in resolving specific carrier-to-carrier disputes.[73]

The resolution of such disputes is governed by the procedural provisions of section 252. The process begins when a competitor asks an

* Parties "aggrieved" by FCC orders may challenge them directly in intermediate federal appellate courts. *See* 47 U.S.C. § 402. Those courts are arranged by circuits, corresponding to the District of Columbia and various regions of the country. Thus, when we refer to a decision of the "Eighth Circuit" or "D.C. Circuit," we are referring to an intermediate appellate court whose decisions are subject to review only by the U.S. Supreme Court.

incumbent to enter into an "interconnection agreement" containing the key terms that will govern the relationship between the two carriers for a period of years. One of two things might happen. First, the two carriers might resolve all relevant issues without regulatory intervention. In that event, they simply file their completed agreement with the relevant state commission, which in turn must approve it so long as it does not harm third parties or otherwise threaten the public interest.[74]

Alternatively, the negotiations might break down. In that event, the state commission arbitrates the disputed issues under procedures spelled out in section 252, applying the rules of the 1996 Act, the FCC's implementing regulations, and any supplemental (and consistent) rules of state law.[75] Either side may appeal the state commission's order by filing suit in the relevant federal district court.[76] The court then reviews the order for compliance with federal law and, if necessary, with state law, too (under its pendent jurisdiction). If it finds problems, it remands the matter to the state commission for further proceedings.

C. The rise and fall of UNE-P

For eight years after 1996, when the FCC issued its massive *Local Competition Order*, adopting the first rules implementing sections 251 and 252,[77] it generally followed a very permissive reading of the "impairment" standard, which (as discussed earlier) governs which network elements are subject to leasing rights. For those eight years, a CLEC could generally lease from an ILEC at low TELRIC-based rates essentially all facilities and network functionalities needed to provide local telephone service. This arrangement was known as *UNE-P*, for "unbundled network elements platform." The key constituent elements in dispute were switching (variable capacity on an ILEC's switch) and shared transport (variable capacity on the ILEC transport links connecting one ILEC switch to another). During this period, the bitterest controversy in telecommunications policy concerned whether these two network functionalities satisfied the "impairment" standard of section 251(d)(2): that is, whether an ILEC's "failure to provide access to such network elements would impair the ability of [a CLEC] to provide the services that it seeks to offer."[78]

The stakes of UNE-P

Preserving UNE-P as a regulatory entry strategy was key to the survival of MCI and AT&T Corp. as independent companies. From late 1999 to 2003, the FCC approved in state after state the Bell companies' applications under section 271 to provide long-distance services to their local exchange customers. To withstand the ensuing Bell onslaught on their long-distance market shares, MCI and AT&T Corp. needed to offer the same local/long-distance bundles that the Bell companies were suddenly authorized to offer, and they therefore needed a strategy for offering local exchange services to ordinary residential consumers. But the revenues associated with residential customers did not justify the considerable expense of building a new loop out to each customer's house. Indeed, those revenues often did not justify incurring even the one-time fee ILECs were entitled to charge CLECs for sending a technician to the central-office switchboard to unplug the existing ILEC loop from the ILEC switch and plug it instead into a different circuit leading to a collocated CLEC switch (a procedure known as a "hot cut"). In short, the only feasible way that MCI and AT&T Corp. could provide local service to residential customers as part of a local/long-distance bundle was UNE-P, which did not require them to deploy local facilities at all.

The ILECs, led by the Bell companies, predictably opposed this regulatory strategy. Among other things, they argued that because UNE-P enabled CLECs to offer local exchange services without reliance on any local exchange facilities of their own, it was tantamount to resale of ILEC services. And the ILECs pointed out that the 1996 Act prescribes a different and more ILEC-friendly rate methodology for CLECs that simply resell an ILEC's retail services. Under sections 251(c)(4) and 252(d)(3), an ILEC may charge CLEC resellers its own retail rates minus the retail-specific costs (of marketing, billing, etc.) that the ILEC will "avoid" by virtue of no longer providing retail service to the customers at issue.[79] In theory, that "top-down" rate methodology entitles a CLEC reseller to a hypothetical margin if it can keep its own retail costs down. In practice, few CLECs found that entry strategy profitable. Instead, they strongly preferred the leasing approach of section 251(c)(3) because, among other considerations, that approach entitled them to pay only the cost-based ("bottom-up") rates derived from the pro-CLEC TELRIC methodology.[80] And for the same reason, ILECs wanted to keep non-facilities-based CLECs from availing themselves of TELRIC-based rates for the functional equivalent of resale.

The FCC, the "impairment" standard, and the courts

In a string of orders from 1996 to 2003, the FCC sided with AT&T Corp., MCI, and the other non-facilities-based CLECs and made the section 251(d)(2) "impairment" findings needed to preserve UNE-P as a regulatory entitlement in the vast majority of circumstances. And three times the courts invalidated those findings—the last time for good.

First, in the 1999 *Iowa Utilities Board* decision discussed earlier, the Supreme Court found that the FCC's initial interpretation of section 251(d)(2) in 1996 was poorly reasoned and unjustifiably permissive. As the Court explained, the FCC could not blithely assume, as it essentially had, "that *any* increase in cost (or decrease in quality) imposed by denial of a network element . . . 'impair[s]' the entrant's ability to furnish its desired services."[81] The Court remanded the matter to the FCC for a better-reasoned analysis of section 251(d)(2). But it did not specifically forbid the Commission, at the conclusion of its inquiry, from once again requiring incumbents to make any given element available at cost-based rates, including those elements that together constituted UNE-P.

With minor exceptions, that is essentially the outcome the FCC reached again in its November 1999 *UNE Remand Order*.[82] In reaffirming UNE-P rights in the vast majority of circumstances, this new order paid at least superficial obeisance to the "impairment" standard as interpreted by the Supreme Court. But that was not enough. In 2002, the U.S. Court of Appeals for the D.C. Circuit invalidated the crux of the *UNE Remand Order* on the ground that the FCC had once more violated section 251(d)(2)'s instruction to impose substantial limitations on unbundling rights.[83]

The premise underlying this 2002 D.C. Circuit decision, now known as *USTA I*, is that "unbundling is not an unqualified good," for it "comes at a cost, including disincentives to research and development by both ILECs and CLECs and the tangled management inherent in shared use of a common resource."[84] The role of the "impairment" standard, the court suggested, is to ensure that the "completely synthetic competition" facilitated by too liberal leasing rights (a thinly veiled reference to UNE-P) does not undermine "incentives for innovation and investment in facilities."[85] The court concluded that the Commission still had not come to grips with this basic statutory concern, had not struck an appropriate balance between too few and too many leasing rights, and had overgeneralized about competitors' "need" for particular elements without conducting any market-specific analysis. Finally, although the Commission had sought to justify its permissive leasing policies on the ground that new entrants would otherwise lack the incumbents' scale economies during the early stages of competitive entry, the court responded that "average unit costs are necessarily higher at the outset for any new entrant into virtually *any* business."[86] The "impairment" standard, the court ruled, prohibits "[a] cost disparity approach that links 'impairment' to universal characteristics, rather than ones linked (in some degree) to natural monopoly"—that is, linked specifically "to cost differentials based on characteristics that would make genuinely competitive provision of an element's function *wasteful*" over the long term.[87]

Like the Supreme Court before it, the D.C. Circuit did not (yet) explicitly *preclude* the Commission from ordering any particular set of leasing rights. Instead, the Commission remained technically free on remand to adopt whatever bottom line it wished so long as it could justify the results under the "impairment" standard as now interpreted by the Supreme Court and the D.C. Circuit. In 2003, the Commission responded with the so-called *Triennial Review Order*, in which a three–two majority decided, over the dissent of Chairman Michael Powell, to preserve UNE-P as an entry strategy in most markets by delegating much of the issue to individual state commissions, which were widely perceived as sympathetic both to that objective and to CLECs in general.[88]

The losers in this struggle—the Bell companies—were soon back before the D.C. Circuit, claiming that the *Triennial Review Order* violated the court's prior mandate in *USTA I*. In *USTA II*,[89] decided in March 2004, the D.C. Circuit vacated the FCC's UNE-P rules, this time on two grounds, one procedural and one more substantive.

The court first held that the Commission had violated its statutory obligations by "subdelegating" much of the market-specific impairment inquiry to the states. "[T]he cases," it found, "recognize an important distinction between subdelegation to a *subordinate* and subdelegation to an *outside party* [such as a state agency]. The presumption that subdelegations are valid absent a showing of contrary congressional intent applies only to the former. There is no such presumption covering subdelegations to outside parties. Indeed, if anything, the case law strongly suggests that subdelegations to outside parties are assumed to be improper absent an affirmative showing of congressional authorization."[90] The court deemed it inconsequential that "the subdelegation in this case is to state commissions rather than private organizations."[91] Congress, it held, is generally free to delegate federal power to the states as it likes, but a federal agency may not subdelegate to the states whatever responsibilities Congress gives it "absent an affirmative showing of congressional authorization."[92] Otherwise, the court concluded, "lines of accountability may blur, undermining an important democratic check on government decision-making," and the states "may pursue goals inconsistent with those of the agency and the underlying statutory scheme."[93]

The court's reaction was in some ways unsurprising. The context of the *Triennial Review Order* suggested to critics of UNE-P that the Commission was punting its section 251(d)(2) obligations to the states as a means of evading legal and political accountability for what in practice amounted to a decision to keep UNE-P alive as an entry option. But the court's rationale swept more broadly than was necessary to its bottom line. Although it suggested (without quite holding) that the FCC could delegate only "fact gathering" authority to the states in the service of the Commission's own policy decisions,[94] that proposition was never likely to emerge as the general rule, and it is unclear that the D.C. Circuit meant to suggest that it should. The FCC cannot be expected to resolve all of the myriad regulatory issues involving local competition as soon as they arise. And, of course, whenever the FCC does not decide such issues, it has effectively "subdelegated" them to the states, whether it says so explicitly or not and whether it even focuses on them or not. The states must often resolve those issues themselves if they are to discharge their own statutory role under section 252 in overseeing the negotiation and arbitration of interconnection agreements.[95] Whatever the uncertain precedential implications in other contexts, however, the court made clear that the FCC's sweeping subdelegation in the *Triennial Review Order* was impermissible.

The court then turned to the merits and rejected the substance of the FCC's "impairment" determination, which had created a nationwide presumption that CLECs are "impaired" if they lack access to the switching element (but had then allowed individual states to rebut that presumption). The court observed that in several passages of the *Order* the FCC itself had acknowledged that any such presumption, "*without* the possibility of market-specific exceptions . . . , would be inconsistent with *USTA I*."[96] The court agreed and, noting that it had just invalidated the Commission's mechanism for making those "market-specific exceptions," invalidated the national "impairment" finding as well. More generally, the court expressed skepticism that the Commission had compiled a strong evidentiary basis for concluding that the switching element was usually essential to CLEC entry.[97]

The collapse of the consumer long-distance carriers

The *USTA II* decision vacated the relevant portions of the *Triennial Review Order* and effectively eliminated UNE-P as an entry strategy,[98] and in the process it fundamentally reshaped the U.S. telecommunications industry. In July 2004, citing the demise of UNE-P, AT&T Corp.

announced that it would stop seeking new customers in the mass market for conventional telephony services. And in 2005, both AT&T Corp. and MCI agreed to be purchased by their former Bell company adversaries: SBC (which bought AT&T and adopted its better-known brand name) and Verizon (which bought MCI and eliminated its name). The acquired companies' prized assets were not their low-margin consumer long-distance operations, but their Internet backbone networks and high-end "enterprise services"—the sophisticated voice and data services purchased by large business customers. Through these mergers, SBC (now AT&T Inc.) and Verizon obtained the assets, expertise, and enterprise customer base they needed to become preeminent communications firms not just in their traditional service regions, but in major metropolitan areas throughout the country and the world.

A number of industry groups and consumer advocates strongly opposed the SBC–AT&T and Verizon–MCI mergers when they were announced in early 2005. The opponents argued that by reconstituting the local and long-distance operations of the original Bell System within SBC's local service region, the SBC–AT&T merger would re-create the discriminatory evils that the 1984 divestiture was designed to prevent. And the opponents likewise resisted the merger of Bell offspring Verizon with longstanding AT&T rival MCI. The merging parties responded that radical changes in the telecommunications marketplace during the ensuing 20 years had erased those discrimination concerns. They argued that the Bell companies, which faced no local exchange competition to speak of in 1984, now confronted greater local competition than ever before—from wireless, cable, and voice over Internet protocol (VoIP) providers—and that such competition would itself discipline any anti-competitive conduct. More generally, they added, regulators in 2005, unlike regulators in 1984, had developed effective regulatory safeguards against discrimination in local exchange markets.

The merging parties essentially won this debate: the Justice Department and the FCC not only approved both mergers in late 2005 but imposed less onerous conditions on the merging parties than many industry analysts expected.[99] Because AT&T and MCI had recently stopped marketing services to new residential customers, the merger-review authorities focused instead on whether these combinations would unduly increase concentration in the business-oriented market for special-access services. In each merger proceeding, the authorities were concerned that in some downtown business areas the merging parties were the only two companies that owned fiber-optic facilities within striking distance of a number of large office buildings. The Justice Department ultimately conditioned its approval on the merging parties' commitment to divest fiber-optic capacity on designated routes to rival special-access providers. The FCC followed up with a few conditions of its own, extracting from the merging parties, among other "voluntary" concessions, a commitment to comply for at least two years with several core net neutrality principles (see chapter 6). When AT&T acquired BellSouth a year later, the FCC extracted further (but still temporary) commitments relating to net neutrality and several other policy issues.[100]

Quite apart from their commercial impact, these mergers marked the end of a 20-year era in regulatory advocacy. Since AT&T's 1984 divestiture of its local exchange operations, the wireline telecommunications industry was characterized by disputes between the regional Bell companies, of which Verizon and SBC had become the largest, and their wireline competitors, of which the traditional long-distance giants AT&T and MCI were the most prominent and politically influential. The elimination of those two companies as independent actors has thus reshaped not merely the commercial landscape, but the political dynamics of telecommunications policy.

D. Wireline broadband facilities and the 1996 Act

When *USTA II* was decided in 2004, it was already apparent to many observers that the ILECs' UNE-P victory was a sideshow. The future of the telecommunications industry lies in mobile wireless services and in packet-switched, Internet-oriented technology, not in circuit-switched wireline telephony. UNE-P is by definition a narrowband, circuit-switched platform and is thus aptly characterized as yesterday's technology. The true threats to ILECs were not long-distance companies, but technological change: both the increasing tendency of American households to "cut the cord" by canceling landline service in favor of cellphones and the convergence of voice and data services over broadband Internet connections, including those of cable companies.

Several years into the new millennium, the nation's largest ILECs—Verizon and postmerger AT&T Inc.—looked for ways to stem the tide by increasing their focus on their affiliated wireless operations and by deploying new broadband networks capable of competing with cable operators in the provision of triple-play services. We address the wireless industry in chapter 4 and the ambiguous regulatory framework for broadband Internet access *services* in chapter 6. Here, we focus more narrowly on the FCC's approach to CLEC rights to *lease* ILEC broadband *facilities* under the 1996 Act.[101]

The 2003 *Triennial Review Order* addressed not only whether UNE-P would remain available for ordinary telephone services, but also whether CLECs should have regulatory entitlements to lease broadband-related ILEC facilities. CLECs claimed that unless regulators gave them such entitlements, the mass market for broadband services would indefinitely remain a duopoly shared between ILECs and cable companies. ILECs responded that such unbundling obligations would depress their incentives to invest in broadband deployment. And they added that such obligations would violate principles of regulatory parity: cable broadband providers generally had a larger share of most residential broadband markets than ILECs did and, because they were not ILECs,[102] were subject to no leasing requirements under section 251.

As a legal matter, this set of facilities-leasing disputes was governed by the same "impairment" standard applicable to all network elements. Again, section 251(d)(2) directs the FCC, when deciding which network elements should be subject to any leasing obligation under section 251(c)(3), to "consider, at a minimum, whether . . . the failure to provide access to such network elements would impair the ability of the [CLEC] seeking access to provide the services that it seeks to offer." For present purposes, there were two major categories of broadband elements subject to this standard: those relating to *line sharing* and those relating to *next-generation networks*.

Line sharing

Like radio broadcasts transmitted over the airwaves, signals traveling over copper wires occupy different frequencies, and equipment on each end can be "tuned" to receive signals only on certain frequencies, much as you can tune your radio to receive only one station at a time. DSL technology involves dividing up the frequencies of a telephone company's copper loop into a "low-frequency" band carrying signals associated with voice telephone services and a "high-frequency" portion associated with broadband Internet access. At your house, a special adapter splits the loop into a voice line leading into your telephone and a data line leading into your computer or home network. At the telephone company central office—or, in many cases, at a roadside metal box known as a *remote terminal*—a separate splitter likewise divides the loop into a voice line leading into a circuit switch and a data line leading into a packet-switched network en route to the Internet. The splitter performs the latter task in combination with a DSL access multiplexer (DSLAM).

In 1999, the FCC—following earlier state initiatives—defined the "high-frequency portion of the loop" as a separate network element and ordered ILECs to provide it on a stand-alone basis to CLECs that specialized in providing competitive DSL services, often to small businesses. An ILEC normally continued providing voice service over these lines and charged end users the standard retail rates. This arrangement, in which an ILEC provided voice service and a CLEC provided DSL over a single copper loop, was known as *line sharing*.[103]

In its 2002 decision in *USTA I*, the D.C. Circuit vacated the FCC's line-sharing rules on the ground that the FCC had inadequately considered the market leadership of cable modem service when finding that the high-frequency portion of the loop met the "impairment" standard.[104] The court reasoned that "mandatory unbundling comes at a cost, including disincentives to research and development by both ILECs and CLECs and the tangled management inherent in shared use of a common resource. . . . [N]othing in the Act appears a license to the Commission to inflict on the economy the[se] sort[s] of costs . . . under conditions where it had no reason to think doing so would bring on a significant enhancement of competition. The Commission's naked disregard of the competitive context risks exactly that result."[105]

As a legal matter, this reasoning raises more questions than it answers. At least on its face, the threshold impairment inquiry asks whether CLECs need access to the ILEC's network in order to compete, not whether ILECs themselves face cross-platform competition. And the court expressed no qualms about the FCC's determination that line sharing met that threshold impairment standard. Nonetheless, in the *Triennial Review Order* issued the following year, the FCC eliminated ILEC line-sharing obligations, albeit at the conclusion of a multiyear transition period designed to cushion the blow for CLECs.[106]

In a bizarre twist, four of the FCC's five members—all but Commissioner Kevin Martin—expressed substantive disagreement with this decision to phase out line sharing. As Chairman Michael Powell explained, "Line sharing rides on the old copper infrastructure, not on the new advanced fiber networks that we are attempting to push to deployment. Indeed, the continued availability of line sharing and the competition that flowed from it likely would have pressured incumbents to deploy more advanced networks in order to move from the negative regulatory pole to the positive regulatory pole, by deploying more fiber infrastructure."[107] The Commission's two Democratic members (Commissioners Michael Copps and Jonathan Adelstein) nonetheless reluctantly agreed to end line sharing anyway as part of a package deal with Republican Commissioner Martin to keep UNE-P available to CLECs in the narrowband voice world. As expected, the D.C. Circuit upheld the gradual elimination of line sharing in its subsequent decision in *USTA II*, but (as discussed) it eliminated the UNE-P regime that had been a critical component of the Commissioners' political compromise.[108]

Next generation networks

The abolition of line-sharing rights did not itself disentitle CLECs from providing DSL over an ILEC's copper loops. A CLEC remains theoretically entitled to lease the whole loop, albeit at the usual TELRIC-based rate (rather than zero for just the high-frequency portion of the loop). In a different portion of the *Triennial Review Order*, however, the FCC essentially foreclosed CLECs from invoking section 251 leasing rights to obtain the broadband functionality of *fiber*-based loop facilities for mass market (i.e., residential and small business) customers.

Very roughly speaking, this portion of the *Order* distinguished between the "legacy" facilities in the ILEC network—that is, the facilities the ILECs built as legacy providers of circuit-switched voice telephone service—and the next-generation fiber-oriented facilities the ILECs build as providers of broadband services for mass-market customers. The FCC largely exempted these fiber-oriented facilities from unbundling obligations under section 251 because, in a nutshell, it feared that imposing such obligations would deter ILECs from building these facilities in the first place and would likewise deter CLECs from building alternative broadband networks of their own.

The Commission thus eliminated section 251 unbundling obligations altogether for packet switches themselves and sharply curtailed such obligations for fiber pipes leading all the way from an ILEC's central office to the premises of mass-market customers—that is, for FTTH loops.[109] Verizon, for example, qualifies for this treatment in the neighborhoods to which it has deployed its all-fiber FiOS network. In "overbuild" situations, where an ILEC replaces existing copper loops with all-fiber loops, the ILEC retains an obligation either to keep the copper loops available for leasing to CLECs or to lease narrowband voice-grade circuits to requesting CLECs over the new fiber loops. In "greenfield" situations, where an ILEC builds these all-fiber loops out to new

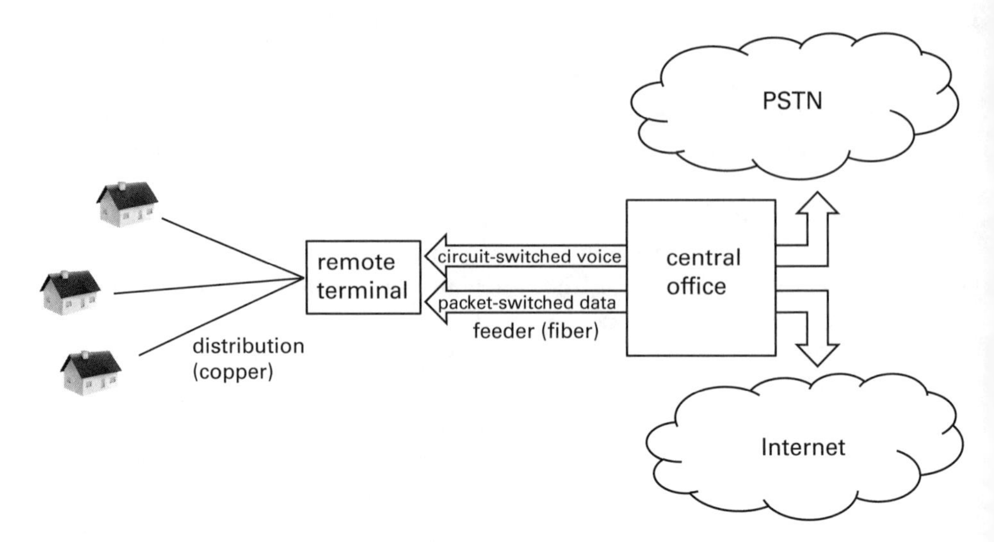

Figure 2.2
Hybrid (FTTN) loop architecture with circuit-switched voice

developments, it owes no unbundling obligations to CLECs at all. The FCC later extended these rules to *fiber-to-the-curb* loops, where the fiber portion of the loop extends from the central office to within a few hundred feet of particular homes.[110]

Somewhat more complex rules apply to *hybrid* (i.e., FTTN) loops, which consist of copper distribution lines that run from individual homes in a neighborhood and converge at remote terminals, where the signals on those lines are multiplexed and connected to fiber-optic feeder links leading to the central office (see figure 2.2).

Hybrid loops occupy a kind of middle ground between pure copper loops and pure FTTH loops in terms of both cost and performance. They are less costly to deploy than FTTH loops because they do not require a telco to install new block-by-block distribution cables, which tend to be the costliest portion of a local network. And because the speed of a DSL connection increases inversely with the length of the copper portion of the loop, hybrid loops have greater bandwidth than pure copper loops of the same length, though less than FTTH loops.

An ILEC that wishes to serve a broadband customer by means of a hybrid copper–fiber loop must install special electronics (splitters and DSLAMs) at the remote terminals where the copper distribution lines are connected to the fiber feeder cables. Although these electronics multiplex the data signals for transmission over the fiber feeder lines back to the central office, this is done by means of a packet-switched technology—not by means of the circuit-switched (TDM) technology traditionally used by different equipment at the same remote terminals for the conventional telephone network.

In the *Triennial Review Order*, the FCC wished to encourage ILECs both to build more fiber into their networks for these purposes and to install the requisite electronics at the remote terminals. To that end, it exempted ILECs from any obligation under section 251 to share the *broadband functionality* of hybrid copper–fiber loops with CLECs. The upshot was that CLECs could typically lease an ILEC's hybrid loops to provide circuit-switched voice services, but they generally had to make significant investments in loop facilities of their own if they wished to provide packet-switched broadband services to their customers.[111]

Of course, as a legal matter, the Commission had to justify these policy decisions under the "impairment" analysis of section 251(d)(2). To that end, it first found that denying CLECs the right to lease capacity on an ILEC's FTTH loops would not "impair" their ability to provide broadband services, reasoning (controversially) that ILECs had no special advantages in the deployment of such loops. The Commission faced a more complex legal challenge in justifying its Solomonic disposition of the unbundling obligations related to hybrid loops, for it acknowledged that denying access to the broadband functionality of those loops *might* somewhat impair a CLEC's ability to provide broadband services.[112]

The FCC first qualified that "impairment" finding by noting that CLECs retain competitive alternatives: for example, they can theoretically build their own fiber links out to remote terminals and lease the copper distribution lines (subloops) from the ILEC for individual customers.[113] Then, taking a page from the D.C. Circuit's rationale for eliminating line sharing, the Commission emphasized that impairment of CLEC business plans is not the end of the statutory inquiry in any event. Instead, it observed, section 251(d)(2) requires it to "consider" such impairment "at a minimum" and thus allows it to consider other factors as well. These factors include the overarching objective of section 706(a) of the 1996 Act to remove regulatory "barriers to infrastructure investment."[114]

Here, the Commission determined that imposing an obligation to unbundle the broadband capabilities of hybrid loops "would blunt the deployment of advanced telecommunications infrastructure by incumbent LECs and the incentive for competitive LECs to invest in their own facilities."[115] In concluding that "the costs associated with unbundling these packet-based facilities outweigh the potential benefits," the FCC predicted that "[t]he end result" of removing those unbundling obligations will be "that consumers will benefit from this race to build next-generation networks and the increased competition in the delivery of broadband services."[116] And, it added, any concerns about the accuracy of this prediction were "obviated to some degree by the existence of a broadband service competitor with a leading position in the marketplace"—namely, cable operators, which then generally surpassed ILECs in market share.[117]

On review, the D.C. Circuit expressed skepticism about some of the FCC's predicate factual findings,[118] but it ultimately upheld the Commission's decision to eliminate the relevant unbundling obligations for next-generation broadband elements. In its core legal holding, the court rejected CLEC arguments that the Commission's "impairment" finding for hybrid loops compelled it to grant CLECs access to the broadband functionality of those loops. Although "'impairment' [i]s the 'touchstone'" of the section 251(d)(2) analysis, the court held, the Act nonetheless compels the Commission to look beyond impairment to a "consideration of factors such as an unbundling order's impact on investment."[119] The court also found it relevant that "robust *intermodal* competition from cable providers . . . means that even if all CLECs were driven from the broadband market, mass market consumers will still have the benefits of competition between cable providers and ILECs."[120] In sum, the court concluded, the Commission's decision "not to unbundle" hybrid loops and FTTH loops "was reasonable, even in the face of some CLEC impairment, in light of evidence that unbundling would skew investment incentives in undesirable ways and that intermodal competition from cable ensures the persistence of substantial competition in broadband."[121]

V. The Twilight of Monopoly-Era Regulation

This chapter has charted the course of telecommunications policy from the first half of the twentieth century, when the local telco monopolized point-to-point communications, to the present, when tens of millions of Americans have canceled their wireline telco service in favor of wireless and cable alternatives. We have discussed how, as competition grew, natural-monopoly regulation gave way to new regulatory models. It yielded first, from the 1970s through the 1990s, to antitrust and regulatory initiatives designed to keep ILECs from suppressing emerging competition in an increasing number of markets. More recently, as ILECs have faced increasingly effective competition, regulators have begun treating them more like their competitors by relaxing monopoly-era regulatory obligations that would otherwise apply only to ILECs. This deregulatory trend presents several topics of continuing debate, two of which warrant brief attention before we close this chapter: retail rate regulation and special-access services.

A. Forbearance from monopoly-era regulation in the mass market

Over the past decade, some state public utility commissions have responded to the growth of competition from wireless and cable telephony providers by relaxing ILEC obligations to serve as carriers of last resort—that is, to provide ordinary telephone service to anyone who requests it. In the same vein, some state commissions have loosened or eliminated regulatory constraints on the intrastate rates that ILECs may charge consumers for basic local telephone service. These initiatives vary enormously from state to state and from service to service within each state, but the long-term trend is toward greater deregulation.

Meanwhile, the FCC has jurisdiction over ILEC retail rates falling within the interstate jurisdiction. As discussed, the two major categories of federally regulated retail rates are interstate long-distance charges and the flat-rated subscriber line charge (SLC), which covers a portion of local network costs. The FCC has largely deregulated ILEC interstate long-distance services.[122] And as discussed in chapter 6, it has never regulated the rates of broadband Internet access services, whether provided by ILECs, cable operators, or anyone else, reasoning that such regulation would be both unnecessary (because the market is at least somewhat competitive) and potentially harmful (because regulation might deter additional broadband deployment). But the FCC has generally maintained the federal scheme of monopoly-style rate regulation for legacy local telephone services on the theory that the market for those services is *not* fully competitive.

In theory, an ILEC can petition the FCC to "forbear" from both federal price regulation (including the SLC rules) and section 251 leasing obligations by showing that competition has deprived it of market power in a particular metropolitan area.[123] And in 2006 and 2007, the FCC granted partial forbearance for two cities where the ILECs faced particularly intense competition: Omaha and Anchorage.[124] Since then, however, the FCC has consistently denied similar forbearance requests and has made it progressively harder for ILECs to show that they have lost market power in particular metropolitan areas.[125]

In 2009, for example, Qwest (now CenturyLink) sought forbearance from facilities-leasing obligations and from dominant-carrier regulation (including SLC rate regulation) of various telecommunications services in Phoenix. Qwest emphasized that over the previous ten years it had hemorrhaged traditional telephone subscribers in Phoenix both to Cox (the local cable company) and to wireless cord cutting. In its 2010 *Phoenix Forbearance Order*,[126] the FCC rejected that forbearance request. Reversing earlier precedent, it decided to disregard wireless cord cutting for purposes of its market-power inquiry because it questioned whether wireless voice services should be treated as price-constraining substitutes for wireline voice services.[127] For example, the Commission speculated that "several classes of customers appear unlikely to drop wireline service [even] in response to a significant price increase," including those who "value the reliability and safety of wireline service," "live in a household with poor wireless coverage," or "desire a service that is more economically purchased when bundled with a local service (e.g., wireline broadband Internet service, or a video service)."[128] Although the Commission disclaimed any "suggesti[on] that mobile wireless [service] must be a perfect substitute for residential wireline services for it to constrain the price of wireline service,"[129] it signaled an abiding reluctance to rely on imperfect intermodal competition as a basis for freeing ILECs from legacy retail regulation for mass-market services.

B. The partial deregulation of special-access services

From a big-picture financial perspective, the regulatory status of residential landline telephone services is much less consequential than the regulatory status of the far more lucrative class of business-oriented services known as *special access*. As discussed, that term encompasses a diverse range of "unswitched" point-to-point transmission services—that is, the provision of high-capacity circuits to deliver signals from one place to another without the involvement of a switch in the middle. The following examples illustrate two of the most important categories of special-access services.

Example 1: enterprise services. Suppose you own a global Internet backbone network and wish to provide service to a large corporate customer—say, a bank—with its headquarters in the downtown business area of a city and branch offices in the surrounding suburbs. And suppose that you have already deployed a ring of fiber-optic cables beneath the downtown streets of that city. It may well be cost efficient for you to build an underground spur, known as a *lateral*, that connects your existing fiber ring to the headquarters building so long as it is reasonably close by. But your bank customer will want you to provide service to all of its office buildings, not just to the headquarters, and you have deployed no facilities outside the urban core. To bridge the gap between your network and the bank's suburban branch offices, you can either dig up many streets and lay miles of new fiber—an expensive proposition—or you can lease high-capacity circuits from a carrier with a more ubiquitous local network. In many cases, that carrier will be the ILEC, and the circuits you lease will constitute special-access services.

Example 2: backhaul for cell sites. As discussed in chapter 4, only a small portion of a typical wireless network is "wireless": namely, the last-mile connection between your cellphone and the nearest cell site. Once the signal from your cellphone reaches the tower, your wireless carrier needs to transmit it—usually via an underground cable—to the carrier's switching station, where it is routed to its ultimate destination; likewise, your carrier needs to transmit incoming signals *from* other parties (people, websites, etc.) *to* the cell site closest to you. This function of transmitting signals between a wireless carrier's cell sites and switches is called *backhaul*, and ILECs have historically supplied many of the special-access circuits needed to perform that function.

For the past dozen years, ILECs and their special-access customers—many of which are telecommunications carriers—have strongly disagreed about how regulated these ILEC special-access services should be. Before we address the specifics of these debates and the mechanics of special-access regulation, we first address the close relationship between special access and UNEs.

What a wholesale customer receives when it buys special-access services—leased capacity on ILEC transmission pipes (high-capacity loops or transport links or both in combination)—is technologically very similar to what it receives if it leases capacity on the same pipes in the form of section 251(c)(3) UNEs. The main difference between these two options relates not to the facilities used, but to the prices paid. Whereas UNEs are subject to the FCC's TELRIC pricing rules, an ILEC typically provides special access as an ordinary interstate service under section 201 of the Communications Act, which prescribes no specific pricing methodology and requires only that rates be "just and reasonable." With rare exceptions, the TELRIC rate is lower—often much lower—than the special-access rate for the same circuit.

Why, then, would any customer ever buy a special-access service rather than the corresponding UNEs? There are two main reasons. First, special-access services are closer to finished services and, for logistical reasons, are often simply easier for a wholesale customer to order and use than UNEs. Second, with the D.C. Circuit's prodding, the FCC has invoked section 251(d)(2)'s impairment standard to limit the availability of high-capacity loop and transport UNEs in several major respects.

For example, the FCC categorically decided in its February 2005 *Triennial Review Remand Order* to bar wireless providers from obtaining UNEs for backhaul. It explained that the wireless industry was thriving and competitive despite its traditional reliance on special-access services and that "whatever incremental benefits could be achieved under the [1996] Act by requiring mandatory unbundling [in the wireless context] would be outweighed by the costs of requiring such unbundling."[130] In the same order, the FCC also narrowed the geographic markets in which *any* provider can obtain business-oriented loops and transport links as UNEs for any purpose. Very roughly speaking, the FCC denies CLECs rights to such UNEs in local areas where there is evidence (in the form of high "business-line density" or significant existing facilities-based entry) that CLECs can economically deploy new facilities of their own or lease them from non-ILEC third parties, although the precise test depends on the type of UNE at issue.[131]

The upshot is that special-access services remain central to the telecommunications industry, accounting for many billions of dollars in annual ILEC revenue. In some but not all contexts, the FCC still subjects ILEC-provided special-access services to *dominant-carrier regulation*, which means, in a nutshell, that the FCC closely regulates their rates and other terms of service.

It may seem anomalous that these high-bandwidth services for *businesses* are still subject to price regulation in any context, given that telcos and cable companies are categorically free of price regulation when they provide broadband Internet access to *residential* customers. The main reason for the discrepancy arises from the historical evolution of the cable industry, which has spent decades overbuilding telco networks in residential neighborhoods throughout the country. As the D.C. Circuit explained in 2009, "The 'last mile' for broadband business customers . . . differs from the analogous last mile for residential customers, who typically have at least two wires into their homes over which they can obtain Internet service (namely, their traditional telephone and cable lines)."[132] Until recently, cable companies did not often build out their networks to pure business districts because that was not where cable television subscribers tended to be. That said, ILECs *have* long faced strong facilities-based competition from CLECs and others in the urban cores of major metropolitan areas, where, as discussed earlier in this chapter, traffic volumes are enormous enough to support multiple facilities-based competitors. But CLECs have deployed fewer fiber loops to reach the less traffic-intensive—and thus less profitable—business districts in smaller cities and the suburbs.

The FCC's regulatory regime for ILEC-provided special-access services is technical and complex, and we outline only the basic contours here. In the 1990s, every major ILEC's special-access services were subject to price-cap rules. As discussed earlier in this chapter, price caps are initially based on rate-of-return calculations but are then mechanically adjusted each year to reflect estimated industry-wide efficiency gains (the X-factor) and the rate of inflation and other macroeconomic variables.[133] In 1995, the Commission set the stage for partial deregulation of these services, explaining that "competition can be expected to carry out the purposes of the Communications Act more assuredly than regulation" can and concluding that price regulation is appropriate only "where and to the extent that competition remain[s] absent in the marketplace."[134]

In 1999, the FCC issued its *Pricing Flexibility Order*, which, as the name implies, relaxed traditional price regulation for ILEC special-access services in certain geographic areas, depending on the degree of local competition.[135] In *Phase I* price-flex areas, which exhibited moderate competition, ILECs were freed from, among other things, antipredation rules restricting them from *lowering* their special-access rates to meet competition. In *Phase II* price-flex areas, which exhibited stronger (and presumptively price-disciplining) competition, ILECs were further entitled to *raise* their special-access rates above price-cap levels and were in general treated no differently from their mostly deregulated competitors. ILEC special-access services in areas that did not qualify for even Phase I treatment remained subject to conventional price-cap regulation.

How did the FCC determine which areas qualified for Phase I or Phase II pricing flexibility and which would remain subject to price-cap rules? In the *Pricing Flexibility Order*, the FCC adopted bright line rules, which it called *competitive triggers*, assigning each Metropolitan Statistical Area (MSA) to one regulatory category or another on the basis of how many competitors were physically collocated in ILEC central offices there. The more collocation there was, the more price-constraining competition there was presumed to be. The collocation-based triggers for Phase II pricing flexibility were more demanding than those for Phase I flexibility. And for both Phase I and Phase II, the triggers for special-access services provided via loops ("channel terminations") were more demanding than those for services provided via transport links, given the greater traffic volumes and thus greater potential competition associated with transport facilities.[136] In 2001, the D.C. Circuit upheld these collocation-based triggers as rough but reasonable proxies for the degree of local competition.[137]

In 2008, the FCC took deregulation one large step further and, granting forbearance petitions filed by leading ILECs, categorically eliminated all dominant-carrier regulation for certain categories of those ILECs' special-access services on a nationwide basis, without regard to the competitiveness of local areas.[138] The services at issue, which the FCC has dubbed *enterprise broadband services*, consist of very high-capacity fiber-based services and any type of packet-switched service, including *gigabit Ethernet* services. The main services *not* affected by this order are legacy TDM-based special-access services, consisting mainly of *DS1* and *DS3* circuits, which provide the circuit-switched equivalents of, respectively, 24 and 672 voice-grade channels. Those services thus remain subject to the locality-specific framework of the *Pricing Flexibility Order.*

Almost every judgment call the FCC has made in this area has been intensely controversial, in part because the financial stakes involved are enormous. In 2005, the FCC opened a proceeding with an eye toward "reforming" special-access regulation,[139] and in the years since then it has repeatedly asked industry participants to "refresh the record" with new data and new proposals for rule modifications. That proceeding remains pending as this edition heads to press eight years later.

Here is a brief sampling of the key disputes that remain unresolved as of this writing. First, CLECs (and some enterprise customers) argue that price caps are set too high; that Phase II pricing flexibility is granted too often; that ILECs therefore earn monopoly profits for special-access services; and that such excessive profits create deadweight losses for the entire economy by raising prices for all downstream goods and services to inefficient levels. Unsurprisingly, ILECs disagree with each of these arguments.[140] Second, CLECs argue that the collocation-based competitive triggers are radically overinclusive proxies for actual competition, whereas ILECs contend that the triggers are underinclusive. Third,

CLECs argue that the volume and term discounts ILECs grant large special-access customers that agree to long-term purchase commitments are anticompetitive schemes to tie up business and exclude special-access competitors. ILECs respond that those discounts are as presumptively pro-competitive as any other measure that moves prices closer to but not below the underlying cost of service.[141]

One unusually important special-access controversy merits discussion before we turn to the wireless industry, precisely because it *involves* the wireless industry. Sprint and other non-ILEC-affiliated wireless providers have long argued to the FCC that, without strong special-access regulation, ILECs can manipulate backhaul rates to skew competition for wireless services. Any wireless carrier traditionally obtained backhaul from ILECs in the form of ordinary TDM-based special-access circuits. Of course, the wireless carrier may be affiliated with that ILEC, as Verizon Wireless is within Verizon's landline footprint and as AT&T Mobility is within AT&T's landline footprint. Sprint and others argue that those two companies dominate the provision of backhaul services within their respective wireline footprints and can leverage that dominance to benefit their own wireless affiliates and harm unaffiliated carriers such as Sprint. This is a key reason why Sprint and its wireless allies claim that, absent aggressive regulatory intervention, the wireless marketplace will devolve into a Bell "duopoly."[142]

The ILECs respond that they cannot leverage backhaul services to undermine wireless competition because, they say, backhaul services are now competitive. With the rise of smartphones and escalating wireless data usage, mobile wireless providers are replacing many of their legacy ILEC-provided special-access circuits with much higher-capacity Ethernet services, provided not only by ILECs, but also by CLECs, including the major cable companies. ILECs claim that they enjoy no special advantages in providing those new backhaul services and thus no ability to leverage those services to harm downstream wireless rivals such as Sprint.

Of course, all these questions are empirical, and as of this writing they remain unresolved. That is true of many other empirical questions that the FCC broached in its 2005 special-access reform proceeding but then placed on the regulatory back burner for a half-dozen years. As a result, the FCC has not yet comprehensively revisited the basic regime it adopted thirteen years ago in the *Pricing Flexibility Order*.[143] That is a curious way to handle a set of regulatory issues that affects billions of dollars in revenue every year. Even if the FCC had gotten all the details of that regime right in 1999, which is doubtful, the telecommunications world has changed radically since then. It would be most surprising, for example, to discover that the collocation triggers adopted in 1999 are still valid indicators of competition; they are very likely either overinclusive (as the CLECs argue) or underinclusive (as the ILECs argue) or both.

In August 2012, the FCC placed special-access reform back on the front burner.[144] It concluded that the existing collocation triggers are poor proxies for competition; that, until the rules are reformed, ILECs can no longer win pricing flexibility on the basis of such triggers; and that ILECs may now obtain such flexibility only by filing fact-specific forbearance petitions, showing that deregulation in particular markets would serve the public interest. Given the FCC's new reluctance to grant such forbearance petitions, as illustrated by the *Phoenix Forbearance Order* discussed earlier, this 2012 special-access order effectively freezes the status quo in place while the FCC elicits competitive data from industry participants. What the FCC will ultimately do with the data it collects and on what timetable remain open questions as this edition heads to press.

3

The Spectrum

We have focused so far on the electronic delivery of information through wires and cables, whether copper, coaxial, or fiber optic. But wires and cables are only part of the telecommunications story, as anyone with a cellphone or a wireless-enabled laptop—or, for that matter, a radio—can attest. And this brings us to the wireless side of the industry.

If this book had been written in 1980, our discussion of the "airwaves" would have focused almost entirely on the regulation of radio and television broadcasting because that was the most significant commercial use to which the electromagnetic spectrum was then put. Engrained tradition had made it seem natural that telephone service involved running wires into people's homes and that television signals should normally be transmitted through the air. But technological change has up-ended this tradition. By 2011, nearly one-third of American households had canceled their wireline phone service in favor of exclusive reliance on their cellphones, and an additional 16%, although subscribing to both wireless and wireline services, used their cellphones for virtually all their calls.[1] Meanwhile, according to FCC estimates, the percentage of American households that received television signals entirely from conventional over-the-air broadcasts (rather than from cable or satellite subscription services) declined from an overwhelming majority in the 1970s to about 10% in 2012.[2]

This transformation in the means of delivering television and telephone services is one more illustration of convergence—that is, the use of different technologies to provide similar services. And in this case, the technologies used to provide these different services have not merely converged but traded places to some extent in a phenomenon sometimes named the *Negroponte switch* (after MIT Media Lab founder Nicholas Negroponte).[3] Because it is more feasible and useful to carry a telephone around than a television, and because the airwaves permit mobility whereas wires do not, the airwaves are in some respects a more natural medium than wires for telephony and other personal point-to-point communications. And as the prevalence of cable television has shown, wires provide greater bandwidth than the available airwaves for broadcasting hundreds of channels of high-definition video signals into large-screen televisions. As in other contexts, however, technology changes much more quickly than regulation, so the policy response to this new reality remains a work in progress.

In this chapter, we address the question of spectrum on the basis of first principles, focusing on how and how much the government should get involved in apportioning the airwaves among different uses and users and what sort of "traffic cop" role it should play in managing interference disputes. We explain the technological and market forces pressing for reform of traditional spectrum regulation, the federal government's recent steps in that direction, and the reasons why fundamental regulatory change has proven so challenging.

These spectrum-reform initiatives have assumed new urgency with the explosive popularity of wireless broadband services. As President Barack Obama remarked in 2010, "Few technological developments hold as much potential to enhance America's economic competitiveness, create jobs, and improve the quality of our lives as wireless high-speed access to the Internet." As he added, however, "This new era in global technology leadership will only happen if there is adequate spectrum available to support the forthcoming myriad of wireless devices, networks, and applications that can drive the new economy."[4] A year later, FCC Chairman Julius Genachowski reaffirmed that the "explosion in demand for mobile services places unsustainable demands" on the finite spectrum resources that have been allocated to those services so far, and "the coming spectrum crunch threatens American leadership in mobile and the benefits it can deliver to our country."[5]

Although the precise nature and extent of that "coming spectrum crunch" are subject to debate,[6] freeing up more spectrum for mobile broadband uses is widely viewed as one of the great public policy challenges of the early twenty-first century. Meeting that challenge will require making the hard political choices needed to reallocate spectrum from federal agencies and broadcasters to increasingly capacity-constrained wireless broadband services. It will depend on innovations in auction theory to persuade spectrum incumbents to cede their holdings voluntarily to more efficient uses. Meeting the spectrum challenge will also necessitate a new generation of engineering innovations, including the further development of smart radio technologies and efficient spectrum-sharing arrangements between private parties and the government. Last but hardly least, it will require *regulatory* innovation in how policymakers conceptualize, resolve, and—where appropriate—head off interference disputes. These are the topics of this chapter, and they are some of the most critical and interesting topics in telecommunications policy today.

I. Overview

In early 1978, while one of us was outside playing baseball, the other spent the better part of his after-school hours breathing in the fumes of burnt solder while assembling, transistor by transistor, a shortwave radio from a build-it-yourself kit. After two or three months, the radio seemed complete, and a visit to the local hardware store produced enough copper wire to string a crude antenna from the side of the house to a tree in the yard. Finally, the radio was turned on: silence. A return visit to the grown-ups at the hardware store revealed that sloppy soldering had shorted out one of the main circuits, an embarrassing but fixable problem. The radio returned home several days later, and, the circuit having been repaired (at a cost of some $100), the sounds of several dozen countries came bursting into the room. It seemed vaguely occult, this discovery that radio signals are continuously flitting through our backyards like a swarm of bats in the night, unknown to us until we reach out to capture them with a copper wire, a properly soldered circuit board, and (at the end of the circuit) a speaker.

For many years, scientists believed that electromagnetic waves were vibrations of the "ether"—a hypothetical substance said to reside invisibly throughout space. In the late nineteenth century, Albert Michelson and Edward Morley conducted experiments that led them and others to abandon the "ether" construct, though the term lives on in colloquial usage. The modern concept of the electromagnetic spectrum, building on these insights and the prior publication of James Clerk Maxwell's equations in the 1860s, soon came into focus. When asked to explain the underlying physical mechanics of the wireless spectrum—that is, how electric and magnetic fields travel in waves that can be sent from point A and received at point B—Albert Einstein is reported to have remarked: "You see, wire telegraph is a kind of very, very long cat. You pull his tail in New York and his head is meowing in Los Angeles. Do you understand this? And radio operates exactly the same way: you send signals here, they receive them there. The only difference is that *there is no cat*."[7]

Several critical points can be gleaned from this enigmatic remark.

First, wired and wireless communications have more in common than many people realize. In both cases, the waves that carry information are electromagnetic signals riding on discrete frequencies. To oversimplify matters a bit, traditional wireless communications involve three basic steps. On one end, a transmitter starts with a *carrier signal* on a given frequency (or frequencies)—represented graphically as a simple sine curve—and "modulates" that signal to represent the information to be transmitted, such as a text message or a song.* Second, as soon as this modulated signal is created, it bursts from the transmitter at the speed of light and promptly runs into many objects—including, if all goes well, the recipient's antenna. This antenna feeds into a receiver, which, when properly tuned, uses electronic filters to accept only signals within the specified frequency range or ranges and to disregard the rest. Third, the receiver decodes the information by retrieving it from the carrier signal in a process called "demodulation." When you tune your radio to a particular station, you are directing the receiver to translate, into sounds or images, one frequency range of electromagnetic signals to the exclusion of all others.

The signals flowing over wires and cables are manipulated in much the same way as radio signals transmitted through the airwaves. A DSL line (see chapter 5) is simply a collection of signals traveling over a copper wire in discrete frequency ranges, some of which are used for voice transmissions and others for data. Coaxial television cables are themselves divided into dozens or hundreds of "channels," each traditionally spanning a slice of frequency six megahertz wide. Some of these

* The *frequency* of an electromagnetic signal is measured by its number of cycles per second—that is, the number of times it moves from crest to trough back to crest during the course of a second. A signal's frequency is inversely proportional to its *wavelength*: the distance from crest to crest or from trough to trough. Thus, a *microwave* signal has a much higher frequency than a conventional AM (or FM) radio signal, with its much longer waves. Specific frequency bands are identified by their *hertz*, or cycles per second, named in honor of the nineteenth-century physicist Heinrich Hertz. A *kilohertz* (kHz) is a frequency of a thousand cycles per second; a *megahertz* (MHz) is a million cycles per second; and a *gigahertz* (GHz) is a billion cycles per second. There are 1,000 kHz in a MHz, and 1,000 MHz in a GHz.

channels correspond to television channels and others to cable modem bitstreams. Like coaxial cable, a strand of optical fiber is, as Thomas Hazlett put it, "spectrum in a tube."[8] The capacity of fiber-optic cables expanded dramatically in the 1990s as engineers found ways to transmit different signals simultaneously over a single strand by means of different colors (light frequencies). In these respects, wires and cables are very much like the airwaves: telecommunications engineers can exploit the division of electromagnetic signals into frequencies to expand the information-carrying capacity of any medium, including empty space itself.

Despite these similarities, the absence of Einstein's "cat"—of a tangible transmission medium such as a wire or cable—carries important consequences for telecommunications policy. We know who owns a wire or cable, and that firm or person generally has the right to control the physical properties of its operation. In contrast, there is no obvious "owner" of the airwaves. Also, a wire or cable is usually wrapped in shielding material and is thus reasonably well protected from interference by external signals. Not so with the airwaves: there is no natural "shielding" in the air that can keep two signals from causing mutual interference in a radio receiver if, for example, they use the same frequency in the same place at the same time.[9] The result of such interference is that ordinary receivers cannot decode the separate information carried by each signal. Although some highly sophisticated receivers might be capable of performing that task, they typically cost more than ordinary receivers, and systems engineers and consumers thus forgo them in many contexts (such as the market for inexpensive transistor radios).

The twin peculiarities of over-the-air transmissions—an absence of obvious property rights in the medium and the prevalent threat of interference—lie at the root of today's spectrum policy. As this chapter explains, the federal government has heavily regulated who may transmit signals over the air and, with recent (and growing) exceptions, has rigidly prescribed the services that may be provided over defined frequency bands.

Since the 1920s, the government has justified such regulation on the grounds that the "airwaves," like the Grand Canyon, are a "public resource" belonging to the whole American polity, that this resource would be quickly exhausted by the unregulated demand for it, and that unpoliced private use of this resource would lead to its despoliation through widespread interference. Taken together, the government's rationales for regulating the spectrum coalesce around the assumption that it is "scarce"—that there is less of it than an unregulated public can use without causing serious interference problems. The Supreme Court first embraced this rationale in 1944 when it rejected First Amendment and other challenges to the FCC's authority to regulate access to the spectrum. In so doing, it concluded not only that the FCC can serve as the traffic cop of the airwaves, but also that it can "determin[e] the composition of th[e] traffic" permitted on particular slices of the spectrum.[10]

By the 1990s, a broad consensus had formed that the FCC's traditional command-and-control licensing regime was exceedingly inefficient and should begin giving way to a nimbler and more decentralized approach to spectrum management. At the highest level of generality, the academic literature describes two basic proposals for such an approach, which are known by the catch-phrases *property rights* and *commons*. Under the first of these approaches, which Congress and the FCC have increasingly embraced over the past twenty years, the government gives private parties alienable property rights (in the form of licenses) to use different portions of the spectrum and allows those rights to be freely bought and sold in a secondary market, much as privately held land is bought and sold. Under the second approach, the government would establish much of the spectrum as a public commons and rely on unlicensed users to avoid interference problems cooperatively—by using only low-power devices, by employing "smart" wireless technology, or otherwise. Under this model, the FCC or some other institution would facilitate coordination by enforcing appropriate restrictions (e.g., on transmission power levels) and developing necessary cooperative protocols (which, for example, would prescribe how devices must "listen before they talk").[11]

We discuss each of these approaches later in this chapter, but two observations warrant emphasis up front. First, popular formulations of the property-rights model tend to overstate the analogy to the law of real property, with its relatively clear-cut rules, and thus understate the continuing need for government oversight of interference issues under a property-rights regime.[12] As we discuss at the end of this chapter, interference disputes can be enormously complex, both technologically and politically, and devising a more coherent policy for resolving them on an industry-wide basis should be one of the FCC's top priorities over the next decade.

Second, although the academic literature has tended to highlight the rivalry between the "property-rights" and "commons" schools, industry participants and many policymakers recognize that any sound spectrum policy will make room for both approaches. The true debate occurs at the margins, as policymakers consider (1) *how much* (not whether) spectrum should be set aside for unlicensed uses such as Wi-Fi and (2) whether, in *licensed* spectrum bands, *unlicensed* devices should be free to transmit low-power (and theoretically noninterfering) signals in, for example, the gaps between the transmissions of licensed spectrum holders.

II. The Nuts and Bolts of Spectrum Allocation and Assignment

When Guglielmo Marconi first made use of the airwaves to transmit information in the 1890s, he envisioned the primary use of spectrum as ship-to-shore communications. During the ensuing 50 years, wireless technology produced the telegraph-like services that Marconi developed, amateur ("ham") radio, and radio broadcasting, which was born in the early twentieth century and gained widespread popularity by the early 1920s. The amateur operators back then, however, included what we might call "hackers" today. Their numbers included some whom in 1910 the United States Navy called "[m]ischievous and irresponsible" in that they took "great delight in impersonating other stations and in sending out false calls."[13] Whatever the extent of this practice or the merits of the navy's call for tighter control of "amateurs," the era of spectrum non-regulation in this country ended in the wake of the *Titanic* disaster in 1912, when "chaos in the spectrum" was said to have confused a potential rescue ship "so it missed the calls of help from the sinking luxury liner."[14] The result was the Radio Act of 1912, which authorized the Secretary of Commerce to license users of equipment that communicated via the spectrum.

Secretary of Commerce Herbert Hoover used his administrative authority under the Radio Act to preclude interference by restricting access to the spectrum. But the courts ultimately balked at Hoover's assertion of authority to regulate use of the airwaves on the basis of a license applicant's merits.[15] In response, Congress enacted the Radio Act of 1927, which established the Federal Radio Commission and gave it broad jurisdiction to regulate access to the spectrum under a general "public interest" standard.[16] The Act further established the regulatory premise that persists to this day: that the spectrum belongs to the public and that, at least as a formal matter, licensees have no property right to continue using it.[17] When Congress passed the Communications Act of 1934, it kept this model of regulation intact but transferred the functions of the Federal Radio Commission to the newly established Federal Communications Commission.

The resulting model of regulation generally involves careful administrative oversight of the spectrum to avoid "harmful interference" between competing uses and users. "Harmful interference," as defined by the FCC, is "unwanted energy" that "endangers the functioning of a radionavigation service or of other safety services or seriously degrades, obstructs, or repeatedly interrupts a radiocommunication service operating in accordance with [international] Radio Regulations."[18] But the federal government has traditionally carried its regulation of the airwaves much farther than needed simply to police against harmful interference. The government has seized upon spectrum "scarcity" as a basis for asserting a comprehensive stewardship of the airwaves, exercising broad authority to determine how particular bands of spectrum shall be used (such as for television or FM radio) as well as precisely who may use them and subject to what restrictions. Here we focus on the two most critical aspects of the FCC's traditional spectrum-management regime: *allocation* of particular spectrum bands for prescribed uses and *assignment* of spectrum within those bands to particular licensees.

A. Allocation

The FCC "allocates" the spectrum by dividing it up into different frequency bands and in many cases specifying the uses to which the frequencies within each band may be put (or at least the technical rules governing such uses). Under the FCC's *band plans*, the Commission "zones" spectrum by, for example, reserving the spectrum between 300 and 535 kHz for aeronautical and maritime communications and the spectrum between 535 and 1605 kHz for AM radio. This allocation function parallels how a zoning authority decides that some land should be designated for residential uses, other land for commercial uses, and so forth. Overall, the FCC has divided the spectrum into scores of different uses, each of which is associated with a particular band (or bands). Interested readers can view the complexity of these allocations by visiting the interactive chart the FCC has posted on its website (http://reboot.fcc.gov/reform/systems/spectrum-dashboard).

Because the physical characteristics of signal propagation vary with the frequency, not all segments of the spectrum can be used for the same purposes. A high-frequency microwave signal, for example, behaves in some respects like visible light: it can carry information successfully only if there is a line-of-sight path between the sender and the receiver. In contrast, signals at the lower frequencies (i.e., the ones with longer wavelengths) can pass more easily through walls or trees and are not subject to "rain fade." This is one reason why the spectrum traditionally reserved for television broadcasters in the UHF band is described as "beachfront property." The spectrum the FCC historically entitled them to occupy—much of the expanse between 470 and 800 MHz—is exceptionally well suited for high-value mobile voice and data services.

The FCC's allocation decisions are complicated by a key institutional consideration: it is authorized to regulate access to the spectrum only for private uses and for state and local governmental uses. Under a presidential delegation of authority, it is the job of the National Telecommunications and Information Administration (NTIA), a subagency of the Commerce Department, to allocate spectrum for use by the federal government, including the military. This is no small point because by one estimate the federal government controls outright nearly 18% of the high-value spectrum between 300 MHz and 3 GHz and shares 52% of that spectrum with other users.[19] To make matters even more complicated, both the FCC and NTIA must confer with the U.S. State Department to develop a unified position for the United States to present at the International Telecommunications Union (ITU), an arm of the United Nations.[20] At regularly scheduled World Radiocommunication Conferences, the ITU develops international spectrum policy on such issues as the reservation of certain frequencies for broad categories of usage around the world. In general, because the ITU's decisions have the force of binding treaty obligations, the FCC accommodates them in developing domestic spectrum-allocation policies.

After allocating the spectrum into bands, the FCC sometimes moves further to determine how much spectrum to make available to particular licensees within a given band (if use of the spectrum in that band is subject to licensing). If, for example, the FCC allocates the spectrum between 535 and 1605 kHz for AM radio, it must decide how to parcel out individual licenses within that overall allocation (e.g., in blocks of 10 kHz), and it must likewise decide how wide of a "guard band" of unused frequency to interpose between different stations within the same geographic area to avoid mutual interference. These licenses are generally subject to additional restrictions that vary from band to band, such as duration of the license, limits on transferability, maximum power levels, requirements to adhere to certain technical standards, and build-out obligations. Like so much of spectrum policy generally, some of these "service rules" seek to limit interference between users.

The amount of spectrum a given licensee needs is a function of the bandwidth required for its service. All else held constant, the greater the frequency range over which signals are transmitted, the more information they can carry—and thus the higher in quality will be the radio signal, television picture, or cellphone call. Policymakers have traditionally assumed that the only feasible way to increase bandwidth is to enlarge the allotment of *contiguous* spectrum a given licensee is entitled to occupy; for example, the FCC has made six megahertz of contiguous spectrum available for television stations. Today, as we discuss at the end of this chapter, new wireless technologies upend that assumption. Additional spectrum remains necessary to address increasing bandwidth needs, but that spectrum need not always be contiguous.

B. Assignment: from comparative hearings to auctions

For many key uses of spectrum—such as television, radio, and mobile voice and data services—the FCC follows up its allocation decisions with exclusive *assignments* of spectrum to particular licensees. We should make clear at the outset that not all spectrum is divvied up this way. In some cases, the FCC does not assign licenses to users on an exclusive basis but allows any qualified user to obtain a license to use a given frequency band, as with the spectrum reserved for amateur radio operators. In other cases, as with the spectrum designated for "citizens' band" (CB) radio, the public may use the spectrum without even obtaining a license (because it is "licensed by rule"). And in its so-called *Part 15 rules*, discussed later in this chapter, the FCC allows wireless devices such as garage door openers and remote controls to operate in unlicensed bands of spectrum, or even in bands of otherwise licensed spectrum, provided they do so at low power levels.

For now, however, we focus on the FCC's traditional regime for licensing the bulk of prime spectrum for the exclusive use of designated private parties. For most of its history, the FCC relied on *comparative hearings* for this purpose: drawn-out, nebulous, inherently subjective inquiries into the relative worthiness of rival applicants for a free spectrum license. In theory, this procedure discharged the FCC's statutory obligation to serve the "public interest" by assigning the use of the airwaves to the "most qualified" users. In practice, however, it tended to favor entrenched incumbents and those with political ties. In one notorious example, the FCC awarded radio and television broadcasting licenses to then-Congressman Lyndon B. Johnson's wife.[21] And in another widely retold finding, not a single newspaper that endorsed Adlai Stevenson over Dwight Eisenhower in the 1952 presidential election received a TV license in a contested proceeding.[22]

Even apart from the appearance of favoritism to political insiders, the comparative hearing process was inevitably expensive, standardless, and time-consuming. Unless only one application was filed for an available license, the FCC had to hold a "beauty contest," and the loser was entitled to appeal.[23] Because there was no procedure for resolving a "tie" between the two applicants, and because the stakes involved were huge, such litigation invariably prolonged and complicated the assignment process. The FCC nonetheless used comparative hearings until the 1980s as its exclusive method of assigning licenses for any commercial use of the spectrum. It had little choice in the matter: the Communications Act of 1934 guaranteed license applicants a right to a hearing and did not provide an alternative assignment mechanism.

The pressure to devise new models for assigning spectrum licenses ultimately grew, particularly when, after many years of delay, the FCC finally allocated blocks of spectrum for cellular telephony in the 1970s and assigned them to specific providers in the early 1980s.[24] The FCC recognized that comparative hearings, originally designed to judge which of several applicants would air television or radio programming in the "public interest," were poorly tailored for determining who should provide common carrier telephone service. And, more generally, such hearings had fallen into well-deserved disrepute for the costs, delays, and arbitrariness they inevitably imposed on the process.

In 1984, Congress first authorized lotteries as a replacement for comparative hearings for cellular telephone licenses.[25] As the name implies, a lottery is a mechanism for assigning licenses among competing applicants for free and, to a large extent, at random. Within a few years, however, it became clear that lotteries were not quite the solution that either Congress or the FCC was looking for. First, the prospect of obtaining a free but commercially valuable license generated so many applications—hundreds of thousands of them—that they caused the partial collapse of the FCC facility used to contain them.[26] And, because the FCC prescreened these applicants to ensure that they met the minimum qualifications for operating a cellular telephone business, the process turned out to be burdensome for the FCC to administer and costly for the parties involved.

But lotteries suffered from an even greater shortcoming, more political than economic. Those who "won" the lottery were under no obligation to keep the licenses themselves; they were generally able to sell them to others on the secondary market so long as the buyers honored all of the initial license restrictions and thus agreed to use the relevant spectrum bands only for their allocated use—cellular telephony. Thus, although the ultimate users of the spectrum often paid enormous sums for these licenses, they paid those sums to the randomly selected private individuals who had won the lotteries rather than to the public treasury. For both fairness and public finance reasons, this struck both Congress and the FCC as wrong. And the front-end delays attributed to the lottery process may well have deprived consumers of prompt access to valued wireless services.

In the early 1990s, to cut down on such delays and raise revenue in the process, Congress authorized government-sponsored spectrum auctions for various types of licenses, including cellular telephony licenses.* To design its auction process, the FCC consulted the branch of mathematics and economics devoted to game theory and its close cousin, auction theory. In some respects, the proto-architect of the FCC's auction process was the now-famous Nobel Laureate John Nash, who developed many of the applicable principles of game theory several years before he was overcome by paranoid schizophrenia. Indeed, Sylvia Nasar devotes several pages of *A Beautiful Mind*, her biography of Nash, to the FCC's auction process in the 1990s and credits him with its apparent success.[27] But the branch of game theory invented by Nash did not definitively resolve such basic questions as whether all goods should be auctioned at once, whether bids should be sealed, and how collusion between firms can best be prevented.[28] The answers to such questions would determine whether the FCC's auctions would succeed—both in terms of generating prodigious sums for the public treasury and in terms of assigning licenses to firms that would make the best use of them for the public benefit.

In conducting its first set of auctions in 1994, the FCC sold rights to use large "blocks" of spectrum at once, enabling bidders to put together mutually dependent spectrum assets—say, licenses in a tristate area such as New York, New Jersey, and Connecticut. The FCC also modified the traditional English bidding model used at art auctions, in which potential buyers openly bid against one another. It decided instead to announce

* Omnibus Budget Reconciliation Act of 1993, Pub. L. No. 103-66, § 6002, 107 Stat. 312, 379–86 (codified at 47 U.S.C. § 309(j)). Not every licensee is subject to an auction requirement. Notable exceptions include public safety agencies and providers of satellite services, including mobile satellite telephony. Also, even where applicable, the auction requirement extends only to the assignment of *initial* licenses, not to applications for renewals or modification (such as those filed by the original cellular licensees who obtained their licenses via comparative hearings or lottery).

the high bid after each round, but otherwise to keep the terms of the bids closed. In so doing, it sought to limit the use of inefficient strategic behavior such as retaliation or collusion. The FCC further adopted a set of anti-collusion rules that, among other things, limited communications among competing bidders.

Juxtaposed against the comparative hearing tradition at least, the FCC's use of auctions has been a reasonably effective means of assigning spectrum licenses to those best able to put them to productive use. From 1993 to 1997 alone, the FCC granted 4,300 licenses via auction, and it did so far more efficiently than under its prior regime. To Congress's delight, the winning bids promised in the aggregate $23 billion to the federal treasury (although not all of that amount was ultimately collected). In 1997, Congress followed up on this fiscal success story by *requiring* auctions for most types of initial spectrum licenses.[29] More recent auctions have underscored the considerable revenue-raising potential of spectrum auctions. For example, the mobile broadband-oriented AWS (for "advanced wireless service") and 700 MHz auctions conducted in 2006 and 2008 generated, respectively, an additional $14 billion and $19 billion for the federal treasury.

No auction process is perfect, and auctions can sometimes lead to outright market failures. Much like the stock market, slices of spectrum are subject to wild swings in market valuation. In Europe, speculative zeal and a poor auction design helped push up the bids for broadband-spectrum rights so high in 2000 that the process eventually bankrupted some of the auction winners and sent the European wireless industry into a temporary tailspin.[30]

Similar problems have occasionally beset the U.S. auction experience as well. One notable example is the NextWave debacle, which illustrates the occasional tension between telecommunications policy and U.S. bankruptcy law. In auctions reserved for small businesses in the 1990s, the FCC decided, with fateful consequences, to allow the winners to pay off the bids for their licenses in a series of installments. But several successful bidders—including NextWave Communications (which bid approximately $4.74 billion for its licenses)—had trouble financing the build-out of their networks. When NextWave ultimately declared bankruptcy and defaulted on its payment obligations to the federal government, the FCC moved to take back its licenses. After several years of litigation, during which this portion of the spectrum went unused, the Supreme Court finally ruled in 2003 that the FCC had to stand in line like any other creditor; it could not simply reach into NextWave's estate, pluck out the company's greatest assets (its spectrum licenses), and sell them to someone else.[31] Meanwhile, NextWave's spectrum lay fallow for the duration of the litigation and beyond. This ruling serves as a cautionary tale: auctions *can* be, but are not inevitably, the most efficient way of assigning spectrum quickly to the parties who can place it to the most productive uses, and success turns on the details of auction-related procedures.

III. Liberating the Airwaves

Over the past decade, spectrum policy has focused on two closely related objectives. First, policymakers have sought to replace the traditional command-and-control regime with a more market-oriented approach to the use of spectrum. The ultimate goal is a world in which licensees, without advance regulatory permission, can place their spectrum holdings to whatever uses will respond best to consumer demand and can flexibly trade spectrum rights on the open market. Second, policymakers have looked for ways to free up new frequency bands to meet the escalating bandwidth demands of mobile broadband services. That means taking spectrum usage rights *away* from broadcasters, federal agencies, and other traditional users. And that challenge raises a host of politically sensitive implementation issues, including whether the government may give TV stations economic incentives to yield their spectrum to mobile broadband uses that the public (which relies overwhelmingly on cable and satellite for TV reception) values more than those stations' over-the-air broadcasts.

A. Beyond command and control

Nobel Laureate Ronald Coase observed in 1959 that the FCC's traditional model of regulation inefficiently limited the possible uses of the airwaves and that the free market offered a more logical system of allocating the spectrum and assigning licenses to use it. As Coase explained, the FCC's traditional rationale—"scarcity" and the ever-present threat of interference—could not alone justify adherence to a command-and-control regime. After all, most other valuable resources (such as land, metal, etc.) are "scarce" in some sense, but the government nonetheless relies on the free market to regulate their use.[32] Similarly, although the stories of "chaos" among early users of the airwaves may support an argument for *defining* property rights clearly and providing for their effective enforcement, they do not justify perpetuating the government's micromanagement of the spectrum's possible uses.

Coase was the first major exponent of what has now come to be known as a "property-rights model" for spectrum management. Under that model, spectrum owners may freely sell or lease patches of spectrum in a robustly competitive secondary market. The government's role is to define the relevant property rights (including the extent of permissible interference) and enforce those rights as well as related contractual agreements, *not* to allocate spectrum for particular uses on "public interest" grounds. For example, subject to limitations necessary to manage interference issues, a television station under a pure property-rights model can sell or lease all its spectrum rights to a mobile wireless carrier, and regulators would play no role in second-guessing whether the public is better served by having one more wireless company and one fewer broadcast television station.

Coase's advocacy for a private-property-like regime of spectrum management rested on his more general proposition—now known as the *Coase theorem*—that, with well-defined property rights, the free market will generally allocate resources to their most efficient uses so long as transaction costs are low enough. For example, a firm using assigned spectrum for its own internal communications would be free to sell its licenses to a wireless carrier and purchase capacity on a fiber-optic network instead. In this example, the spectrum would be more valuable to the buyer than the seller; both parties would be better off if they made the trade, and so would the wireless carrier's many subscribers. Granted, Coase's property-rights formulation gave only a partial roadmap for spectrum reform because, as discussed later in this chapter, defining property rights in the spectrum is no easy task and is far more complex than, say, defining rights to real property.[33] But despite its incompleteness, Coase's insight was enormously influential: over the long term, it would fundamentally reshape how policymakers conceptualized the government's role in licensing spectrum.

We say "over the long term" because in fact it took many years for Coase's 1959 critique to affect the FCC's actual spectrum policies. For decades, parties entered into all sorts of private transactions for the exchange of spectrum rights, but these transactions bore only a distant resemblance to those that would be permitted under a genuine property-rights model. With FCC approval, one company could purchase another company along with the spectrum licenses that it held, but the buyer generally had to continue using the spectrum to provide the same service specified in the license. That model thus locked in legacy uses of spectrum either by precluding efficient transactions altogether or by reducing their likelihood by interposing substantial transaction costs in the form of bureaucratic hurdles. The command-and-control model thus foreclosed alternative and more socially valuable uses of spectrum that would be available under a market-based system.

Thomas Hazlett has gone one step further, arguing that the origins of the command-and-control regime lay largely in a desire to protect incumbent licensees, and he has cataloged the government's long history of enabling those incumbents to keep competition at bay through various pretexts dressed up as "public interest" concerns.[34] Of course, the FCC did not view its congressionally assigned role in this pejorative light; instead, it traditionally subscribed to the "'wise man' theory of regulation," under which it is deemed "capable of deciding what [uses of spectrum are] best for the public."[35] These separate but complementary interests of incumbents and regulators helped confine Coase's views to the classroom for many years. Indeed, almost twenty years after the publication of his article, two FCC commissioners colorfully remarked that the odds of using auctions or some other market-based system for assigning spectrum licenses were "about the same as those on the Easter Bunny in the Preakness."[36]

In the late 1980s and early 1990s, the FCC finally took a few steps in a market-based direction, but more by happenstance than by design. In one instructive story, Morgan O'Brien, the founder of the wireless company that became known as Nextel (now part of Sprint), parlayed his nine years' experience as an FCC lawyer into creating a competitor to the existing cellular providers. O'Brien's insight was that the spectrum allocated to "specialized mobile radio" (SMR), a wireless dispatch service for taxis and other service vehicles, could just as easily be used for cellular telephone service. The main difference between the licenses for these two services was that the dispatch licenses came with various restrictions designed to keep them from becoming a source of competition for cellular telephone providers. Those restrictions, as Hazlett recounts, produced an enormous discrepancy in the value of the two types of licenses in the secondary market: "the same amount of spectrum sold for just $100,000 with a dispatch license and $2 million with a cellular license."[37]

For O'Brien, the discrepancy in cost between the two licenses presented a striking arbitrage opportunity. He acquired dispatch licenses and—knowing from his FCC experience that "he could not succeed in a straight-up rule making to re-allocate SMR bands to cellular"—devised a "below-the-radar-screen approach" that sought various waivers from the FCC's restrictions on those licenses, emphasizing the need to "upgrade dispatch service, not compete with cellular."[38] After much time and effort, he eventually succeeded in freeing up this spectrum for the wireless telephony services that would create the greatest consumer value (although regulatory controversies arising from Nextel's spectrum strategy resurfaced in the early 2000s, as discussed later). Although this first chapter in Nextel's founding has a happy ending, the convoluted path that O'Brien had to take and the fact that this regulatory discrepancy existed in the first place underscore the gross inefficiencies of the government's traditional spectrum policy.[39]

Since the mid-1990s, the FCC has taken more deliberate steps toward a market-based model. One example of this trend, discussed earlier, is the use of auctions rather than comparative hearings for the assignment of many licenses. A second example is the FCC's allocation of frequency bands for flexible commercial uses. On key spectrum bands, the licensee may use its spectrum holdings to provide "any mobile communications service," as well as "fixed services" if provided in combination with mobile ones (but not broadcasting services under any circumstances).[40] For the most part, carriers have used such licenses to provide mobile voice and broadband services. Unlike traditional licenses, however, these licenses are like private property in that, generally speaking, they neither restrict allowable uses nor dictate a particular technology. In 2003 and 2004, the FCC took additional steps toward making these licenses even more like private property. Until then, licensees of flexible-use spectrum had flexibility to use their spectrum as they wished, but they could not lease it to third parties without first jumping through burdensome and time-consuming bureaucratic hoops.[41] In its *Secondary Markets Order*, the FCC streamlined its review of certain spectrum license transfers[42] and followed up a year later with further reforms of the same type.[43] These moves, although long overdue, were still controversial (and opposed by one commissioner) because they departed from the traditional conception of public interest regulation by leaving certain spectrum licensees with a comparatively free rein.[44]

B. Reclaiming spectrum for mobile broadband

As discussed, the physical properties of electromagnetic signals make some frequency bands more amenable than others to particular uses.[45] All else held equal, signals travel farther and permeate walls more effectively at lower frequencies. But the antennas needed for wireless devices are smaller and thus more portable at higher frequencies. At very low frequencies (such as the so-called "medium band" at 520 to 1710 kHz), signals interact with the earth's atmosphere in such unpredictable ways that they can be used effectively for conventional AM radio broadcasts but not, for example, mobile communications services. And at frequencies higher than several GHz, signals attenuate so rapidly with distance and are so susceptible to blocking by intervening objects that they are best used for fixed, high-power, line-of-sight transmissions (such as satellite TV broadcasts or tower-to-tower microwave beams), not mobile services.

With these considerations in mind, the Obama administration announced an ambitious initiative in 2010 to redeploy additional spectrum—about 500 megahertz in the aggregate—to mobile broadband services within ten years.[46] This initiative focuses mainly on spectrum between 225 MHz and 3.7 GHz (i.e., 3,700 MHz) because that spectrum range—in particular the lower-middle portion of it—is technologically best suited for mobile broadband services.[47] As the FCC's *National Broadband Plan* observed, however, only a small fraction of that range today, accounting for 547 megahertz in the aggregate, "is currently licensed as flexible use spectrum that can be used for mobile broadband."[48] That amount, the *Plan* found, is not nearly enough to keep pace with the escalating bandwidth demands of mobile broadband services, given that "[m]ore bandwidth begets more data-intensive applications," which in turn "beget[] a need for more bandwidth."[49] As the *Plan* observed, "the cost of not securing enough spectrum [for mobile broadband] may be higher prices, poorer service, lost productivity, loss of competitive advantage and untapped innovation."[50]

Much of the usable "new" spectrum will have to be drawn from one of two classes of spectrum incumbents: federal agencies and television broadcasters. In each case, the task of spectrum reallocation presents a distinct set of policy challenges.

Reclaiming government-held spectrum and options for spectrum sharing

As discussed, NTIA is legally responsible for spectrum currently allocated to federal agencies, including the military. But it lacks any generalized authority to compel those agencies to use the spectrum more efficiently. And history has shown that federal agencies do not readily part with their spectrum holdings, which collectively account for hundreds of thousands of frequency assignments in widely scattered locations across the electromagnetic spectrum.

The root of this problem is that federal agencies do not internalize the opportunity costs of their continued use of this spectrum—that is, they have no particular incentive to abdicate to others whatever portions of their assigned spectrum they do not need, no matter how commercially valuable those portions might be in other hands. It is thus quite likely that many of the assigned frequencies are either barely used or used inefficiently.[51] Federal policymakers are acutely aware of this systemic inefficiency in the allocation of spectrum resources, and they have floated far-ranging proposals for freeing up hundreds of additional megahertz for private-sector use.[52] So far, however, those proposals are longer on ambition than results.

The frequency band between 1755 and 1850 MHz, which sits between two other bands currently used for mobile broadband, illustrates the immense complexity of such reform measures. In 2012, NTIA mapped out a strategy to wean federal agencies off the 1755–1850 MHz band and free up much of it for commercial uses.[53] The challenge is that those agencies use the band for critical national-security purposes, including the operation of complex wireless systems that in some cases would be extremely difficult and expensive to retool for use on other bands. But the solution cannot be to acquiesce in perpetual monopolization of this entire band by federal agencies. Hoping to strike a middle ground, some policymakers have proposed a complex regime in which mobile providers would to some extent share this band with federal agencies along several dimensions, including geography, time, and available frequencies. Mobile wireless carriers have reacted to this proposal with ambivalence. Although the industry welcomes the release of new spectrum, it expresses concern about the details of the proposed sharing. For example, it says, some of spectrum-sharing techniques anticipated in this plan are speculative and commercially untested.[54] The industry has thus encouraged the government in the shorter term to move federal agencies completely off a specific portion of this band—the portion between 1755 and 1780 MHz—and make it available for the exclusive use of private licensees.

Over the longer term, however, the private sector will have to devise creative new solutions to sharing with the government as it exhausts the bands the government is willing to relinquish altogether. Indeed, the 500 megahertz of new spectrum that the Obama administration notionally hopes to redeploy for mobile broadband uses, even if it can be redeployed instantaneously and at low cost, cannot by itself meet the escalating bandwidth needs of the mobile broadband industry using today's technology. Part of the solution must come instead from spectrum-sharing arrangements, including with federal spectrum holders.

In a highly publicized July 2012 report, the President's Council of Advisors on Science and Technology (PCAST) opined that comprehensive "clearing and reallocation of Federal spectrum is not a sustainable basis for spectrum policy due to the high cost, lengthy time to implement, and disruption to the Federal mission," and thus "the norm for spectrum use" on such spectrum "should be sharing, not exclusivity."[55] PCAST expressed optimism that technological innovations had made it far more feasible than before for network engineers to achieve such sharing through a combination of smaller cells, lower-power transmitters, and smarter receivers.[56] Here, too, the wireless industry responded with studied ambivalence, faintly praising the report for "highlight[ing] a range of forward-looking options, some of which are not yet commercially available," and insisting that "the gold standard for deployment of ubiquitous mobile broadband networks remains cleared spectrum."[57]

Reclaiming spectrum from broadcasters

Apart from spectrum used by federal agencies, the spectrum currently held by broadcasters presents the best opportunity for reallocating large amounts of spectrum for mobile broadband uses. That is particularly true of the frequencies used today by UHF broadcasters in the 600 MHz band, loosely defined as the spectrum stretching from around 500 MHz to just below 700 MHz. Because of the breadth of this band and its excellent propagation characteristics, much of the 600 MHz band lies in the sweet spot for mobile broadband services. As prior experience confirms, however, the process of inducing broadcasters to cede their spectrum to mobile providers is prolonged, complex, and fraught with political controversy.

In the late 1990s and early 2000s, the FCC oversaw the digital television transition, in which all major U.S. TV stations began broadcasting in digital format and ultimately, in 2009, stopped transmitting in analog format. As part of that transition, the FCC assigned new frequencies for the TV stations that had been broadcasting in UHF channels 52–69, and it reallocated the spectrum encompassing those former TV channels—the "lower" and "upper" 700 MHz bands—for mobile voice and broadband services. This band-clearing initiative created immense consumer value. Whereas consumers needed more spectrum to accommodate bandwidth-hungry mobile data services, the vast majority of people who watched programming on channels 52–69, just like people who watch any TV channel today, watched it by means of subscription cable or satellite services rather than conventional over-the-air broadcasts. After a decade of planning and implementation, the spectrum was mostly auctioned off by 2008, and the first mobile broadband services using this spectrum were introduced a few years after that. (In the next chapter, we discuss why it takes so long to bring such "new" spectrum to market.)

Policymakers then turned their attention to the UHF channels in the 600 MHz band. The challenge of freeing up that spectrum for wireless broadband services is as complex as it is fascinating to game theorists and public choice scholars. The central irony in this story is that the FCC arguably could have exercised its wide statutory discretion to reallocate ("reband") all of this broadcast spectrum unilaterally and auction it off to mobile providers, at least once the incumbent broadcasters' licenses expired. The catch was that, until 2012, the FCC lacked any lawful mechanism to *compensate broadcasters* for their loss of spectrum. And it was widely understood that if broadcasters could not be compensated, the reallocation of this spectrum was a political nonstarter, given the strength of the broadcasting lobby.

In 2012, Congress broke the impasse by directing the FCC to hold voluntary *incentive auctions* for the 600 MHz band.[58] In a nutshell, incentive auctions encourage incumbent licensees to cede spectrum voluntarily by entitling them to a portion of the proceeds that the government obtains by auctioning the spectrum off to the highest bidders. With respect to the UHF broadcast spectrum in particular, the statute directs the FCC to conduct a one-time set of two distinct but tightly intertwined auctions: a *reverse auction* involving broadcasters and a *forward auction* involving mobile wireless providers. As of this writing, the FCC has not yet set a date for these auctions.

In the reverse auction, broadcasters that choose to participate will confidentially disclose to the FCC what compensation they will accept in order to give up some or all of their spectrum rights in a given market. They can offer to cede such rights in one of three possible ways. First, a broadcast station can offer to "go dark" and give up all of its spectrum (a six megahertz channel). Second, it can agree to relocate from its UHF channel to a new VHF channel, situated lower than 216 MHz.* Third, two stations in a market can become spectrum partners, agree to cede

* Although VHF channel placements were previously considered more attractive in an analog broadcasting world, they now are considered to have less favorable signal characteristics than UHF in a digital environment. VHF frequencies are also less favorable than UHF for mobile wireless services, in part because they require the use of longer antennas in mobile devices.

one of their six-megahertz frequency allotments, and share the other six-megahertz allotment for continuing broadcast operations. In the last scenario (as in the second), each station will remain an FCC licensee, and each will maintain its regulatory entitlement to be carried by local cable operators (under the "must carry" rules discussed in chapter 9).

Within any given geographic market, the "winners" of these reverse auctions will be the broadcasters who, roughly speaking, accept the lowest compensation for the rights they are ceding, although the FCC has yet to specify many of the details for that reverse-bidding mechanism, including the procedure for determining how much each winning bidder will receive.[59] The broadcasters that actively participate in this auction are likely to be those with the least-watched stations in a given market (i.e., those with the most tenuous business plans). The incentive auction gives such marginal broadcasters a one-time opportunity to realize the opportunity costs of their spectrum holdings—that is, to share in the far greater value that mobile broadband providers can extract from this spectrum.[60]

The overall value created by this reverse auction will depend not only on how many broadcasters participate and how many megahertz they cede overall, but also on whether the FCC can free up large *nationwide blocks* of spectrum. Mobile devices are designed to receive signals in a handful of defined frequency bands across the electromagnetic spectrum. Each band is more valuable to a carrier and its subscribers if it is available for mobile uses in all geographic areas in which those subscribers live and travel, an objective that requires clearing television broadcasters from the same band in each such area. Broadcasters know this, and, in the absence of FCC oversight, any given broadcaster in City X would have incentives to hold on to its spectrum block until after broadcasters in Cities Y and Z cede the same spectrum block to wireless providers; at that point, the hold-out broadcaster would be able to extract an inefficiently high price for this missing piece in a wireless provider's spectrum puzzle.

In the 2012 legislation, Congress addressed this hold-out concern in two ways. First, it directed the FCC to hold a single set of incentive auctions simultaneously, lest broadcasters otherwise stay on the sidelines during early bidding in the hope of a greater future reward. Second, it reaffirmed the FCC's authority to alter the frequencies on which (and locations from which) local broadcasters may transmit their signals if they remain on the air after the auctions are implemented, thereby maximizing the value of the remaining spectrum for mobile broadband uses.

This *repacking* process is more complex than it might first appear. In the densely populated Northeast, for example, television stations are arranged in an intricate daisy chain of interdependent channel assignments from city to neighboring city; for example, there is a channel 3 in Philadelphia, but not in New York. To address interference concerns, every change in a channel assignment in one city will need to be coordinated closely with channel assignments in neighboring cities. And the FCC will thus need to elicit a high level of participation among stations in the Northeast in order to free up many blocks of nationally cleared spectrum. Even then, however, there will likely be some bands that are cleared of broadcasters in some markets but not others—a fact that will decrease the value of those bands throughout the U.S. for mobile providers seeking a national 600 MHz footprint. And there may well be major metropolitan markets such as Detroit and San Diego where new mobile broadband spectrum, even once it is cleared of all U.S.-based broadcasters, will be subject to substantial interference from Canadian or Mexican TV stations, which the FCC has no authority to repack.

The second phase of this auction process—the forward auctions—involves its own set of challenges. These forward auctions are like traditional commercial spectrum auctions, but with two critical twists. First, auction participants will probably have far less information than usual about what they are bidding for. For example, depending on how the FCC structures the auction process, wireless bidders may not know at the time of their initial bids exactly what type and quality of spectrum they will receive after the repacking process is complete.

Second, the forward auction is complicated by its complex interrelationship with the reverse auction. By congressional design, this entire spectrum-freeing initiative can succeed—and spectrum can be reallocated—only if the amount that winning wireless participants agree to pay the government in the forward auction exceeds (1) the amount that the FCC must pay winning broadcast participants for the spectrum rights they ceded in the reverse auction, plus (2) the estimated costs (which the FCC must reimburse) of repacking the remaining broadcast spectrum as well as certain administrative costs.[61] If, for a given target level of cleared spectrum, forward-auction bids fall short of that statutory benchmark, the FCC may not reallocate any spectrum. Instead, the FCC can try again with more modest reallocation goals, decreasing the amount of compensation due to broadcasters by reducing the number of winning broadcaster bids it will award in the reverse auction and thus the amount of spectrum it will clear. That adjustment, however, will also likely reduce the aggregate bids that forward-auction participants will offer for the total (now smaller) amount of spectrum being auctioned. And the FCC thus may need to re-run the bidding several times, for decreasing target amounts of cleared spectrum, before reaching the point (if any) where forward-auction revenues meet the statutory criteria for concluding the auction and reallocating spectrum. This intricate relationship between the reverse and forward auctions creates one of the most mind-bendingly complex problems in game theory that any regulatory authority has ever been asked to resolve.

If this incentive-auction initiative succeeds, however, it will produce considerable benefits for broadcasters, broadband providers, and the public.[62] Mobile broadband providers and consumers will receive significant additional spectrum to help support increasingly bandwidth-hungry services. And broadcasters will receive billions of dollars as well for agreeing to vacate spectrum that the overwhelming majority of their viewers do not rely upon to receive TV programming anyway.

C. Controversy in the transition

Some observers find it objectionable for the government to pay off the broadcasting industry for use of "the public airwaves," as contemplated by this new plan for incentive auctions. Recall that the FCC did not traditionally hold spectrum auctions to assign broadcasting licenses, and almost all broadcasters have therefore paid nothing to the federal treasury for their licenses (although, to be sure, many broadcasters have paid large sums to *prior broadcast licensees* through mergers and acquisitions). Some policymakers have thus made the essentially political objection that if an incumbent licensee did not pay "the public" for its spectrum at auction, it should be denied the "windfall" it would receive if it were permitted to sell its license for millions of dollars on the newly privatized market.[63]

Despite its intuitive appeal, this "windfall" objection suffers, in our view, from two basic conceptual flaws.[64] First, as Eli Noam explains, any claim that the public "owns" the spectrum, such that it "deserves" compensation for its use by private actors, is arguably no stronger than the claim (which no one makes) that the public deserves special compensation from the airlines for the right to fly planes through lanes in the public airspace.[65] Auctions are most persuasively justified not as mechanisms for compensating the public for the use of "its" airwaves, but as a means of assigning spectrum rights as quickly as possible to those who would make the most efficient use of them. Second, from a consumer-welfare perspective, granting spectrum incumbents a "windfall"—if that is the only quick way to free up the spectrum at issue for more efficient uses—is usually superior to letting the incumbents tie up that spectrum in perpetuity with the less efficient uses specified in their licenses.[66]

Similar "windfall" objections have also greeted previous FCC initiatives to free up spectrum for mobile voice and broadband services. We briefly discuss two here: the FCC's 2004 Nextel rebanding decision and the FCC's more recent initiatives to let two satellite telephony providers use their non-auctioned spectrum to provide ordinary terrestrial cellphone service.

The 2004 Nextel spectrum-relocation controversy

Recall that Nextel's founder Morgan O'Brien purchased inexpensive "dispatch" spectrum on the secondary market and then quietly won enough regulatory flexibility in the use of this spectrum, most of it in the 800 MHz band, to provide cellular telephone service. A dozen years later Nextel's well-subscribed cellular operations threatened widespread interference with the transmissions of hundreds of public safety authorities that still used adjoining frequencies within the 800 MHz band for traditional dispatch operations. Exacerbating the problem was a basic engineering reality: high-power, geographically expansive transmissions (dispatch) and low-power, smaller-cell transmissions (cellular) cannot readily coexist on the same spectrum band.

After years of mutual discord, Nextel and public safety officials forged a deal under which Nextel would vacate some of its existing spectrum assignments in exchange for others in the 1.9 GHz band and would underwrite the considerable costs to the public safety authorities of modifying their equipment to operate in different frequencies. The only catch was that the FCC would have to approve the deal and grant Nextel the new and valuable 1.9 GHz spectrum it wanted—without an auction.[67] In July 2004, after several years of intense lobbying, the FCC finally announced its approval of a modified version of Nextel's plan.[68] Although the FCC traditionally has enjoyed wide latitude in managing the spectrum, the movement toward an auction-based property-rights model greatly complicated the legal picture, making the Nextel proceeding "by far one of the most complex matters" that Chairman Michael Powell had faced at the FCC.[69]

The FCC's ultimate disposition was complicated, but its basic terms required Nextel (now part of Sprint) to relinquish some of its spectrum in the 800 MHz band and pay (1) the transition costs incurred by the public safety agencies in modifying their equipment to operate on new frequencies plus, potentially, (2) additional sums to the federal treasury as an "antiwindfall payment" (if relocation costs fell short of the estimated value of the spectrum). Verizon Wireless and other opponents claimed that this plan violated section 309(j) of the Communications Act, which (as discussed) requires auctions for most new spectrum assignments. Verizon then leveled additional arguments against the plan just days before the FCC approved it, warning that "proceeding on this course would place *the Commission's members themselves* in direct violation of federal laws governing the *personal* accountability of federal officials for the disposition of federal resources[,] . . . some of which are criminal in nature."[70] Unfazed by this extraordinary warning, the FCC unanimously approved the plan, and Verizon let the matter rest.

Mobile satellite service controversies: LightSquared and DISH

Several years later the FCC triggered similar controversy when it sought to ease limits on when *satellite*-based telephony providers can use their spectrum licenses to compete with *terrestrial* mobile providers such as Verizon Wireless, AT&T, and Sprint. Beginning in the 1990s, a number of providers with names such as Iridium, Globalstar, and MSV (later SkyTerra and then LightSquared) began offering so-called *mobile satellite services* (MSS). These MSS providers use fleets of satellites to provide connectivity in the many remote areas worldwide that lack conventional cellular network coverage. To offer such services, an MSS provider needs two basic regulatory entitlements: orbital assignments for the satellites in question and spectrum licenses for the frequencies used to transmit signals between those satellites and customers' handsets. Because these satellite-oriented spectrum licenses are subject to international agreements, they are assigned differently from the licenses used by conventional terrestrial mobile providers. Although federal law generally *requires* the FCC to use auctions to assign new spectrum licenses to terrestrial providers,[71] it *forbids* the Commission to use "competitive bidding" to assign "spectrum used for the provision of international or global satellite communications services."[72] Thus, unlike conventional wireless providers, MSS providers have received their spectrum licenses for "free."

If MSS providers had to rely entirely on satellites to offer service, they would pose little competitive threat to terrestrial wireless providers, and no windfall issues would arise. In any developed area, the per customer network costs of a satellite-based telephony network are far higher than those of a terrestrial network, and handsets that are fully equipped for communications with satellites are generally bulkier and more expensive than conventional cellphones. True satellite-based telephony is thus an expensive niche service used mainly by a few corporate customers and the military.

But why *should* the holders of MSS spectrum be forced to rely solely on satellites to provide service? Suppose an MSS provider builds (or partners with others to build) a conventional terrestrial cellular network that transmits and receives signals over the same spectrum used for satellite transmissions. Such services are far less costly to provide than satellite-based services using the same spectrum, and these new terrestrial networks can thus be used to provide low-priced service to millions of ordinary consumers, thereby meeting the larger objective of freeing up more spectrum for mobile broadband. Conventional wireless providers have sometimes argued that it is unfair to allow MSS providers, who paid nothing for their licenses, to compete directly with terrestrial providers, who have collectively paid tens of billions of dollars at auction (at least for all licenses allocated after the auction regime was created). But if policymakers banned all terrestrial uses of MSS spectrum in order to avoid such "unfairness," they would essentially waste enormous spectrum resources—at least for the duration of these MSS licenses—by consigning the relevant spectrum bands to less cost-effective satellite-based uses for which there is comparatively little demand.

For nearly a decade, the FCC has struggled with this set of competing concerns and has gradually chosen efficiency over "fairness."* Starting in 2003, it gave MSS providers ever-greater leeway to integrate terrestrial functionality with satellite-based functionality so long as the "terrestrial components" of their services remain in some sense "ancillary" to their "principal" satellite-based offering.[73] In 2011, the FCC conditionally allowed MSS provider LightSquared to go one step further, side-step these *ancillary terrestrial component* (ATC) restrictions, and become a full-blown wholesaler of spectrum to ordinary retail providers of terrestrial wireless services, subject to some limitations.[74] Had this plan

* Similar issues arise in the television broadcasting context, where licensees are free to use a portion of their spectrum for "ancillary or supplementary" purposes so long as they air at least one free video signal. *See* 47 U.S.C. § 336. But Congress required licensees who exercise this option to pay the federal treasury a percentage of any revenues attributed to such services—in essence, a small anti-windfall tax. *Id.* § 336(e). No analogous requirement applies to the MSS licensees that benefit from new flexibility in the use of their non-auctioned spectrum.

succeeded, a customer could have ordered a regular phone and service plan from, say, Best Buy and made calls on LightSquared's spectrum bands using Best Buy's phones.

The FCC essentially rescinded its 2011 order a year later, however, amid intractable concerns about the interference of LightSquared's terrestrial operations with global positioning systems (GPS), scuttling the company's plans to become a major wireless wholesaler. But that 2011 order set an important precedent for the regulatory treatment of another company, DISH, with MSS spectrum ambitions of its own.

In 2011, DISH, best known as a satellite TV company, acquired insolvent satellite telephony providers DBSD and TerreStar from bankruptcy and, in early 2012, received regulatory approval to acquire those companies' considerable MSS spectrum holdings in the 2 GHz band.[75] In December 2012, after an intensive rulemaking proceeding, the FCC eliminated all ATC restrictions and authorized DISH to use the spectrum for the provision of terrestrial mobile services, subject to various interference-related rules designed to protect mobile operations in adjacent bands.[76] The Commission reasoned that the 2 GHz band will provide much-needed new capacity for mobile broadband services; that maintaining ATC restrictions would irrationally subvert that objective; and that DISH, as the existing MSS licensee, can most efficiently exploit the full terrestrial potential for that spectrum as well. That said, the FCC wanted assurances that DISH, which had no terrestrial network and had never operated a mobile wireless service, would actually put this new spectrum to immediate use rather than letting it lie fallow. To that end, the FCC imposed specific build-out milestones that DISH would have to meet—deadlines by which DISH must offer service to increasing percentages of the relevant population.[77] And the FCC announced license-cancelation penalties to deter DISH from falling short of those milestones.[78]

The FCC noted that it had imposed similar build-out requirements in past proceedings involving spectrum rights.[79] Whether such requirements are necessary to promote efficient outcomes is subject to debate. In theory, market forces should give spectrum holders adequate incentives to put any valuable resource to prompt use. But that theory has not deterred policymakers in a range of fields from establishing "use it or lose it" requirements to ensure that those authorized to lease public resources are making productive use of them.[80] In practice, moreover, some spectrum licensees have held spectrum in limbo for many years before finally selling it to parties intent on extracting its consumer value. In 2006, for example, a consortium of cable companies known as SpectrumCo acquired considerable spectrum at the AWS auction, but then let it lie fallow for five years until, in 2011, Verizon Wireless finally offered to buy it at a sufficiently attractive price (see chapter 4). The FCC appears to have taken that experience into account in adopting build-out requirements for DISH.

Avoiding the sunk cost fallacy in the assignment of spectrum rights
As this series of orders illustrates, the FCC has concluded that it cannot focus single-mindedly on fairness concerns arising from the prices originally paid for spectrum rights without disserving the public's more general interest in the efficient use of spectrum. Arguments to the contrary are sometimes infected by the *sunk cost fallacy.* In competitive industries, firms set prices on the basis of their *opportunity costs*—the cost of forgoing alternatives—not on the basis of their historical costs, which are "sunk" (generally unrecoverable) and thus irrelevant to their current business decisions. For example, if you put your house on the market today, the ultimate sales price will not depend on whether you inherited the house from your parents or bought it at the top of the market six years ago just before the market crashed. You wouldn't charge less than the market rate if you inherited the house, and you wouldn't be able to charge more if your own purchase price exceeded the current valuation. But despite the cold market realities, many individuals do not act rationally in such cases, instead indulging the sunk cost fallacy—that is, a belief that it is sensible to compromise one's present and future self-interest because of past economic decisions.

On occasion, the FCC itself has fallen prey to this fallacy. The D.C. Circuit once invalidated an FCC order because the Commission, in justifying the disparate treatment of two different classes of spectrum licensees, had presumed that carriers that obtained their spectrum at auction would have incentives to return a profit on that spectrum more quickly than carriers that obtained it for free. As the court observed, with an intemperate bluntness characteristic of its longstanding treatment of many FCC orders, "This is a foolish notion that should not be entertained by anyone who has had even a single undergraduate course in economics. Failing that advantage, a moment's reflection would bring one to the realization that the use to which an asset is put is based not upon the historical price paid for it, but upon what it will return to its owner in the future. Would anyone be less interested in earning a return on money he had inherited than on money he had worked for? Of course not! Are radio licensees not as alert as inheritors?"[81]

Several years later, in a different context, a chastened FCC recognized the sunk cost fallacy for what it is and rejected claims that liberalizing the uses of spectrum by providers that had obtained it outside the auction process would give them an anticompetitive advantage in competing with providers that had paid at auction: "[T]he telecommunications experience in the U.S. has . . . been consistent with the theory that historic costs don't alter pricing. For example, within a given market, the prices charged by cellular operators who obtained their licenses via comparative hearings [or] lotteries are not lower than the prices of those firms that purchased their cellular licenses in the secondary market, or firms that obtained . . . licenses in an auction. Similarly, where a U.S. cellular license has been bought at a significant cost from a party that obtained it at no cost, we have not observed any increase in consumer prices."[82] These observations, the FCC concluded, confirm what economic theory suggests: "[L]icensees do not have an incentive to forgo recovery of the value of spectrum and price below competitive levels merely because the spectrum was obtained without an auction. Pricing that does not include recovery of the market value of an asset such as spectrum represents a loss (compared to the price that could be sustained in the marketplace) that [these] operators would have to bear regardless of how much, if anything, they spent on acquiring the asset initially."[83]

D. Commons: Einstein's cat in the age of the mouse

Policymakers have long understood that freeing up spectrum for its most socially beneficial uses will require heavy reliance on Coase's property-rights model. But like the counterpoint in a Bach fugue, a second school of thought has grown up alongside the property-rights model, both rivaling and complementing it. In its purest, most academic form, this *commons* model would dispense with any scheme for licensing exclusive rights to private entities across very broad swaths of spectrum. Everyone operating devices within those swaths would be free to exploit that spectrum simultaneously for any number of uses, just as everyone is free to exploit Central Park simultaneously for any number of uses (commercial and otherwise), subject in each case only to the most basic rules of the road.

One technological underpinning of this approach is the emergence of "smart" transmitters and receivers with advanced microchip technology.

These devices are capable of avoiding mutual interference by transmitting at low power over wide expanses of spectrum and exploiting the intelligence of next generation receivers to identify and decode the transmitted signals.[84] The commons school calls not only for more unlicensed spectrum in which these smart devices can operate, but also for rights of unlicensed users to operate such devices in *licensed* bands of spectrum as well.

These unlicensed uses in licensed spectrum can take several forms. For example, *underlay* transmissions operate at such low power levels as to become theoretically indistinguishable from background noise and thus pose a theoretically low threat of interference with the primary uses of the relevant frequency bands. Alternatively, unlicensed users can employ higher-power transmissions to fill the *white spaces* (unused frequencies) within a licensee's spectrum.[85] These white spaces sometimes fluctuate from frequency to frequency at any given moment, and sometimes they are relatively fixed, as in the portions of the UHF spectrum that otherwise remain unused at all times (because, for example, they serve as "guard bands" designed to prevent interference between adjacent licensees). In its 2010 *White Spaces Order*, the FCC took cautious but significant steps in authorizing the use of white-space-exploiting devices within the television broadcast bands and in establishing a common database that will help such devices identify the specific frequencies on which they may operate in particular locations.[86]

The "commons" approach is not quite as novel as it might sound.[87] Under its Part 15 rules, the FCC allows certain wireless devices—such as Wi-Fi-enabled laptops and tablets, cordless phones, TV remote controls, and garage door openers—to use unlicensed blocks of spectrum or licensed spectrum at very low power levels without prior authorization.[88] Operators of these "Part 15 devices" have no guarantee of protection against interference by others and bear responsibility for avoiding interference with licensed users.[89] Moreover, the FCC has long required that all Part 15 devices be "certified" as compliant with certain operating rules—such as restrictions on transmission power levels—that reduce the probability of interference with other users of the spectrum.[90]

The first step in the modern wireless revolution was the deployment of *spread-spectrum* technology, first authorized for commercial use in unlicensed bands in the late 1980s.[91] This technology enables paired devices, such as a wireless router and a laptop, to exchange signals over a range of frequencies within a band rather than over a single frequency. The best-known spread-spectrum applications today are cordless phones and Wi-Fi, which have now filled up much of the unlicensed spectrum that the FCC once regarded as its "garbage bands." Since the mid-1990s, spread-spectrum technology has also proliferated in licensed bands, as illustrated by the success of the proprietary code division multiple access (CDMA) technology used widely in cellular telephone service (see chapter 4).

Like spread-spectrum technology, *cognitive radio* devices (including so-called software-defined radios) use advanced microchip technologies to find and exploit underutilized spectrum within a given band. For example, a transmitter and receiver, acting in tandem, might hop dynamically from frequency to frequency during the course of a call or data session, avoiding frequencies along the way that at given instants are occupied by unrelated signals or excessive noise. By "intelligently" filling in the gaps of unused spectrum within a given band, such devices can use available spectrum much more efficiently than can traditional "dumb" devices. Also like spread spectrum, cognitive radio technology can be used not just in unlicensed bands, but also in licensed ones, where they can facilitate shared access to spectrum through (for example) secondary market arrangements.

Understanding these new technologies requires some familiarity with a principle of information theory known as *Shannon's Law*. In 1948, Bell Labs engineer Claude Shannon observed, in mathematically precise detail, that the information-carrying capacity of a communications channel increases in direct relation to the breadth of the frequencies employed and the "signal-to-noise" ratio.[92] That ratio reflects the power of the transmission as compared to the background electromagnetic radiation, whether emitted by other wireless devices, by "unintentional radiators" such as car motors, computers, and hairdryers, or by natural sources of noise such as lightning, cosmic radiation, and so forth. Because no wireless channel is free of such noise, conventional transmissions must use significant wattage to enable receivers to identify and decode the relevant signal, just as you must raise your voice over the "noise floor" of a crowded restaurant to be heard.

Shannon's Law largely explains why television broadcasters consume such vast tracts of spectrum. TV broadcasters transmit enormous amounts of information in the form of high-quality real-time video to television sets on static, preassigned frequencies. The conventional way to convey all that information to ordinary television receivers is to allow broadcasters to blare their signals at enormous wattage over broad swaths of spectrum. That arrangement works out well for the television station and its passive viewers, but only at the expense of crowding out many other wireless applications for dozens of miles in every direction, regardless of whether the channels are being watched or even used (as they sometimes are not—late at night, for example).

Ultra-wideband technologies represent in some ways the technological opposite of conventional television broadcasting, conveying large volumes of information *both* at low power levels *and* without monopolizing particular spectrum bands. Imprinting information either on many different carrier signals or in millions of short pulses of radiation, ultra-wideband devices transmit signals over such an enormous expanse of spectrum that, under Shannon's Law, they can convey information using power levels so low as to fall beneath the "noise floor" for conventional uses of the affected spectrum. Like the other technologies mentioned earlier, ultra-wideband systems rely on an intelligent receiver that translates the signals by listening for a familiar pattern sent by the transmitter. Although some (nonengineer) commentators have argued that these advances might theoretically eliminate interference concerns altogether, that argument misunderstands the relevant engineering principles. It is more accurate to say that ultra-wideband and other spread-spectrum technologies enable network engineers to exploit given spectrum resources more efficiently and that in the process they often allow more simultaneous exchanges of information than would otherwise be possible.[93]

In February 2002, after some delay, the FCC finally permitted the use of ultra-wideband devices under restrictive conditions designed to minimize interference risks to the devices used by spectrum incumbents.[94] Unfortunately, regulatory approval of this new technology—like any other—is only the first step. The industry must also develop common standards before the technology can gain widespread adoption and scale economies, and efforts to design such standards can be highly contentious. That has been the case with ultra-wideband technologies, and, as a result, they remain more interesting in theory than in application. By contrast, Wi-Fi technologies have succeeded in the marketplace precisely because the industry has coalesced around common standards—the 802.11 family of protocols forged by the Institute of Electrical and Electronics Engineers (IEEE), a private engineering association that operates as a standard-setting authority.

In 2002, commons advocate Yochai Benkler cited these technologies as harbingers of what he predicted would be the next revolution in telecommunications technology: a widespread alternative to "carriers" as providers of last-mile access. If end users were allowed to treat spectrum as a commons, he claimed, they would create "open wireless network[s] that no one owns," that are "[b]uilt entirely of end use devices," and that can be set up "simply by using equipment that cooperates, without need for a network provider to set up its owned infrastructure as a precondition to effective communication."[95] These devices, Benkler speculated, would spontaneously converge into "mesh" networks, repeating one another's signals as needed from origin to destination while avoiding interference by observing constraints on the power of their respective transmissions. In such an environment, telecommunications regulators would shift their focus from regulating *service providers* to regulating *equipment providers* to ensure cooperation among devices in the commons.[96] But commons model advocates such as Benkler emphasize that developments like these will not reach their potential unless the government first frees up large swaths of spectrum for unlicensed uses.

For the commons regime to work as advertised in any frequency band open to unlicensed uses, however, it must overcome the fabled "tragedy of the commons." The problem lies in the temptation for users of resources held in common—whether those resources are something tangible such as a grazing field or intangible such as the electromagnetic spectrum—to spoil them through overuse or neglect.[97] As Stuart Benjamin has explained, that temptation would be overwhelming if large portions of the spectrum were opened up to individuals to exploit as they choose. In that event, each user would have strong incentives to "cheat" by, for example, employing high-power devices that would interfere with the devices of other users, and those other users would then have escalating incentives to raise *their* power levels to make themselves heard over the increasing din, and so forth.[98] Such problems might be controlled by regulation, but Benjamin contends that "the level of regulation involved [would be] significant."[99] Indeed, commons advocates often acknowledge the need for such regulation in one form or another but devote little explanation to exactly how it would work.[100] Critics of the commons approach similarly argue that the advantages of observing private-property rights in spectrum management are similar to the advantages of observing those rights throughout the rest of the economy: the purchase and sale of such rights permit efficient private ordering through the mechanism of price signals. Those price signals would be largely absent in a "commons" world without enforceable property rights.[101]

The academic back and forth between opposing theorists obscures a central point: few people advocate either a pure property-rights or a pure commons approach to spectrum policy.[102] There is a fairly broad consensus today that the government should maintain some bands for unlicensed uses but should also continue relying heavily on its auction-based property-rights regime to govern the beachfront spectrum that is used increasingly for mobile voice and broadband services. Indeed, that is precisely what Congress did in the 2012 incentive-auction legislation discussed earlier. Although Congress focused on freeing up broadcast spectrum for the exclusive use of designated licensees (mobile providers), it simultaneously authorized the FCC to maintain narrow guard bands between licensed spectrum assignments and to "permit the use of such guard bands for unlicensed use[s]" so long as those uses do not "cause harmful interference to licensed services."[103] That said, the legislation appeared to provide for less unlicensed spectrum than commons advocates had hoped for, a setback they attribute both to their limited political influence (tomorrow's innovators tend to have less clout than today's spectrum incumbents) and to the political allure of auction revenues for licensed spectrum.

In sum, the real debate between property-rights and commons advocates arises at the margins. The two sides argue in particular about how many megahertz should be allocated to unlicensed uses and about how sensitive policymakers should be to concerns that underlay or white-space technologies will interfere with licensed spectrum uses.[104] That debate is important, and it will play out in many different settings for years to come, but it is more circumscribed in its significance than the academic literature might suggest.

IV. The Future of Interference Policy

If you asked the typical property-rights advocate to describe his ambition for spectrum policy, he might answer: "To make spectrum rights as much like land-ownership rights as possible." He would analogize between bidding wars for land and auctions for spectrum as well as between private leases of land and private leases of spectrum rights. He would argue that just as landholders have strong incentives to sell leases or easements to third parties if that will produce the most efficient use of their property, so too will private "band managers" have every incentive to lease unused portions of their spectrum (or sell rights to use noninterfering underlay or white-space technologies) when *that* is the best way to squeeze maximum value from their holdings. The purest of the private-rights advocates would go one step further, arguing that the FCC will create maximal value in the use of the airwaves if, after auctioning them off, it walks away forever while Adam Smith's invisible hand works its magic.

This last step in the argument is unrealistic, and the reason lies in the unpredictability of interference across spectrum bands and the intractable complexity of many interference disputes. Indeed, if and when pervasive competition makes all traditional forms of telecommunications regulation obsolete, there will still remain a critical need for government oversight of interference policy. The government's enduring challenge will be to devise guidelines that (1) maximize the value of spectrum by allowing diverse technologies to flourish and permitting neither "too much" nor "too little" interference, (2) increase predictability in the outcome of interference disputes, and (3) take account of the quite different interference issues that arise in different spectrum bands and different industry contexts.[105]

As discussed, interference is often an attribute of radio equipment, not of radio waves; it arises when, for example, a radio receiver tuned to one frequency cannot filter out unwanted signals in nearby frequencies. A cheap transistor radio is the classic case in point. It is much less adept than a high-end stereo system at tuning in to the precise frequencies used in one station's radio broadcasts and screening out the frequencies used by other stations. Most categories of receivers—from FM radios to GPS devices—exhibit a range of vulnerability to interference. And as a general rule, the more expensive the device is, the better it can withstand interference.

Throughout its history, the FCC has used a poorly defined standard of "harmful interference" that largely accepts as given a spectrum licensee's subjective claims about how much spectrum it needs in order to guard against interference.[106] Pressured by Congress, the FCC has historically been deferential to the subjective claims of radio and television broadcasters in particular. To some broadcasters, any new use of frequencies near an existing station's channel assignment would pose an unacceptable risk of interference if a single listener cannot receive a broadcast that she might otherwise receive—even if many other people would benefit from new uses on those adjacent frequencies. Also, in judging whether that single listener faces any interference, broadcasters often presume that she owns the lowest-quality and most interference-prone receiver on the market. Citing the interests of this lone listener with a cheap radio, the broadcasting lobby persuaded Congress to subvert much of the FCC's initiative to license new low-power FM stations in 2000.[107] Broadcasters have more recently invoked similar interference concerns in opposing—with less success—the "white spaces" initiative noted earlier, which permits high-tech devices to exploit the interstices in the signals used by television broadcasters.[108]

Of course, policies that err on the side of disallowing any interference to incumbents present enormous opportunity costs—the costs of forgone stations, forgone technologies, and forgone bandwidth. As Coase explained more than 50 years ago, policymakers should ensure that "the gain from [allowing additional] interference more than offsets the harm it produces."[109] In other words, spectrum policy should seek to prevent not only *too much* interference, but also *too little*. And the federal government's command-and-control regime is particularly ill suited to that objective. By prescribing allowable uses of spectrum and limiting commercial transactions, that regime has kept many spectrum incumbents, including broadcasters and federal agencies, from fully internalizing the opportunity costs of inefficient uses of spectrum.[110] Where the regime persists, therefore, it gives spectrum incumbents no incentive to strike an efficient balance between too much and too little interference; *any* interference will appear undesirable to them.

Excessive conservatism in interference policy is also perversely self-reinforcing: it suppresses the very technologies that might increase spectral efficiency by making interference less likely. As Thomas Hazlett explains, when the government declines to allow new entry on the ground that it would disturb cheap, interference-prone receivers, "[i]nterference becomes a self-fulfilling regulatory prophecy" because "radio sets need not upgrade performance" (to conserve on their use of spectrum), and, worse yet, "there is no demand for equipment to receive additional signals (which fail to be licensed by the FCC)."[111]

This is *not* to say that interference concerns are always overblown. The more interference produced by one set of spectrum users, the greater the costs imposed on third parties to upgrade receivers to avoid that interference. In other words, there are costs on both sides of the equation, and balancing them can be unavoidably subjective and laden with controversy.

The property-rights model can help perform that task efficiently where the interference in question is all within a single band managed by a single "landlord" who leases out spectrum rights to third parties and has strong incentives to maximize the value of the band as a whole. But the property-rights model can provide no easy solutions to interference problems arising between holders of two mutually exclusive licenses. Such interference can arise from either spectral or geographic proximity. For example, it can arise when transmissions within one licensee's band interfere with the reception of signals on an adjacent band used by another licensee. And interference can arise whenever the FCC assigns a given frequency to one licensee within one geographic area while assigning the same frequency to a different licensee in a neighboring geographic area. To be sure, licensees in some contexts have strong incentives to reach solutions on their own; one example is the mobile wireless industry, where the main players are well established and can satisfy their customers only if they form stable, technologically cooperative relationships. In other contexts, however, the property-rights model offers no straightforward solutions to interference disputes, and the government may therefore need to play a critical dispute-resolution role for the indefinite future.[112]

The LightSquared–GPS controversy exemplifies the complexities of this challenge. As discussed earlier, LightSquared benefited from a series of FCC orders granting certain MSS licensees increasing flexibility in the use of their spectrum. And in 2011, the FCC conditionally authorized LightSquared to complete its evolution from a niche provider of satellite-oriented telephony services to an emerging spectrum wholesaler for providers of conventional terrestrial wireless services.[113] But there was a glitch. Tests showed that if LightSquared made full use of its spectrum for terrestrial transmissions, it would cause widespread interference to GPS receivers used for navigation and military purposes in neighboring frequency bands. LightSquared downplayed the significance of these test results, though, and argued that any inference problems were the fault of the GPS industry anyway. As it observed, the FCC had put the GPS industry on notice many years earlier that MSS providers such as LightSquared could make "ancillary" terrestrial use of their ostensibly satellite-oriented licenses. LightSquared argued that the GPS industry should have produced more sophisticated receivers less prone to interference from transmissions in LightSquared's adjacent spectrum.[114]

The GPS industry responded that although it had known that the FCC would permit genuinely *ancillary* terrestrial uses, it had not known that the FCC would make it as easy as it ultimately did in 2011 for LightSquared to devote its spectrum almost entirely to terrestrial uses.[115] In other words, the GPS industry had bet that the FCC would continue imposing such onerous restrictions on LightSquared's licenses that LightSquared spectrum would lie nearly fallow. If so, there would have been no need for the GPS industry to produce costlier, less interference-prone receivers; in that event, consumers (for example) would have paid more for navigation systems but with no commensurate benefit. But the FCC initially defied the GPS industry's prediction. Confronted with the explosive demand for mobile broadband services, it conditionally relaxed restrictions on the use of LightSquared's spectrum much more than expected. And that decision threatened extensive interference for the installed base of inexpensive GPS equipment used in car and airplane navigation devices, military systems, and agricultural equipment sold by staunch LightSquared adversary John Deere. In early 2012, amid widespread opposition to LightSquared's plans, the FCC signaled that it would reverse course and maintain the restrictions it had previously seemed poised to eliminate.[116]

The LightSquared controversy encapsulates several key lessons about interference policy. First, when such controversies erupt, policymakers must balance competing priorities—in this case, between (1) the need for greater efficiency in the use of broadband spectrum and (2) the need to keep critical infrastructure operable and consumer-electronics equipment affordable. The interests on each side of the balance are often incommensurable. For example, there is no way to assign objective values to a healthy broadband economy on the one hand and national security on the other. For that reason, picking winners and losers in these disputes is as much a political exercise as a regulatory one.

Second, both the nature of these controversies and their ultimate resolution are difficult or impossible to predict years in advance. Third, that unpredictability carries great costs. Equipment manufacturers must sometimes decide years before interference problems arise what technological steps they should take to alleviate those problems—and at what extra cost. As the LightSquared–GPS episode reveals, they must sometimes make those highly consequential decisions in the dark. And for their part, spectrum owners will be reluctant to sink billions of dollars in next-generation wireless networks if they cannot reasonably predict what interference-related restrictions the FCC will impose on the use of those networks. Such spectrum-interference issues are particularly difficult to resolve when the relevant receivers are produced by independent firms and not coordinated with service providers that supply them to customers. That is the case, for example, with GPS devices and over-the-air TV sets, but not with mobile devices sold by commercial wireless providers.

Although the FCC cannot eliminate all uncertainty, it can do more to mitigate it by adopting at least high-level guidelines for acceptable interference levels in different frequency bands. That, however, is not how the

FCC typically proceeds in this area. Consider the case of Qualcomm's MediaFLO product, designed to provide video delivery to mobile devices over a spectrum band that Qualcomm obtained at auction.* Qualcomm recognized that this service would pose interference risks for third-party services on adjacent bands, and it petitioned the FCC for advance guidance on how to proceed. More than twenty months later, the FCC finally issued an order detailing the basic rules that would govern Qualcomm's use of its spectrum.[117] That order was extremely specific to Qualcomm's particular service and to some extent directly regulated the inputs to that service (i.e., the technical details of how Qualcomm could transmit signals). For those reasons, the order provided little or no guidance to any future provider of innovative services.

To date, the FCC has proposed no industry-wide framework for resolving similar interference issues and appears likely to continue deciding these interference issues on an ad hoc basis. As a result, the next innovator seeking advance guidance on permissible uses of its spectrum will have to do what Qualcomm did: hire lawyers and lobbyists and then wait, potentially for years, as the FCC exercises its case-by-case discretion. But that process inflicts delays and expenses that many innovators cannot afford, and it therefore inevitably suppresses innovation in the use of spectrum resources.

What might a comprehensive interference policy look like? One of us has coauthored a "zoning code" proposal designed to ensure greater predictability and to facilitate more innovation than the FCC's current approach.[118] Under this proposal, the FCC would adopt distinct interference policies for different frequency bands, given the different propagation characteristics of those bands and thus the different interference issues that can arise on them. For many of those bands, the FCC would rely on a property-rights model and would avoid specifying which technologies or services could be used on any given band. And it would adopt interference standards based on *outputs*—that is, the interference levels actually caused by a service. That approach would stand in stark contrast to the Commission's frequent reliance on case-by-case regulation of the *inputs* tied to particular licensed services, as exemplified by the Qualcomm MediaFLO example.[119] In the words of an FCC-sponsored working

* Qualcomm eventually concluded that, for reasons generally unrelated to FCC interference policy, this new service would be a commercial failure. In 2011, Qualcomm agreed to sell the relevant spectrum to AT&T for use in its mobile broadband services.

group, an output-oriented approach would "provide desired flexibility while protecting the reasonable expectations" of other licensees,[120] and it would allow market forces to put the spectrum as a whole to its most valued uses.

A second approach to interference policy would focus less on mediating interference disputes than on avoiding such disputes altogether by calling for interference-protection standards years in advance.[121] Consider the case of TV manufacturing. Whereas mobile providers work closely with equipment manufacturers to ensure spectral efficiency in the handsets designed for particular frequency bands, broadcasters have no similar relationship with TV manufacturers. Given the FCC's traditional solicitude for broadcast viewers, TV manufacturers have never had strong incentives to install sophisticated, more costly tuners that can screen out interference from nonbroadcast wireless uses—just as GPS manufacturers had few incentives to incur greater costs to install more sensitive devices capable of screening out LightSquared's transmissions.

The problem is that once millions of consumers have bought unsophisticated, interference-prone TVs or GPS devices, they can be easily rallied to oppose alternative spectrum uses that their inexpensive receivers cannot screen out. That outcome is inefficient if the social value of the foreclosed uses exceeds what the incremental cost would have been of building the more sophisticated receivers that could have peacefully coexisted with those alternative uses.[122] Under one approach to this problem, policymakers could simply tell spectrum incumbents in advance how much interference they must put up with and hope that the market will produce whatever equipment is needed to accommodate those rules.[123] This approach is more likely to succeed the shorter the interval is between the announcement of the rules and the appearance of potential interference. The longer that interval, the more likely it is that the incumbents or manufacturers will ignore the future problem, sell millions of consumers cheap receivers, and hope that political pressure will induce a change of regulatory policy once interference arises. To avoid that outcome, regulators could alternatively prescribe *minimum equipment standards*, specifying more directly just how accommodating transmitters and receivers must be. These approaches are not always mutually exclusive, and context is key to determining which of them (or both or neither) is necessary to produce optimally efficient uses of spectrum.

To date, such big-picture thinking about interference policy has appeared more often in staff-level white papers and academic scholarship than in FCC orders.[124] The demise of the FCC's *interference temperature initiative* in the early 2000s suggests that comprehensive reform may still be many years off.

In November 2002, the FCC's Spectrum Policy Task Force offered a variety of recommendations for the future of spectrum policy and in the process introduced the concept of "interference temperature."[125] Under that concept, the FCC would define an objective level of tolerable radiation on particular frequencies in a particular geographic area and would assure the licensees for those frequencies that they will not face interference above that "temperature." In effect, this approach would define the property rights conferred on spectrum "owners" and recognize that some level of interference is unavoidable. In so doing, it would give the industry new and powerful incentives to design receivers capable of accommodating certain interference levels by, for example, exploiting advances in microchip technology.

The FCC opened an official inquiry on this interference temperature proposal in 2003, and spectrum incumbents announced their strong opposition.[126] Television broadcasters warned of widespread degradation of their signals. Mobile voice and data providers argued that they had already taken extraordinary technological steps to squeeze all available bandwidth out of the sharply limited spectrum the government had assigned them. And, they said, permitting additional uses within the same bands would simply degrade the quality of their service without creating any corresponding consumer benefits. It may be that any interference temperature concept would have more utility in bands, such as those occupied by federal agencies, in which spectrum incumbents have less pressing incentives to ensure efficient spectrum usage.

In 2007, the FCC quietly pulled the plug on this initiative "without prejudice to its substantive merits," noting that "[c]ommenting parties generally argued that the interference temperature approach is not a workable concept," and "no parties provided information on specific technical rules that we could adopt to implement it."[127] It is unsurprising, however, that industry commentary would so lopsidedly favor shelving any regulatory initiative that might benefit *future* innovators at the potential expense of *today's* incumbents. Incumbents have very specific interests in preserving the value of their investments, have millions of existing customers, and often command immense political resources. In contrast, future innovators may not even recognize themselves as such; they may still be teenagers experimenting in a garage. If the FCC is to promote long-term innovation in this area, it will have to write its own policies (subject to technological reality checks from industry), not rely on today's interested parties to write those policies for it.

* * *

Simply coming up with the "right" theoretical answers to spectrum policy is complex and contentious enough. As we have noted, however, it is only one challenge among several that policymakers must meet if there is to be true reform in this area. To begin with, policymakers must fill in the practical details of whatever regime they choose, defining (for example) the interference rules applicable in various contexts, identifying opportunities for unlicensed uses, facilitating market-based flexibility for licensed spectrum, and implementing a resource-intensive enforcement scheme to administer those reforms.[128] They must further devise a comprehensive transition plan that overcomes the political and legal obstacles that lie in the path of liberalization. And they must confront formidable coordination problems among the many different governmental authorities involved, including not just the state and local authorities who use spectrum for important public safety functions, but also the Defense Department and other federal agencies that do not answer to the FCC and have few incentives today to part with their allotted spectrum, no matter how inefficiently they may use it. Taken together, these factors conspire to make genuine spectrum reform as complicated and controversial as it is important to the future of telecommunications.

4 Mobile Wireless Services

Although the electromagnetic spectrum is used for many commercial purposes, the one that has most pervasively transformed person-to-person communications over the past quarter-century has been the family of mobile voice and data services characterized by the use of "cellular" technology. In 1981, virtually no one in the general public owned a cellular telephone, and trips to a telephone booth were time-wasting detours of necessity for anyone who needed to stay in touch while away from home or the office. Thirty years later the overwhelming majority of adult Americans had a cellphone, and the total number of wireless connections—including cellphones, tablets, netbooks, laptop cards, and the like—exceeded the total number of Americans.[1]

The mobile services marketplace is generally considered more competitive than the residential marketplace for fixed-line communications services such as wireline broadband Internet access. Most people live in localities covered by several alternative providers of mobile wireless services; customers routinely switch from one carrier to another; and prices per unit of consumption have fallen precipitously since the 1990s. For example, the average per minute price of a wireless phone call has fallen from approximately 43 cents in 1995 to less than five cents in 2010.[2] And in the most revolutionary development of all, skyrocketing numbers of Americans now use their wireless devices not only to make calls or send text messages, but also to reach the Internet at increasingly high bandwidths.

Policymakers have generally subjected mobile wireless services to much lighter forms of regulation than conventional wireline telephone service. This divergence in treatment is attributable largely to the different economic characteristics of these two industries. As discussed in chapter 2, the sheer expense of digging up the streets to install physical cables throughout individual neighborhoods created what regulators traditionally viewed as natural-monopoly conditions in the "last mile." The last-mile connections between a wireless carrier's transmission sites and its subscribers, however, consist of signals sent at different frequencies through the air. The main limits on the availability of the spectrum to multiple wireless providers are regulatory, not economic. There was a time, as we discuss later in this chapter, when the FCC enforced an artificial duopoly, assigning spectrum rights to just two mobile telephony firms in each geographic market—the incumbent telephony company and one independent carrier. Since the 1990s, however, the FCC has set aside additional frequency bands for wireless services, and this more liberal spectrum policy, combined with the proliferation of digital technology, has produced the competitive wireless services market that prevails today.

Given this competition, regulators have hewed closely to a deregulatory policy since the mid-1990s. But they have not left the wireless industry completely unregulated. First, although that industry does not have natural-monopoly characteristics, it remains a *network* industry, with the associated interconnection and standards-setting challenges familiar to students of the wireline marketplace. Regulators have concluded that although many of those challenges can be resolved through market dynamics, others require governmental intervention. Second, the mobile wireless industry has become increasingly consolidated since the first years of the new millennium. That consolidation has triggered unusually close scrutiny of industry mergers and acquisitions, as highlighted by AT&T's rejected attempt to buy T-Mobile in 2011. This chapter examines these issues in depth. We begin, however, with a basic overview of how wireless telephone service works and how it traditionally has been regulated.

I. The Basics of Cellular Technology

A mobile device is a sophisticated two-way radio that relies on a network of transmission antennas operated by a wireless carrier. In some cases, particularly in rural areas or along highways, these antennas are attached to towers specially built for radio communications. In other cases, wireless carriers attach their antennas to rooftops, utility poles, or other preexisting structures.[3] For the most part, only the link between the cellphone and the nearest transmission site—the analogue to the loop in a wireline network—is "wireless." Once the signals reach that site, they are normally routed through a wireline connection (a backhaul link, as discussed in chapter 2) to one of the carrier's centralized switches and then through more wireline connections en route to their ultimate destination. Of course, if you are calling someone else on her cellphone, the last link of the transmission on her end will be wireless as well, but most of the transmission in between will be channeled through wireline facilities.

The key to mobile voice and data services is the division of a wireless service area into many small geographic *cells*, each served by a single transmission site. This "cellular" approach is best introduced by contrasting it to the quite different method used for ordinary radio broadcasting. When you turn on your car radio, select a station, and drive twenty miles down the highway, you will generally receive the station's signals from the same transmission site during the entire trip. The station's transmitter, placed atop a tall tower, hill, or skyscraper, blasts out signals at enormous wattage to reach everywhere in a metropolitan or multi-county area. Until the final decades of the twentieth century, most mobile person-to-person telecommunications services were also provided by means of essentially this same technological model. Taxi companies, the police, and others used (and still use) their assigned slices of spectrum for radio "dispatch" services, in which signals are transmitted over wide geographic areas from centrally located antennas (sometimes with the aid of repeaters that boost the signal strength in areas far away from the transmission site).

This centralized transmission approach worked well enough so long as the airwaves were used mostly for "one-to-many" transmissions such as radio broadcasts to large audiences or generalized alerts from police dispatchers ("calling all cars"). But it is ill-suited for mobile voice and data services. Precisely because a conventional radio transmitter pumps out signals at high power to reach listeners throughout a broad geographic area, it tends to crowd out different uses of the same spectrum. This same approach would be immensely wasteful if the high-powered transmitter were tying up a given frequency throughout an entire metropolitan area *not* to broadcast radio signals to hundreds of thousands of passive listeners, but simply to send voice or data signals to a single mobile customer.

Cellular technology helps avoid such waste by dividing a broad geographic area into small discrete cells, which generally range from a few blocks wide (in downtown urban areas) to several miles wide (in more rural areas). These cells are often portrayed on a map in a honeycomb arrangement. The transmission antennas serving each cell emit relatively low-power signals that soon fade out once they pass beyond the cell's boundaries. As a result, wireless carriers can use the same set of frequencies for calls in one cell that they are simultaneously using for other calls in geographically distant cells.* "Recycling" the same frequencies in this manner enables these carriers to exploit their assigned spectrum much more efficiently than if they used a single high-powered transmitter for the entire area. Because mobile devices require commensurately lower power to communicate with these nearby transmission sites, they can operate for several hours on smaller batteries than they would need if they had to transmit to a single tower much farther away. Small batteries, along with advances in microengineering, allow for very small devices. That fact helps explain the rapid growth of the wireless industry, for consumers are more likely to carry a cellphone that slips inconspicuously into a pocket than one that, like the earliest cellular phones, has the size and weight of a brick.

These efficiency advantages, however, come with two basic trade-offs. A mobile wireless business requires both (1) deployment of an extensive and sometimes aesthetically controversial network infrastructure and (2) investment in extremely sophisticated technology to keep track of and handle the calls of millions of mobile customers.

First, a wireless carrier must incur the costs of establishing transmission sites and installing all of the antennas that compose a cellular network. Even apart from the considerable expense of this infrastructure, the need for many antennas can also present significant bureaucratic obstacles to the build-out of a cellular network because many localities exercise their traditional zoning authority to limit the placement of unsightly towers. Section 332(c)(7)(B) of the Communications Act, added in 1996, balances the interests of zoning authorities with those of wireless carriers by limiting the substantive bases on which localities can exclude transmission facilities from particular areas and permitting aggrieved parties to seek review in either federal or state court.[4] This provision requires localities to base any denial of a siting request on "substantial

* Modern cellular technology is enormously complex, and our technological summary here is necessarily simplified. For example, a cellular network today is typically divided not only into cells, but also into arc-shaped *sectors* within those cells, each corresponding to one of several directional antennas placed on a cell site. Moreover, network engineers increasingly cope with congestion in populous urban areas by supplementing conventional cell sites with even lower-power *distributed antenna systems* installed on utility poles and lamp posts.

evidence," an amorphous standard that, as one court explains, "requires balancing two considerations. The first is the contribution that the antenna will make to the availability of cellphone service. The second is the aesthetic or other harm that the antenna will cause. The unsightliness of the antenna and the adverse effect on property values that is caused by its unsightliness are the most common concerns. . . . But adverse environmental effects are properly considered also, and even safety effects: fear of adverse health effects from electromagnetic radiation is excluded as a factor, but not, for example, concern that the antenna might obstruct vision or topple over in a strong wind."[5]

The second basic trade-off presented by the efficiencies of cellular technology is the need to invest in considerable "intelligence" within the network, both to identify where a subscriber is at any given moment and to hand off calls seamlessly from one cell site to the next as the subscriber changes location. The intelligence of today's cellular systems resides in the base-station equipment at the bottom of each antenna, in a carrier's centralized switches, and, to an increasing extent, in the mobile device itself.

When you turn your cellphone on, it "listens" for a signal announcing the presence of your wireless carrier, constantly transmitted by each base station in the form of a carrier-specific "system identification code." In turn, your cellphone transmits two pieces of information about itself: a ten-digit telephone number your carrier has assigned you, plus a non-public electronic serial number that was installed in the phone at the time of manufacture. So long as you keep your phone turned on, it will stay in contact with the cellular network. If someone calls you, the network will already know where you are, it will instantaneously send the call to the appropriate cell site, and it will transmit the call to you from there, whether that site is in the same city as your home address or on the other side of the continent.

Now suppose that when you turn your cellphone on, it hears only the signal of some *other* carrier's base station because you are out of the range of your own carrier's network. Your phone will still send its identifying information to that other carrier, which will check to see whether it has a *roaming* agreement with your carrier (discussed more fully later in this chapter). If it does, and if your phone is compatible with the other carrier's network technology, that other carrier will stand in for your carrier and process your calls upon receiving confirmation from your carrier that you are a current subscriber.

The easiest way to understand how a wireless network hands off calls from one cell to the other as you drive through them is to consider what happens in conventional GSM systems using "time-slicing" technology, even though the details are different in non-GSM systems.* When you place or receive a call over a GSM phone, the network assigns your phone a pair of channels among the several dozen it has devoted to the cell you are in; you transmit in one channel ("uplink") and receive in the other ("downlink"). During the call, the telephone and the base station adjust their power levels to maintain a certain level of quality, and when the telephone moves to the edge of a cell, the power level reaches the maximum allowed. When the ratio of signal strength to background noise falls below a predetermined level, a new pair of channels is assigned in a neighboring cell. Once the system determines that you will receive better reception using the channels in the new cell, it tells both your phone and the new base station to resume the transmission using the new channel. These handoffs are sometimes fumbled and calls are dropped because, for example, all channels in the adjacent cell are in use or because there is no adjacent cell at all—that is, there is a "hole" in the coverage area.

Given the explosive popularity of wireless services, particularly in metropolitan areas, each carrier must struggle to squeeze enough bandwidth out of its limited spectrum to serve its customers' anticipated demands. Part of the solution lies in so-called *cell splits*, which further "recycle" spectrum by dividing up service areas into yet smaller cells with lower-power antennas. On top of that, carriers must also find ways to use the same frequencies for the transmission of multiple calls in the *same* cell or sector. To do this, they use one of the several time- or code-based multiplexing mechanisms discussed later in this chapter. Carriers similarly use various digital compression techniques that, as discussed in chapter 5, reduce the amount of total information that needs to be transmitted in order to convey a reasonably accurate representation of a sound or image. Cell splits and other capacity-increasing measures may require large capital investments, and wireless providers thus face complex economic trade-offs on the margins between taking those mea-

* As discussed later in this chapter, GSM ("Global System for Mobile") became the world's most common second-generation (2G) mobile technology standard around the turn of the millennium, although it has always had technological rivals in the United States and is now giving way worldwide to more advanced 3G and 4G standards.

sures and acquiring additional spectrum (via auctions or commercial deals).

II. The Regulatory Landscape

A. A brief history of mobile wireless regulation

Cellular technology was conceived in the 1940s, but it stood in line for several decades while policymakers debated in fits and starts whether and how much spectrum should be allocated to this quixotic new technology instead of to broadcasting and other conventional spectrum uses.[6] By one estimate, the multiple delays in this process cost the U.S. economy in the aggregate tens of billions of dollars.[7] In the early 1980s, after monitoring an experimental system in Chicago and satisfying itself that consumers would actually order cellular telephone service,[8] the FCC finally allocated 40 (later raised to 50) megahertz of spectrum in the 850 MHz band for that purpose throughout the United States.[9] The Commission obtained the spectrum for these original cellular licenses by reallocating spectrum that had previously been allocated for UHF channel 70 and above on the television dial.

The FCC initially conceptualized the service areas of cellular providers as "local," perhaps by analogy to local radio stations, and it doled out cellular licenses to each of 734 *cellular market areas* (CMAs). Although 50 megahertz was not enough spectrum to support full-blown competition within each of these markets, it was enough to support two rival carriers. The FCC thus created an official duopoly in each market, assigning one license (for a frequency band that was ultimately 25 megahertz wide) to the incumbent LEC and another to an independent provider chosen, for the most part, by lottery.[10]

When the FCC granted these initial cellular licenses in the early 1980s, the telecommunications world was preoccupied with the impending break-up of AT&T's wireline Bell System, and few in the broader industry recognized just how commercially significant mobile telephony would become. Indeed, at a 1982 news conference announcing AT&T's divestiture of its local exchange operations, the company's own CEO was reportedly unable to answer a question about which set of companies would inherit the cellular licenses: AT&T proper (soon to be a standalone long-distance company) or the newly independent Bell companies. The consent decree ultimately assigned the licenses to the Bell companies, along with some initial restrictions on their ability to offer service across exchange area boundaries.[11] The immensity of that decision became clear ten years later when AT&T made up for its lack of cellular licenses by purchasing McCaw Communications' extensive wireless network for $11 billion.*

In each local market, the wireless industry remained a regulated duopoly throughout the 1980s and early 1990s, and cellphone service remained spotty and expensive.[12] In the mid-1990s, the FCC finally created more competition by exercising its newly granted auction authority to assign new wireless licenses—technically called *PCS* ("personal communications services") licenses—covering an additional 120 megahertz of spectrum. This spectrum was divided into a number of frequency blocks, some of which were 30 megahertz wide, and some 10 megahertz wide. These licenses were originally assigned on a local or regional basis, and they remain so today; two of the blocks are assigned to (geographically larger) "major trading areas," and the rest are assigned to (geographically smaller) "basic trading areas."[13]

Unlike the first generation of mobile wireless carriers, which had built all-analog networks in the 1980s, these new second-generation (2G) carriers built out their networks from the ground up with digital technology. That technology enabled them to provide higher-quality service and to squeeze more calls into the available spectrum, in part through digital compression techniques. Within several years, all major wireless providers had made that technological leap by offering digital service to their customers, and no major provider offers analog service today.

In the first decade of the new millennium, the FCC promoted the growth of third- and fourth-generation (3G and 4G) wireless technologies by allocating additional spectrum bands for commercial mobile services, including the *AWS* and *700 MHz* bands. The FCC auctioned off the main AWS spectrum blocks in 2006, after reallocating bands that, under the FCC's command-and-control regime, had previously been consigned in part to an obsolescent class of television-oriented "wireless cable" services (along with various other uses). And in several different proceedings between 2002 and 2008, the FCC also auctioned off licenses in the "lower" and "upper" 700 MHz bands, which had previously been occupied by UHF television channels between 52 and 69.[14] As noted in

* That network became the basis of AT&T Wireless, which was spun off from AT&T in 2001 and purchased in 2004 by Cingular, then a joint wireless venture between Bell companies SBC and BellSouth. History came full circle when, after buying the AT&T wireline company in 2005 and taking its brand name, SBC/AT&T bought BellSouth in 2006 and gave Cingular, which it now owned outright, the brand name "AT&T" as well.

the previous chapter, these auctions collectively raised tens of billions of dollars for the federal treasury.

Wireless providers use all of these bands and several others to provide mobile voice and data services today. There are four basic facts to understand about this hodgepodge of spectrum bands, which are scattered in various locations across the radio spectrum. First, the licenses to use these bands are among the most flexible of any the FCC issues: the FCC generally permits carriers, within broad limits, to use them to provide any type of service using any type of technology. Second, the services provided over these bands are market substitutes. For example, the mobile voice and data services that Sprint provides on one set of spectrum bands compete with the similar services that Verizon provides on its own set of spectrum bands.

Third, handsets are typically designed to work on some, but not all, of these various spectrum bands—PCS, AWS, 700 MHz, the original "cellular" (850 MHz) band first allocated in the 1980s,* and miscellaneous other bands. In particular, to keep handsets compact and their batteries long-lived, manufacturers typically design any given handset to operate only with particular technologies on specified frequencies. Suppose you own a handset designed to operate only with particular 2G and 3G wireless technologies and only on the PCS and AWS spectrum bands in which your provider holds its licenses. You cannot use that handset to exploit *either* (1) a new 4G network on any band *or* (2) spare capacity on otherwise compatible 2G or 3G networks operating on the 700 MHz band (or any other band besides the PCS and AWS bands). For that reason, it may take a few years before a carrier can put a new spectrum band to use with a new generation of technologies. The basic standards for a new technology must first be developed and specified so that the industry can achieve scale; then manufacturers must design handsets for the technology and whatever spectrum bands it will be used in; and then, before rollout for a particular handset model, the carrier may wish to satisfy itself that it can sell a critical mass of those handsets to its subscribers.

Fourth, these various spectrum bands have different propagation characteristics. Recall from chapter 3 that the lower a signal's frequency is,

* Although the term *cellular* is still used today to denote the licenses granted (and spectrum allocated) in those first wireless-era proceedings, its use in that specialized sense occasionally causes confusion. To be clear: all mobile wireless providers, whether they operate in the "cellular" bands or any other, use cellular *technology* to provide service.

the more strength that signal will retain over distance, and the easier it will be for that signal to penetrate walls and thus work well indoors (all else held equal). Those propagation characteristics confer some advantages on networks using frequencies in the upper and lower 700 MHz bands and in the original "cellular" bands, which operate in the vicinity of 850 MHz. And they confer some propagation disadvantages on networks using frequencies in the PCS band (1850–1990 MHz) and AWS bands (1710–1755 MHz and 2110–2155 MHz). Low-band spectrum presents the most significant advantages in sparsely populated rural areas, where its superior propagation characteristics enable providers to build fewer cell towers that cover larger cells. That advantage is absent, however, in more populous urban areas, where network congestion requires providers to "recycle" spectrum—whether low band or high band—by using very small cells anyway. In those areas, *capacity* is more important than the extent of geographic *coverage*, and low-band spectrum holds no special capacity benefits over higher-band spectrum. Indeed, the FCC has found that, with current technologies, "higher-frequency spectrum may be just as effective, or more effective, for providing significant capacity, or increasing capacity, within smaller geographic areas."[15]

The relative advantages of lower-band (lower than 1 GHz) versus higher-band (higher than 1 GHz) spectrum have played an important role in regulatory policy debates. The carriers with the most low-band spectrum are Verizon and AT&T, in part because they inherited many "cellular" licenses assigned to their local exchange affiliates during the duopoly era of the 1980s, and in part because they were the most aggressive bidders in the 700 MHz auctions. Their rivals contend that the FCC should take those advantages into account in various regulatory contexts. For example, they urge the FCC to keep AT&T and Verizon from obtaining more low-band spectrum at auction, including the new 600 MHz auction authorized by the 2012 spectrum legislation (see chapter 3). Verizon and AT&T respond that the advantages of low-band spectrum are overstated, at least in populous areas, and that, to the extent such spectrum presents advantages, they paid a premium to obtain those advantages in the 700 MHz auctions and should not be penalized now.[16]

As of this writing, the FCC has left this general set of issues unresolved, and so has Congress. For example, Congress split the baby in the 2012 legislation itself. On the one hand, it prohibited the FCC from adopting any auction-specific ban on any given company's ability to participate in an auction. On the other hand, it added that this prohibition does not "affect[] any authority the Commission has to adopt and enforce rules of general applicability, including rules concerning spectrum aggregation that promote competition."[17] This second provision appears to authorize the FCC, as appropriate, to impose new "spectrum caps" (see next section) in an industry-wide rulemaking proceeding and use them to constrain or condition the involvement of the largest carriers in future auctions, including the 600 MHz auction. The precise relationship between these two provisions will be a source of lively debate in the years to come.

B. The general deregulation of wireless telephony

Mobile wireless services fall into two major statutory categories: "commercial" and "private." The term *commercial mobile radio services* (CMRS) encompasses "any mobile service . . . that is provided for profit," is "interconnected with the public switched network," and is provided "to the public or . . . to such classes of eligible users as to be effectively available to a substantial portion of the public."[18] It includes conventional mobile telephony as well as commercial paging and various other services. In contrast, *private mobile services* have traditionally consisted of dispatch-oriented services that are not interconnected with the public switched network: for example, a taxi driver cannot call home on his car radio.[19] Much turns on this statutory classification. Unlike private mobile services, commercial mobile services are subject to basic common carriage requirements, albeit only in the most skeletal form. As discussed later in this chapter and in chapter 6, the proper statutory classification of mobile broadband Internet access is a matter of controversy, whose resolution affects a range of regulatory initiatives.

In the decade after assignment of the first cellular licenses, the FCC imposed common carriage obligations on wireless providers by cobbling together various sources of authority, both under Title II, which had long governed wireline "common carriers," and under Title III, which covers "radio" (which, in this context means the electromagnetic spectrum in general rather than broadcast radio services in particular). In 1993, Congress revamped section 332 of the Communications Act to formalize the basic substance of this approach and to harmonize the regulatory treatment of the different classes of wireless telephony services.

Congress began with the presumption that such services would be subject to common carriage regulation under Title II. But foreshadowing its broader, industry-wide grant of "forbearance" authority three years later,[20] it authorized the FCC to *exempt* wireless carriers from any Title

II requirements that the Commission deems unnecessary, except for the skeletal "reasonableness" and "nondiscrimination" obligations of sections 201 and 202 and the FCC complaint procedure of section 208.[21] In 1994, the FCC exempted wireless carriers from the tariffing obligations of section 203 and the market entry and exit regulations of section 214, reasoning that competition made most forms of traditional common carrier regulation superfluous at best and counterproductive at worst.[22] The practical upshot of the detariffing determination is that, although section 202 still theoretically prohibits a narrow category of "unjust and unreasonable discrimination in charges and service," wireless carriers are generally free in practice, under the so-called *Orloff* doctrine, to strike dissimilar bargains with similarly situated customers.[23]

Two FCC policies—relating to equipment bundling and resale, respectively—illustrate the deregulatory trajectory of wireless policy in this age of competition. First, as anyone who has purchased a discounted cellphone with a two-year service contract is aware, the FCC has exempted wireless carriers from generally applicable restrictions on a common carrier's right to sell customers bundles of services and equipment on more favorable terms than if it sold the two separately. As discussed in chapter 2, such restrictions, which derive from the *Computer Inquiries*, sometimes make economic sense when the discounted bundle includes the products of a dominant firm in a monopolistic market, particularly one that is subject to price regulation.[24] The firm may have both the incentive and the ability to leverage its monopoly in that market by discriminating against competitors in adjacent markets and, when subject to traditional rate-of-return regulation, by engaging in anticompetitive cross-subsidization as well.

But such restrictions make little sense when, as in the wireless sector, the markets for both the service and the equipment are competitive; in that context, bundling is both commonplace and pro-consumer.[25] In the early 1990s, the FCC agreed with this general conclusion (after first coming out the other way) and excluded wireless carriers from the scope of its general bundling prohibition.[26] This is plainly the right policy call. Consider the consumer who is unwilling to pay $400 up front for a smartphone but is willing to pay $150 for the phone and commit to paying $60 each month for service for the next two years. Wireless companies routinely accommodate such customers by giving them the option of paying a low "subsidized" price for their phones if they agree to purchase service for a designated period. As a general matter, few would preclude wireless companies and their customers from entering into such service contracts. Competition in the markets for both wireless service and mobile devices, together with the absence of price regulation for either, leaves firms with neither the incentive nor the ability to exploit such bundling to the detriment of consumers.

That said, the *early termination fees* (ETFs) that wireless providers use to enforce these contractual commitments have come under scrutiny by the FCC, which has effectively cajoled the industry into giving consumers better notice about the threat of such fees when they sign up for service.[27] ETFs have also drawn the attention of plaintiffs' lawyers, who have filed class-action lawsuits seeking invalidation of various ETF provisions under state law.[28] Not surprisingly, the wireless industry views substantive legal constraints on ETFs as a threat to the economic logic of the current handset-subsidy regime, and it argues that any state-level action violates the preemption provision of section 332 (which we will discuss shortly).

As with the bundling of equipment and service, the FCC has also taken a deregulatory approach to the subject of wireless resale. Suppose that a facilities-based wireless carrier charges individual subscribers $40 a month for a particular wireless service but gives a 20% volume discount to any large business that purchases that service in bulk for at least 100 people—say, a sales force that needs to stay in constant contact with the home office and with clients. A non-facilities-based reseller might then seek the same 20% volume discount for 100 different wireless accounts, which it would in turn sell, at a slightly less discounted rate, to *individual* subscribers. The facilities-based carrier might well reject that request if it believed that a large number of those 100 subscribers would otherwise pay full price by subscribing directly to it.

During the early years of the industry, the FCC generally required wireless carriers to offer their services for resale without restriction, just as it had required AT&T to accommodate resellers such as MCI in the early days of long-distance competition.[29] The difference, however, was that AT&T was a monopolist in the wireline long-distance market, whereas competition was beginning to blossom in the wireless telephony market. In 1996, the FCC found that as the new PCS licensees built out their networks and gave consumers a fuller choice of facilities-based providers, there would be progressively less consumer-welfare justification for the government to intervene in the market to preserve resale opportunities. And it thus decided to "sunset" its wireless resale rules by 2001.[30]

The Commission reasoned that as competition in a market increases, efficient resale opportunities tend to arise through market dynamics. In our example, the facilities-based carrier in a reasonably competitive market might well have strong incentives to sell those 100 accounts to the reseller at the discounted rate because otherwise it would have no assurance of selling those accounts at all. Also, government prohibitions on resale restrictions can impose significant costs of their own. Even apart from the administrative costs of enforcement, which can be considerable, such prohibitions may operate as an economically inefficient flat ban on all price discrimination among classes of consumers. For example, there are many circumstances in which a firm will wish to target its discounts narrowly to a particular class of customers that might otherwise take their business elsewhere. Requiring such a firm to accommodate resale arbitrageurs in extending those discounts to consumers at large may discourage it from offering such discounts to anyone in the first place, resulting in a net loss of consumer welfare. Of course, this concern arises only in markets with less than perfect competition, but that characterizes virtually all competitive markets in the economy.[31]

The FCC has likewise taken a more deregulatory approach to spectrum management, as discussed in the previous chapter. Mobile telephony licenses are no exception to this general trend.[32] For example, the FCC initially precluded any single firm from holding licenses accounting for more than 45 megahertz of spectrum in any metropolitan area.[33] At the time, it viewed this *spectrum cap* as an important safeguard against excessive consolidation during the initial development of competition in wireless telephone service. In 2001, however, the FCC decided to raise and ultimately eliminate the cap "in light of the strong growth of competition in [mobile wireless] markets" and move instead "to case-by-case review of spectrum aggregation," conducted mainly in the merger context.[34] As noted, however, Congress recently took pains in its 2012 incentive-auction legislation to preserve "any authority the Commission has to adopt and enforce rules . . . concerning spectrum aggregation that promote competition," and the FCC responded by opening a new inquiry into whether it should reimpose spectrum caps in one form or another.[35]

To this point, we have addressed the minimalistic approach of *federal* wireless regulation in the wake of Congress's amendment of section 332 in 1993. In a similar deregulatory vein, Congress sought to insulate the industry from unnecessary regulation of all kinds by preempting most *state* efforts "to regulate the entry of or the rates charged by any commercial mobile service."[36] This preemption provision comes with assorted qualifications, all set forth in section 332(c)(3). First, under the most important qualification, states may "regulat[e] the other terms and conditions of commercial mobile services."[37] Second, a state may "petition the [FCC] for authority to regulate the rates" of a mobile wireless service if (1) "market conditions with respect to such services fail to protect subscribers adequately" or (2) "such market conditions exist and such service is a replacement for land line telephone exchange service for a substantial portion of the telephone land line exchange service within such State."[38] Such petitions are rarely filed and even more rarely (if ever) granted.[39]

Unsurprisingly, lawyers have argued at length about how to draw the statutory line between (preempted) "entry" and "rate" regulation and (unpreempted) regulation of "the other terms and conditions" of service. For example, may disgruntled customers sue wireless carriers under state tort, contract, or consumer-protection law for false advertising, misleading billing practices, or simply allowing too many dropped calls after promising to provide seamless coverage throughout a given area?[40] In 2000, amid widespread disagreement in the courts, the FCC stepped in and opined that such state law claims generally may proceed. It added, however, that "this is not to say that such awards can never amount to rate or entry regulation"; instead, whether a particular remedy constitutes "rate or entry regulation prohibited by Section 332(c)(3) would depend on all facts and circumstances of the case."[41]

Running parallel to these litigation-oriented disputes are similar controversies about the proper role of state consumer-protection laws designed specifically to ensure the high quality of wireless services. Some state commissions have enacted prophylactic "wireless bill of rights" initiatives despite concerns that such market intervention is unnecessary in a competitive industry subject to generally enforced rules of contract and tort law.[42] Whatever the ultimate fate of such initiatives, the industry's trade association, the Cellular Telecommunications and Internet Association (CTIA), has tried to address the underlying discontent by instituting a consumer code for wireless services, to which individual companies have voluntarily committed themselves.[43] And in 2011, CTIA persuaded the FCC to hold off on proposed federal "bill shock" rules by updating its consumer code to give wireless subscribers advance warning before their usage subjects them to high overage or roaming fees.[44] That self-regulatory initiative gave the wireless industry what it wanted—a flexible, industry-friendly regime for billing practices—and the FCC what it wanted: effective consumer protection without the political fallout it would have incurred if it had sought to regulate these retail practices directly.

III. Competition Policy for the Wireless Broadband Ecosystem

As discussed, the FCC has forborne from most forms of *retail* regulation of mobile services. In addition to its close scrutiny of wireless mergers, however, it has charted a more interventionist course in several key areas of competition policy, including roaming, number portability, and net neutrality. We address the first two issues in this section and conclude this chapter with a discussion of the complex competitive issues raised by the proposed AT&T/T-Mobile merger in 2011; we defer our discussion of net neutrality issues until chapter 6.

We begin the discussion with a brief overview of the core economic differences that have historically led regulators to regulate wireless providers more lightly than their wireline counterparts. In chapter 1, we explained that the wireline side of the industry was long subject to natural-monopoly treatment because it exhibited an unusual confluence of network effects, immense scale economies, and widespread monopoly-leveraging opportunities. The second and third of those concerns have always been somewhat attenuated in the wireless context: there are no wireless monopolies to worry about, "natural" or otherwise, and thus few anticompetitive "leveraging" opportunities.

There are, to be sure, large fixed costs and scale economies in the wireless market. To serve even a few customers, facilities-based wireless firms must invest large sums to acquire spectrum rights, either at auction or in a secondary market; to build out an infrastructure of transmission antennas, base stations, and switches; and to deploy "backhaul" facilities (or procure backhaul services) to connect its cell sites to its switches (see chapter 2). Within a given metropolitan area, the average cost per customer of those investments is obviously much lower for the millionth customer than for the first. Scale economies affect equipment prices as well; the one millionth copy of a smartphone tailored to a particular technology costs less per unit than the first copy and can thus be profitably sold at lower prices to consumers. But these facts do not make wireless telephony a natural-monopoly market, just as the high fixed costs and low unit costs of the automobile industry do not make car manufacturing such a market. In the wireless ecosystem, a firm's long-run average costs do not decline "over the entire extent of the market"[45] with every increase in output, as the persistence of multiple competitors in this industry suggests.

All this said, the wireless industry, no less than its wireline counterpart, is highly susceptible to network effects. That fact gives rise to a rich assortment of issues concerning whether and when the government should intervene in the wireless market to force cooperation among different networks.

Some of these are familiar issues of *interconnection*: the terms on which one network must cooperate with another to enable its subscribers to call or be called by the other network's subscribers. Wireless telephony could never have developed without rights of interconnection with wireline networks. Because wireless networks are costly to build and operate, consumers have always paid substantial fees to use them, and few consumers would have paid those fees if their cellphones had been able to exchange calls only with other cellphones on the same network and not with the vast embedded mass of conventional landlines. Few doubted, therefore, that wireline ILECs would be required to interconnect with wireless carriers on *some* terms. In the *Carterfone* decision, discussed in chapter 2, the FCC required such interconnection with mobile radio networks even before the dawn of cellular telephony, but on less than favorable terms.[46] When the FCC first authorized cellular service in the early 1980s, it took the interconnection requirement to the next logical step, conditioning an ILEC's own cellular licenses on its commitment to interconnect with unaffiliated cellular operators on more favorable, nondiscriminatory terms.[47]

The enduring interconnection disputes between wireless carriers and their wireline counterparts have generally involved the compensation that such carriers owe or are owed when they handle part of a call that originates or terminates on some other network, including ordinary wireline telephone networks. That compensation issue can be enormously complicated, in part because a wireless carrier might interconnect *directly* only with a large regional ILEC such as a Bell company (i.e., Verizon, AT&T, or CenturyLink). When a wireless carrier handles calls sent to or from the subscriber of some other carrier—such as a wireline CLEC, a small rural ILEC, or another wireless carrier—it often interconnects *indirectly* with that carrier by relying on an intermediary to route calls between the other two carriers' networks. That intermediary may be either the regional Bell company or an independent third-party provider such as Inteliquent (Neutral Tandem). Whether and how such *transit* services should be regulated are matters of intense controversy. We defer that set of issues until chapter 7, which addresses questions of intercarrier compensation generally. In this section, we address the proper scope of government involvement in promoting *roaming* and *interoperability* among wireless networks: features that enable one carrier's subscribers to use their wireless devices when traveling through geographic regions covered only by other carriers' wireless networks.

A. Roaming

Although there are several national wireless networks in operation today, there were none at all in the early days of wireless telephony because, as we have seen, the FCC assigned the first "cellular" licenses on a strictly local basis. Although the FCC later allowed different carriers to consolidate these geographically disparate networks, the local character of the initial license assignments created an acute need to ensure that subscribers could roam on other networks—that is, use their cellphones when they traveled away from their own carrier's geographically limited network and into the territory of unaffiliated wireless carriers. Without roaming, cellular service would have been much less valuable to consumers, and the wireless industry might never have gotten off the ground. A consumer can roam on other carriers' networks, however, only if (1) those other carriers agree to let her use those networks and (2) her cellphone is compatible with those networks in terms of both frequency bands and technology.

Carriers address the first issue by negotiating carrier-to-carrier *roaming agreements*, which allow the mutual use of each carrier's network by the other carrier's subscribers. Each carrier pays the other for the privilege in cash or in kind and passes the costs on to its own subscribers, either directly in the form of per minute roaming fees or as part of a flat monthly rate for a national calling plan. Roaming should not be confused with resale, although the two concepts are similar and shade into each other at the margins. Roughly speaking, roaming describes an arrangement where a *facilities-based* carrier relies on network capacity from other carriers to supplement its own network, mainly (though not exclusively) for the benefit of end users who are traveling out of their carrier's home region and need network coverage. In contrast, resale describes an arrangement where a *non*-facilities-based carrier such as TracFone—often known as a *mobile virtual network operator* or "MVNO"—outsources all of its network needs to wholesale providers, even in areas where it actively markets services to customers.

The FCC originally required carriers only to provide "manual" roaming on a pay-per-use basis—usually by means of a credit card—to *individual cellphone users* in the absence of a carrier-to-carrier agreement. These cumbersome arrangements usually arose in remote rural areas and involved small local carriers that had not reached direct roaming agreements with the cellphone users' own carriers. In the first years of the new millennium, the FCC appeared generally reluctant to intervene in the market for such carrier-to-carrier ("automatic") roaming agreements, having apparently concluded that market forces would give each carrier adequate incentives to negotiate efficient roaming arrangements in the absence of regulation.[48]

Beginning in 2007, however, the Commission reversed course and concluded that sections 201 and 202 of the Communications Act compel carriers to negotiate "reasonable" and "nondiscriminatory" automatic roaming arrangements with other carriers to the extent the underlying wireless services are, like conventional voice telephony, classified as Title II common carrier services.[49] In 2011, the Commission went one critical step further and in its *Data Roaming Order* required carriers to negotiate roaming agreements even where the underlying wireless services are, like mobile Internet access, classified as Title I "information services" instead—a critical statutory distinction we introduced in chapter 2 and explore more fully in chapter 6.[50]

This *data-roaming* requirement is controversial. It targets and is generally opposed by the two largest wireless providers: AT&T and Verizon. Those providers argue that mandatory roaming arrangements undermine their incentives (and those of their roaming partners) to invest in next-generation technologies for their wireless networks, much as they had argued that aggressive network-leasing obligations under sections 251 and 252 would thwart incentives to deploy next-generation wireline facilities (see chapter 2).

In the *Data Roaming Order*, however, the FCC concluded that such rules are needed to jump-start competing data services by smaller carriers. Such carriers include not only the other two "national" providers (Sprint and T-Mobile) with nationwide networks in varying stages of development, but also the dozens of regional carriers (such as U.S. Cellular, C Spire, and Leap/Cricket) that have no plans to deploy network facilities at all beyond the local markets where they sell service. Without data-roaming rules, the Commission reasoned, such carriers might be unable to secure the roaming arrangements needed to offer nationwide data coverage, which many customers demand as a precondition to signing up for service.

The FCC further dismissed the concerns expressed by AT&T and Verizon about the anti-investment incentive effects of mandatory sharing requirements. The Commission observed that the data-roaming rules, unlike the mandatory leasing rules of sections 251 and 252, do not actually prescribe the terms of these network-sharing arrangements; they do not mandate particular rates, nor do they require roaming providers to treat all roaming buyers indiscriminately. Instead, they create a forum in which those buyers may claim that the terms of their roaming deals are so one-sided as to be "commercially unreasonable." The FCC concluded that this nebulous standard strikes the right balance between the sellers and buyers of roaming services and preserves everyone's incentives to build out multiple next-generation networks. Only time will tell whether that empirical judgment is sound.

Verizon Wireless challenged the data-roaming rules in the D.C. Circuit, focusing less on their policy merits than on the FCC's disputed statutory authority to adopt them. Among its other arguments, Verizon contended that section 332 of the Communications Act bars the FCC from imposing common carrier requirements on wireless services that, like mobile data services, are not "interconnected" with the circuit-switched telephone network and are thus "private radio services" rather than "commercial mobile radio services."[51] The FCC responded that no matter what the statutory classification of data services, its data-roaming rules "do not amount" to common carrier requirements because they "will allow individualized service agreements," "will not require providers to serve all comers indifferently on the same terms and conditions," and include no obligation to make these roaming deals public.[52] In December 2012, the D.C. Circuit rejected Verizon's challenge and upheld the data-roaming rules, essentially on the basis of the FCC's rationale.[53]

B. Interoperability

Roaming requires *interoperability* between two networks. Two carriers cannot enter into a meaningful roaming agreement in the first place if they have built out their networks with incompatible technology; in that case, one carrier's customers cannot use their handsets to communicate with the radio equipment on the other carrier's towers. That fact gives rise to basic questions about the proper role of government in prescribing technological standards in this quickly evolving industry.

From the advent of cellular service, the FCC specified a single analog standard—Advanced Mobile Phone System (AMPS)—to ensure interoperability among geographically disparate cellular systems throughout the country. This uniformity provided the technological basis for all customers to use the same cellphones when roaming in parts of the country not served by their own wireless carriers. When wireless carriers received PCS licenses in the mid-1990s, however, they began building out 2G networks that, from the beginning, used purely digital technology—an advance that allowed them to provide better signal quality than their first-generation analog rivals and more voice conversations per increment of spectrum. But the FCC did not require all of them to use and they did not in fact use the same *type* of digital technology. Instead, they have employed different and incompatible standards that often preclude one carrier's customers from roaming on another carrier's network.

This technological divergence, which relates to 2G and 3G technologies, remains relevant to today's policy debates even though 4G wireless technologies are quickly converging worldwide on a single family of standards known as *LTE* ("Long Term Evolution"). As you read the pages that follow, keep in mind that many wireless networks still support several generations of technologies at once and divide up their spectrum holdings and radio equipment accordingly. Such carriers use an increasing portion of their assets to provide 4G LTE services to their high-end customers with the latest and most expensive handsets and tablets. But they simultaneously use other assets to support the 2G and 3G technologies used by millions of customers with less expensive (or simply older) devices.

In the long term, most carriers will shut down these older networks, beginning with the 2G facilities, in order to rededicate all of their spectrum holdings to newer technologies such as 4G LTE. That is in everyone's interest. Each generation of wireless technologies not only provides faster data speeds than its predecessors, but is far more efficient in its use of available spectrum. For example, if a carrier has 40 MHz of spectrum in a given city, it can support far more usage by its customers if it dedicates all that spectrum to LTE than if it divides up the spectrum to support an assortment of 2G, 3G, and 4G technologies. But before a provider can shut down a legacy wireless network, it must first induce all the customers using that network to trade in their older devices for newer ones; otherwise, those customers will simply be stranded with devices that no longer work at all. Such customer *migrations* are multiyear undertakings fraught with commercial, technological, and policy challenges. For example, long after the first 2G phones came on the U.S. market in 1995, many customers were still using analog phones. Indeed, the FCC *required* all licensees of "cellular" spectrum to maintain analog networks until 2008 in order to support that gradually dwindling base of analog customers.[54]

With that background, we now examine the technological divergence of legacy digital technologies within the United States. All these standards pursue the same goal: squeezing additional bandwidth out of available spectrum by using the same frequencies in the same places for more than one call at a time. But the 2G standards used to accomplish that goal diverged into two general families: those that allocated bandwidth on the basis of *time slicing* and those that used *spread-spectrum* technology (see chapter 3) to allocate bandwidth on the basis of a sophisticated *code.* We address each in turn.

The time-slicing family of 2G standards is analogous to the TDM techniques we discussed in chapter 2 when surveying the technology of wireline networks. This approach aggregates, within the same frequency, the signals associated with several different calls. It "samples" each call's signals many times a second and transmits those samples—for example, digital representations of the sounds of your voice—within precisely calibrated time slots. If the samples are taken often enough, the system can convey an accurate representation of the sounds coming out of your mouth. Of course, like wireline telephony companies, any wireless carrier saves bandwidth by disregarding many of the natural frequencies associated with the human voice, so the transmitted signal will never approximate what you would sound like if the listener were sitting next to you. The main 2G time-slicing standard today is GSM. In the United States, GSM is used by AT&T, T-Mobile, and a variety of smaller regional providers.

The second major family of 2G standards is CDMA (introduced in chapter 3). In the 1980s, building on military applications developed during World War II, Qualcomm's Irwin Jacobs pioneered CDMA for commercial uses, and it is now used by such carriers as Verizon Wireless and Sprint in the United States. CDMA relies on spread-spectrum technology to disperse the signals associated with each call over many different frequencies. Some have likened CDMA to permitting ten pairs of people to converse simultaneously across a dinner table in ten different languages and counting on each pair to pick out and focus on the language they understand. To perform this feat, a CDMA system exploits advances in digital-processing technology to enable the devices on each end of a transmission to understand which signals belong to which calls and decode them for end users.

Although CDMA is more efficient than the time-slicing standards at squeezing more bandwidth out of available spectrum, its equipment (because of the necessary intelligence in it) is more sophisticated and was initially more expensive. Nonetheless, it won many converts throughout the industry, for reasons that are instructive about the trends in high-technology markets more generally. As *The Economist* reported in 2003, "Perhaps the greatest lesson from the story of CDMA is that what seems impossibly complex today may well seem simple tomorrow—thanks to the relentless advance of Moore's law," which "states, roughly, that the cost of a given amount of computing power halves every 18 months. . . . Without it, Dr Jacobs admits, CDMA handsets would have been too large, and the base-stations would have been too expensive. 'Our argument was that Moore's law will take care of cost, size and power for us,' he says"[55]—and he was right.

Whereas U.S. regulators let these digital access standards proliferate, European regulators—not known for their faith in unregulated markets—took the opposite course.[56] In the late 1980s, anticipating the transition from the first (analog) to the second (digital) generation of wireless services, they adopted GSM as a universal standard and required all European providers to adopt it. This decision was widely hailed as a principal reason why Europe jumped ahead of the United States in the 1990s in overall cellphone penetration. A European consumer could invest in a cellphone and expect compatible digital service from any transmission site she passed, no matter where in Europe she traveled. For their part, manufacturers were assured of a giant market for digital cellphones of a particular type. Precisely because more phones could be designed for a single standard, both manufacturers and consumers benefited from scale economies that were unavailable where, as in the United States, the proliferation of incompatible standards balkanized the cellphone market.

European regulators followed a similar course in the early 2000s by centrally managing the transition to 3G wireless services, which offer broadband Internet access in addition to plain vanilla mobile telephony. They required every auction winner of a 3G license to provide service using a spread-spectrum standard commonly known as *UMTS* ("Universal Mobile Telecommunications System"). The main U.S. carriers that were using GSM for 2G service, including AT&T and T-Mobile, independently chose UMTS as their 3G technology for voice and data, often in combination with technological enhancements known as *HSPA* ("High Speed Packet Access"). In contrast, Verizon, Sprint, and the other U.S. carriers that were using CDMA for voice adopted a related set of technologies for 3G data services known as *EvDO* ("Evolution—Data Optimized"). Although all of these 3G technologies were significant advances in their time, they are less efficient and generally slower than today's 4G LTE technologies.

Which regulatory approach was superior: Europe's decision to mandate specific wireless standards or America's decision to let the marketplace experiment with mutually incompatible standards? There is no clear answer, and either policy approach presents a different mix of costs and benefits. The laissez-faire U.S. approach created uncertainty about which standards would prevail and thus made it riskier for manufacturers to develop products designed for any particular standard. Heterogeneity among the standards that reached the market also raised the unit costs for products designed for any particular standard. And it has limited roaming opportunities by reducing the number of transmission sites that a typical consumer's handset can communicate with. For example, a Sprint customer with a CDMA/EvDO handset cannot roam on AT&T's network, and an AT&T customer with a GSM/UMTS handset cannot roam on Sprint's network.

All this said, the balance may still tip in favor of the U.S. laissez-faire approach. If U.S. regulators had followed the lead of their European counterparts in the late 1980s and adopted GSM as the universal standard, they would have eliminated the U.S. market for CDMA products and thereby suppressed innovation in spread-spectrum technologies more generally. The question will eventually become academic now that the international standards-setting community, without any government compulsion, has adopted LTE as the leading 4G standard worldwide. But wireless technology is a richer field because of the free-wheeling experimentation facilitated by U.S. authorities during the second and third generations of wireless technology.

To this point, we have addressed the interoperability questions posed by carriers' use of different *technologies*. We briefly pause to consider the interoperability questions posed by carriers' use of different *frequencies*. Just as one cannot use a CDMA cellphone on a GSM network, neither can one use a cellphone that transmits and receives in one set of frequency bands (those used by carriers A and B) on a network that transmits and receives only in other frequency bands (those used by carriers C and D). You might reasonably ask why handset manufacturers do not simply make devices capable of operating on all frequencies used by all carriers. The answer, as noted earlier, is that such versatility would come with many drawbacks. All else held equal, the more frequencies a device supports, the larger and more expensive it will be, among other trade-offs. Manufacturers thus typically make handsets capable of operating on a few discrete spectrum bands used for commercial mobile services, but not all of them. In more technical terms, your mobile device comes equipped with a finite number of *radio-frequency filters* designed to accept signals throughout defined frequency bands while screening out the extraneous and potentially interfering signals on all other bands.

The interoperability issue assumed new prominence once the FCC opened up the 700 MHz bands to mobile wireless services. Questions immediately arose about which frequencies within those bands would be supported by the new generation of LTE handsets. In 2009, an international standards-setting body known as 3GPP ("3d Generation Partnership Project") addressed that issue with standards that tended to advantage some carriers over others, and the disadvantaged carriers urged the FCC to intervene.

Some context is necessary to understand this dispute. Private standards-setting initiatives, operating through generally open processes, can serve an indispensable role in the communications ecosystem. They reassure manufacturers that their devices will work on a critical mass of carrier networks and thus induce those manufacturers to make such devices available more quickly and in greater numbers than they otherwise would. And for much the same reason, these private standards-setting initiatives increase scale economies—and thus lower per unit costs—by enabling manufacturers to make similar products not just for one carrier's customers, but for many carriers' customers. But private standards-setting initiatives sometimes trigger allegations that self-interested participants have manipulated the process to secure anticompetitive advantages for themselves over their rivals. The result can be years of acrimonious litigation or proposals for regulatory intervention.[57]

Such proposals greeted 3GPP's standards in 2009. With the participation of AT&T and Verizon Wireless (among many others), 3GPP defined four discrete categories of equipment functionality, each corresponding to a discrete set of frequency blocks within the 700 MHz band. In the wake of this decision, device manufacturers began making LTE handsets that operate only with certain 700 MHz frequencies used by AT&T and Verizon Wireless but not also with the lower 700 MHz "A Block" frequencies used by smaller carriers such as C Spire and U.S. Cellular. An industry consortium calling itself the "700 MHz Block A Good Faith Purchasers Alliance" petitioned the FCC to reverse that outcome by imposing an "interoperability" rule on carriers and manufacturers.[58] That rule would require all 700 MHz devices in the United States to operate on all 700 MHz bands and would ban all devices that operate only on the frequencies used by AT&T and Verizon.[59] Without such a rule, the consortium argued, smaller carriers operating in the A Block could not sell their customers the same cutting-edge handsets made for AT&T and Verizon customers, and handsets made specifically for A Block carriers would be more expensive because they would lack the scale economies of handsets made for those much larger carriers. The consortium also accused AT&T and Verizon of deliberately manipulating the standards-setting process to achieve these outcomes.

AT&T and Verizon rejected that accusation and opposed the consortium's request for an interoperability rule.[60] They argued that equipment manufacturers, not the carriers, had originated the relevant standards-setting choices in an open process. They added that those choices made technological sense because wireless devices with filters that can receive A Block signals (in addition to other 700 MHz frequency blocks) are subject to greater interference from several sources, including Channel 51 television stations broadcasting on adjacent UHF spectrum in many metropolitan areas. Such interference, they explained, would degrade operations not only on the A Block itself, which AT&T and Verizon do not use, but also on the other 700 MHz blocks that they do use. AT&T and Verizon further asserted that they paid more at auction for their non-A Block spectrum so that they could avoid such interference problems. And they claimed that the proposed interoperability rule would impose major costs on the industry and expose millions of 700 MHz handsets to greater interference.

The FCC first put the proposed interoperability rule out for public comment in early 2010.[61] It sought a second round of comment in early 2012, this time signaling that it was more interested in achieving some type of interoperability solution if the interference problems could be resolved.[62] That proceeding remains pending at press time.

C. Number portability

A standard ten-digit telephone number is "portable" if a customer can take it with her when she cancels service with one provider and orders service from another.[63] On a technological level, such portability requires each carrier to cooperate with national databases that peg any given number to a particular carrier and instruct all other carriers processing calls destined for that number to send them to that carrier's network.

Under section 251(b)(2) of the Communications Act, as revised in 1996, all local exchange carriers must provide number portability for their subscribers. Although wireless carriers are not formally classified as LECs, the FCC exercised independent authority in 1996 to compel all wireless carriers to ensure number portability among themselves within several years.[64] The Commission's policy objective, as in the wireline context, was to reduce the *switching costs* that customers incur—in the form of printing out new business cards or telling friends about a new telephone number—when they move from one carrier to the next in the hope of better service. Portability rules can ameliorate such switching costs, but usually with a trade-off: providers can comply only by incurring extra engineering and administrative costs.[65] Here the FCC reasoned that imposing those extra costs on the industry was necessary because without number portability some consumers would be locked into their existing service, no matter how mediocre, and would be unable to benefit from competition on the merits from superior services.[66] That would be bad not just for them individually, but for wireless consumers generally: the more captive a wireless carrier's customer base is, the more complacent that carrier will be about seeking to meet its customers' needs.

The wireless industry fiercely resisted the Commission's wireless-to-wireless number portability requirement for many years, arguing that customer turnover was great enough already, that portability rules would add only marginally to the already intense competition among carriers for existing wireless customers, and that the significant implementation costs would outweigh the benefits. Industry opposition delayed but did not derail the eventual imposition of these wireless-to-wireless number portability obligations. In June 2003, the D.C. Circuit rebuffed the final legal challenges to the Commission's rules.[67] The industry's entreaties for a legislative override fell flat when Verizon Wireless, breaking ranks with its fellow carriers, announced its support for number portability upon realizing that, with its general reputation as the most reliable wireless network, it stood to gain more than to lose from greater customer churn. Thus, in November 2003, the FCC's portability rules finally took effect.[68]

Despite deep divisions on wireless-to-wireless portability, the wireless industry stood united behind a related and ultimately successful proposal: if that industry had to implement such portability for itself, the *wireline* industry should also have to accommodate a cord-cutting customer's request to port her conventional wireline telephone number to her new *wireless* carrier. In late 2003, the FCC agreed and imposed rules to that effect.[69] The obligation theoretically runs both ways: another customer can cancel her existing *wireless* account and take the number with her when signing up with a *wireline* carrier. But wireline–wireless number portability clearly operates to the net benefit of the wireless industry because customers increasingly view mobile wireless phones as a potential replacement for stationary landline phones and not vice versa.

D. Market concentration and the AT&T/T-Mobile merger controversy

When we began writing the first edition of this book in 2003, there were six "national" wireless carriers, defined as carriers that own network facilities across the United States.* These were Verizon Wireless, a joint venture between majority shareholder Verizon and U.K.-based Vodafone; Cingular, a joint venture between Bell companies SBC and BellSouth, both now part of AT&T Inc.; AT&T Wireless, previously affiliated with the legacy AT&T Corp. long-distance company; T-Mobile USA, a wholly owned subsidiary of Deutsche Telekom; Sprint; and Nextel. Two years later mergers had reduced that number from six to four: Cingular purchased AT&T Wireless in 2004 and later assumed AT&T's brand name,[70] and Sprint merged with Nextel in 2005.[71] Under a longstanding telecommunications industry practice discussed in chapter 10, the parties in each transaction sought and obtained merger approval in two formally independent proceedings—one before the FCC, and another with the Department of Justice.

Such merger proceedings are complicated and often controversial. As in any other industry, wireless consolidation presents a complex mix of costs and benefits for consumers. All else held equal, the fewer competitors there are in any given market, the less discipline each competitor will impose on the others' retail prices, and the more likely it is that they can successfully collude (either explicitly or tacitly) to keep prices high. Depending on market structure and the total number of competitors, the negative effects of consolidation can range from negligible to very pronounced. On the other hand, consolidation can create substantial *synergies* (i.e., merger-related efficiencies) that exert an opposing, downward pressure on consumer prices. As the Department of Justice and the Federal Trade Commission (FTC) explain in the *Horizontal Merger Guidelines*,

> A primary benefit of mergers to the economy is their potential to generate significant efficiencies and thus enhance the merged firm's ability and incentive to compete, which may result in lower prices, improved quality, enhanced service, or new products. For example, merger-generated efficiencies may enhance competition by permitting two ineffective competitors to form a more effective competitor, e.g., by combining complementary assets. . . . [I]ncremental cost reductions may reduce or reverse any increases in the merged firm's incentive to

---

* No wireless network is truly national in geographic scope. Although the four "national" carriers have facilities across the United States, each has significant holes in its coverage, and each must thus strike roaming arrangements with other carriers.

elevate price. Efficiencies also may lead to new or improved products, even if they do not immediately and directly affect price.[72]

Merger review proceedings thus often turn on a complex economic comparison of the consumer benefits of merger-related efficiencies against the consumer harms of reduced competition. Under established policy, "[t]he greater the potential adverse competitive effect of a merger, the greater must be the cognizable efficiencies, and the more they must be passed through to customers, for the Agencies to conclude that the merger will not have an anticompetitive effect in the relevant market."[73]

In approving the Cingular–AT&T and Sprint–Nextel mergers in 2004 and 2005, policymakers essentially concluded that, at least in those proceedings, the efficiencies of greater wireless consolidation outweighed any anticompetitive impact. As a federal report concluded in 2010, wireless consolidation had "enabl[ed] large national carriers to exploit economies of scale" and thus "create[d] greater productivity and economic efficiency. This industry consolidation may have especially improved the efficiency of the large national carriers, allowing them to offer more wireless services for similar or lower prices." Indeed, the report found that "the overall average price (adjusted for inflation) for wireless services declined each year from 1999 to 2008."[74] The FCC may have also taken some comfort in the fact that even after these mergers, the U.S. mobile wireless marketplace remained less concentrated than almost all of its counterparts in western Europe and the Asia-Pacific region.[75]

The government's acquiescence in ever-greater consolidation ended abruptly in 2011, when the FCC and the Department of Justice teamed up to reject AT&T's proposed $39 billion acquisition of T-Mobile, announced earlier that year. That merger attempt ranks among the most controversial in recent memory, and we close out this chapter with a concise summary of the main issues in that proceeding.

As its main affirmative argument, AT&T stressed that the merger would be critical to its efforts to cope with its mounting network-capacity problems and continue providing high-quality service at affordable rates.* AT&T began with the FCC's own premise that, with the rise

* Although network capacity was a key theme of AT&T's advocacy, AT&T also committed that if the merger were approved, the company would build out LTE functionality to areas covering 97% of the U.S. population. And it argued that this deployment initiative, along with the new mobile applications LTE makes possible, would create many new jobs in the affected communities. The merger-review authorities ultimately concluded, however, that competition would likely force AT&T to build out LTE aggressively whether it merged with T-Mobile or not.

of smartphones, mobile wireless usage is increasing exponentially and will quickly overwhelm existing network capacity. The FCC had been stressing this theme to persuade Congress to enact incentive-auction legislation over the opposition of the broadcast lobby (see chapter 3), but AT&T found the same theme useful for merger advocacy as well. AT&T claimed that because it was the first carrier to offer the iPhone, its capacity crisis was cresting early and that it could not wait half-a-dozen years for new spectrum to become available at auction; it needed more spectrum immediately. It added that the merger partners had uniquely complementary spectrum and network assets: T-Mobile operated on largely the same spectrum bands as AT&T, and, alone among other major wireless providers, the two carriers used the same family of GSM/UMTS/HSPA technologies.

Of course, AT&T was not proposing simply to take T-Mobile's spectrum away from T-Mobile's customers and use it for its own customers. That proposal would hardly have impressed the merger-review authorities, particularly given that T-Mobile had capacity issues of its own in a number of markets. Instead, AT&T and T-Mobile claimed that by pooling their spectrum and network assets together, they could achieve key network synergies and thereby accommodate far more customer traffic than the two companies could serve separately. And they asserted that by exploiting these merger-related synergies, the combined company could reduce its incremental costs of expanding capacity and ultimately sell consumers more for less. The question, however, was how strong this incremental price-lowering pressure would be. Merger opponents argued that the pressure would be far too weak to counterbalance the price-increasing effects of reduced market competition, whereas AT&T and T-Mobile argued that it would overwhelm those effects. Each side produced elaborate cost models, designed by prominent economists, to predict the net economic impact of this merger.[76]

This economic debate ultimately boiled down to an empirical dispute on two broad sets of issues. First, the two sides argued about the strength of merger-related network synergies and whether AT&T could cost-efficiently address its capacity constraints through less radical measures, such as by acquiring additional spectrum from other sources or by making greater use of cell-splitting and distributed antenna systems. Second, the parties argued about how much the elimination of T-Mobile as an independent company would harm wireless competition. Merger opponents characterized the wireless market as national in scope, noted that T-Mobile was one of four national wireless providers, and characterized the proposed combination as a classic "four-to-three" merger. AT&T responded that the relevant market was not national, but local (as the FCC and Department of Justice had long found), and that consumers in many localities could buy service not only from a national provider, but from one of several facilities-based regional providers that offered nationwide coverage (through roaming agreements). AT&T also noted that some of these regional providers, such as U.S. Cellular and MetroPCS, were prospering in several metropolitan markets where T-Mobile was declining. The two sides disagreed, however, about whether the services offered by these regional providers were close substitutes for T-Mobile's services. The two sides also presented starkly different images of T-Mobile itself. Merger opponents viewed T-Mobile as an innovative and indispensable "maverick," whereas AT&T characterized it as a moribund also-ran.

In addition to these "horizontal" competition issues, merger opponents also raised a set of "vertical" concerns that deserve brief mention even though the merger-review authorities never cited them as an important basis for rejecting this merger. Sprint and other merger opponents asserted—and still assert today—that the U.S. wireless marketplace is veering toward a "duopoly" dominated by the Bell-affiliated AT&T and Verizon, which together account for a strong majority of wireless subscribers nationwide. These opponents argued that an AT&T/T-Mobile merger would cement that duopolistic market structure by entrenching the scale and scope advantages that AT&T and Verizon enjoy over smaller providers. In particular, they argued that the merger would increase AT&T's incentive and ability to harm its smaller rivals' access to key inputs, such as handsets, roaming, spectrum, and reasonably priced backhaul (see chapter 2). AT&T contested each of these input-market arguments on the merits and claimed that the arguments were for the most part non-merger-specific—in other words, that the merger could not itself make any of the putative input-market problems worse. AT&T thus contended that the FCC should address such concerns only in connection with industry-wide rulemaking proceedings, such as those for handset interoperability, data roaming, and special-access pricing.

In August 2011, the Department of Justice filed suit in federal district court to enjoin the proposed merger, focusing entirely on horizontal competition concerns and ignoring the input-market arguments entirely.[77] For a few months, AT&T and T-Mobile began gearing up for trial while simultaneously trying to persuade the FCC to approve the merger, albeit with divestitures and conditions. In November, however, the FCC announced that it would soon issue an order identifying its competition-related concerns about the merger and referring the matter to a prolonged fact-finding proceeding before an administrative law judge. Such "hearing designation orders" are normally considered the death knell for any FCC merger application.

Although AT&T and T-Mobile managed to keep the FCC from issuing that order by formally withdrawing their merger application, the FCC nonetheless issued essentially the same document in the form of a 157-page staff report, which accepted the merger opponents' key horizontal theories of competitive harm.[78] The report found that "by combining these two nationwide providers, the proposed transaction would result in an increase in both subscriber and spectrum concentration that is unprecedented in its scale."[79] It further expressed concern that by extinguishing T-Mobile as an independent company, the merger would deprive the market of "a disruptive competitive force" that could "play a special role in counteracting the exercise of market power."[80] The staff report acknowledged that the merger might have generated some network synergies, which, if all else were held equal, might have increased output and imposed some downward pressure on price. But it concluded that "the predictions of the Applicants' economic model are not sufficiently robust or credible to carry the Applicants' burden" of proving "that the benefits of the proposed transaction would outweigh the harms."[81] A month later AT&T and T-Mobile threw in the towel, abandoning any prospect of merging.

Meanwhile, AT&T's arch-nemesis Verizon Wireless was preparing to announce a quite different means of relieving its own capacity constraints: it would purchase all the spectrum that various cable companies had acquired in the 2006 AWS auction but had never put to use (see chapter 3).[82] That deal generated its own set of controversies, including concerns about an asserted "spectrum gap" between the top two wireless providers (AT&T and Verizon) and everyone else.[83] But the FCC ultimately allowed Verizon to acquire this spectrum (with some conditions), if only because the spectrum might otherwise have lain fallow for many more years.[84] The contrast between that transaction and the benighted AT&T/T-Mobile proceeding illustrates a likely theme of twenty-first-century wireless policy: policymakers will go to great lengths to free up more capacity for use by mobile wireless providers, but probably not at the expense of losing a major national carrier.

5

A Primer on Internet Technology

Most of our discussion to this point has addressed technological and regulatory issues confined to the *physical* layer of wired and wireless communications services. We now turn to the higher layers of those services—the protocols and applications that, together with the physical-layer technologies, compose the modern Internet. This chapter provides a brief history of the Internet and a basic technological primer on what it means to conceptualize the Internet in terms of "layers." Chapter 6 then turns to the most important modern controversy surrounding the relationships *among* those layers—a controversy dubbed "net neutrality."

I. The Basics

A network is defined in part by its physical infrastructure—copper wires, fiber-optic cable, radio spectrum. But it is also defined by the logic embedded within the streams of 1s and 0s flitting across it. To understand the distinction between those two "layers" of a network and thus to understand the nature of the Internet, it is first necessary to understand the nature of digital technology.

A. From analog to digital

When Alexander Graham Bell spoke into the first telephone in the 1870s, the sound came out of his mouth in continuous pressure waves of air particles. The transmitter's role in that telephone was to convert those waves into analogous variations of electrical current. Once that electrical current reached the receiver (i.e., Mr. Watson's handset), they were converted back into continuous pressure waves of air particles in the form of audible sounds. The common denominator in each of these steps was the use of continuous waves (of air particles or electrical current) to convey information. The term *analog* describes the various methods of transmitting information in such continuous waveforms.

The term *digital*, in contrast, describes the quite different way in which computers operate and communicate. The silicon chip at the heart of your laptop or smartphone is an intricate network of microscopic transistors whose basic function is to open and close circuits. At the most elemental level, each circuit has two possible states: on and off. The 1s and 0s in a digital transmission—known as *bits*, short for "binary digits"—correspond to those two states and can be used as a mathematical shorthand for describing anything, from the sound of a voice to a video clip to a thousand-page document. Digital technology came of age in the computer world in the 1950s and 1960s. Once computers entered the telephone network in the 1970s and 1980s, this technology swept through the telecommunications world, transforming it forever.

A digital communication often begins with an analog signal, such as the human voice. A device on the speaker's end "samples" the properties of that voice at regular intervals, many times a second. These samples are represented as a particular configuration of 1s and 0s, which collectively describe the sounds the voice makes in roughly the same way that the topography of a mountain range can be described by a series of numbers reflecting longitude, latitude, and altitude. If enough samples are taken in rapid succession, the resulting bitstream can convey a complex enough mathematical representation to capture all the important nuances of the human voice—or any other sound or image.

In a digital transmission, what gets sent is this abstract mathematical representation, not (as in an analog transmission) a direct and continuous representation of the wave characteristics of the sound itself. The device at the receiving end then decodes the stream of 1s and 0s and translates them back into a continuous analog sound. How does the receiving device "know" how to decode the 1s and 0s? It "knows" because it and the transmitting device share an agreed-upon *protocol* for the exchange of encoded information, much as two telegraph operators can encode and decode otherwise inscrutable messages because both have learned the Morse Code. We discuss the question of protocols more fully later in this chapter.

Why do telecommunications engineers go to the trouble of converting analog signals into digital form and back again? One important efficiency made possible by digital technology is *compression*, a means of conserving bandwidth when transmitting information. Consider an evening news broadcast in which the background image behind the anchor remains constant for extended periods. In a conventional analog transmission, the same information about that background image must be wastefully transmitted anew many times a second, even though the information is unchanged each time. In a digital transmission, the information need be transmitted only once, along with short subsequent placeholders indicating to the receiving device that the background image has not changed and should continue to be shown in its original form. Other compression techniques use sophisticated algorithms to represent, say, repeating patterns in a music file.

An additional benefit of digital technology is also one of the most obvious: greater signal clarity. If you are old enough to remember placing long-distance telephone calls before the mid-1980s, you probably recall that the voice on the other end sounded more distant and less distinct than the voice of your neighbor on the other end of a local call. That was an unfortunate side effect of analog technology. On conventional telephone facilities, signals tended to fade over distances. The only way to transmit an ordinary voice call from New York to Los Angeles through analog technology is to amplify the continuous wave signal transmitted over the telephone network, imperfections and all. The result is background static and some distortion. Digital technology avoids that problem. Because a digital signal is just a mathematical representation of an underlying analog signal, its constituent 1s and 0s can be regenerated from one device to the next with no resulting loss of signal fidelity and no increase in background noise. The difference between analog and digital technology is illustrated by the difference in quality between the tenth successive photocopy of a document (which will be much fuzzier than the first) and the tenth successive email exchange of a document (which may be exactly the same as the first).[1] This is the primary reason why long-distance calls today, unlike those placed in the 1970s, sound just as good as local calls.

The distinction between analog and digital technology is relevant to but different from the distinction (discussed in chapter 2) between circuit-switched and connectionless packet-switched technologies. To review, ordinary telephone networks have traditionally been circuit switched, which means that when a call is placed between two points, a fixed increment of transmission capacity (the "circuit") is held open for the duration of the call on a static, predetermined path, regardless of whether any information is passing through the circuit. In contrast, connectionless packet-switched technologies do not dedicate fixed capacity to a given communication, generally relying instead on a system known as *dynamic routing*. That system economizes on transmission capacity by subdividing the information contained in the communication into millions of packets and sending them off in different increments over whatever paths might be most efficient at any given instant. When these packets reach their ultimate destination, they are then reassembled from start to finish and decoded.

Packet-switched networks, by definition, *must* be digital. Circuit-switched networks can be *either* analog *or* digital. In the old days, the circuit that was held open for the duration of a telephone call in both wireline and wireless networks was used exclusively for the transmission of analog signals. Today, the circuit is still held open, but now it is host to long streams of 1s and 0s. Put more concretely, when you now place a call on a conventional landline network, your voice may travel down a copper wire in the form of continuous (analog) electrical impulses, but those impulses will have been converted into digital bits by the time they pass through the first computerized switch that will help direct the call to its ultimate destination.

B. Modularity and layering

To explain what "the Internet" is, we begin with the related concepts of *modularity* and *layering*, which have spurred tremendous entry and innovation in the markets for Internet-related products and services. Modularity is a means of managing complexity by enabling different products to work together through well-understood sets of rules.[2] Think of a Chinese menu that allows one choice from column A, one from column B, one from column C, and one from column D. A more restrictive menu would contain a precombined list of offerings that could not be ordered separately from one another. Like the Chinese menu, modularity allows mixing and matching among different technologies in adjacent markets. And in the Internet sphere, what gets mixed and matched are the data technologies at different "layers" of Internet communications.

To demystify that point, we introduce the layering concept with an example from the nineteenth century: a Union army telegraph transmission from Gettysburg to Washington in 1863. There are four basic levels on which we can describe this transmission. First, we can describe the physical properties of the medium over which the signal was sent. In this case, it was a copper wire—as opposed to, say, a fiber-optic cable or the airwaves. Second, we can describe the shared code, or "protocol," that enabled the sender and receiver to know what words were being sent over that copper wire. In our example, the protocol was the Morse Code: dot-dash, dash-dot-dot-dot, and so on. Both the message's sender and the immediate recipient needed to agree on that protocol (or some other shared protocol), or else no message could be effectively transmitted. Third, we can describe the telegraph operator, who possessed the relevant "intelligence" to understand that dot-dash signifies the letter *a*, dash-dot-dot-dot signifies the letter *b*, and so on. If the operator at either the sending or the receiving end lacked proficiency in the Morse Code, the message would get garbled. Fourth, we can describe the content of the message itself, as translated by the telegraph operator on the receiving end. In our example, the message (at the end of Day 3) would have been: "Confederates retreat after disastrous charge up hill."

Each of these four attributes of the "call" from Gettysburg to Washington describes a layer of the transmission: a physical layer (the copper wire), a logical layer (the Morse Code), an applications layer (the telegraph operators), and a content layer (the message about the Confederate retreat). The important point to grasp for now is that each layer is largely independent of the others, meaning that, for the most part, we can vary one element without varying the others. For example, if Guglielmo Marconi had invented the radio in 1862 rather than 1895, the same message, in the same Morse Code format, could have been sent over the airwaves rather than through a copper wire. In other words, one can vary the physical medium of delivery in our hypothetical transmission but leave the logical, application, and content layers untouched. Second, if Samuel Morse had not developed his system and some other protocol had been developed to convert letters into simpler elements (similar to the dots and dashes of a Morse Code transmission) understandable to a trained operator of a telegraph machine, the same Union army message could have been sent through the same physical medium (the copper wire) and interpreted by the same people. Third, the same message could have been delivered if another set of operators had been at the helm and were also proficient in the Morse Code. Finally, our Civil War combatants could have used the same physical medium (copper wire), the same logical protocol (Morse Code), and the same application (the telegraph operators) to transmit a different message content altogether if the battle had gone differently: "Confederates take hill, begin unimpeded march toward Washington."

To summarize now and move our discussion into the digital era, there are several mutually independent layers in any Internet communication. Engineers often refer to seven layers, but for conceptual simplicity (and at the risk of some oversimplification), we group them into just four.* At the bottom is the physical layer. This denotes the physical characteristics—for example, copper wires, fiber-optic cable, or the airwaves—of the medium over which the information is transmitted. Next up is the logical layer. It includes the basic protocols for a transmission: the digital signal formats that enable the electronic devices on each end to cooperate in transmitting information successfully. On top of this layer is the applications layer, which includes, among a great many other things, Web browser software such as Explorer, Safari, or Chrome and the corresponding server software used by websites. And on the very top is the content layer, which describes the actual words in an email, the music video shown by a media player, or the images and text contained in a webpage.

This division of technology into self-contained, mutually independent layers makes the Internet environment highly modular. This means that, at least in theory, firms can compete independently at each layer without worrying about entering the market for services at other layers. For example, Internet modularity enables a website such as eBay or an applications software provider such as Apple's iTunes to provide service to anyone on the Internet without becoming an ISP in its own right and without incurring the prohibitive expense of building a physical network.

Without the Internet's careful separation of layers, consumers today, like Prodigy or CompuServe subscribers in the 1980s, would be at the mercy of their service provider for their online content. As discussed in more detail later in the chapter, the Internet is not only modular but

* As an example of how our discussion somewhat oversimplifies Internet technology, we have lumped into a single "logical" layer what telecommunications engineers would consider two different layers. In 1978, the International Standards Organization introduced a now standard seven-layer "open systems interconnection" (OSI) model for understanding layering hierarchies in digital environments. The protocols that define Internet communications appear on layers 3 and 4 of that "OSI stack"—"network" (IP) and "transport" (TCP, User Datagram Protocol (UDP), etc.), respectively. And in any given transmission, those protocols often ride "on top of" a layer 2 "data link" protocol such as Ethernet or ATM. For a comprehensive yet highly readable discussion of the Internet's constituent layers, see James F. Kurose & Keith W. Ross, Computer Networking: A Top-Down Approach (5th ed. 2010).

open,* in the sense that no one owns the core protocols at the logical layer, and anyone can develop complementary products at the adjacent physical and applications layers. This tradition of modularity and openness provides much of the explanation for the Internet's phenomenal success in generating consumer value. Because the Internet stimulates entry and disaggregated competition at each layer, and because it allows end users to mix and match the best technologies at each layer, it is a uniquely hospitable platform for innovation of all kinds.

C. The logic of the Internet

At the *physical* layer, examined later in this chapter, the Internet includes millions of networked computers and smart devices joined together by routers, fiber-optic pipes, cellular networks, and other transmission media into a worldwide network of networks. At the *logical* layer, the Internet consists of a common computer "addressing" scheme—the *Internet protocol* (IP)—and a set of protocols for the accurate and efficient transmission of packet-switched data across different computer networks, of which the best-known is the *transmission control protocol* (TCP). Those protocols, known collectively as the *TCP/IP suite*, enable each packet in a transmission to "tell" the packet switches it encounters where it is headed and enables the computers on each end to confirm that the message has been accurately transmitted and received.[3] "The Internet" is defined as the combination of these characteristics: the IP-based addressing system and the interconnected network of networks that rely on IP-related protocols as a common logical-layer standard.[4]

Together, these elements of the Internet enable a computer in one corner of the world to find a different computer in another corner of the world and exchange information that can be understood by the applications software loaded onto the computers at each end of the transmission.† Indeed, the two computers may use entirely different hardware and operating systems. One might be a powerful server running on the

* The term *open* can mean different things in different contexts; here we use it simply to mean nonproprietary. Our use of the term does not mean that all relevant Internet standards are built on "open-source" software (which has a different set of implications) or that they are open in name but proprietary to an individual firm, such as the Java standard pioneered by Sun Microsystems.

† We are using the term "computer" here in its broadest sense to include not only servers and PCs, but any smart device capable of Internet connectivity, from tablets and smartphones to advanced meter readers.

UNIX operating system, and the other might be a Windows-based PC, an iPad running iOS, or an Android smartphone. The critical point is that all "computers" connected to the Internet speak the same IP-based logical-layer language. As noted, this language is used by all in the Internet world and, like English, is owned by no one.

We can make this point more concrete with a simplified example relating to one major category of Internet applications: the World Wide Web. Suppose that you wish to download a particular webpage—that is, load it into your computer's memory and call it up on your screen. To do that, you type in the webpage's *domain name*, such as "www.amazon.com." Domain names—valuable commodities in the world of e-commerce—are allocated by private companies such as Verisign under the general supervision of a U.S.-based nonprofit entity: the Internet Corporation for Assigned Names and Numbers (ICANN).[5]

A domain name such as www.amazon.com is just a user-friendly shorthand for the real information needed to reach the website: the *IP address* of the computer hosting that site. The IP address consists of numerical sequences separated by dots, such as 72.21.194.1. That address performs much the same function as the number you dial in an ordinary phone call: it designates the computer you are trying to reach. Like the number assigned to a mobile phone, an IP address is not location specific. It can be accessed from anywhere, and it can be located anywhere. In our simplified example, your computer finds the IP address for the website by transmitting the domain name to a special type of computer on the Internet called a *domain name server*, whose job is to keep track of which domain names correspond to which IP addresses.

The message transmitted in a typical Web inquiry is broken down into discrete packets—strings of 1s and 0s—which may fly off individually in several directions in search of the fastest, least congested route to the computer running that website. Messages are compartmentalized this way for the sake of efficiency. By analogy, as John Naughton explains in his masterful history of the Internet,

> [Nobody] would contemplate moving a large house in one piece from New York to San Francisco. The obvious way to do it is to disassemble the structure, load segments onto trucks and then dispatch them over the interstate highway network. The trucks may go by different routes—some through the Deep South, perhaps, others via Kansas or Chicago. But eventually, barring accidents, they will all end up at the appointed address in San Francisco. They may not arrive in the order in which they departed, but provided all the components are clearly labelled the various parts of the house can be collected together and reassembled.[6]

In an Internet packet (more technically known as a *datagram*), the labeling function is performed by an address *header*—the 1s and 0s that appear in preassigned slots near the beginning of each packet and convey information about the packet's destination. The related packets in a message will ideally end up in the right place in short order because various packet switches throughout the Internet's telecommunications infrastructure—the *routers*—are constantly exchanging information about the most efficient way to reach a particular destination. Additional preassigned slots within an Internet packet contain other standard information, such as where the packet originated, how many bits it contains, and how it relates to the other packets within the same transmission. The TCP/IP suite is, in essence, the set of rules governing which slots within a packet will contain what information about a packet's destination, source, length, and so forth. Together, those rules enable the computers on each end to ensure the efficient transmission of the data "cargo"—the substance of the transmitted information—across the Internet. The "TCP" part of the TCP/IP suite governs the assembly and reassembly of the data at each end (including checking for errors such as missing data), and the "IP" part is responsible for routing data from one node to another.*

The high-capacity computer hosting the website you contact might be in the same city where you live or halfway across the globe; you do not know, and it most likely makes no difference to you. When it receives your inquiry, that computer, known as a *server* (in that it "serves" you, the "client"), sends a burst of digital packets back to you. Again, the 1s and 0s in the header of each packet contain addressing information to ensure that the return message reaches your computer. Other 1s and 0s identify the content of the webpage using protocols specific to the World Wide Web. Your computer is able to translate those 1s and 0s into pictures and words only because it is outfitted with client software (a *browser* such as Explorer, Safari, or Chrome) designed to understand the meaning of 1s and 0s transmitted from distant websites. This is a critical point: the telecommunications facilities of the Internet itself—and, more generally, the Internet's physical and logical layers—do not generally "know" what those 1s and 0s mean; they simply send the 1s and 0s your way and let your computer software figure out the rest.

* Even the TCP/IP standard is modular in the sense that other protocols besides TCP can govern the assembly and reassembly of data sent on IP networks. UDP, for example, is an alternative to TCP that provides fewer error-checking services and is used for, among other things, domain-name look-ups and VoIP services.

The Internet's effectiveness depends on universal agreement on the nonproprietary protocols to be used to translate information into 1s and 0s and back again. This agreement is the legacy of the government's early sponsorship of the Internet's antecedents combined with the power of network effects once the Internet assumed public stature. The important point for now is that because the Internet's core logical-layer standards are not owned by any firm, any operator of a data network can connect to the Internet. Similarly, because the Internet's intelligence is provided primarily by the devices connected to it and not by centralized switches, any applications developer is free to make her work available via the Internet and give all Internet users access to it. As a consequence, the creator of new media content, such as a short film, can rely on the Internet to distribute her work, thereby displacing such traditional intermediaries as movie theaters and television networks. To summarize, these two related features of the Internet's open architecture—the openness of its protocols and the ability of anyone to develop applications and content for it—help explain the Internet's spectacular growth.

Under traditional accounts of the Internet's origins, this open architecture is no accident: the engineers who developed the basic protocols promoted an *end-to-end* design principle that gave maximal control to the users and devices on each end and minimized the intelligence necessary to operate the Internet itself.[7] When applied strictly, this principle means that packets are delivered on a first-come, first-served basis without regard to their content, origin, or destination and are free from any intermediate error checking or filtering. In contrast, the intelligence in a conventional telephone network resides in centralized switches, which tightly control such applications as call waiting and caller ID. The Internet has thus been described as the circuit-switched telephone network "turned inside out."

But these same characteristics make the Internet, as originally conceived, an imperfect medium for real-time applications such as voice and videoconferencing. As discussed in chapter 2, telephone company engineers guarantee bandwidth by dedicating a circuit to each call, even when no one is talking, while conserving on the network's overall switching and transport needs by limiting the bandwidth assigned to each circuit. In the Internet world, by contrast, the principal limit on bandwidth—if no one else is using the network—is the overall capacity of the routers and pipes between point A and point B. But if the network is busy, users face degradation of whatever application they are trying to run. Such degradation is generally acceptable for applications that are not highly performance sensitive. You might not care or even notice if, because of congestion, a conventional webpage takes half a second longer to load. But you would find it highly annoying to join a videoconference where every impromptu exchange of words takes an extra half-second to "load." In short, in a highly congested network operating just below its capacity limits, real-time applications proceed far more smoothly in a circuit-switched environment, where bandwidth is guaranteed, than in a connectionless packet-switched environment, where thousands of users compete for the same bandwidth from instant to instant.

Because of the Internet's comparative drawbacks in handling performance-sensitive, real-time applications, broadband service providers tend to provide voice (phone) and video (TV) services in "managed" data streams that are at least logically separate from and sometimes prioritized over the packet streams associated with regular Internet access during periods of congestion. "Logically separate" does not necessarily mean "physically separate." For example, AT&T and some other service providers have begun transmitting the data packets associated with "voice," "video," and "Internet" services over the same last-mile transmission pipes in a unified IP format. In these networks, phone conversations and TV shows take the form of IP packets, just like Web-browsing sessions, and all of those packets coexist on the same wires leading into a home or business. And in some of these networks, a key technological difference between "voice" and "video" services, on the one hand, and "Internet" services, on the other, is that network engineers mark the voice and video packets for special handling to distinguish them from "Internet" packets when the network is busy.* As discussed in the next chapter, one critical question in the net neutrality debate is whether service providers should have analogous flexibility, *within* the category of services they market as "Internet access," to prioritize some data streams over others during periods of network congestion, depending on their varying quality-of-service needs.

* All relevant versions of the Internet protocol contain header fields that allow network operators to mark time-sensitive packets for priority during periods of congestion. *See* Kurose & Ross, Computer Networking, at 367. To date, however, the operators of the Internet's constituent IP networks have developed no common system for honoring one another's priority markings. For the most part, therefore, network operators today can prioritize some IP packets over others only if they keep them on a single IP network, in what the FCC has called "managed" or "specialized" IP services. These points are subtle but critical, and we return to them in chapters 6 and 7.

D. Killer apps: email and the Web

To launch a successful network standard, one must persuade a critical mass of users to adopt it. In the case of the Internet, the advent of digital technology and digital networks, which businesses had previously adopted for their own purposes, provided an important building block that made adopting Internet technologies easier from an engineering perspective. But the question remained: Why would anyone *want* to use the Internet? The answer lay in two killer applications—email and the World Wide Web—that led the total number of Internet users to double each year through the late 1990s.

The basic protocols that facilitate email were developed in the 1970s and early 1980s on a nonproprietary basis. The Simple Mail Transfer Protocol (SMTP) provided an effective means for delivering email messages and is still used today, although it is often supplemented by various complementary protocols.[8] Although the original email systems were somewhat difficult for nonspecialists to use, ordinary consumers began exchanging emails in increasing numbers in the 1990s with the development of user-friendly software programs such as Qualcomm's Eudora, Microsoft Outlook, and Lotus Notes. Thus, whereas very few business cards in 1990 contained a line for an email address, very few business cards in 2000 *lacked* such a line.

The driving force behind email's explosive popularity during that decade was the familiar phenomenon of network effects. The technology quickly reached a tipping point: because more and more people began to use email, it became correspondingly more valuable to each user, and the *absence* of an email address became a serious liability within many professions. Another reason for email's quick adoption cycle was its use of an open standard. Users did not worry about being locked into a proprietary standard, and businesses (such as Microsoft and others) could enhance email's appeal by easily developing extensions for the open standard.

The Internet's second mass-market application, the World Wide Web, was conceived by Tim Berners-Lee in 1989 at CERN, the Swiss particle research laboratory.[9] Like email, the Web relies on a set of nonproprietary protocols for formatting webpages on computer screens ("hypertext mark-up language," or HTML), establishing transmission procedures between a Web server and its clients ("hypertext transport protocol," or HTTP), and identifying the server address and file location where a particular webpage can be found (the "uniform resource locator," or URL). The Web's defining characteristic, a core feature of the HTML protocol, is the use of hyperlinks. These are the embedded codes in a webpage (often associated with highlighted words) that, when clicked, tell your computer to retrieve another webpage, either from the same website or another one, thereby enabling users to move quickly from one webpage to the next.

The best way to think of the Web is as one application among many that rides on top of the Internet's lower-layer TCP/IP protocols and serves as a platform in its own right, supporting still higher-layer applications such as Web-based streaming video or instant messaging. Some people still confuse the Internet with the Web, but this is a bit like confusing the Windows operating system with the Microsoft Word program that runs on top of it. That the Internet itself and the Web operate at independent layers explains why you can use your computer to run applications other than the Web (such as VoIP or email) over the Internet and why you can use Web-oriented software for functions unrelated to the Internet (such as searching the files in a closed corporate database).

Like email, the Web was not an immediate popular success. To view websites—collections of files posted on Web servers—a user needs a browser that translates her requests into code that those servers can understand and then translates the code sent back by those servers into sights, words, and sounds she can understand. The first browsers were text-based, pictureless, and nonintuitive, at least from a layperson's perspective, and their audience was largely limited to university settings. That changed in 1993 when Marc Andreessen and Eric Bina, then working at the National Center for Supercomputing Applications at the University of Illinois, released to the public (for free) the first multimedia browser with a point-and-click graphical user interface, which they called "Mosaic." Along with Jim Clark, Andreessen then founded Netscape. There he improved upon Mosaic to produce the more popular Navigator browser and became an Internet tycoon overnight by taking Netscape public in 1995.[10] The spectacular success of Netscape's initial public offering helped awaken the general public to the Internet's transformative significance for the economy at large. In retrospect, it also marked the beginning of the late 1990s Internet gold rush, which largely ended with the NASDAQ crash of 2000–2001.

The development of user-friendly browsers in the early 1990s followed the federal government's decision to privatize the Internet and allow it to support electronic commerce (more on that later). By 1996, the Web had entered the popular consciousness and begun a period of explosive growth. Why did executives, in a relatively short period, turn from asking "what is a website?" to "when can we get our business's website up and running?" As with email, the short answer lies in the network-effects phenomenon.

In the early years of the Web, ordinary consumers had no real incentive to devote the time and resources needed to use it because too few of their peers were using it, and no one had developed user-friendly software to exploit its commercial potential. The tipping point arrived with the invention of Mosaic and Navigator, which made the Web easy for the masses to surf. As John Naughton explains, the Web presented the same "chicken and egg story" as the slow initial growth of the telephone—but with one key difference: "[W]hereas the spread of the telephone depended on massive investment in physical infrastructure—trunk lines, connections to homes, exchanges, operators, engineers and so forth—the Web simply piggy-backed on an infrastructure (the Internet) which was already in place. By the time Mosaic appeared, desktop PCs were ubiquitous in business and increasingly common in homes. . . . The world, in other words, was waiting for Mosaic."[11]

Once the Internet reached critical mass, it could rely on network effects to keep it—and its most successful applications, such as email and the Web—from fragmenting into mutually unintelligible systems. In that respect, the development of the Internet, email, and the Web is somewhat like the development of spoken languages—which, indeed, are the most fundamental of all "standards." As words change meaning and new words come into use, individuals adjust their own linguistic practices to ensure that they are understood by others. The language changes, but it remains mutually intelligible to everyone within a single linguistic community.

The same is true of the protocols that constitute the Internet itself—email, the Web, and other Internet applications. As the standards change, individuals follow because that is the only way they can continue exploiting the Internet's prodigious network effects. But whereas language normally evolves without much guidance from any recognized decisionmaking body,[12] the same is not true of the Internet. Instead, since 1986 the key standards associated with the Internet have evolved largely under the close supervision of the Internet Engineering Task Force (IETF). In its own words, the IETF is an open and "loosely self-organized group of people who contribute to the engineering and evolution of Internet technologies."[13] And the World Wide Web Consortium (W3C) performs a similar standard-setting function for Web-oriented protocols.

Significantly, the IETF and the W3C have no formal regulatory authority of any kind. Thus, once the IETF introduces a new standard—say, IPv6, an upgrade to the Internet's core addressing scheme that, among other things, provides for more Internet addresses than its predecessor (IPv4)—it has no means, beyond its powers of persuasion, of ensuring that firms employ this upgrade. In many cases, including this one, the self-interest of firms will coincide with the collective interest of the entire Internet community, and firms will voluntarily adopt the new standard. But where those interests are not aligned, the IETF cannot enforce compliance with its official standards.

E. VoIP

We close this discussion of key Internet applications by briefly addressing an IP-based service that did not become widespread until several years into the twenty-first century but is now supplanting circuit-switched telephony and will someday lead regulators to declare the "sunset" of the conventional telephone system. This service is VoIP, a diverse family of voice and videochat applications that reduce voice conversations to exchanges of IP packets, often by means of an open signaling standard known as the *Session Initiation Protocol* (SIP).[14]

The FCC has defined two basic categories of VoIP services: *interconnected* services, which are subject to various regulatory obligations, and *non-interconnected* services, which are subject to few such obligations. Very roughly speaking, an "interconnected" VoIP service allows its end users to communicate not only with others online, but also with regular telephone users on the circuit-switched PSTN (public switched telephone network).[15] These interconnected VoIP services in turn fall into two general subcategories: *fixed* and *nomadic*. A fixed VoIP service is so named because, like a conventional circuit-switched telephone service, it is tied to a specific last-mile network; a customer can use it at home and nowhere else. If you order VoIP as part of a triple-play package from your cable or wireline telephone company, you will receive a fixed VoIP service.

In contrast, *nomadic*—also known as *over-the-top*—VoIP services, such as those offered by Vonage or Skype, are not tied to any particular broadband connection. In that respect, a nomadic VoIP service, unlike its fixed VoIP counterpart, is just one Internet application among many riding over end-user transmission platforms that do not "know" what the associated packets are used for and do not prioritize them over other

Internet packets. Because no one network provider manages smooth routing for those packets from end to end, nomadic VoIP services lack the quality assurances of fixed VoIP services, although technological advances have narrowed the performance gap. In 2004, the FCC deemed these quintessentially Internet-based services categorically "interstate" and thus beyond state common carrier regulation insofar as it is infeasible to determine the endpoints of a nomadic VoIP provider's calls.[16] In 2007, a federal appeals court upheld that determination while suspending judgment on whether the same result would apply to fixed VoIP services, an issue that the FCC had stopped just short of resolving (and assured the court that it had not resolved).[17]

Although VoIP services use a wide variety of different technologies, we can illustrate how a typical nomadic VoIP call interconnects with the PSTN by looking at Vonage's service. A Vonage subscriber originates a voice call over the Internet by plugging an ordinary telephone into a special Vonage-provided adapter, which is associated with an IP address. The subscriber connects that adapter, in turn, not to a telephone jack, but to whatever broadband connection he has separately purchased from, say, a cable modem or DSL provider. Once connected, the adapter communicates with Vonage's server. If a Vonage subscriber calls someone on the PSTN, Vonage, in partnership with various telecommunications wholesalers, arranges to drop the call off on the PSTN after first converting it from IP packets into the standard time-division multiplexing ("TDM") format understood by circuit-switched networks (see chapter 2). Similarly, Vonage (through its wholesale partners) obtains telephone numbers managed by the North American Numbering Plan Administrator—a federal authority that works closely with the FCC—and assigns specific numbers to its subscribers so that they can be called by regular PSTN subscribers.

Because a Vonage subscriber's telephone number is, from Vonage's perspective, just a proxy for the IP address associated with the adapter, it liberates the subscriber from the geographical constraints usually associated with landline telephone numbers. Suppose, for example, that a subscriber resides in Denver and obtains from Vonage a number with the 303 area code, yet he often commutes to New York City during the workweek. When he travels to New York (or to Brussels, Tokyo, or anywhere else), he plugs his adapter into a broadband connection there—say, in his hotel room. When a CenturyLink wireline subscriber back in Denver calls him, she can reach him through CenturyLink's ordinary circuit-switched network by making a local call to his 303 number without incurring toll charges. One of Vonage's telecommunications wholesalers has made this possible by establishing a physical point of presence in the Denver area and associating Vonage's 303 numbers with it. A shared database tells all telecommunications carriers, including CenturyLink, to drop off calls to Vonage's 303 numbers at that point. From there, each call is sent on its way, ultimately in the form of IP packets, to the subscriber's adapter, wherever on the Internet it might be plugged in. What looks like a local call to the friend back in Denver is in fact a long-distance call with a local stop at an Internet gateway. It is a bit like visiting a distant Web server through a dial-up Internet connection, except that the call is processed seamlessly to the called party without any discernible set-up delay, and the digital application being run is a voice conversation rather than a session of Web browsing.

Such VoIP services enable end users to treat voice telephone calls and their accompanying features as just another set of applications they can run over any broadband connection at the edge of diverse packet-switched networks. They free person-to-person calling from the control of telephone company software locked in centralized circuit switches. In that respect, these VoIP services invite end-user innovation for voice and videochat services in the same way that the Internet facilitates such innovation for communications in general: again, it turns the circuit-switched telephone network "inside out."

II. The Internet's Physical Infrastructure

To this point, we have focused on the Internet's logical-layer characteristics and some of its key higher-layer applications. Now we shift our attention to the physical layer.

A. Beginnings

A brief synopsis of the Internet's origins helps explain its close relationship to the traditional telecommunications infrastructure. In the usual telling, the Internet is described as the brainchild of the same military–industrial complex that brought us the theory of mutually assured destruction.[18] In the 1960s, at the height of the Cold War, some of America's brightest minds were fixated on a macabre question: If the Soviet Union wiped out much of the United States in a preemptive nuclear strike, how could the president convey a "launch" order to America's own assembled nuclear forces? For Paul Baran, then a technologist at Rand, the answer to this telecommunications problem seemed obvious: build a network with a series of broadly interconnected *nodes* (routers) so that no one node is critical to the functioning of the network as a whole.

The problem was that the military's long-distance communications infrastructure was largely contained within AT&T's circuit-switched network. As we discussed in chapter 2, an efficient circuit-switched voice network conserves on switching capacity by economizing on the number of switches that need to be occupied for the duration of a voice call. This efficiency requires a highly centralized network with a rigid hierarchy of switches. The efficiency of that hierarchical arrangement, however, is also its greatest military vulnerability. Precisely because it was centralized and hierarchical, AT&T's circuit-switched network was exceptionally susceptible to destruction. If the Soviets had hit just a few central switches, they might have been able to prevent Washington from sending messages to missile silos in Nebraska, Arizona, and Montana.

The solution was a digital and distributed (decentralized) network developed by a branch of the Defense Department known as the Advanced Research Projects Agency (ARPA). Like the nodes in Baran's imagined distributed network, each of the nodes in the ARPANet was linked to several other nodes in a generally nonhierarchical way that maximized the number of routes a signal could take from Point A to Point B. One node would hand off a message to another with the destination information attached, the next node would hand it off to yet a third node with the same destination information, and so on down the line until the destination was reached. If one node along the way was out of commission, the message could be rerouted to another, still-active node. Thus, even though the ARPANet still leased its transport links from AT&T, it was much less vulnerable than AT&T's own circuit-switched network. Just as Baran had envisioned, if half of the ARPANet had been wiped out in an attack, signals would still have arrived at missile silos or any other critical governmental location, albeit in a more roundabout way.

For the ARPANet to work, however, the messages had to be transmitted in digital form, in part because that is the only feasible way of preserving the clarity and integrity of messages after frequent repetition at many nodes. And the network could operate nimbly and efficiently only if it was packet switched, for, as we have explained, packet-switching technology enables the portions of a "call" (or "session") to hop flexibly through a large number of nodes in small packets without needlessly tying up capacity on each of those nodes in the form of dedicated circuits.

The ARPANet also accomplished a number of relatively mundane objectives, such as linking together government-supported labs for purposes of aggregating the processing power of the mainframe computers in each lab. Indeed, some observers submit that the government funded the ARPANet project more to achieve such conventional objectives than, as in Baran's vision, to preserve communications in the event of a catastrophic attack.[19] Either way, the distributed and digital character of this network, conceived amid the Cold War paranoia of the 1960s, forms the basis for the current Internet.

Today's Internet does, however, differ from the original ARPANet in several important respects. The ARPANet started out as a single unified network overseen by the government and affiliated research institutions. The main applications from the 1970s through the birth of the modern Internet in the early 1990s were file exchanges (employing the now outdated but still used File Transfer Protocol), TELNET (a means of logging on remotely to a mainframe computer), newsgroups, and email.[20] The most influential users were academics who relied on government funding to develop the basic protocols on which the Internet still relies. And the federal government's sponsorship was critical throughout these early years. ARPA funded a Berkeley program to incorporate TCP/IP into the UNIX operating system; the Defense Department required its contractors to adopt TCP/IP; and the National Science Foundation provided grants to the IETF and invested more than $200 million to support "NSFNet," a TCP/IP network that linked the networks of various universities.[21]

This official sponsorship finally ended in the early 1990s, when the federal government proposed to privatize the NSFNet and for the first time open it up to commercial uses.[22] The government thereby sought to enlist "the enthusiasm of private sector interests to build upon the government funded developments to expand the Internet and make it available to the general public," as two Internet pioneers have observed.[23]

The result of this privatization decision is the wildly successful modern Internet. As to the physical and logical layers, the Internet is a decentralized network of dissimilar networks, most of them private, with otherwise incompatible computers and other smart devices, all joined together by the common use of the TCP/IP set of protocols. As to the higher layers, it is an engine for economic growth and an unrivaled resource for electronic applications and content, whose variety is bounded only by the limits of the human imagination. By 1997, it had become conventional wisdom that, in the words of Bill Clinton and Al Gore, "[t]he private sector should lead" the Internet's continued growth and that although the government had played a critical role in the Internet's development, "its expansion has been driven primarily by the private sector."[24]

B. The Internet's Constituent IP Networks

Of the many thousands of IP networks that together constitute the Internet, most are "edge" networks ranging from home Wi-Fi configurations to massive corporate IP networks. In this section, however, we focus on the commercial IP networks that offer Internet transport or content delivery services—the function of connecting one place or user on the Internet with another. The most important of these networks fall into three basic categories: access/ISP networks, backbone networks, and specialized content delivery networks (CDNs). Keep two caveats in mind as you read the following discussion. The distinctions between these network categories can blur at the edges, and any given company may perform two or all three of these roles. For example, both AT&T and Level 3 provide access, backbone, and CDN services.

Access networks/ISPs

Most end users rely on *access networks* such as Comcast, Verizon, and Sprint to bridge the "last-mile" gap between them and the rest of the Internet. These access networks are typically (though not invariably) operated by *Internet service providers* (ISPs), and we use the terms *access provider* and *ISP* interchangeably here.* An ISP is your liaison to the broader Internet. Whenever you use the Internet, you expect your ISP to provide access to the entire Internet, including all public websites and email addresses. To satisfy that expectation, your ISP enters into various *peering* and *transit* arrangements with other IP networks, as discussed later in this chapter.

At least in populous areas, an end user may use up to several access networks during the course of the day: one at home, one at the office, and one or two for mobile broadband devices such as smartphones and tablets. This is a recent development. When the first edition of this book was published in early 2005, mobile broadband services were still in their infancy, and they did not come of age until Apple introduced the first iPhone in 2007. We discuss the explosive popularity of mobile broadband services in the previous two chapters and focus here on fixed-line broad-

* Some smaller telephone companies provide last-mile transmission services ("access" in the conventional telco sense) to formally distinct ISPs, which in turn arrange for connectivity with the broader Internet and sell bundled broadband Internet access services to end users.

band services. Even fixed broadband services were still a bit of a novelty in early 2005, when residential Internet subscriptions were almost evenly divided between broadband and "dial up."[25]

Those dial-up services warrant brief mention here, if only because they accounted for how essentially all Americans gained access to the Internet until the late 1990s. In a dial-up connection, an end user places an ordinary telephone call though the telephone company's circuit switch to an ISP, which converts the call into IP format and communicates on the caller's behalf with the wider Internet. The narrowest bottleneck in that arrangement is the circuit switch. Again, telcos have always economized on the capacity of circuit switches by squelching the information they can process in any given call, a step that leaves voice conversations clear enough but slows bandwidth-hungry Internet applications to a crawl. In particular, the traditional design of telco networks limits dial-up Internet access to about 56 kilobits per second, which severely constrains the kinds of Internet applications that end users can run.

To provide broadband Internet access, therefore, telephone companies need to route Internet traffic *around* the bandwidth-constraining circuit switch. Under the DSL technology introduced in chapter 2, the telco splits the signals on copper telephone wires into separate voice and data frequencies. It sends the voice signals through the circuit switch as usual but shunts the data signals off through a multiplexer and packet switch en route to an ISP. Unlike the voice path, the data path is "always on": your DSL modem is in constant communication with the ISP.

Because the transmission capacity of the copper wires themselves declines with their length, many telcos are deploying fiber-optic cable deep into individual neighborhoods to minimize how far the data signals must travel over those wires. Alone among the major telcos, Verizon has gone one step further, deploying fiber all the way to individual homes in some areas. But Verizon began winding down new deployments in 2010 amid concerns about the enormous costs (see chapter 2).

The other major type of fixed-line residential broadband access is cable modem service, provided over the same facilities that deliver cable television signals into people's homes. Cable companies traditionally assign Internet data to discrete "channels" (frequency blocks) within their cables in the same way that they assign television stations to such channels. Just as DSL providers split data from voice, cable modem providers split data from conventional television signals and in two places: at the customer's home and typically at the *headend*, which is the cable company's counterpart to a telco central office (and is where the company collects television programming signals from fiber-optic and satellite links). Like the telcos, cable companies have spent the past decade deploying fiber-optic cables deep into residential neighborhoods to replace traditional coaxial cable, with its more limited bandwidth.

By deploying extensive fiber and adopting a new transmission technology, DOCSIS 3.0, cable companies can now provide broadband performance that in almost all cases is superior to what telcos can provide when they rely on conventional all-copper loops equipped with DSL. As noted in chapter 1, some industry analysts have predicted that cable operators will enjoy a growing bandwidth advantage over telcos in the years to come and may ultimately exercise quasi-monopoly power in the marketplace for fixed-line residential broadband services.[26] At this point, however, these predictions remain speculative, particularly given new advances in the DSL technologies used on hybrid fiber–copper loops.[27] Over the long term, different broadband platforms—including cable, telco, and mobile—may well be imperfect substitutes, but they will likely continue to compete among several dimensions, including not only raw bandwidth, but also price and (in the case of wireless) mobility.

Backbone networks and the basics of peering and transit

Internet backbone networks—such as those operated by AT&T, Level 3, and Sprint—use long-distance fiber-optic cable to connect other, geographically dispersed networks, including the networks of large businesses, ISPs, and other backbone providers. As noted in chapter 2, there is enormous overlap between long-distance telephone networks and Internet backbone networks. For example, the fiber strands that carry Internet backbone traffic often coexist along the same intercity routes as the fiber strands that carry long-distance telephone traffic. It is thus no coincidence that the largest conventional telcos—AT&T, Verizon, and CenturyLink (formerly Qwest)—own some of the largest Internet backbone networks. These backbone providers help unite the Internet by interconnecting with one another and the Internet's other major constituent IP networks, ensuring that each computer or smart device on the Internet can talk to any other.

Interconnection arrangements between IP networks are forged in private bilateral negotiations and take two basic forms: *peering* and *transit*. These arrangements are best conceptualized as forms of direct and indirect interconnection, respectively. Through peering, two networks interconnect *directly* to connect their respective customers, whereas in transit one network hires an intermediary to connect it *indirectly* to any number of other networks.

The prototypical peering arrangement involves two networks of roughly equivalent market stature. Each "peer" exchanges Internet traffic with the other peer for ultimate delivery to their respective customers, which can range from other commercial IP networks to individual end users. Until the past decade or so, peering rarely involved any exchange of money; instead, each peer compensated the other in kind by agreeing to route traffic that the other originated. This arrangement is called *settlement-free peering*.

In recent years, however, it has become increasingly common for one IP network to pay another to peer with it, particularly if it is handing off far more traffic *to* the other network than it is receiving *from* that other network.[28] These arrangements are known as *paid peering*. For example, in 2005 Level 3 forced smaller backbone provider Cogent to agree that it would begin paying compensation if the two networks' "traffic ratios" began falling too far "out of balance"—that is, if the amount of data traffic Cogent funneled into Level 3's network greatly exceeded the amount of data traffic it received in turn.[29] In late 2010, Level 3 found itself on the other end of a similar impasse when Comcast began charging it for the use of direct-peering links to send high volumes of streaming-video traffic to Comcast's residential broadband customers. Level 3 contended that Comcast's paid-peering policy violated net neutrality principles, whereas Comcast argued that it was simply applying the same traffic-ratio principles that Level 3 had invoked its earlier dispute with Cogent. We return to this dispute in the next two chapters because it illuminates a deep conceptual conundrum at the heart of the FCC's net neutrality regime.

Peering negotiations can become a high-stakes game of chicken if one network operator threatens to *depeer* (stop exchanging traffic with) the other. Indeed, in the 2005 dispute Cogent agreed to a compensation arrangement only after Level 3 briefly depeered it.[30] In the worst-case scenario, abrupt depeering between two high-level backbone networks can leave one set of Internet users temporarily unable to communicate with another, as illustrated most recently by another brief impasse between Cogent and Sprint in 2008.[31] The practical consequences of such brinkmanship are important but complex, and we return to them and their policy dimensions in chapter 7.

Whereas peering may or may not involve an exchange of money, transit always does. The difference between the two interconnection arrangements is subtle. In a peering relationship, each peer hands off data traffic that is exchanged *only with the other peer's customers* (either third-party networks or end users). The defining attribute of transit relationships is that they are not so limited. As Michael Kende explained in a widely cited FCC white paper in 2000:

In [the figure here], backbone A is a transit customer of backbone C; thus, the customers of backbone A have access both to the customers of backbone C as well as to the customers of all peering partners of backbone C, such as backbone B. If backbone A and backbone C were peering partners, ... backbone C would not accept traffic from backbone A that was destined for backbone B.[32]

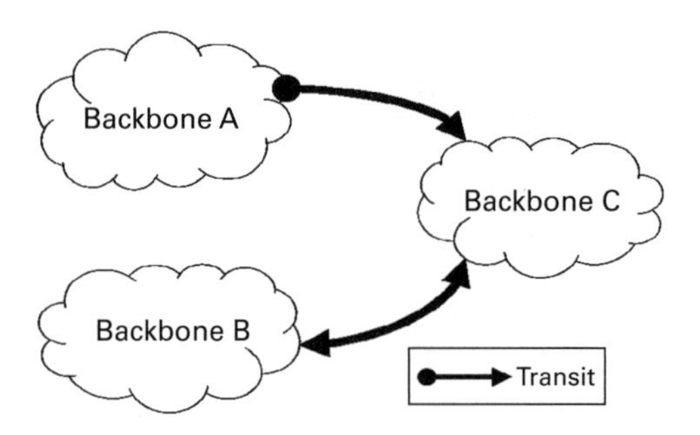

Of course, if A does not wish to pay C for transit, it would have other options for connecting its customers to B's customers. For example, it can negotiate for peering privileges directly with B, or it can hire any number of other transit providers as alternatives to C.

A concrete example helps illustrate how these relationships work in practice. Suppose that you are a small ISP that offers broadband Internet access to rural communities. You cannot expect to establish peering relationships with each of the innumerable IP networks worldwide that collectively constitute the Internet. Instead, you will likely buy transit services from a large global backbone provider such as Level 3. Level 3 will interconnect with your network and act as an intermediary between you and (1) all of Level 3's customers (both end users and transit customers) and (2) all of Level 3's peers, which in turn will connect your customers with *their* transit customers via the peering link with Level 3. The very largest backbone providers, known as *Tier 1* networks, have global networks of such commercial importance and comprehensive reach that they never need to purchase transit services from other backbones. As of this writing, the U.S.-based Tier 1 networks are AT&T, Verizon, Sprint, CenturyLink (formerly Qwest and Savvis), and Level 3 (which recently acquired fellow Tier 1 operator Global Crossing). They compete with foreign-based Tier 1 networks such as NTT, Deutsche Telekom, and Tata.

Since the privatization of the Internet in the early 1990s, all these peering and transit arrangements have been free from any type of prescriptive regulation. Neither the FCC nor any other U.S. authority limits the prices that a larger backbone network may charge a smaller one for transit services or mandates that backbone providers interconnect at all.* To date, the federal government's intervention in this marketplace has been confined to review of proposed mergers. For example, when WorldCom acquired MCI in 1998, the Justice Department and the European Union compelled the two companies, as a condition of their merger, to divest MCI's backbone affiliate—which, at the time, was second only to WorldCom's UUNet in market share. If not for this divestiture, the combined company would have controlled 50% of the total Internet backbone market, presenting a risk "that it would attempt to tip the market by charging existing peers for interconnection or by degrading the quality of interconnections."[33] For similar reasons, together with concerns about excessive concentration in the voice long-distance market, the Justice Department and the European Union effectively blocked the proposed merger of MCI–WorldCom and Sprint in 2000.[34] On occasion, the FCC has used its own merger-review authority to impose temporary constraints on the ability of newly merged companies to reduce their number of settlement-free peering arrangements.[35] Otherwise, however, it has avoided regulating the peering and transit marketplace altogether.

By most accounts, transit services are highly competitive today. One reason is that no one backbone provider of such services has ever grown large enough to dominate its rivals. Another equally important reason is that conventional backbone providers now compete not only with one another, but also with alternative mechanisms for getting data from Point

* As discussed in chapter 7, some foreign governments have urged the International Telecommunication Union, an agency of the United Nations, to create a regime in which individual nations could regulate international peering and transit arrangements. The U.S. government and most of the U.S.-based Internet ecosystem oppose such initiatives. As they observe, such foreign intervention would fundamentally alter Internet economics, creating regulated money flows from networks carrying (largely U.S.-based) content to foreign networks receiving that content.

A to Point B on the Internet. Those alternative mechanisms include the CDNs that we discuss next.

CDNs and the disintermediation of Internet backbone networks
In the mid-1990s, Internet data transport was far more hierarchical than it is today, and backbones were thus much more central to Internet data transport.[36] If you owned a website in 1995 and wished to send a webpage to an end user subscribing to a distant ISP, your ISP would typically route your data *up* to whatever backbone provider your ISP bought transit services from; that backbone provider would then route the data *across* to the (usually separate) backbone provider serving the end user's ISP; and the data would then proceed *down* through that ISP en route to the end user. Often this hierarchical routing would redundantly take place *each time* any given end user from the same ISP wished to download your webpage.

Today, instead of relying on backbone networks to carry your website's data across the Internet each time someone asks for it, you can hire (or build) a *content delivery network* to accomplish the same task much more efficiently. CDNs operate by arranging for the transport of data content to cache servers dispersed throughout the Internet, close to end users in many different locations. Each time an end user in one of those locations wishes to view your content, she generally communicates only with a nearby CDN server, not with your central database. And if the CDN you hire has interconnected directly with that end user's ISP, the localized communication from end user to cache server never needs to involve a backbone middleman at all.

Some of the largest Internet content providers (such as Google and Amazon) have built proprietary CDNs for their own use. Thus, when you click on Google News, you are communicating with one of Google's many cache servers, not with some centralized database in Mountain View, California (Google's headquarters). Many Internet content providers, however, obtain the efficiencies of this functionality by hiring third-party CDN providers such as Akamai and Limelight. Because these CDNs circumvent points of traffic congestion on the Internet, they offer not only low-cost alternatives to conventional backbone services, but higher-quality connectivity as well. For example, as Akamai explained to prospective customers in 2002, CDN services can make all the difference in how end users experience Internet content: "Let's assume someone has ten minutes to spend at your Web site: some are able to access 10+ pages, while some can't stand the wait and give up after two requests. If page speed were to be increased by as little as five times, these visitors would have the ability to view 50+ pages during the same short session, ensuring a better user experience—critical to your efforts to acquire and retain customers and partners."[37]

In addition to the rise of CDNs, other changes in the Internet's architecture have also contributed to the decline of conventional backbone providers as the central hubs and spokes of the Internet transport ecosystem. For example, through exchange points operated by Equinix and other private companies, many ISPs now peer directly with each other rather than hiring backbone providers to connect them via transit services.[38] And the price effects of such alternative interconnection arrangements are enormous. In part because Internet participants can often forgo conventional backbone services in favor of direct interconnection, the price of transit services has plummeted year after year since 1995.[39]

* * *

This chapter has covered much terrain, including the layered and modular nature of the Internet; the traditional end-to-end premise of Internet communications; the limitations that this premise imposes on performance-sensitive, real-time applications; the mechanics of peering and transit arrangements; the games of chicken that can arise from peering impasses; and the role of CDNs and other disruptive forces within the Internet's transport ecosystem. As we discuss next, these technological features take center stage in today's most profound telecommunications policy controversies: net neutrality (chapter 6) and IP-to-IP interconnection (chapter 7).

6

Net Neutrality and the Regulation of Broadband Internet Access

Until fairly recently, all telecommunications services were joined at the hip with the particular facilities on which they were provided. When AT&T's Bell System was split up in 1984, for example, few people expected telephone networks to offer widespread video services, nor did they expect cable television networks to offer voice or data services. The Internet, however, upsets this established order by uncoupling particular *services* from the physical *networks* over which they have traditionally been provided. Every single form of content ever conveyed over any electronic communications system—voice (telephony), audio (radio), video (television), documents (faxes), and so forth—can be instantly converted into 1s and 0s and transmitted as the cargo in digital packets flying over the Internet, no matter what the underlying physical medium.

This phenomenon, *technological convergence*, is attributable to the Internet's openness and modularity—the way that it divides computer-enhanced communications into separate physical, logical, application, and content layers. Competition and product diversity have flourished on the Internet's higher layers because the universality of the Internet's core protocols frees applications and content providers (such as Pandora and Facebook) from any need to become ISPs (such as Comcast or Verizon).

Nothing guarantees, however, that market forces alone will indefinitely preserve the traditionally modular and layered nature of the Internet. The basic policy challenge is to determine whether, when, and how the government should intervene to prevent a dominant provider in any one layer of the Internet from acting in ways that stifle competition and innovation in adjacent layers with the effect of reducing overall consumer welfare. That concern, among others, motivates the modern *net neutrality* movement, which took its current form around 2003 and, in 2010, produced the FCC's *Open Internet Order*, which remains on review in the D.C. Circuit as this edition heads to the printer.[1]

Roughly speaking (and with key exceptions), the FCC's net neutrality rules restrict the ability of broadband ISPs, insofar as they provide "Internet access service," to treat IP packets differently on the basis of their content or to charge content providers for transmitting those packets to the ISPs' end users.* A world of complexity is embedded within that seemingly simple description, and our task in this chapter is to make that complexity somewhat understandable. As we discuss, the debate about net neutrality rules is ultimately far more complicated and far more interesting than is popularly understood.

This chapter is divided into three main sections. Section I discusses the historical pedigree of the modern net neutrality movement in the FCC's *Computer Inquiry* rules of the 1970s and 1980s and the "open-access" debate of the late 1990s and early 2000s. Section II explores the major issues within today's net neutrality debate and examines how the FCC addressed them in its *Open Internet Order.* Section III then turns to the legal dimensions of the debate, examining the FCC's disputed authority to adopt net neutrality rules and the still-simmering controversy about whether, to bolster that authority, the Commission can and should treat broadband ISPs as common carriers under Title II of the Communications Act.

I. The Historical Origins of the Net Neutrality Movement

A. The *Computer Inquiries*

The roots of the net neutrality debate can be traced to the *Computer Inquiries*, a series of orders the FCC adopted in the 1970s and 1980s to govern the relationship between traditional "common carriers" (telephone companies) and the emerging data-processing industry. For present purposes, the orders known collectively as *Computer I* now have only historical significance. The chief legacy of the *Computer Inquiries* derives instead from the set of orders called *Computer II*, finished in 1981, and the later set of orders known as *Computer III*, completed in the decade before the Telecommunications Act of 1996.[2]

* For simplicity, we use the term *Internet content providers* to encompass all providers of over-the-top Internet content and applications; the category thus includes applications providers such as Skype and BitTorrent as well as content providers such as Netflix.

In a nutshell, the rules adopted in *Computer II* and *Computer III* were designed mainly to control the physical-layer monopoly power that AT&T's Bell System and its progeny (see chapter 2) then exercised in providing the links over which distant computers could "talk" to each other. It was a more straightforward matter in that monopoly era than in today's broadband world to keep market dominance at the physical layer from threatening competition at the higher layers. For example, in the dial-up era, customers could "call" one of many ISPs over the conventional telephone network just as they could call another person. The telephone company could not tell you which ISP to call because it was expected to operate the network as a common carrier. Just as it could not keep you from calling your next-door neighbor, it could not keep you from calling the ISP of your choice.

In *Computer II*,[3] the FCC expanded on traditional common carrier concepts to foster the growth of independent providers of "enhanced" services, whose ranks included the forerunners of today's ISPs. It began by distinguishing between (1) *basic* services, defined to include a "pure transmission capability over a communications path that is virtually transparent in terms of its interaction with customer supplied information,"[4] and (2) *enhanced* services, defined to include "services, offered over common carrier transmission facilities used in interstate communications, which employ computer processing applications that act on the format, content, code, protocol or similar aspects of the subscriber's transmitted information; provide the subscriber additional, different, or restructured information; or involve subscriber interaction with stored information."[5]

The "basic" versus "enhanced" distinction, although technical and legalistic, is central to today's regulatory debates about the Internet. Roughly speaking, a carrier providing a basic service delivers voice or data signals to their intended destination without providing additional data-processing functions beyond those needed for internal network management. The category of basic services covers everything from ordinary voice telephone calls to a telephone company's lease of private lines to large business customers, even when these businesses use their own equipment to perform a variety of data-processing functions over those lines. A basic service offered to the public at large is a *common carriage* service. An enhanced service provider, in contrast, sells content or data-processing services to the public by means of underlying transmission facilities. Early examples included "dial-a-joke," voicemail providers, Lexis-Nexis, and Westlaw; recent ones include the full range of Internet applications and content. At the risk of some oversimplification, a basic service, from the end user's perspective, operates primarily at the physical layer, whereas an enhanced service involves significant provision or manipulation of data at the higher layers as well.

The 1996 Act codifies the distinction between basic and enhanced services with different but essentially synonymous terms, derived from the 1984 AT&T consent decree.[6] For all relevant purposes, the term *telecommunications* means a basic service; *telecommunications service* means a basic service offered at common carriage; and *information service* means an enhanced service.[7]

We examine these statutory definitions in detail toward the end of this chapter, but it is important to keep several points in mind from the outset. First, what Internet companies such as Google and Amazon provide their customers are generally viewed as "information services," not "telecommunications services." Second, whereas a "telecommunications service" is subject to traditional common carriage obligations under Title II of the Communications Act, an "information service" is *not* subject to those obligations. Information services are generally subject to regulation only under Title I, which gives the FCC some ill-defined residual jurisdiction over interstate communications. As discussed later, any major FCC action predicated on an exercise of Title I authority typically sparks multiyear litigation about whether the Commission has overstepped the bounds of that nebulous statutory authority. Third, the line between "telecommunications services" and "information services" is indistinct and often disputed, and the legal status of broadband Internet access services is a key case in point.

The origins of the government's deregulatory approach to information services go back to the *Computer Inquiries*. In *Computer II*, the FCC reaffirmed its policy of encouraging the growth of long-distance data-processing applications—the precursors of today's Internet—by shielding information service providers from common carriage regulation under Title II. In a similarly deregulatory vein, the Commission recognized that telephone companies—and specifically AT&T's then-integrated Bell System, with its prodigious resources—could play a valuable role in developing such applications. But the Commission took various steps to keep the telcos from leveraging their monopoly power in last-mile transmission to harm unaffiliated information service providers, which relied on the telcos' transmission services for delivery of their own offerings.

The FCC originally excluded telephone companies altogether from the market for enhanced services. In *Computer II*, however, it authorized telephone companies to enter that market subject to two critical conditions. First, it imposed a regime of *structural separation*, under which the largest telephone companies (the Bells and GTE) could provide enhanced services only through a formally separate corporate affiliate. Second, it directed each telephone company to separate out—unbundle—the raw transmission functions (such as high-speed circuits) underlying any information service from higher-layer enhancements; tariff those transmission functions as a stand-alone telecommunications service; purchase that service for its own use from that tariff; and sell the same telecommunications service on a nondiscriminatory basis to all unaffiliated information service providers that request it. This requirement is sometimes known as the *Computer II* "unbundling" rule.[8] Together, these *Computer II* requirements were intended to prevent the telephone company from discriminating in favor of or cross-subsidizing its own information service operations.

In the mid-1980s, in a series of orders known as *Computer III*, the FCC simultaneously relaxed some *Computer II* restrictions and expanded others. First, the *Computer III* rules eliminated the structural separation requirement and substituted a more flexible scheme of "nonstructural" safeguards against discrimination and cross-subsidization. The short life of the structural separation requirement reflected both the deregulatory climate of the mid-1980s and, more generally, an abiding ambivalence about how to balance the efficiencies of vertical integration against the potential dangers.[9] In the end, the FCC concluded that those efficiencies were great enough to allow telephone companies into the market for information services, such as voicemail, so long as various nonstructural safeguards provided some protection against monopoly abuses.[10]

Second, even as the FCC eliminated the structural separation requirements, in *Computer III* it retained the *Computer II* unbundling obligation and expanded on it, at least in theory. Specifically, the nonstructural safeguards imposed on the major carriers included a new set of affirmative nondiscrimination obligations: *comparably efficient interconnection* (CEI) and *open network architecture* (ONA). These obligations bear a distant family resemblance to and were a source of inspiration for the facilities-unbundling rules of the 1996 Act. Although the details of these new *Computer III* obligations generated much controversy at the time,[11] they had limited competitive significance even when they were adopted and have almost none today. Indeed, few people today can confidently explain what exactly these arcane rules ever required as an engineering matter. Perhaps the most enduring legacy of *Computer III* is the FCC's decision to maintain federal jurisdiction over enhanced services under Title I of the Communications Act and to preempt state regulation of most such services to ensure a deregulatory environment for the fledgling Internet industry.[12]

B. The "open-access" debate

Congress left the *Computer Inquiry* rules essentially untouched when it overhauled the Communications Act in 1996. The various nondiscrimination requirements of *Computer II* and *Computer III* thus continued to ensure that telephone companies could not leverage their then-clear dominance in the market for last-mile transmission services to preclude robust competition in the adjacent market for higher-layer Internet services. In 1996, moreover, there were essentially no residential broadband services. Residential consumers who used the Internet almost invariably got there by means of dial-up connections over common carrier telephone lines. Again, the telcos could not influence your choice of dial-up ISPs any more than they could influence your choice of dial-up friends. And because consumers had a choice among many dial-up ISPs, a typical ISP had strong incentives to permit full access to any application or content provider on the Internet; otherwise, its disappointed subscribers could and would choose one of many alternative ISPs. The Internet, in short, was exceptionally competitive—and exceptionally slow.

Then broadband services, led at first by cable companies, transformed everything. Not until the mid-1990s did anyone think that cable television pipes, originally designed to carry the same TV signals past everyone's house, could support the same range of two-way information services as telephone networks with their dedicated, customer-specific loops. But cable wasted no time in proving the skeptics wrong and secured an early head start in the race for the residential broadband market, leaving the telcos to play catch-up with their first-generation DSL services. Unlike the telcos, moreover, these cable companies never had any *Computer II* obligation to "unbundle" the transmission component of their broadband Internet access and sell it to unaffiliated ISPs. Instead, cable providers (such as Time Warner Cable) offered their customers Internet access through an affiliated ISP (such as Road Runner). For cable modem customers then and for most broadband customers today, the distinction between a broadband transmission provider and an ISP has little commercial significance.

By the turn of the millennium, many industry observers had shifted from skepticism about cable's technical capabilities to concern that the largely unregulated cable networks might replace common carrier telephone networks as bottleneck last-mile facilities for residential Internet access and in the process put dial-up ISPs out of business. This concern spurred an incongruous coalition of telephone companies, dial-up ISPs, and consumer advocates to propose regulations requiring cable systems to provide *open access* to unaffiliated ISPs. In a nutshell, these advocates of "open access" asked the government to ensure that consumers of cable modem service could choose among ISPs and enjoy the same sort of independent relationship with their chosen ISPs that consumers traditionally enjoyed in the dial-up context. They hoped that these unaffiliated ISPs would serve as logical-layer buffers between the Internet's physical layer, on the one hand, and its applications and content layers, on the other, keeping cable companies from trying to leverage their presumed dominance in the broadband transmission market to discriminate against unaffiliated providers of content and applications.[13]

The high-water mark of the open-access push came with two regulatory initiatives around the turn of the millennium. First, when a national cable company sought to acquire a cable franchise in Portland as part of a nationwide merger, local franchising authorities conditioned their approval on the company's agreement to live up to vaguely worded guarantees of "non-discriminatory access to the . . . cable modem platform."[14] The Ninth Circuit later invalidated this Portland initiative, reasoning that under the Communications Act any authority to impose open-access requirements lies with the FCC rather than with localities.[15] Second, when Time Warner merged with AOL in 2000, the FTC conditioned its merger clearance on Time Warner's agreement to give at least three nonaffiliated ISPs "access" to its cable broadband platform.[16]

These and similar open-access initiatives tended to be vague on the technical details governing how cable companies were expected to integrate nonaffiliated ISPs into the physical architecture of their broadband networks. From an engineering perspective, this was unexplored territory. It was far less straightforward to "open" a cable modem network to multiple independent ISPs than to continue applying the longstanding requirement of traditional common carrier telephone companies to accommodate multiple dial-up ISPs. These engineering complexities soon led key "openness" advocates to shift their emphasis from rights of physical access by unaffiliated *ISPs* to rights of nondiscriminatory treatment for higher-layer *applications and content providers*. The modern net neutrality movement was thus born.

The most enduring legacy of the open-access debate is the series of legal precedents it produced, which remain highly relevant to the ongoing argument over net neutrality. Open-access advocates generally argued that cable broadband services are (or include) "telecommunications services" subject to common carrier regulation under Title II of the Communications Act. This was a provocative claim. Unlike telcos, cable companies had always exercised extensive editorial control over the content transmitted over their networks (chiefly one-way cable programming), and the prospect of common carrier regulation was alien and disagreeable to them. From 1999 to 2005, therefore, the two sides in the open-access debate waged a legal battle about how cable broadband service should be "characterized" under the Act: as a "cable service" subject to regulation under Title VI, as a "telecommunications service" subject to regulation under Title II, or as an "information service" subject to (minimal) regulation under Title I.

Of course, the Communications Act is the FCC's organic statute, and one might have expected the FCC to weigh in promptly on this legal controversy. For several years, however, the FCC remained mute because, given the perceived enormity of the stakes, opposing political pressures kept it in a state of near paralysis. As a result, the Commission stood by and watched as the courts took the lead in defining national telecommunications policy. Indeed, the Ninth Circuit began its opinion in the *Portland* case by noting that the "FCC has declined, both in its regulatory capacity and as amicus curiae, to address the issue" before the court.[17] The Court thus proceeded to the merits of that issue, concluding that cable modem service was neither a Title VI cable service nor a pure Title I information service and that it instead included a "telecommunications service" subject to federal-level common carrier regulation.[18]

In 2002, the FCC itself finally addressed that issue in its long-awaited *Cable Broadband Order*,[19] rejecting any open-access regime for cable broadband service and in the process concluding that the service fell outside of Title II (and Title VI) and within the presumptively unregulated sphere of Title I. The Commission began by finding that cable broadband service at least *includes* a Title I information service "regardless of whether subscribers use all of the functions provided as part of the service, such as e-mail or web-hosting, and regardless of whether every . . . service provider offers each function that could be included in the service."[20] The FCC next reaffirmed prior suggestions that the categories "information service" and "telecommunications service" are, by statutory definition, "mutually exclusive."[21] The term "telecommunications service," it observed, is defined as "the offering of telecommunications for a fee directly to the public . . . regardless of the facilities used," and "telecommunications" in turn is defined as "the transmission . . . of information of the user's choosing, without change in the form or content of the information as sent and received."[22] The Commission reasoned that when a broadband ISP offers an "information service," it cannot simultaneously offer "transmission . . . *without* change in the form or content of the information as sent and received" because such changes are the essence of an information service.[23]

The upshot, which persists (controversially) to this day, is that cable broadband services are unitary "information services" without any "telecommunications service" component. In plain English, this means that they are categorically beyond the reach of Title II common carrier rules and are, indeed, completely unregulated unless the Commission affirmatively acts to regulate them under its residual Title I authority. Moreover, courts tend to take a narrow view of that Title I authority, as discussed later in this chapter. Thus, every effort by the FCC to regulate these services, through net neutrality rules or otherwise, has triggered a court challenge and given rise to legal uncertainty.

One further aspect of the *Cable Broadband Order* warrants attention here because had the FCC come out the other way on that issue, its "statutory characterization" of cable broadband service would have assumed merely academic significance for the open-access debate. In particular, the FCC rejected proposals to apply to cable companies the same *Computer II* unbundling rule it had long applied to telcos: that is, an obligation to strip out the transmission component of this information service and offer it on a common carriage basis to all would-be purchasers, such as unaffiliated ISPs. As it explained, "for more than twenty years, *Computer II* obligations have been applied exclusively to traditional wireline services and facilities"—that is, to telephone companies, not to cable companies.[24] The Commission declined to extend the requirement beyond that industry context or, more generally, "to find a telecommunications service inside every information service, extract it, and make it a stand-alone offering to be regulated under Title II of the Act."[25]

The *Cable Broadband Order* was promptly challenged and temporarily nullified. In 2003, the Ninth Circuit, relying on its earlier decision in the Portland case, rejected the FCC's unitary "information service" characterization of cable broadband service and concluded that such services are properly characterized as simultaneously a Title I information service *and* a Title II telecommunications service.[26] But in 2005, in its landmark *Brand X* decision, the Supreme Court reversed the Ninth Circuit and, by a vote of six to three, reinstated the Commission's unitary "information service" classification of cable broadband service.[27] The majority reasoned that the statutory definitions of "telecommunications service" and "information service" are ambiguous; that the Ninth Circuit should have deferred to any reasonable FCC resolution of such ambiguity; and that the FCC's statutory construction was in fact reasonable. This decision was not quite the end of the story, though: in 2010, the FCC came very close to reversing the very statutory interpretation the Supreme Court had upheld. We return to that controversy in the final section of this chapter.

In the meantime, the *Brand X* decision settled matters for cable broadband providers, but not for the more heavily regulated telcos. In a 2002 notice of proposed rulemaking, the FCC had already suggested that whenever a telco bundled Internet access together with broadband transmission (such as DSL), the bundled product was an "information service" without a "telecommunications service" component—just like cable broadband service.[28] Even under such a regime, however, the telcos, unlike the cable companies, would remain subject to the *Computer Inquiry* rules unless the Commission took the extra step of repealing those rules. Thus, when the telcos offered broadband Internet access to consumers, they were required to offer the underlying transmission component (such as DSL transmission) at tariff to independent ISPs.

In its *Wireline Broadband Order* of 2005, the Commission followed through on its *Brand X* victory by eliminating this regulatory asymmetry and extending to wireline telcos the same deregulatory policies it had applied to cable broadband service from the beginning. In particular, it concluded that the broadband marketplace was sufficiently competitive and dynamic that continued application of common carrier regulation to any broadband Internet access providers, including traditional telephone companies, would serve no purpose beyond the destruction of healthy investment incentives.[29] On that basis, the Commission categorically exempted wireline telcos from the *Computer Inquiry* rules when they provide broadband Internet access services.

II. The Net Neutrality Debate

By the time the Supreme Court decided *Brand X* in 2005, the open-access debate had become antiquated.[30] That debate focused on the rights of independent ISPs such as AOL and Earthlink, who were unaffiliated with last-mile transmission providers. It had become clear by the early 2000s, however, that broadband technology makes such independent ISPs much less central to the consumer market for Internet access. Unlike dial-up customers in the 1990s, who subscribed separately to telcos and dial-up ISPs, a broadband subscriber today essentially equates her last-mile transmission provider (typically a telco, cable company, or mobile provider) with her ISP. And even if competition among non-facilities-based ISPs—the goal of open-access mandates—were meaningful in a broadband environment, such competition would not have resolved the core concerns of open-access advocates about the broadband practices of cable companies and other non–common carriers. As Tim Wu pointed out in 2003, such competition would "not necessarily mean that broadband operators will simply retreat to acting as passive carriers in the last mile,"[31] allowing uncompromised access to the full range of Internet content and applications.

In the first years of the new millennium, therefore, Wu and Lawrence Lessig, among others, sought to address the same underlying concern by developing a new regulatory proposal that they dubbed "net neutrality."[32] Whereas open-access proposals would have granted independent *ISPs* such as Earthlink and NetZero access rights to broadband transmission platforms, net neutrality rules grant such rights to *applications and content providers* such as Netflix, Skype, and BitTorrent.

These early advocates based the net neutrality concept loosely on the *end-to-end* design principle formulated in a famous 1984 white paper by Internet pioneers Jerome Saltzer, David Clark, and David Reed.[33] The authors of that paper originally described that principle in engineering rather than policy terms. As they explained, error correction and similar tasks required for most forms of inter-network data exchanges are generally more efficient if performed by devices at the network "edge" rather than by routers in the network core. Twenty years later net neutrality proponents developed that engineering design principle into a normative policy judgment that the Internet's constituent IP networks should remain "dumb" in the sense that they should not "know" what content IP packets contain, should treat every packet exactly the same, and should leave all content-aware "intelligence" to end users at the network edge.

Although populist advocacy sometimes phrases the net neutrality principle in such absolute terms, no one familiar with engineering realities would seriously argue that all IP networks should always be oblivious to the types of content contained in IP packets; some content requires special handling to function properly and some does not. The problem is that absolutist rhetoric often eclipses nuance in this debate. Our objective in sections II.A and II.B is to penetrate the rhetoric and explain as concretely as possible what people are arguing about when they argue about net neutrality. In the process, we discuss how the FCC addressed or (in some cases) ignored the key elements of this debate in its 2010 *Open Internet Order*. In subsection II.C, we then turn to the core economic elements of the net neutrality debate, assessing both the purported need for such rules and the purported costs of imposing them.

At the highest level of generality, net neutrality advocates support—and, with important exceptions, the *Open Internet Order* adopts—two different types of substantive rules: (1) a ban on "blocking" or "degrading" lawful content over an Internet access platform and (2) a ban on, or at least close regulation of, contractual deals between broadband networks and Internet content providers for favored treatment over that platform. These two types of requirements are analytically distinct, although they are often blurred together, and we address them in subsections II.A and II.B, respectively.

Before we address those core substantive rules, however, we first note the *Open Internet Order*'s third main rule: transparency. The *Order* requires all broadband ISPs to "publicly disclose accurate information regarding the network management practices, performance, and commercial terms of its broadband Internet access services," except that they need not disclose "competitively sensitive information or information that would compromise network security or undermine the efficacy of reasonable network management practices."[34] Although the implementation details can be contentious, this transparency requirement is far less controversial than the *Order*'s rules restricting actual network conduct. As one of us has written, moreover, this transparency norm is as fundamental to the protection of consumer interests as it is nonintrusive: while avoiding unnecessary constraints on experimentation, it enables consumers to make informed choices among competing services.[35] As Justice Brandeis observed, "sunlight is said to be the best of disinfectants,"[36] and that is as true of broadband network practices as any other practice.

A. The anti-blocking rule and "reasonable network management"

The Madison River and Comcast–BitTorrent controversies

In February 2004, FCC Chairman Michael Powell became the first major federal policymaker to embrace the anti-blocking principle when he "challenge[d] the broadband network industry" to honor several "Internet Freedoms" for consumers. These "Freedoms" included "access to [consumers'] choice of legal content," subject to "reasonable limits . . . placed in service contracts," and a right "to run applications of their choice," except where doing so "exceed[s] service plan limitations or harm[s] the provider's network."[37]

The next year, after Powell had left the FCC, the Commission followed his lead by issuing, in conjunction with the *Wireline Broadband Order*, a theoretically nonbinding *Policy Statement* that embraced Powell's "Internet Freedoms" in substance.[38] The *Policy Statement* provided, among other things, that consumers are "entitled to run applications and use services of their choice," such as VoIP or video, "subject to reasonable network management" and "the needs of law enforcement."[39] At the time, the only documented violation of these principles had occurred in 2005, when a small rural telephone company named Madison River Communications blocked its subscribers' access to VoIP services. Some observers alleged, and the FCC apparently concluded, that Madison River had blocked these services not for any legitimate network-management purpose, but simply to protect the lucrative access charges it earned for handling long-distance calls over the conventional telephone network. Madison River quickly suppressed the ensuing controversy by paying a small fine and pledging to stop this practice.[40]

Although the FCC stressed in its *Policy Statement* that it was "not adopting rules," it soon forced two of the nation's largest broadband providers—SBC (now AT&T Inc.) and Verizon—to accept the *Statement's* principles as binding (though temporary) conditions on the Commission's approval of their pending mergers with, respectively, AT&T Corp. and MCI.[41] For the ensuing two years, however, the debate about whether the FCC should convert its anti-blocking "principles" into industry-wide rules remained quiescent. Although the FCC opened a new proceeding on the topic,[42] it showed little inclination to adopt any formal net neutrality rules, perhaps because most broadband providers accepted the anti-blocking principle in the abstract, at least to the extent it applied to fixed (rather than mobile) broadband networks.[43]

That period of regulatory quiescence ended when in late 2007 independent tests suggested that cable operator Comcast had manipulated Internet packet headers to suppress its customers' use of BitTorrent, a peer-to-peer file-sharing application.[44] The ensuing controversy vaulted the anti-blocking principle once more to the forefront of the FCC's policy agenda. In its August 2008 *Comcast Order*, the FCC asserted Title I jurisdiction over this controversy; condemned Comcast for degrading "disfavored" applications; announced that such applications-specific degradation would be unlawful unless it "further[s] a critically important interest and [is] narrowly or carefully tailored to serve that interest";

found that Comcast's type of "network management" could not survive such scrutiny; and concluded that Comcast had thereby violated the Commission's *Policy Statement* in particular and the purposes of the Communications Act in general.[45] As discussed later in this chapter, the D.C. Circuit vacated the *Comcast Order* in 2010 on the ground that the FCC had articulated no valid theory of Title I authority that could justify imposing those requirements on broadband Internet access services, and so the FCC began looking for alternative legal rationales for such requirements.[46]

The Open Internet Order *and the differential treatment of fixed and mobile broadband platforms*

In late 2010, following the reversal of the *Comcast Order*, the FCC issued its *Open Internet Order*, which invoked a different Title I theory to impose a comprehensive net neutrality regime, including the basic anti-blocking restriction endorsed in the *Comcast Order.** In so doing, it instituted an anti-blocking rule that applies differently to fixed providers and mobile providers. Fixed broadband ISPs "shall not block lawful content, applications, services, or non-harmful devices" and may not "impair[] or degrad[e] particular content, applications, services, or non-harmful devices so as to render them effectively unusable"—except that they *may* engage in "reasonable network management," an elusive concept we will examine shortly.[47]

In contrast, a *mobile* broadband ISP is subject to a more lenient no-blocking rule. Like its fixed counterparts, it "shall not block consumers from accessing lawful *websites*," nor shall it "degrad[e]" such a website to the point of unusability, except that it, too, may engage in "reasonable network management."[48] But the Commission radically narrowed this no-blocking/degradation rule when it turned from websites to mobile applications. It prohibited mobile ISPs only from blocking or severely degrading "applications that *compete with the provider's voice or video telephony services*," subject again to the "reasonable network management" caveat.[49] It explained the far narrower protection for mobile apps

* The regulations adopted in the *Open Internet Order* apply only to "mass-market" services. The FCC defined that term to include services "marketed and sold on a standardized basis to residential customers, small businesses, and other end-user customers such as schools and libraries," but not "enterprise service offerings, which are typically offered to larger organizations through customized or individually negotiated arrangements." Report & Order, *Preserving the Open Internet*, 25 FCC Rcd 17905, ¶ 45 (2010) ("*Open Internet Order*").

on the ground that "accessing a website typically does not present the same network management issues that downloading and running an app on a [mobile] device may present."[50]

More generally, the FCC cited several rationales for subjecting mobile broadband providers such as Sprint and Verizon Wireless to far narrower net neutrality restrictions than fixed broadband ISPs such as cable operators and wireline telcos. It found that "[m]obile broadband is an earlier-stage platform" that "is rapidly evolving" and is thus more difficult to control with prescriptive rules; that "most consumers have more choices for mobile broadband than for fixed"; that mobile broadband networks face greater "operational constraints" because of spectrum shortages; and that they thus have a stronger need to engage in creative network management than their fixed counterparts.[51] These considerations led the FCC not only to apply the anti-blocking rule more leniently to mobile providers, but also to exempt them altogether from the FCC's new "nondiscrimination" rule, discussed in subsection II.B.

The concept of "reasonable network management"

As noted, all of these prohibitions on blocking or degradation are subject to the overarching prerogative of any broadband ISP (fixed or mobile) to engage in "reasonable network management." What is that, and how do we know it when we see it? The FCC explained that "[a] network management practice is reasonable if it is appropriate and tailored to achieving a legitimate network management purpose, taking into account the particular network architecture and technology of the broadband Internet access service."[52] Such practices, the FCC added, can include not only "ensuring network security and integrity," but also "reducing or mitigating the effects of congestion on the network."[53] In a passage that upset advocates of a more aggressive approach, the Commission rejected the "strict scrutiny" standard it had announced in the *Comcast Order*. It found instead that network practices, to be "reasonable," need only be "appropriate and tailored to the network management purpose they seek to achieve, but they need not necessarily employ the most narrowly tailored practice theoretically available to them."[54] Given "[t]he novelty of Internet access and traffic management questions, the complex nature of the Internet, and a general policy of restraint in setting policy for Internet access service providers," the FCC also decided that it would adopt principles of reasonable network management only "on a case-by-case basis, as complaints about broadband providers' actual practices arise."[55]

No matter what the precise legal formulation, regulatory adjudication of "network management" disputes will be an amorphous exercise that can place regulators in the unenviable position of second-guessing the technological choices of network engineers. Most IP networks are efficiently designed to handle reasonably foreseeable traffic loads over some defined time horizon. To conserve on costs, however, engineers do not typically "gold plate" best-effort residential broadband networks to accommodate moments of extreme congestion.* As a result, virtually all such networks contain potential bottlenecks of shared capacity. During peak usage periods, which can fluctuate widely in timing and degree, congestion in these bottlenecks can degrade basic Internet access for all subscribers. Such congestion poses a challenge for network engineers, who must cope with the explosive popularity of high-bandwidth Internet applications such as high-definition video streaming and peer-to-peer video file sharing, all while conserving on costly capital investments.[56] Network engineers will sometimes try to maximize bandwidth for the greatest number of users by manipulating how a network accommodates different users and uses.

The FCC's oversight of those decisions will inevitably involve highly fact-specific inquiries into the case-by-case "reasonableness" of particular network-management practices from an engineering perspective. Because the subject matter of these inquiries is esoteric and technologically dynamic, one of us has proposed that the FCC rely in part on *self-*

* Gerald Faulhaber and David Farber (former chief economist and chief technologist of the FCC, respectively) explain:

One possible solution to the network management issue is that ISPs should simply expand capacity, so that congestion never occurs. . . . Is that a reasonable option? If demand for Internet traffic capacity were relatively level, and the variance of traffic were low, then this might well be an attractive option, as the amount of capacity required to avoid congestion altogether might be some small multiple (say 1.5) of average demand. But the reality is that Internet traffic varies by time of day and is highly variable, or "bursty." Installing capacity sufficient to carry all demand all the time could well involve providing capacity dozens of times larger than average demand, with a concomitant increase in costs to customers to pay for capacity that sits idle for all but an hour a year. It is the nature of stochastic "bursty" traffic that peak demand will be much larger than average demand, so providing for the peak would be very expensive, and certainly against good engineering and economic principles. "Just add capacity" is a recipe for a very expensive Internet, primarily because of the bursty nature of Internet traffic. (Gerald R. Faulhaber & David J. Farber, *The Open Internet: A Customer-Centric Framework*, 4 Int'l J. of Communication 302, 324 (2010) (footnote omitted).)

regulatory organizations (SROs) to provide the engineering community's perspective on disputed network-management practices.[57] Although the FCC has taken no formal steps in that direction, it has welcomed the role played by one new SRO in particular—the Broadband Internet Technical Advisory Group (BITAG). Conceived in the months leading up to the *Open Internet Order* and composed of a broad cross-section of Internet stakeholders, BITAG describes its mission as "bringing transparency and clarity to network management processes as well as the interaction among networks, applications, devices and content."[58] But because BITAG has no official relationship with the FCC or with any other federal agency, the FCC itself will be left to make the hard judgment calls needed to decide what bandwidth-conserving practices are "reasonable."[59]

"Application agnosticism" and usage-sensitive pricing

You might wonder why broadband ISPs, instead of trying to suppress high-bandwidth *uses*, do not instead discourage inefficient bandwidth consumption by charging higher prices to their most bandwidth-consumptive *users*. That question brings us to an economic peculiarity about the retail market for Internet access in the United States. Although many *mobile* broadband plans come with finite buckets of data usage, most *fixed-line* broadband plans have traditionally included nearly "all you can eat" connectivity. Under all-you-can-eat plans, consumers pay a flat fee for a particular "speed" (data rate) but do not pay any incremental per bit price for causing extra data traffic to cross shared network facilities. For example, they pay the same for a 5 Mbps connection whether they use that connection once a day to download a static webpage or all day to download and upload high-definition video files. Traditionally, therefore, no price signals deterred a minority of subscribers from overconsuming network capacity at the expense of the majority.

The question in the *Comcast* controversy was whether, given the absence of retail price signals for data consumption, it was "reasonable" for a broadband ISP to treat the use of certain lawful applications (such as BitTorrent) as a proxy for undue consumption of finite and shared network resources and thus to limit the bandwidth consumed by those applications in order to ensure adequate network capacity for the majority of its subscribers. Net neutrality advocates argued and the Commission essentially agreed that such application-specific judgments are presumptively problematic. Indeed, the FCC pointedly suggested (over

Comcast's strong denials) that Comcast may have suppressed the use of BitTorrent not for any genuine engineering reason, but because Comcast wished to preclude the threat that this file-sharing application poses to its own lucrative cable television services, including video on demand.[60]

Similarly, in the *Open Internet Order*, the Commission suggested that, for the most part, network-management practices will be deemed reasonable only if they are "use- (or application-) agnostic."[61] Such application-agnostic practices might include throttling "the bandwidth available" to high-volume subscribers during periods of congestion to "ensure that heavy users do not crowd out others."[62] Or they might include retail pricing plans under which customers pay more if they consume more data. As the FCC explained, "prohibiting tiered or usage-based pricing and requiring all subscribers to pay the same amount for broadband service, regardless of the performance or usage of the service, would force lighter end users of the network to subsidize heavier end users."[63]

That finding was somewhat controversial. In early 2008, Time Warner Cable intended to become the first major U.S. broadband provider to implement a form of "metered pricing" and announced that it would offer, on a trial basis, a new tiered pricing scheme under which customers would pay a flat fee for a designated level of Internet traffic per month and usage-sensitive fees for all traffic beyond that level. But it promptly canceled this early experiment in the face of protests by the same advocacy groups that were pushing for net neutrality regulation. Some of those groups contended that as consumers had begun streaming more high-quality video content over the Internet, Time Warner Cable had adopted usage-sensitive Internet pricing to discourage consumers from canceling their conventional cable TV subscriptions.[64]

Many broadband Internet providers ultimately adopted a variation on the same theme, however, by charging consumers extra for exceeding defined monthly data usage limits or in some cases by imposing monthly data caps.[65] The main difference was that these standard tiers or caps were set at a much higher level than Time Warner Cable's original experimental tiers. As a result, only a small percentage of customers faced a realistic prospect of hitting the caps or incurring data overages. But these arrangements, despite their endorsement by the FCC, have reignited the debate about whether data tiers anticompetitively suppress over-the-top alternatives to broadband providers' own subscription video services. We address that debate later in this chapter.

B. Restrictions on favored treatment of particular content

So far, we have addressed only the net neutrality rules governing the *blocking* or *degradation* of particular applications and content. A theoretically more interesting net neutrality debate concerns whether regulators should restrict commercial deals providing for *superior* access to a broadband Internet access platform for certain applications and content.[66]

To take one concrete example, the provider of an Internet video-streaming service may wish to ensure quality of service for its customers by paying broadband Internet access providers to mark its IP packets for priority. Whenever those packets hit a broadband provider's router during periods of congestion, the provider would move them to the front of the queue for immediate transmission to the next router. The result would be lower latency and jitter and thus higher quality of service (see chapter 5). To take another example, the provider of an online videogame application might wish to pay broadband providers for whatever similar prioritization techniques might be needed to run graphics-intensive, real-time gaming applications involving the simultaneous participation of game participants across the globe. The policy question is whether the government should prohibit or tightly restrict such *priority-tiering* agreements or other forms of preferential treatment, as the FCC did in its *Open Internet Order* with respect to fixed (but not mobile) broadband Internet access services. This section summarizes the FCC's rules on that issue, and the following section turns to the underlying economic debate about the need for such rules.

Background

Two threshold points are needed to place the priority-tiering debate and the FCC's attempted resolution in their proper context. First, regulation of priority-tiering arrangements has nothing to do with whether broadband ISPs may impose *unilateral* charges on content providers for *best-effort* access to the ISPs' subscribers. The anti-blocking rule already forecloses such unilateral charges by prohibiting a fixed-line broadband ISP from blocking lawful Internet packets on the basis of which content provider is sending them. And without that greater power to block content, the ISP logically has no power to block (or threaten to block) those packets if a content provider does not pay involuntary "tolls" to avoid such blocking.[67]

Ever since Michael Powell's "Internet Freedoms" speech in 2004, moreover, there has been an overwhelming industry consensus that a fixed-line broadband ISP cannot validly threaten to block Internet content because a content provider refuses to pay for ordinary, best-effort access to end users (i.e., non-prioritized routing over last-mile broadband networks).* The academic literature sometimes disregards that consensus when it analyzes the merit of "zero-price rules" that keep broadband ISPs from financing broadband networks by charging both end users and (noncustomer) content providers for access.[68] In general, fixed broadband ISPs disavow any intent to charge a nonzero price to any content provider simply for transmitting its content to the ISPs' end users. The main question instead is whether these ISPs may enter into agreements to charge a limited subset of content providers for specialized treatment on that platform.

Second, as discussed in chapter 5, broadband providers increasingly *do* prioritize some IP-based applications over others on unified IP transmission platforms—and they often do so without controversy.[69] In particular, some broadband providers offer voice, video, and Internet services all over the same pipes in a unified stream of IP packets but mark the IP packets associated with "voice" and subscription "video" services for priority over the IP packets associated with regular "Internet" services. These so-called *managed* or *specialized* IP services have not traditionally raised net neutrality concerns because net neutrality has focused on how providers treat applications available through broadband Internet access, not how they treat applications *across* all classes of service provided over the same last-mile pipes by means of the Internet Protocol.

This point is subtle but critical, and we illustrate it by means of a concrete example. Suppose that you subscribe to a triple-play voice–

* One short-lived but notorious deviation from that consensus appeared in a *Businessweek* interview with Ed Whitacre, the CEO of Bell company SBC, just before it closed its acquisition of AT&T in 2005. Asked how he intended to address "Internet upstarts like Google, MSN, and Vonage," Whitacre responded: "How do you think they're going to get to customers? Through a broadband pipe. Cable companies have them. We have them. Now what they would like to do is use my pipes [for] free, but I ain't going to let them do that because we have spent this capital and we have to have a return on it. So there's going to have to be some mechanism for these people who use these pipes to pay for the portion they're using. . . . The Internet can't be free in that sense, because we and the cable companies have made an investment and for a Google or Yahoo! or Vonage or anybody to expect to use these pipes [for] free is nuts!" *Online Extra: At SBC, It's All about "Scale and Scope,"* Bloomberg Businessweek Magazine (Nov. 7, 2005), http://www.businessweek.com/stories/2005-11-06/online-extra-at-sbc-its-all-about-scale-and-scope. Of course, broadband providers have long recovered "returns" on their "pipes" from their own retail subscribers, and SBC's government-relations executives quietly disavowed Whitacre's remarks.

video–Internet service bundle from a broadband provider that has unified all three services within a single IP platform—as, for example, AT&T has done in its U-verse system.[70] Now suppose that you want to watch a live sports event in high definition as part of the "multichannel video" service in this bundle. The IP packets containing the images and sounds of that sports event will travel over the same pipes as the IP packets containing your Internet data. But because the "multichannel video" packets are subject to special handling from one end of the transmission to the other, you can almost always expect the same high quality of service as you would receive if your provider had transmitted that program over separate dedicated pipes.

In contrast, you could *not* expect such uniformly high quality of service if instead you tried to stream the same live sports event over a "best-effort" Internet connection, where no IP packets are prioritized over others. Instead, your experience would vary radically with the degree of network congestion. If the network is relatively congestion free, the real-time sports program you stream over the Internet will arrive smoothly. But if some portion of the Internet transmission path *is* congested, the non-prioritized packets associated with that program will have to wait in queue along with all other "Internet" packets at various routers before reaching you. The result will be latency and jitter—split-second delays and other transmission anomalies. These anomalies would generally be unproblematic for traditional Internet applications such as Web browsing or email, but they would be potentially quite disruptive for live streaming of a high-definition sports event. Repeated half-second delays are far less noticeable and problematic in bulk-file transfers than in coverage of the final two minutes of a football game.

In our example, a key difference between the live sports event transmitted as part of the "multichannel video" service and the live sports event transmitted "over the Internet" lies in whether the IP packets associated with each are marked for priority from origin to destination. But that distinction is unstable. Suppose that a triple-play broadband provider arranges to accept IP packets from a content provider at a predetermined hand-off point (such as a CDN) and marks the packets for prioritization from that point to the broadband provider's own end-user customers. Suppose now—and this is the twist—that the broadband provider's customers can obtain this same content by entering a publicly accessible Internet address into a Web browser and that if they do so from the broadband provider's network, the content will still be prioritized from end to end. Finally, suppose that these customers can obtain this content simply by ordering stand-alone "broadband Internet access" without ordering any separate "video" service. In that case, they will perceive that they are obtaining the content over "the Internet" even though it is functionally equivalent to "managed" IP video signals.*

Such arrangements would blur the distinction between "over-the-top Internet video" and "cable television." Some would welcome such blurring as the logical culmination of technological convergence—the elimination of artificial service distinctions based on regulatory rather than engineering distinctions. But net neutrality advocates worry that if broadband providers are permitted to blur these lines and start prioritizing some "Internet" packets over others, they "will destroy, once and for all, the egalitarian vision of the Internet."[71]

Before we dive into that debate, it is worth pausing to consider why end-to-end packet prioritization is not more widespread on the Internet already. Today, such prioritization is generally limited to "managed IP" traffic that remains on a single network. Examples include videoconferencing applications over a corporate IP network and (as discussed) subscription video on a residential broadband network. To date, end-to-end prioritization has been rare for IP packets on the so-called "public Internet"—that is, the system for exchanging IP packets across multiple *unaffiliated* IP networks from origin to destination. Why is this? One can at least imagine a world where each of the Internet's thousands of constituent IP networks honors *every other* network's prioritization decisions and thus prioritizes, on its own network, packets that other networks have marked for priority.

Such arrangements are uncommon mainly because of a basic collective-action problem. Consider Backbone Network X, which serves as an ISP to content providers (such as Netflix and Hulu), aggregating their content and handing it off to other backbone networks for delivery to individual end users. At the point of hand-off, Network X has strong incentives to present all (or at least an inefficiently high percentage) of its content-provider customers' packets as "high priority." Network X is paid by and loyal to its own content-provider customers, and it therefore has every reason to favor their packets over those of content providers

* There are additional variations on these technological themes: for example, in 2012 Comcast introduced an IP-based video-on-demand service (Xfinity Xbox) that is similar to our hypothetical service except that the content is not available via a publicly accessible website, and consumers can obtain it only by ordering Comcast's subscription cable service. We discuss the net neutrality debate surrounding that service later in this chapter.

that pay other backbone networks to serve as their ISPs. In economic terms, Network X does not *internalize the opportunity costs* of prioritizing its traffic over other networks' traffic when it hands that traffic off. Of course, each other network knows that this is true of Network X (and of all other interconnecting networks), so each adopts a default policy of treating all incoming traffic the same—and thus disregarding any priority markings.

In theory, IP networks might be able to overcome this collective-action problem by arranging to pay (in cash or in kind) to honor one another's priority markings, so long as they can work out the engineering details of precisely what that means in practice. In that event, each network *would* incur the opportunity costs of its prioritization decisions, and it would pass those costs through to the content-provider customers that want such end-to-end prioritization. Such arrangements might theoretically lead to efficient outcomes, in which content would be prioritized (or not) depending on consumer demand and the extent to which given content actually needs prioritized treatment in order to perform effectively.

For years, various players in the Internet community have tried to devise a universally recognized mechanism for such "QoS-aware" traffic exchanges across multiple networks.[72] Such initiatives remain very much a work in progress, and their prospects are anyone's guess. In the meantime, QoS-aware arrangements for end-to-end packet prioritization are more likely to appear in the form of multilateral agreements among given IP networks, such as enterprise-oriented ISPs that provide high-end videoconferencing services to corporate customers,[73] or between a CDN and a residential ISP. Of course, any inflexible regulatory ban on Internet packet prioritization could suppress such arrangements altogether, which brings us back to the net neutrality debate.

The FCC's "nondiscrimination" rule

In the years leading up to the *Open Internet Order*, the major proposals for regulating priority-tiering arrangements fell into two categories, which we call, respectively, the "flat ban" and the "common carrier approach." The flat ban would categorically prohibit broadband ISPs from entering into paid commercial contracts with Internet content providers and charging them for QoS-enhancing prioritization.[74] In contrast, the common carrier approach would permit a broadband ISP to enter into such contracts but would subject it to a core common carrier obligation when it does so: the ISP would be required to offer the same deal on the same contractual terms to *other* willing Internet content providers.[75]

The FCC's *Open Internet Order* addressed these issues by adopting what it called a "nondiscrimination" rule, which is separate from the no-blocking rule and applies only to fixed (not mobile) broadband ISPs serving mass-market customers. This new regulation provides (subject to the "reasonable network management" caveat) that fixed broadband providers "shall not unreasonably discriminate in transmitting lawful network traffic over a consumer's broadband Internet access service."[76] At first glance, this language suggests adoption of a common carrier approach rather than a flat ban. In other industries, common carriers have long offered faster or more reliable transport to "shippers" of goods (analogous to Internet content providers) who are willing to pay a premium, and those shippers then receive priority over other shippers who agree to pay less in exchange for less speedy and reliable service.[77] Such agreements comport with traditional common carriage principles so long as the carrier offers to sell the same higher-tier service to and does not "unreasonably discriminate" among all similarly situated parties that wish to buy that service.

Despite its use of "unreasonable discrimination" nomenclature, however, the FCC's rhetoric appears far closer to a flat ban on priority-tiering arrangements than a common carriage rule. Although the FCC avoided any categorical pronouncements, it cautioned that "a commercial arrangement between a broadband provider and a third party to directly or indirectly favor some traffic over other traffic in the broadband Internet access service connection to a subscriber of the broadband provider (*i.e.*, 'pay for priority') would raise significant cause for concern. . . . [A]s a general matter, it is unlikely that pay for priority would satisfy the 'no unreasonable discrimination' standard."[78] As this passage reveals, the FCC is using the term *discrimination* in a far looser, more vernacular sense than is typical in regulatory policy or economics. As used in the *Open Internet Order*, the term appears to mean "differential handling of IP packets" rather than "disparate treatment of similarly situated business partners."[79] And the *Order* suggests that such differential packet handling is likely to be "unreasonable" if it is the product of a paid commercial agreement between a broadband ISP and an Internet content provider. Later in this chapter, we analyze the FCC's disputed rationale for, and the potential unintended consequences of, this strong presumption against paid priority-tiering agreements.

The "specialized services" exception, IP video, and data tiers

The FCC stressed that its new "nondiscrimination" rule does not prohibit what it called *specialized services*—managed VoIP, subscription video, home monitoring, and any other *non*-"Internet" IP services whose packets are prioritized over the "Internet" packets delivered on unified IP transmission platforms.[80] The Commission explained that these managed services "provide end users valued services, supplementing the benefits of the open Internet," and "may drive additional private investment in broadband networks."[81]

At the same time, the Commission announced that it would "closely monitor" the development of new "specialized services" to keep this regulatory exception from swallowing the nondiscrimination rule. It expressed particular concern that broadband ISPs may "constrict or fail to continue expanding network capacity allocated to broadband Internet access service to provide more capacity for specialized services."[82] This concern manifests a larger fear expressed by advocates of net neutrality regulation: that if broadband ISPs may prioritize some content sources over others, they will artificially constrain bandwidth on their best-effort Internet platforms in order to force all content providers to "pay for priority." As discussed in subsection II.C, reasonable people can disagree about how strong this concern is. To the extent the concern is valid, however, one potential policy response is to scrutinize any actions by broadband ISPs that cause the quality of their best-effort Internet services to drop below industry norms.[83] In 2011, merger-review authorities adopted essentially that approach by requiring Comcast, as a condition for merging with NBC Universal, to offer a service that typically provides 12 Mbps in downstream bandwidth (subject to potential adjustments) wherever it has deployed DOCSIS 3.0 technology.[84]

The "specialized services" exception assumed new prominence when, in 2012, Netflix attacked Comcast's policy of exempting its own IP-based video-on-demand services from the monthly usage restrictions it applied to Internet services. At the time, Comcast's standard tier of Internet service entitled users to use up to 250 gigabytes (GB) of capacity per month. As Netflix had acknowledged, that was more than enough for the overwhelming majority of users,[85] but potentially not for a very small but growing percentage who stream unusually large volumes of over-the-top Internet video. In early 2012, Comcast announced that it would exempt from this 250 GB limitation any video programming that its users watch when they stream IP-based video on demand over Comcast's proprietary "Xfinity Xbox 360" app, which Comcast cable subscribers can select if they have an Xbox gaming console connected to their televisions and subscribe to Microsoft's Xbox Live service. No similar exemption applied to over-the-top services such as Netflix's.

Some technological background is necessary to understand the ensuing dispute. Like most conventional cable companies, Comcast traditionally assigned all of its subscription-video services to separate frequency channels within its cable distribution infrastructure using a technology known as *QAM* ("*quadrature amplitude modulation*"). Like AT&T in its U-verse system, Comcast was now finding ways to use "managed IP" to deliver its video services, and it began streaming these Xfinity Xbox packets along the same physical pipes that it used to transmit Internet packets. But Comcast was quick to point out that, unlike AT&T, it was not "prioritizing" the Xfinity Xbox packets over Internet packets, and it had configured its network to ensure at all times the same Internet bandwidth for any given customer whether the Xbox service was in use or not.[86] Nonetheless, Comcast *was* counting its customers' use of Netflix video-on-demand services toward the 250 GB data restriction, but not their use of Comcast's own competing Xfinity Xbox service.

Was this differential treatment a problem? Netflix and its supporters maintained that it constituted a serious threat to net neutrality principles in general and to over-the-top video competition in particular.[87] Netflix CEO Reed Hastings wrote on his public Facebook page:

> I spent the weekend enjoying four good internet video apps on my Xbox: Netflix, HBO GO, Xfinity, and Hulu. When I watch video on my Xbox from three of these four apps, it counts against my Comcast internet cap. When I watch through Comcast's Xfinity app, however, it does not count against my Comcast internet cap. For example, if I watch last night's SNL episode on my Xbox through the Hulu app, it eats up about one gigabyte of my cap, but if I watch that same episode through the Xfinity Xbox app, it doesn't use up my cap at all. The same device, the same IP address, the same wifi, the same internet connection, but totally different cap treatment. In what way is this neutral?[88]

As Netflix's general counsel argued to Congress a few months later, "When you couple limited broadband competition with a strong desire to protect a legacy video distribution business, you have both the means and motivation to engage in anticompetitive behavior."[89]

Comcast responded that its new initiative was both lawful and pro-consumer. The Xfinity Xbox app, it claimed, "essentially acts as an additional cable box for [its subscribers'] existing cable service," which is distinct from Comcast's Internet service both because the company charges separately for it and because it maintains control over the relevant packets from content source to destination.[90] Comcast claimed that, given the specialized services exception, treating this new non-Internet service differently from ordinary Internet content could hardly be unlawful under the *Open Internet Order.**

Comcast's supporters further maintained that it would make no policy sense for the government to rule for Netflix in this debate. As they observed, no one raised net neutrality objections when Comcast provided all of its video-on-demand services the old QAM way—over segregated channels, but also without any of the signals counting toward the 250 GB Internet usage threshold. They contended that if new nondiscrimination obligations are triggered whenever cable companies shift their video-on-demand signals from segregated QAM channels to the same IP platform used for Internet services, they will simply deter providers from making that shift in the first place. That outcome, they concluded, would benefit no one and would simply discourage the efficient evolution of traditional cable architecture toward more converged models similar to the one AT&T has developed through its triple-play U-verse platform.

This dispute reveals a deep fault line at the heart of the *Open Internet Order*. Over the long term, it will be efficient for all broadband companies to provide all of their services—both Internet access and specialized services such as subscription video—over unified IP platforms. And over the long term, the distinction between "broadband Internet access" and "specialized services" may blur. Does it make sense to create "neutrality" rules that apply to only one service category but not the other when services from one category compete so directly with services in another? If not, should policymakers abolish the "specialized services" exception or, less radically, require broadband providers to offer high enough data tiers and large enough allocations of Internet bandwidth to permit over-the-top video to compete meaningfully with the providers' own subscription video services? Should such requirements apply only to cable incumbents such as Comcast or to all triple-play broadband providers, including new video entrants such as AT&T? And if policymakers impose "neutrality" rules specific to the IP platform, how can they avoid giving

* As discussed in chapter 9, Comcast's adversaries in this dispute focused not only on the FCC's general net neutrality rules, but also—indeed, primarily—on the merger conditions that Comcast undertook when it merged with NBC Universal. We are focusing here, however, on the broader industry implications of this dispute for the FCC's net neutrality regime.

cable operators perverse incentives to preserve their legacy QAM architectures simply to avoid incurring new regulatory obligations? These questions will likely preoccupy policymakers for some time to come, as will several related questions prompted by the rise of over-the-top video services, to which we return in chapter 9.

We conclude our discussion of this issue with a critical caveat about the debate over specialized services: unless and until the FCC fundamentally alters its rules, that debate will relate only to *fixed-line* broadband services. As noted, the FCC imposed no nondiscrimination obligations at all on *mobile* broadband providers, and the specialized service "exception" to those obligations therefore never arises for those providers. Thus, if a mobile broadband provider applies its data tiers differently for some applications rather than others, that disparate treatment is presumptively permissible under the *Open Internet Order*.

Nondiscrimination and paid peering

In the *Open Internet Order*, the FCC did not analyze a potential conflict between its net neutrality rules and its traditional refusal to regulate Internet peering and transit arrangements (see chapter 5). On the one hand, the FCC announced a strong presumption against any "commercial arrangement" between a broadband ISP and Internet content provider "to directly or indirectly favor some traffic over other traffic in the broadband Internet access service connection to a subscriber of the broadband provider (*i.e.*, 'pay for priority')."[91] But it simultaneously asserted, in a one-sentence footnote, that "[w]e do not intend our rules to affect existing arrangements for network interconnection, including existing *paid-peering arrangements*."[92] These two passages are in deep tension.

As discussed in the previous chapter, paid peering is an arrangement in which one IP network pays another for the right to exchange traffic *directly* with it rather than indirectly via transit links. For example, a CDN might pay for the right to send traffic directly into an ISP's network, thereby providing more efficient and reliable connectivity between the CDN's content-provider customers and the end users served by that ISP. Such arrangements are voluntary and completely unregulated. If the CDN refuses to pay the ISP's price for direct interconnection, it must find an alternative, indirect way to deliver its packets to the ISP's customers. For example, it may hand off those packets to a third-party transit provider that independently interconnects with the ISP. The content will reach the ISP's end users, but perhaps less quickly and predictably than if the two networks had peered directly (because the associated packets may now have to make more hops from origin to destination).

You would be right to perceive a strong economic similarity between paid prioritization and paid peering. Like paid prioritization, paid peering helps ensure high-quality performance for particular content (as well as greater network efficiency). And both paid prioritization and paid peering involve payments by content providers or their agents (CDNs or backbone providers) to ISPs for efficient access to the ISPs' end users. It is therefore unclear how it makes any policy sense to prohibit one of these business arrangements but endorse the other. In both cases, either the ISP has some form of market power that warrants a regulatory response, or it does not. The tension here is particularly acute because there is not even a stable distinction between content providers (parties to "paid prioritization" arrangements) and CDNs (parties to "paid peering" arrangements). A content provider such as Netflix can outsource CDN functionality to third-party providers such as Level 3, or it can build (or buy) its own CDN, as Google and Microsoft have done. Indeed, Netflix recently announced that it is now deploying its own CDN, which will "eventually" serve "most of [its] data."[93] The upshot of the *Open Internet Order* is that when Netflix uses its own CDN, it can pay ISPs for superior access to their customers via direct-peering links but not via packet prioritization.*

This distinction makes little apparent policy sense, and the *Order* makes no attempt to justify it. Perhaps the FCC did not fully appreciate the deep relationship between its net neutrality rules and the complex but unregulated world of Internet peering and transit. Whatever the explanation, this yet unresolved tension in the *Open Internet Order* is likely to animate policy disputes for many years to come. Significantly, the *Order's* criticism of paid prioritization does not appear in the FCC's codified regulations, which provide merely that fixed broadband providers "shall not unreasonably discriminate in transmitting lawful

* To give the Commission the benefit of the doubt, we are assuming that the *Open Internet Order*'s presumptive opposition to paid arrangements that "directly or indirectly favor some traffic over other traffic" (¶ 73) refers only to paid prioritization in the most literal sense—the marking of packets for priority in individual routers within an ISP's last-mile network. Of course, if the FCC meant to encompass *all* paid arrangements that "favor" traffic, including those for direct interconnection, the *Order* would lapse into outright self-contradiction: it would both forbid and permit a CDN-equipped Netflix to enter into paid-peering arrangements with ISPs.

network traffic over a consumer's broadband Internet access service."[94] As a technical matter, the relevant passages in the *Order* simply *predict* how the FCC will rule on claims that paid prioritization constitutes "unreasonable discrimination," and the FCC retains discretion to conclude that such prioritization is often acceptable after all, at least so long as ISPs offer it on reasonably comparable terms to all similarly situated content providers who wish to buy it. It is also conceivable that the FCC will more closely scrutinize paid-peering arrangements in the years to come. In chapter 7, we consider that question and, more generally, whether the government should oversee Internet peering and transit arrangements.

C. The economic elements of the net neutrality debate

Commenters approach the net neutrality debate from many different perspectives. Our approach in this section is mainly economic: we analyze the debate through roughly the same analytical lens that antitrust theorists use to view monopoly-leveraging disputes in other industries whenever vertically integrated firms are said to leverage (or preserve) their monopoly in one market by thwarting competition in others. In general, antitrust evaluates such disputes by defining the relevant markets and then evaluating whether the firm in question is dominant in one of those markets, whether it has both the incentive and the ability to exploit (or protect) that dominance by harming competition in adjacent markets, whether the costs of any remedy outweigh the benefits, and so forth.[95]

When we say we are viewing the net neutrality debate through the lens of substantive antitrust policy, we are not saying that antitrust institutions are superior to administrative agencies as policymakers in this context.[96] One can support an antitrust-oriented economic perspective on the net neutrality debate without concluding that the debate should be consigned to the antitrust *process*.[97] For institutional reasons, antitrust courts are far more reluctant than administrative agencies to constrain unilateral conduct by firms with market power, and they have crafted various rules reflecting that policy of judicial restraint.[98] In contrast, administrative agencies can apply the same basic economic analysis and reach less conservative conclusions about what remedial steps might best promote consumer welfare on the facts of a given controversy. Indeed, the FCC routinely applies antitrust-oriented competition principles to justify forms of market intervention that antitrust enforcement agencies and antitrust courts would reject under current law.[99]

Framing the debate

Our economic perspective on the net neutrality debate is not the only one, and some participants in the discussion dispute that it is the right one. Advocates of aggressive net neutrality regulation do not generally analyze Internet-related vertical leveraging issues from the industry-agnostic perspective of traditional competition policy, or at least they would weigh the costs and benefits of market intervention very differently from the way traditional antitrust policy would weigh them.[100] Some of these advocates might favor strict net neutrality rules even if there were little risk that, in the absence of such rules, broadband platform operators would discriminate among Internet applications providers in *anticompetitive* ways that would harm overall economic welfare as measured by substantive antitrust analysis.

To these advocates, virtually any "content-aware" differentiation among Internet packets is problematic even where it is economically efficient. They believe that such differential treatment threatens the core attribute that makes the Internet a uniquely valuable global resource: the equal opportunity that the end-to-end principle promotes, at least in theory, for all fledgling innovators at the edge of the network. Skeptics of regulation respond that although the end-to-end ethic for best-effort Internet traffic is important, it is unthreatened by new business models that ensure greater performance for QoS-sensitive applications, and the government will do more harm than good if it forbids such experimentation. And they stress that if such business models do ultimately pose a threat to the Internet's structural integrity, policymakers can take appropriate action at that point. This fundamental difference in perspective helps to explain why the two sides of the net neutrality debate tend to talk past one another and why that debate often generates more heat than light.

Some advocates of aggressive net neutrality regulation also argue that the government should strictly enforce "the egalitarian vision of the Internet"[101] not only or even primarily to maximize consumer welfare, but mainly to protect First Amendment values—to ensure a robust marketplace of ideas by guaranteeing strictly equal access to the broadband platform by anyone with ideas to share.[102] Portions of the *Open Internet Order* tap into this strain of advocacy, suggesting that net neutrality rules are needed to preserve "the Internet's role as a platform for speech and civic engagement," "an informed electorate," "the health of a functioning democracy," and "the basis for informed civic discourse."[103]

It is important to place this free expression rationale in its proper context. The Internet has immensely enriched the marketplace of ideas because, among its many other benefits, it greatly reduces the costs to any given speaker of communicating her views to the world at large. But the Internet has never been an electronic town hall where everyone has guaranteed rights to speak as loudly as anyone else, and no one suggests that it should be. For example, as discussed in chapter 5, content powerhouses such as Google or Netflix that spend millions or billions on CDN functionality will vastly outperform underfunded websites or dorm-room bloggers because the powerhouses' packets will make fewer hops en route to content recipients and will short-circuit congested peering points.

Moreover, although "free expression" concerns might justify a narrow ban on blocking particular content sources, they cannot logically support the very broad forms of market intervention that they are often invoked to justify.* For example, if (as the FCC suggested) Comcast acted anticompetitively when it throttled BitTorrent peer-to-peer applications in 2007, its actions were still content neutral: Comcast did not "discriminate" against *viewpoints* at all, much less in ways that could directly threaten the "marketplace of ideas" (as opposed to economic efficiency). If network-management practices ever do give rise to a discernible free

* Basing regulation on free expression concerns is also difficult to square with First Amendment law, which the Supreme Court has generally invoked to restrict rather than expand government intervention in media contexts. In *Miami Herald Publishing Co. v. Tornillo*, the Supreme Court held that the First Amendment invalidated a state law that required newspapers to give political candidates an opportunity to reply to unfavorable editorials, reasoning that the marketplace of ideas will prosper best if the government does *not* act as a referee of "fair" access to privately owned means of public expression. 418 U.S. 241 (1974). The notable exception to this rule involved conventional television and radio broadcasting. In its controversial (and now highly suspect) *Red Lion* decision in 1969, the Supreme Court rejected a First Amendment challenge to the original fairness doctrine: a requirement that broadcasters give equal time to opposing viewpoints. *See* Red Lion Broad. Co. v. FCC, 395 U.S. 367 (1969). But the Court upheld that rule only because the broadcast spectrum, long considered a public resource, was viewed as so inherently "scarce" that the government *had to* grant limited rights of private access to it in order to ensure genuine public debate. No one would seriously argue that the Internet has any of the "scarcity" properties that underlay the *Red Lion* decision. Any Internet connection allows end users to reach millions of information sources worldwide, not the three or four broadcast television channels available locally when *Red Lion* was decided.

expression problem, there will presumably be time to formulate appropriately tailored solutions to it.

For the same reasons that a flat ban on last-mile packet discrimination would not ensure an absolutely egalitarian town hall, it also would not create a perfectly level playing field for all commercial innovators at the "edge" of the Internet. To begin with, a rigidly application-blind approach to packet handling on the Internet would itself be "discriminatory" in one sense: it would disfavor the providers of performance-sensitive applications. As Tim Wu explains, "the Internet's greatest deviation from network neutrality" has ironically consisted of its traditional "favoritism of data applications, as a class, over latency-sensitive applications involving voice or video."[104] But even if all applications were equally performance sensitive, "neutrality" would remain unattainable in an Internet environment long dominated by CDN-based and other network services designed to give content providers who pay for the privilege a competitive edge over their rivals.

These considerations cast doubt on the most absolutist advocacy for a flat prioritization ban, whether rooted in First Amendment ideals or solicitude for commercial innovators. For example, one Senate sponsor of proposed net neutrality regulation argued in 2006 that a flat ban on packet "discrimination" is needed to "allow[] folks to start small and dream big" because it would help out the "small mom and pop businesses that can't afford the priority lane" and who would otherwise have "no hope of competing against the Wal-Marts of the world."[105] But the government cannot ensure that "small mom and pop businesses" will reach consumers as effectively as large companies do unless it starts issuing vouchers for subsidized CDN services and otherwise regulates commercial arrangements far deeper into the Internet. Otherwise, start-ups will have to keep doing what the Internet ecosystem has demanded of them from the beginning: they will have to persuade the capital markets to finance their ambitions for improved performance, whether through CDN services or any other method.

This does not mean, of course, that the last-mile practices of broadband ISPs are unworthy of any regulatory concern. Net neutrality advocates claim that although CDNs require massive capital investments, the market for CDN services is inherently more competitive than the market for last-mile broadband services, less likely to generate market failures, and thus less likely to require regulatory intervention. Whether or not that is true, however, it is not an argument about abstract rights to free expression or to a completely level playing field for all commercial actors.

It is instead an *empirical* argument about market power and the potential for market failures. These are traditional antitrust concepts and should be analyzed as such.

Market power and residential broadband competition

Reduced to its essentials, most economic advocacy for net neutrality regulation begins with the argument that there is inadequate competition in the residential broadband marketplace and that the government should step in to prevent abuses of the resulting market power. If each American consumer had a choice of ten broadband Internet access providers, there would probably be no strong basis for such intervention because competition would presumably ensure each provider's responsiveness to consumer choice.[106] Instead, the root fear is that the provision of fixed Internet access to residential and small-business customers is a distinct product market and in essence a duopoly dominated by incumbent cable and telephone companies; that it will remain so indefinitely; and that each provider has an incentive to abuse its market power in ways that harm the Internet and consumer welfare.* Net neutrality advocates are particularly concerned about the risk that any given broadband provider, to the extent it vertically integrates broadband transmission with the provision of particular applications or content (such as voice or video), will leverage its power in the broadband market to discriminate anticompetitively against unaffiliated over-the-top providers of competing applications or content.[107]

In response, opponents of regulatory intervention claim that the retail Internet access market is more competitive and dynamic than net neutrality advocates contend and that the potential for further intermodal

* In the *Open Internet Order*, the FCC speculated in passing that even if the broadband marketplace were fully competitive, any given broadband ISP, no matter how small, could and would extract "inefficiently high fees" from applications and content providers if it were generally permitted to charge such providers for prioritized handling of packets bound for the ISP's end users. *Open Internet Order* ¶ 24. The Commission reasoned that once an end user has signed up for an ISP's service, the ISP "is typically an [Internet applications or content] provider's only option for reaching [that] end user," and the provider would thus have no choice but to accede to such fees. *Id.*; *see also id.* ¶¶ 27, 32. As discussed in the next chapter, this *terminating access monopoly* concern arises mainly in connection with the public switched telephone network (PSTN), where the FCC had encountered it previously, and it arises there for reasons peculiar to the PSTN, with its pervasive interconnection rules and tariffs. In contrast, the case for adopting prescriptive regulation of Internet traffic exchanges, let alone a flat ban on a whole class of commercial QoS agreements, has not been made.

competition keeps all providers in check. They also argue that because of the unique cost characteristics of fixed-line broadband services, competition among even a very small number of rivals may suffice to protect consumer interests as effectively as competition among larger numbers of rivals protects consumer interests in other markets. In particular, they reason that the high fixed costs and negligible marginal costs in the broadband market give providers unusual incentives to keep and recruit as many customers as possible—and thus to accommodate any significant consumer concerns—because each customer represents almost pure profit in that no costs are avoided if any customer defects to an alternative provider.[108]

As noted earlier in this chapter, the FCC had cited the potential for new and disruptive broadband entry as a key basis for deregulating cable broadband providers in 2002 and telco broadband providers in 2005. When it issued its *Open Internet Order* in 2010, however, it reversed course. It found that regulatory intervention (in the form of net neutrality rules) had become necessary in part because, just as in 2002, "most residential end users today have only one or two choices for wireline broadband Internet access service" and "[t]he risk of market power" and its associated inefficiencies "is highest in markets with few competitors."[109]

One interesting question in this debate is whether and to what extent *mobile* (or other wireless) broadband services will become close enough substitutes for *fixed-line* broadband services to curb any market power that fixed providers may otherwise exercise. In the *Open Internet Order*, the FCC essentially treated fixed and mobile broadband services as separate markets and, as discussed earlier, subjected the latter to less regulation than the former. As of this writing, mobile broadband is a partial but quite imperfect substitute for fixed-line broadband. Mobile broadband connections are less reliably robust than their fixed-line counterparts, and because mobile network capacity is far more constrained, mobile pricing plans are generally more expensive at the very high levels of usage to which consumers have become accustomed for fixed-line connections. That said, these two types of broadband services may become closer substitutes over time, depending on how mobile broadband networks develop and how capable they are of meeting consumers' service needs.

Vertical integration and the ICE principle

Skeptics of strict net neutrality rules argue that even if a broadband provider faced no competition, and even if it theoretically had the *ability* to harm competition in the content and applications markets (a core fear of net neutrality proponents), it would still often have only limited *incentives* to discriminate against unaffiliated providers of complementary applications and content in ways that would harm consumer welfare. This argument, which we first examined in chapter 1, is complex and warrants further elaboration here.

Since the emergence of the Chicago School of antitrust economics in the 1970s, antitrust law has taken a skeptical view of claims that vertically integrated, non-price-regulated firms will try to "leverage" their monopoly status in one market to harm competition in adjacent markets.[110] Even a broadband *monopolist*, for example, has powerful incentives to deal evenhandedly with unaffiliated content providers; after all, anticompetitive discrimination against unaffiliated content would devalue the broadband platform for end users. To be sure, there will be contexts where such discrimination would enable the platform provider to extract higher overall profits.[111] In other cases, however, Antoine Cournot's "one monopoly profit" observation will hold true (see chapter 1): monopolization of application markets would not enable the platform provider to earn overall profits greater than what it could otherwise earn simply by setting a profit-maximizing price for the underlying platform. And in those cases, the platform provider will have incentives to maintain an open and nondiscriminatory platform regardless of regulatory policy.

Where applicable, this principle, known as the *internalization of complementary efficiencies* (ICE), does *not* hold that broadband ISPs will never favor their own affiliated content and services over those of independent companies. For example, they may favor their own affiliates in order to capture the efficiencies that vertical integration permits[112] or to attract consumers through efficient product differentiation.[113] But the ICE principle (where it applies) *does* hold that platform providers will have no rational incentive to favor their affiliates in ways that distort efficient competition and harm consumers. And it is a broadly accepted axiom of U.S. competition policy that, with rare exceptions, economic regulation should be designed to promote *competition* in the interests of consumers rather than individual *competitors*.[114]

The ICE principle is nonetheless subject to a number of important exceptions: contexts in which vertical integration might give dominant firms incentives to discriminate in anticompetitive ways against unaffiliated Internet content or service providers.[115] Under one of these exceptions—Baxter's Law, discussed in chapter 1—a vertically integrated company that is subject to price ceilings on its platform services may well have incentives to thwart rival content providers in order to recover the monopoly profits that those price ceilings keep it from recovering in the platform market.[116] That is why the predivestiture, heavily price-regulated Bell System tried to undermine MCI and other unaffiliated carriers that were undercutting Bell's ability to continue extracting supra-competitive long-distance profits (see chapters 1 and 2). But no analogous concern arises in the broadband context, where retail broadband services are generally *not* subject to price regulation.

Another important exception to the ICE principle arises when a platform provider believes that an applications provider poses a competitive threat to the underlying platform. For example, Microsoft, as a monopoly provider of PC operating systems, may not normally have incentives to discriminate anticompetitively against unaffiliated applications software. But as the Justice Department successfully argued at the turn of the millennium, Microsoft did have and was found to have acted upon incentives to crush an applications provider (Netscape) that was thought to have threatened the market position of the Windows platform itself.[117]

In the Internet access context, an analogous question arises about whether broadband providers that face inadequate broadband competition might likewise have incentives to thwart applications (such as VoIP and streaming video) that threaten any service traditionally offered by a given broadband provider (voice for telcos and video services for cable companies). That is one reason why so much scrutiny greeted Madison River's treatment of VoIP services and Comcast's treatment of a peer-to-peer technology used for sharing large video files. However, this concern—anticompetitive discrimination by dominant broadband providers against Internet-based threats to legacy monopoly profits—can by itself justify only targeted governmental safeguards against such discrimination. It could *not* justify keeping, say, CenturyLink from agreeing to prioritize Netflix video traffic on CenturyLink's broadband network because that arrangement could not possibly enable CenturyLink to exclude platform rivals. To justify such a ban, therefore, the FCC would need to invoke and substantiate different and more broadly applicable rationales, such as the "manufactured scarcity" concern discussed next.

We do not wish to make too much of the ICE principle in the net neutrality context, however, because it can take the analysis only so far. The principle is subject to several additional exceptions beyond those we have discussed, and economists may disagree about whether the exceptions are more prevalent than the rule in some industries.

For example, the ICE principle may not deter a dominant platform provider from inefficiently discriminating against unaffiliated applications providers if its own entry into the applications market makes it easier to engage in profitable *price discrimination*. Price discrimination is the practice of selectively charging more to customers who are less likely than other customers to switch to competitive alternatives when a firm raises its prices; familiar examples include the airline industry's practice of charging more for tickets bought just before departure. For firms in industries with large fixed costs and low marginal costs, price discrimination is often an important strategy for recovering massive infrastructure investments. And the prospect of these extra revenues can help justify marginal investments and can thus foster greater broadband deployment. Nonetheless, "the platform monopolist's desire to price discriminate can outweigh ICE and lead it to exclude efficient innovation or price competition in complementary products. In the classic case, the monopolist does so more or less intentionally because control of the complementary market allows it to maximize profits through large markups on complementary goods . . . . Even where price discrimination itself *enhances* efficiency, the platform monopolist may impose highly inefficient restrictions on applications competition in order to engage in price discrimination."[118]

The ICE principle likewise may not deter a dominant firm from engaging in anticompetitive conduct if, for example, a content market is to some extent independent of the platform market and is itself subject to scale economies or network effects[119] or if the dominant firm is simply irrational and ignores the ICE principle's exhortation to deal evenhandedly with unaffiliated providers of complementary applications.[120] In sum, our essential but limited point about ICE is this: if broadband providers enjoy significant market power at the physical layer, they might, but do not inevitably, have incentives to discriminate against unaffiliated Internet content providers.

Concerns about "manufactured scarcity"

As discussed, the *Open Internet Order* announced a presumption against any "pay-for-priority" arrangement in which an Internet content provider compensates broadband ISPs to prioritize its packets over last-mile networks during periods of congestion. Supporters of this prohibition sometimes justify it on the ground that packet prioritization is a zero-sum game: simply as a logical matter, they say, a broadband ISP cannot prioritize some packets by advancing them to the front of a router queue without deprioritizing other packets by holding them an instant longer in that queue.

Although that argument may have some rhetorical appeal, it is technologically naive. Some Internet applications, such as real-time high-definition video, *need* QoS enhancements to function properly. But many other applications, such as bulk-file transfers and static webpage downloads, do not. It may often be possible to prioritize the packets associated with applications that do need such enhancements without, in any practical sense, degrading the performance of the applications that do not. As one engineering text explains, "in many multimedia applications, packets that incur a sender-to-receiver delay of more than a few hundred milliseconds are essentially useless to the receiver," and such "characteristics are clearly different from those of elastic applications such as the Web, e-mail, FTP, and Telnet," for which even "long delays" are "not particularly harmful."[121]

Quite apart from this zero-sum point, the FCC and net neutrality advocates also offer a more sophisticated "manufactured scarcity" rationale for banning paid-prioritization arrangements. They argue that if such arrangements become widespread, they will become highly profitable to broadband ISPs, and those ISPs will look for ways to force as many content providers as possible to pay for them. According to this argument, the appeal of these prioritization fees will give broadband ISPs strong incentives to underinvest in broadband capacity and thereby degrade the performance of all *non*-prioritized Internet content delivered over the best-effort transmission platform. That feared dynamic, in the words of Lawrence Lessig and Robert McChesney, would bifurcate Internet content providers into separate castes depending on their ability and willingness to pay and would enable ISPs "to sell access to the express lane to deep-pocketed corporations and relegate everyone else to the digital equivalent of a winding dirt road."[122]

In the *Open Internet Order*, the FCC essentially agreed with this rationale as a basis for aggressively regulating, if not categorically prohibiting, paid-prioritization arrangements involving fixed broadband ISPs. The FCC found that "if broadband providers can profitably charge edge providers for prioritized access to end users, they will have an incentive to degrade or decline to increase the quality of the service they provide to non-prioritized traffic. . . . [B]roadband providers might withhold or decline to expand capacity in order to 'squeeze' non-prioritized traffic, a strategy that would increase the likelihood of network congestion and confront edge providers with a choice between accepting low-quality transmission or paying fees for prioritizing access to end users."[123]

The FCC offered no empirical justification for this manufactured-scarcity rationale for presumptively banning paid-prioritization agreements, and that rationale is subject to debate. A broadband ISP that "withholds" capacity from its best-effort platform in order to "squeeze" providers of non-prioritized content would leave its customers dissatisfied whenever they seek access to such content. The sources of non-prioritized content would probably outnumber the sources of prioritized content, given the millions of Internet content sources in the world. And any ISP that artificially constrains bandwidth in order to "penalize" nonpaying content providers would thus risk losing customers to ISP rivals that do not degrade their best-effort platform and offer faster speeds for the vast bulk of Internet content. The risk incurred by the bandwidth-constraining ISP would be greatest in markets where broadband competition is strong and rivals compete explicitly on the basis of speed, as they often do today.

Also, to the extent that the "manufactured-scarcity" concern is valid and requires a regulatory response, policymakers can address it without imposing a flat ban on prioritization services. One of us has encouraged policymakers to consider a middle course on these issues, permitting broadband providers to charge content providers for enhanced network services subject to FCC oversight, while giving those providers various incentives to provide basic and growing levels of bandwidth on the best-effort Internet access platform.[124] As noted, the nation's merger-review authorities adopted essentially this approach when they conditioned approval of the Comcast–NBC Universal merger on Comcast's commitment to offer, in its DOCSIS 3.0 markets, a service that typically provides 12 Mbps in downstream bandwidth (subject to potential adjustments).[125]

Weighing the costs and benefits of intervention

We have focused so far mostly on the asserted *benefits* of net neutrality regulation—claims that such regulation is needed to ward off perceived threats to the open Internet and consumer welfare. But any form of regulatory intervention carries potential costs, too, and those costs must be weighed against the benefits.

In this case, any rule or strong presumption against paid-prioritization agreements will limit consumers' ability to use their Internet access platforms to run certain types of QoS-sensitive applications, such as the real-time high-definition sports programming discussed earlier in this chapter. If consumers nonetheless want to use such applications, the broadband ecosystem will have to create workarounds to meet that demand, and the question is whether these workarounds will be as efficient and pro-consumer as the prohibited alternative.

Four potential workarounds help illustrate the possible trade-offs. First, a ban on paid prioritization might induce broadband providers to keep doing what they have long done: segregate any real-time, QoS-sensitive application from their Internet services and always offer it instead as part of a separate "specialized service" such as subscription video. In other words, the regulatory ban might artificially prolong today's sharp distinction between "Internet access" and "cable TV" and cloud prospects for efficient convergence between the two. As discussed earlier, some net neutrality advocates appear willing to postpone such convergence indefinitely in order to keep the service marketed as "Internet access" as egalitarian as possible.

Second, a ban on paid prioritization might theoretically induce broadband ISPs to accommodate QoS-sensitive content over their best-effort platforms by building extra bandwidth in their last-mile networks in the form of routers and transmission lines with greater capacity than the ISPs would otherwise need if they could assign traffic to priority tiers. Net neutrality advocates sometimes cite the prospect of this outcome—higher-capacity networks—as a desired side effect of net neutrality regulation. But it presents a trade-off. Any broadband ISP will incur extra costs in deploying this extra capacity and could avoid incurring those costs, without materially degrading its service, by entering into commercial agreements to prioritize the QoS content that *needs* to be prioritized in order to function properly without prioritizing the content that does *not* need to be prioritized. If ISPs must choose the first option because the second is foreclosed, they will ultimately pass through to consumers some portion of those costs in the form of higher retail rates, which in turn may potentially reduce broadband subscribership at the margins.[126] Or the ISP might simply forgo the inefficiently costly upgrades, thereby diminishing the supply of any high-QoS Internet-based applications consumers might wish to purchase.

Third, broadband ISPs might try to achieve the same efficiencies of *paid*-prioritization agreements by prioritizing some forms of Internet content over others *without* charging content providers for that function (to the extent that such unpaid differentiation remains permissible under the FCC's rules). But the lack of price signals would pose high hurdles to that plan. Broadband networks have not traditionally "known" in real time what content payload a given IP packet is carrying, nor can they easily assess by themselves how much special handling any given content needs in order to satisfy the expectations of a content provider's customers. If broadband ISPs may not charge for priority treatment and must rely on content providers' packet markings to determine the QoS needs of particular content, the content providers would have every incentive to "cheat" by marking all of their packets for priority. That, as discussed, is why interconnecting IP networks have not traditionally honored each other's priority markings. The solution to that collective-action problem is to attach price signals, in the form of prioritization fees, to decisions to mark packets for prioritization. Only then will each network internalize the opportunity costs of prioritizing their packets over other networks' packets, and only then will priority be allocated efficiently to the content that needs such priority in order to function properly in the manner that consumers most value. Those fees, however, are precisely what a paid-prioritization ban seeks to suppress.

Fourth, broadband ISPs might try to work around the regulatory ban by charging their content *recipients*—that is, their own subscribers—rather than content *providers* to prioritize those providers' QoS-sensitive content. But it is unclear why it would make sense to fashion a regime in which such payments are permitted, but direct payments from content providers to ISPs are not. Suppose that for a content provider's service to function well, its packets need to be marked for priority from source to destination. Even if the content recipient rather than the content provider is paying the ISP to honor those priority markings, the content provider will still need to arrange for its packets to be marked for priority from the outset of the transmission to any ISP subscriber who has agreed to pay to receive those marked packets. Now suppose that the content provider charges each content recipient $10 a month to access its content on that basis and that each content recipient pays her broadband ISP $2 to honor the content provider's priority markings. From an economic perspective, this arrangement is essentially equivalent to one in which the content recipient makes a single payment of $12 to the content provider, and the content provider pays the ISP $2 to honor its priority markings. In each case, the same money ends up in the same places, and the same functions are performed.

This is not to say, of course, that the two payment regimes would produce the same economic behavior or consequences in all cases. But these observations raise serious questions about why policymakers should want to create regulatory barriers to the way such payment regimes are structured if, at least some of the time, the parties can achieve functionally equivalent outcomes by jumping through different and potentially less efficient hoops. Scott Hemphill explains:

> Although an access provider is prohibited from charging content providers, it is free to charge consumers under the leading network neutrality proposals. As a result, the access provider may charge the consumer for premium service—prompt delivery of video, for example—in the expectation that the content provider, in turn, will compensate the consumer for the extra expense. When indirect extraction replaces (prohibited) direct extraction, private bargaining tends to undo the effect of the government regulation. The shift to indirect extraction also imposes a social cost, making a ban on direct extraction not only ineffective, but counterproductive as well.[127]

Supporters and opponents of net neutrality regulation also argue at length about the effects of such regulation on broadband investment incentives. As we have seen, a key rationale for regulatory intervention is a fear of market power arising from the small number of rival broadband ISPs in most geographic markets. But regulatory choices not only *reflect* the existing level of competition in a market but also *affect* future levels of competition by increasing or reducing incentives for new competitive entry. For example, Christopher Yoo argues that if the root problem is an undersupply of broadband access providers, the proper solution is to maintain deregulatory policies that encourage new entry into the broadband market by allowing each broadband provider to differentiate itself from others.[128] Net neutrality rules, he claims, would stifle such differentiation, deter new entry, and perversely solidify the competitive problem that gave rise to net neutrality proposals in the first place. In contrast, Tim Wu argues that the fixed-broadband marketplace will remain a duopoly for the foreseeable future no matter what regulatory steps are taken and that regulators must therefore focus on preventing the duopolists from harming innovation at the "edge" of the Internet.[129]

One final dimension of the net neutrality debate warrants discussion here and is more procedural and institutional than substantive. Opponents of prescriptive net neutrality regulation claim that no matter how the underlying economic issues should be resolved in the abstract, policymakers should adopt a cautious case-by-case approach to the resolution of particular net neutrality disputes, including arrangements for paid prioritization. And they submit that if and when market failures arise, policymakers should opt for after-the-fact remedies rather than prophylactic rules, which, they say, grow obsolescent quickly in this dynamic environment and thus inevitably create unintended consequences. In contrast, advocates of prescriptive regulation assert that unregulated broadband ISPs pose threats to the Internet that dwarf any costs of regulatory intervention and that unless the government takes bold steps now to foreclose those threats, those ISPs will undermine the Internet's openness in ways that cannot easily be undone later.[130] These advocates generally support regulating broadband ISPs as common carriers and subjecting them to prescriptive regulation under Title II of the Communications Act.

This last point brings us back to the *Brand X* controversy about whether broadband ISPs should be viewed as providers of Title I information services or of Title II telecommunications services (i.e., common carrier services). That controversy, seemingly resolved by the Supreme Court in 2005, reemerged with ferocity in 2010, roiled the telecommunications policy world for several months, and continues to simmer today.

III. Title I, Title II, and the Limits of FCC Power

A. Overview of Title I and the FCC's "ancillary authority"

In Internet policy debates, engineers and economists have a clear advantage over lawyers: the questions they answer, although complicated, are at least sensible questions to ask. The same cannot always be said of the legal questions that permeate those policy debates. Since the mid-1990s, telecommunications lawyers have battled over how to pigeonhole various Internet-related services—from VoIP to IP video to broadband Internet access—within the obsolescent framework of the Communications Act of 1934. Those legal disputes can be mind numbing in their scholastic complexity, and they are increasingly unhinged from the underlying economic and engineering realities that should be driving the policy debate. But they are critical to understand because the courts will invalidate any policy initiative, no matter how sensible, that ends up on the wrong side of the law.

These legal disputes arise mainly because the Communications Act, written in a preconvergent era, assumes that particular types of facilities will normally be associated with particular services and that those services should be individually classified and regulated pursuant to mutually distinct statutory regimes. For example, conventional telephone companies are subject to common carrier regulation under Title II, and their circuit-switched voice services are subject to the traditional division of regulatory authority between the FCC and the states (see chapter 2). And multichannel video services are regulated under Title VI, which essentially divides regulatory responsibility between the FCC and local franchising authorities (see chapter 9). In revising the Communications Act in 1996, Congress left intact each of those statutory silos, along with the markedly different rules contained in each for governing the corresponding physical-layer platform.

But Congress did not clearly foresee the rise of higher-layer applications that mimic traditional services yet can ride on top of any physical-layer Internet access platform. The result has been profound legal uncertainty. For example, because Congress did not anticipate the advent of VoIP, it gave no clear guidance about which types of VoIP services, if any, should be regulated like regular telephone services under Title II and which should remain presumptively unregulated.[131] And because Congress did not foresee the advent of online video services such as Hulu, it gave no clear guidance about whether the providers of those services are subject to the rights and obligations of multichannel video programming distributors (MVPDs) under Title VI.[132]

Similar uncertainty surrounds the legal status of residential broadband Internet access, which had not quite emerged in 1996. As noted earlier, the dispute about that legal status ultimately turns on the obscure statutory definitions of "telecommunications service" and "information service," added by the Telecommunications Act of 1996. Those definitions, based on the FCC's older "basic" and "enhanced" service concepts, were conceived during the dial-up era, without broadband Internet access in mind. Indeed, before the late 1990s, most policymakers often seemed to assume that residential consumers would indefinitely continue accessing Title I "information services" such as AOL and Internet websites by means of a single mechanism: ordinary telephone calls (Title II "telecommunications services") to ISP modem banks over the PSTN.

When Congress wrote the 1996 Act, it did not anticipate that cable companies, free of conventional common carrier requirements, would invest heavily in broadband, tightly integrate ISP functionality with last-mile transmission, and offer the resulting service bundle to consumers. If Congress had waited just two or three years past 1996 to codify its core definitional concepts, it presumably would have been clearer about exactly where such "broadband Internet access" fits within the larger statutory framework. But because it acted when it did, Congress bequeathed a legal and political quagmire to the FCC and the federal courts.

In its 2005 *Brand X* decision (discussed earlier in this chapter), the Supreme Court appeared to resolve that dispute when, by a vote of six to three, it upheld as "reasonable" the FCC's conclusion that cable broadband services are pure Title I "information services" without any Title II component.[133] Most observers believed that although the FCC had thereby disclaimed traditional common carrier regulation for broadband ISPs, it still had Title I authority to adopt basic net neutrality rules. The *Brand X* Court did much to encourage that way of thinking by obliquely suggesting that to some unspecified extent the FCC "remains free to impose special regulatory duties on facilities-based ISPs under its Title I ancillary jurisdiction."[134]

What, though, is this "Title I ancillary jurisdiction," and what kinds of rules does it support? On its face, Title I is unremarkable. It contains very little beyond general pronouncements about the purpose of the FCC, the scope of the Communications Act, assorted statutory definitions, and the "forbearance" provision we examined in chapter 2. The first two provisions of Title I—sections 1 and 2 respectively of the Communications Act—establish the FCC "[f]or the purpose of regulating interstate and foreign commerce in communications by wire and radio" and recite that "[t]he provisions of this [Act] shall apply to," among other things, "all interstate and foreign communications by wire or radio."[135] And section 4(i), sometimes called the FCC's "necessary and proper clause,"[136] authorizes the Commission to "perform any and all acts, make such rules and regulations, and issue such orders, not inconsistent with this [Act], as may be necessary in the execution of its functions."[137] But it does not specify the permissible scope of those "functions."

Together, these unremarkable provisions constitute what people are talking about when they discuss regulation of Internet-related services "under Title I." To say that a given communications technology—broadband Internet access or higher-layer applications such as VoIP or instant messaging[138]—should be regulated under Title I is to embrace two conclusions. The first is that the service in question slips through the cracks of the substantive titles of the Communications Act (II, III, and VI) and is thus immune from the industry-specific regulations contained in those titles. The second is that the FCC nonetheless may assert general regulatory jurisdiction over the services (insofar as they are "interstate") in order to promote its core regulatory responsibilities under those substantive titles and, just as important, to preempt states and localities from regulating the services themselves.

Although the provisions of Title I may seem a slender basis for substantive FCC regulatory authority, it is difficult to imagine how the FCC could function without such authority in an industry that, like this one, spawns new technologies and thus new regulatory issues far more quickly than Congress can legislate to address them. When the cable television industry arose in the 1960s, the FCC exercised its Title I authority to protect the local advertising revenues of the nation's over-the-air broadcasters by, for example, limiting the ability of cable operators to transmit the signals of distant television stations (see chapter 9). In its 1968 decision in *United States v. Southwestern Cable*, the Supreme Court upheld those regulations as ancillary to the Commission's undisputed responsibility to preserve the broadcasting industry, even though Congress had never formally authorized the FCC to regulate the previously nonexistent cable industry.[139] Indeed, as the Court later remarked, *not* to give the FCC latitude in meeting such new regulatory problems "would place an intolerable regulatory burden on the Congress—one which it sought to escape by delegating administrative functions to the Commission."[140]

That said, the FCC lacks unbounded authority to impose under Title I whatever regulations it likes on anyone involved in the interstate or international transmission of electronic communications. Instead, the courts have confined the FCC's Title I authority to regulations that are "reasonably ancillary to the effective performance of"[141] or "necessary to ensure the achievement of"[142] the Commission's responsibilities under the Act's other substantive titles. The scope of such "ancillary" authority has always been murky. Several years after its initial decision in *Southwestern Cable*, the Supreme Court upheld the Commission's adoption of more extensive regulations for the cable industry, including a new obligation for cable operators to serve their communities by transmitting programming of their own in addition to the signals independently aired by broadcasters.[143] This time, however, the Commission prevailed only by the narrowest of margins, and Chief Justice Burger, who cast the tie-breaking vote, observed that the Commission had "strain[ed] the outer limits" of its ancillary jurisdiction.[144]

The Supreme Court finally delineated those limits when in 1979 it balked at the FCC's assertion of Title I jurisdiction to require cable operators to provide (among other things) "public access" channels on their systems.[145] The Court based its decision on what it perceived as a tension between the Commission's jurisdictional theory and the substance of its regulations. The basic problem was that although the Commission had predicated jurisdiction on its underlying Title III authority to regulate broadcasters in the same video programming industry, it had exercised that jurisdiction by imposing, in the form of these public access channels, the very form of common carrier regulation that the Act would prohibit if the regulated parties had been broadcasters rather than cable companies.[146]

In the first decade of the new millennium, the D.C. Circuit issued several decisions that further constrained the FCC's Title I authority, culminating in the court's 2010 invalidation of the *Comcast Order* discussed earlier in this chapter. First, in 2002 the D.C. Circuit rejected the Commission's invocation of Title I authority to require television broadcasters to include "video descriptions" in their shows—that is, aural descriptions of the show's visual content for the benefit of the visually impaired.[147] The court limited its holding to the particular concerns presented in that case, including both its desire to "avoid potential First Amendment issues" arising from this regulation of broadcast content and the fact that "[a]fter originally entertaining the possibility of providing the FCC with authority to adopt video description rules, Congress declined to do so."[148]

Three years later, the D.C. Circuit again invoked the limits of Title I in vacating the FCC's *broadcast flag* rules. These rules, designed to limit unauthorized consumer digital recording of broadcast content, required equipment manufacturers to configure digital televisions and other consumer devices to honor copying restrictions encoded in digital television broadcasts. The court viewed these rules, too, as a misapplication of Title I ancillary authority. It found that, "at most, the Commission only has general authority under Title I to regulate apparatus used for the receipt of radio or wire communication while those apparatus are engaged in communication," whereas the broadcast flag rules were designed to suppress post-broadcast copying.[149]

Despite these legal setbacks, the FCC still believed it was acting on solid ground when it issued the *Comcast Order* in 2008, finding that Comcast's throttling of BitTorrent had violated the general purposes of the Communications Act, if not any actual statutory requirement. Although the FCC was acting under Title I (because broadband Internet access remained classified as an "information service"), it had some reason to feel confident in its authority. Until then, the courts had seemed more solicitous of the FCC's efforts to use Title I as a mechanism for dealing sensibly with emerging and congressionally unanticipated technologies than for supplementing the statutory schemes applicable to the more established communications media. *Southwestern Cable* was one case in point; another was the D.C. Circuit's 1982 decision to uphold the Commission's use of its ancillary authority in *Computer II* to regulate the provision of Title I enhanced services by Title II common carriers.[150] That, however, is not how the *Comcast* litigation played out.

B. The *Comcast* decision and the "third way" proposal

In early 2010, just after the FCC had opened the proceeding that led to the *Open Internet Order*, the D.C. Circuit invalidated the *Comcast Order* as another example of Title I overreach. As the court noted, the FCC's legal analysis had asserted expansive ancillary authority to promote broad congressional policy statements through whatever regulatory steps the Commission found most convenient. For example, the Commission had predicated its order in part on a finding that Comcast's suppression of BitTorrent threatened Congress's general aspiration, expressed in section 1 of the Communications Act, to promote "rapid" and "efficient" communications to the American public.[151] The D.C. Circuit held, however, that such "policy statements alone cannot provide the basis for the Commission's exercise of ancillary authority," stressing that the FCC's contrary view would "virtually free the Commission from its congressional tether."[152] Indeed, the court added, "we can think of few examples of regulations that apply to Title II common carrier services, Title III broadcast services, or Title VI cable services that the Commission, relying on [such] broad policies . . . , would be unable to impose upon Internet service providers" if its expansive theory of ancillary authority were valid.[153]

The court thus reaffirmed a key limiting principle: the FCC may invoke ancillary authority only "to support its exercise of a *specifically delegated power*" under a substantive provision of the Communications Act rather than "to pursue a stand-alone policy objective."[154] The Commission's appellate lawyers had also argued that even if ancillary authority were so constrained, the *Comcast Order* could still be justified as necessary to support the exercise of "specifically delegated powers." But the D.C. Circuit largely refused to consider those arguments on their merits because, for the most part, the Commission had neglected to include them in the *Comcast Order* itself.[155]

The D.C. Circuit noted that all of these esoteric Title I issues had arisen only because, "in its still-binding 2002 *Cable Modem Order*, the Commission ruled that cable Internet service is [not] a 'telecommunications service' covered by Title II of the Communications Act."[156] As if on cue, the Commission responded to the *Comcast* decision by promptly announcing that it might reverse its unitary "information service" characterization of broadband Internet access service and reclassify it (or some component of it) as a "telecommunications service" under Title II.

The first salvo in the FCC's response appeared in a highly publicized FCC blog post by General Counsel Austin Schlick in May 2010. He argued that reclassifying broadband Internet access this way would place the FCC's net neutrality agenda on a more "sound legal footing" than would any revised theory of Title I ancillary jurisdiction.[157] He stressed that because the underlying statutory definitions are ambiguous, the Commission had broad discretion to reverse course on the threshold legal classification of broadband Internet access. After all, he noted, the *Brand X* majority had found that "'the Commission is free within the limits of reasoned interpretation to change course if it adequately justifies the change,'" and the three dissenters, led by Justice Antonin Scalia, "expressed the view that the agency *must* classify a separable telecommunications service within cable modem offerings. As many as all nine Justices, it seems, might have upheld a Commission decision along the lines Justice Scalia suggested."[158]

In a contemporaneous statement, FCC Chairman Julius Genachowski asserted that he was not proposing to subject broadband ISPs to the full suite of Title II common carrier rules designed for the telephone monopolists of the mid–twentieth century, including tariff requirements and retail price regulation.[159] He proposed what he called a "third way" approach, under which the FCC would reclassify broadband Internet access as a Title II service (in whole or in part) but exercise its forbearance authority (see chapter 2) to insulate broadband providers from most Title II requirements. Such providers, he said, would instead be subject only to a small handful of core Title II requirements, such as the prohibitions in sections 201 and 202 on "unjust," "unreasonable," and "discriminatory" practices. The FCC soon formalized Genachowski's proposal by issuing a Notice of Inquiry seeking public comment on this "third way" proposal.[160]

The FCC's supporters viewed reclassification as legally essential to support a range of FCC initiatives, including not only net neutrality but also FCC subsidies for broadband deployment (see chapter 8). They agreed with Schlick's assessment that retaining the "information service" classification while devising new Title I jurisdictional theories would be "a recipe for prolonged uncertainty. Any action the Commission might take in the broadband area—be it promoting universal service, requiring accurate and informative consumer disclosures, preserving free and open communications, ensuring usability by persons with disabilities, preventing misuse of customers' private information, or strengthening network defenses against cyberattacks—would be subject to challenge on jurisdictional grounds because the relevant provisions of the Communications Act would not specifically address broadband access services."[161]

These supporters also claimed that this Title II treatment should be easy to defend on appeal not only because so many Supreme Court Justices appeared receptive to that approach in *Brand X*, but also because, in the supporters' view, broadband Internet access is in fact a "telecommunications service."[162] After all, they said, that service's main function is to provide a clear transmission path from end users to any Internet content of their choice, much as a standard telephone line's function is to provide a clear transmission path between two people or, in the case of dial-up Internet access, a person and an ISP modem bank. Reclassification supporters further pointed out that many rural telcos had for years offered DSL transmission as an explicit Title II service, mainly to qualify for universal service funding (see chapter 8).

Opponents of reclassification disputed each of these claims. First, they noted that the Title II DSL services offered by rural telcos were pure last-mile transmission services unbundled from Internet access and that the relevant question is how to classify the retail *Internet access* service offered by broadband ISPs qua ISPs.* And they argued that core elements of that retail service, such as domain name system (DNS) look-up functionality (see chapter 5), fall within the definition of "information service" because they involve "retrieving, utilizing, or making available information via telecommunications."[163] To this last point, reclassification proponents replied that DNS look-up functionality cannot place broadband Internet access within the "information service" definition because, they claimed, it falls within that definition's explicit carve-out for "any use of any [information-processing] capability for the management, control, or

* Theoretically, the FCC could have bifurcated the inquiry by reinstating a form of the *Computer Inquiry* requirement that telcos separate their ISP "information services" from pure last-mile transmission services, and it could have regulated only the latter under Title II. But it is unclear how that approach would have supported the Commission's net neutrality rules because any violation of those rules arguably occurs on the ISP side of the line. Moreover, because the *Computer Inquiry* rules had never applied to any non-telco information service providers, it is also unclear how, simply as an engineering matter, the FCC would have distinguished between the "ISP" and "transmission" components of, for example, cable broadband networks.

operation of a telecommunications system or the management of a telecommunications service."[164] The two sides thus disputed how broadly this definitional carve-out extends; how it relates to a pre-1996 regulatory antecedent known as the *adjunct-to-basic doctrine*;[165] and whether it encompasses network functionalities that, like DNS look-up, are designed for the convenience of end users as much as or more than for the convenience of network engineers. This sampling of issues illustrates how exceptionally dry the "statutory classification" debate can become on its purely legal merits.

On a policy level, the opponents argued that reclassification would be both over- and underinclusive as a means of achieving the FCC's broadband objectives. First, even if accompanied by forbearance from most provisions of Title II, "third way" reclassification would subject broadband ISPs at a minimum to the provisions of sections 201 and 202, which require "just" and "reasonable" practices and prohibit "unreasonably discriminatory" ones. Would those provisions subject broadband ISPs to retail rate regulation for the first time? Perhaps so, although proponents of regulation did not advocate that outcome: they focused on relations between ISPs and content providers, not between ISPs and their retail subscribers. Second, as noted earlier, common carriage principles have traditionally allowed paid-prioritization arrangements so long as carriers offer them on the same terms to all similarly situated buyers, whereas the *Open Internet Order* expresses hostility to such arrangements as a general matter. Opponents thus argued that merely reclassifying broadband ISPs as "common carriers" would not support a key component of the net neutrality agenda that had motivated the reclassification initiative in the first place.

C. Denouement and litigation

All of these legal issues became academic when in late 2010 the Genachowski FCC bowed to intense political pressure and shelved its "third way" reclassification initiative indefinitely. Nearly all major broadband ISPs—cable, telco, and wireless—opposed that initiative, and they cited similar statements of opposition from key congressional allies. As discussed in chapter 10, the FCC, as an "independent" federal agency outside the Executive Branch, is highly vulnerable to such congressional pressure, and it must pick its political battles wisely. In this case, it decided to conserve its political capital for other important objectives, such as spectrum reform (see chapter 3).

To justify the *Open Internet Order*, therefore, the FCC's lawyers had to find new, post-*Comcast* theories of authority under Title I or elsewhere. They left no stone unturned in that pursuit. What follows is a brief and noncomprehensive sampling of the Commission's many alternative jurisdictional rationales, which remain under judicial scrutiny at press time.

First, the *Open Internet Order* relied heavily on section 706(a) of the 1996 Act. This provision exhorts the FCC to "encourage the deployment" of "advanced telecommunications capability" (which, very roughly speaking, means broadband) by exercising its regulatory authority to "promote competition in the local telecommunications market" and "remove barriers to infrastructure investment."[166] As the *Comcast* court noted, the FCC had previously interpreted this provision not as an independent source of authority, but as mere guidance on how it should exercise whatever authority it otherwise had.[167] The *Open Internet Order* reversed that Commission precedent and invoked section 706(a) as an affirmative basis for regulating broadband ISPs in order to promote infrastructure investment and "the deployment of advanced services,"[168] although the *Order* was a bit vague on how the net neutrality rules would promote that particular objective.

The FCC independently invoked a variety of more conventional theories of ancillary jurisdiction. For example, it relied on section 201 of the Communications Act, which authorizes the FCC to ensure that the rates and terms of conventional telecommunications services are "just and reasonable." Although that provision does not directly apply to broadband ISPs so long as they remain classified under Title I, the FCC reasoned that its net neutrality rules were necessary to keep these broadband ISPs from interfering with the over-the-top VoIP providers that, although also arguably Title I providers, nonetheless "contribute to the marketplace discipline of voice telecommunications services" that *are* "regulated under Section 201."[169] In that respect, the FCC argued, its net neutrality rules were ancillary to the execution of its statutory responsibilities under section 201. Opponents of regulation argued that although section 201 might justify very narrow protections for emerging competition to traditional Title II services (as in the *Madison River* case discussed earlier), such provisions could not justify the much more expansive net neutrality regime the FCC proposed and ultimately adopted, including the presumption against paid-prioritization arrangements. In other words, the opponents said, that regime was much broader than necessary to promote the Commission's actual responsibilities under Title II and was therefore unlawful.

To support the *Open Internet Order*, the Commission not only needed to identify affirmative sources of regulatory authority but also had to rebut its opponents' claims that it was impermissibly regulating broadband ISPs "as common carriers." Section 3(51) of the Communications Act provides that "a telecommunications carrier shall be treated as a common carrier under this [Act] *only to the extent that it is engaged in providing telecommunications services*."[170] The FCC's opponents claimed that any ban on "unreasonable discrimination" in particular impermissibly applied the classic hallmarks of common carrier regulation on broadband ISPs when, by hypothesis, they are not "engaged in providing telecommunications services." The FCC responded that it was not really subjecting ISPs to such regulation because even though it was banning them from unreasonably discriminating among *Internet content providers* (for example, when offering paid-prioritization services), ISPs remained free to deal with *retail* customers however they wished.[171] To this argument, the ISPs responded that section 3(51) bars the FCC from treating them as common carriers with respect to either class of customers on either side of this potentially two-sided marketplace.

Many parties challenged the *Open Internet Order* in various courts, and the usual forum-selection games began. The parties opposing any regulation—Verizon and MetroPCS—sued in the D.C. Circuit, which had ruled against the FCC in the *Comcast* litigation. In contrast, the public interest groups seeking greater regulation for mobile broadband providers sued in several other circuits, hoping to avoid D.C. Circuit review. The case was nonetheless assigned by lottery to the D.C. Circuit. All the public interest groups subsequently withdrew their appeals, leaving only Verizon and MetroPCS as petitioners.[172] The challengers argue that the FCC lacks statutory authority to issue the rules in question, that the rules lack a sufficient empirical predicate and are thus "arbitrary and capricious," and that the rules violate the free speech and takings clauses of the U.S. Constitution.[173]

The FCC appeared none too eager to let the *Open Internet Order* ripen for judicial review. Soon after it issued the *Order*, it invoked an obscure Office of Management and Budget review process to delay the effective date of its new rules and thus any judicial scrutiny of those rules by nearly a year. When that process was completed, the FCC asked the D.C. Circuit to hold the case in abeyance while tangentially related regulatory proceedings remained pending before the Commission. After the court denied that request, the Commission and the other litigants agreed to a lengthy briefing process that would not conclude until late 2012, two years after the *Open Internet Order* was initially issued, with oral argument and the eventual decision pushed into 2013.

* * *

After the FCC issued its *Open Internet Order* in late 2010 and suspended its "third way" reclassification initiative, it turned in 2011 to a hornet's nest of obscure technical issues concerning "intercarrier compensation," which we address in the next chapter. Those issues triggered far less emotion and publicity than the net neutrality controversies of the prior year, but they are arguably far more important in their overall industry impact.

That said, there is no clear demarcation between the net neutrality rules discussed in this chapter and the intercarrier compensation issues discussed in the next. In fact, these two sets of rules are intertwined at the deepest levels. We noted a key aspect of that relationship earlier: the tension between (1) the FCC's ostensible presumption against paid agreements to favor particular Internet content and (2) its simultaneous disavowal of any intent to regulate paid-peering arrangements, which are often designed to favor particular Internet content. We now take a closer look at network-to-network compensation arrangements more generally and the policy issues they raise, both on the conventional telephone system and on the Internet.

7

Interconnection and Intercarrier Compensation

There are many thousands of telecommunications networks in the world, and each day millions of voice or data communications cross from one network to another. On the circuit-switched PSTN, regulators have grappled for decades with a complex maze of rules governing when and where voice-oriented telephone networks must interconnect and who must pay what to whom when they exchange calls. On the Internet, many thousands of IP networks likewise interconnect for the exchange of data traffic, but *their* interconnection arrangements—based on the peering and transit models we introduced in chapter 5—are completely unregulated. In general, the Internet has functioned efficiently anyway, without regulatory oversight. Why has it made sense to regulate interconnection on the PSTN but not on the Internet? As voice services become just another IP application and the circuit-switched telephone system draws closer to its eventual sunset, will regulators need to oversee interconnection arrangements in the emerging all-IP ecosystem?

These are some of the most critical emerging questions in telecommunications policy today, and they are the core questions we address in this chapter. In section I, we begin by examining the historical legacy of *intercarrier compensation* regulation on the PSTN, the rules that govern how interconnecting telephone companies recover the costs attributed to calls between their respective subscribers. We then turn in section II to the economic debate about how those cost-recovery rules should be structured. In section III, we turn from theory to practice, addressing the FCC's decision, in its landmark *USF–ICC Reform Order* of 2011, to shift the PSTN from one cost-recovery model to another, including the complex legal dimensions of that shift.[1] Finally, in section IV, we focus once more on the Internet and ask whether and when any form of government regulation will be needed to govern interconnection in the all-IP environment of tomorrow, whether for voice traffic or any other kind.

I. The Historical Crazy Quilt of Intercarrier Compensation Schemes

A. Introduction to PSTN interconnection rules and intercarrier compensation

Title II of the Communications Act requires every "telecommunications carrier" to interconnect with every other for the exchange of traffic generated by one carrier's customers and bound for the other's. That obligation arises from several different provisions. First, since the original passage of the Communications Act of 1934, section 201 has granted the FCC plenary authority over "common carriers"—a synonym under the Act for "telecommunications carriers"[2]—insofar as they are engaged in the provision of "interstate" services. The FCC invoked that provision to justify essentially all of its interconnection orders before the Telecommunications Act of 1996, including the interstate access charge and "expanded interconnection" regimes discussed in chapter 2. But this section 201 power is limited to interstate services and in 1994 was found to grant the FCC insufficient authority to order physical collocation.[3] In its comprehensive rewrite of the statute in 1996, therefore, Congress added the new interconnection provisions of section 251.

As noted in chapter 2, the section 251 interconnection obligations vary depending on a carrier's regulatory status. Telecommunications carriers that are not ILECs* are subject to a general obligation under section 251(a) to "interconnect directly *or indirectly*"—that is, through a third-party network—with other carriers.[4] In contrast, under section 251(c)(2), ILECs must agree to interconnect directly—"at any technically feasible point" and on highly regulated rates and terms—with any requesting "telecommunications carrier . . . for the transmission and routing of telephone exchange service and exchange access."[5]

Again, these interconnection mandates apply only to networks that are subject to regulation as "telecommunications carriers"—that is, as common carrier providers of "telecommunications services"—under Title II of the Act. Those networks, which in the aggregate constitute the PSTN, are our focus in the first three sections of this chapter. These interconnection mandates do *not* apply to the many network providers that elude characterization as "telecommunications carriers," such as

* As discussed in chapter 2, ILECs (incumbent local exchange carriers) are legacy telco monopolists (*see* 47 U.S.C. § 251(h)), whereas CLECs (competitive local exchange carriers) are wireline competitors to ILECs in local markets.

Title I "information service" providers or "private carriers."[6] In particular, these Title II interconnection mandates have never been found to apply to ISPs, Internet backbone providers, content delivery networks, or the Internet's other constituent IP networks. As discussed in the final section of this chapter, the interconnection arrangements among those non–Title II networks have been subject to market-based contracts and no prescriptive regulation.

When regulators do find it necessary to create interconnection obligations, as they have for PSTN-based networks, they do not simply issue a one-line directive to interconnect and leave it to the parties to work out all the details. Implementation disputes are routine, and regulators have found it necessary to adopt a complex web of rules to resolve them. First, regulators must resolve disputes about the physical particulars of where and how one network must interconnect with another.[7] To take just one example, some ILECs have long wished to require CLECs and other interconnecting carriers to establish many different *points of interconnection* (POIs) with each ILEC network so that every traffic exchange is geographically proximate to the ILEC customer involved in initiating or receiving a voice call. That multiple-POI approach would enable ILECs to conserve on the costs of transporting calls throughout their ubiquitous networks, but it would also raise the costs of the CLECs that would have to arrange to drop off and receive traffic at all those POIs (often by purchasing special-access services from the ILECs). For many years, the FCC has addressed this issue by entitling each non-ILEC carrier to avail itself of a single POI per LATA ("local access and transport area"), a geographic area derived from the 1984 AT&T consent decree.[8]

Second, when regulators create interconnection obligations, they must also resolve disputes about the price terms of the physical interconnection facilities involved. Suppose, for example, that a CLEC or wireless carrier wishes to lease a high-capacity transmission circuit on an ILEC's network in order to connect its network with the ILEC's. What rate should it pay for that circuit, known as an *entrance facility*? That was a long-simmering dispute for many years until, in 2011, the Supreme Court at least temporarily resolved it in *Talk America Inc. v. Michigan Bell Telephone Co.*[9] Affirming a state commission decision, the Court ruled that under section 251(c)(2) an ILEC can be required to lease such entrance facilities to interconnecting carriers at low regulated rates even though the FCC had previously ruled that those facilities do not meet

Figure 7.1
Call "origination" and "termination"

the "impairment" test for purposes of section 251(c)(3)'s main facilities-leasing provision (see chapter 2).[10]

Most important of all, when regulators create interconnection rights, they must decide highly complex questions of *intercarrier compensation*—the amount that Network X must pay Network Y for the usage-sensitive costs that Y incurs when handling calls originated by X's customers and bound for Y's customers.

Suppose you pick up your U.S. Cellular mobile phone and call a few miles down the road to your mother's house, served by CenturyLink's wireline network. In the industry lingo, U.S. Cellular *originates* the call and hands it off to CenturyLink. CenturyLink then *terminates* (i.e., completes the circuit for) the call by routing it to the switch nearest to your mother's house and dedicating network capacity to the call for as long as it lasts (see figure 7.1).

Viewed in isolation, this particular call to your mother is easily accommodated by existing network capacity. But the cumulative effect of all such calls placed during peak traffic periods is to impose substantial demands (and associated costs) on the networks involved. For example, CenturyLink must use, on the margin, larger switches and fatter transport pipes to accommodate the extra incoming telecommunications traffic from U.S. Cellular. Now assume, simply for the sake of illustration, that the majority of calls exchanged between these two networks are placed *by* U.S. Cellular's customers *to* CenturyLink's customers. This means that U.S. Cellular incurs proportionately lower incremental costs to accommodate the smaller number of calls that are placed by CenturyLink's customers and bound for U.S. Cellular's customers than CenturyLink incurs to handle the larger volume of U.S. Cellular traffic bound for CenturyLink's customers.

This example illustrates the basic question in intercarrier compensation policy: Who should pay for the additional network costs that a circuit-switched telephone company incurs because of its duty to terminate calls originated on *other* telecommunications networks by callers with whom the terminating carrier may have no direct relationship? There are two basic answers. One, roughly speaking, would require each terminating carrier to absorb these extra costs itself and pass them on to its own subscribers in the form of higher retail rates. This approach is called *bill-and-keep* (because each carrier "bills" its customers for its costs and "keeps" the proceeds rather than transferring any of them to the other carrier). The other solution would require the *calling* party's carrier, whose customers originate the calls that "cause" these extra costs, to compensate the terminating carrier that incurs the costs. This latter approach is formally known as *calling party's network pays* (CPNP); for simplicity, we refer to it simply as "calling-network-pays." U.S. telecommunications policy followed this latter approach for many decades until 2011, when the FCC announced that it would gradually transition the industry to a bill-and-keep approach.

The choice between these two approaches, which we examine in sections II and III of this chapter, has immense financial consequences for carriers. As noted in chapter 2, the traditional flow of intercarrier compensation payments, which still accounts for many billions of dollars each year, reflects basic regulatory choices made around the time of the 1984 AT&T breakup and then in the passage of the 1996 Act. Under the dichotomous system that the FCC's 2011 order seeks to eliminate, a long-distance carrier pays *access charges* to the local carriers on each end of a long-distance call; and a local exchange carrier typically pays lower *reciprocal compensation* rates whenever it hands off local calls to be terminated on another carrier's network. Each of these two schemes follows basic calling-network-pays principles, albeit in quite different ways. Competition and technological innovation have undermined the local/long distance distinction underlying these two schemes and, more generally, have drawn into question the stability of any calling-network-pays approach. Dissatisfaction with that approach sparked considerable interest in the bill-and-keep alternative, which, as noted, the FCC finally adopted in 2011. The order embodying that intercarrier compensation decision and related universal service fund reforms—the *USF–ICC Reform Order*—is one of the most important FCC initiatives in recent memory, and it will occupy much of our attention in this chapter and the next.

We begin, however, by surveying the regulatory backdrop for that order and in particular the incoherent patchwork of mutually inconsistent intercarrier compensation schemes that proliferated after 1996, each of which applied only to an arbitrarily compartmentalized class of calls or carriers. The result was arbitrage and competitive distortion, as we demonstrate through several case studies.

B. Access charge arbitrage scandals—and their origin in regulatory artificiality

In July 2003, the *New York Times* ran a front-page story concerning obscure allegations against MCI, a prominent stand-alone long-distance carrier that was later absorbed into Verizon.[11] The story reported that MCI may have "defrauded other telephone companies of at least hundreds of millions of dollars over nearly a decade. . . . The central element of MCI's scheme, people involved in the inquiry said, consisted of disguising long distance calls as local calls."[12] The particular allegations against MCI may or may not have been true, and MCI denied them. Our objective here is not to assess the factual legitimacy of those allegations, but to explain why any carrier might have had an incentive to engage in such arbitrage and why such schemes exposed radically unstable fault lines at the core of the traditional intercarrier compensation regime.

Suppose that a telephone company in Los Angeles wishes to place a long-distance call from its local customer to a called party in New York served by Verizon's local exchange network, as shown in figure 7.2. If the Los Angeles carrier delivers the call directly to Verizon (the terminating LEC), it must pay *terminating access charges* to Verizon. Under the traditional intercarrier compensation regime, however, it would save money if it sent the call instead to a CLEC accomplice in the New York area and arranged for that accomplice to deliver the call to Verizon as a "local" traffic exchange, as shown in figure 7.3.

Why would the Los Angeles phone company add this seemingly unnecessary middleman to an otherwise uncomplicated call from Los

Figure 7.2
A call from a long-distance carrier directly to a LEC

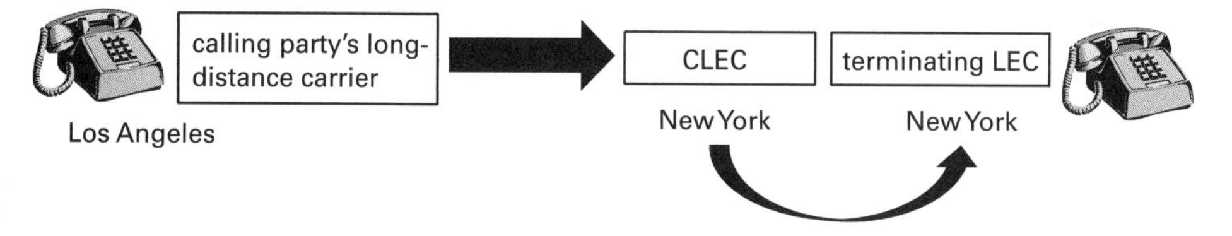

Figure 7.3
A call from a long-distance carrier to a LEC via a CLEC partner

Angeles to New York? The answer has to do with regulatory arbitrage. Terminating LECs have been traditionally entitled to charge much higher rates for completing "long-distance" calls than "local" calls. As a result, the Los Angeles phone company would do better if it reimbursed the CLEC for the lower "local" call-termination rate it paid to Verizon in New York than if it paid Verizon the higher rate for a completed "long-distance" call to the same destination, even if that company also had to pay the CLEC additional consideration (beyond simple reimbursement) for its trouble. For this scheme to work, of course, the CLEC accomplice would need to strip off the signaling information that identifies Los Angeles as the origination point for the call; by doing so, it can trick Verizon's switch into perceiving the incoming transmission as a "local" call originated by the CLEC rather than as a "long-distance" call originated by the Los Angeles company. For many years, allegations of such "arbitrage" schemes have roiled the telecommunications industry, and many billions of dollars have been at stake.

From a theoretical perspective, the most interesting thing about these allegations is that such arbitrage opportunities arise in the first place. Why, in particular, is there this tempting discrepancy in Verizon's (or any terminating carrier's) call-termination rates? After all, the cost of terminating the call—of routing it through Verizon's own New York switch en route to the called party—is the same no matter where the call originally came from. Indeed, that is why Verizon can be so easily fooled into thinking that these long-distance calls are local. Technologically speaking, Verizon's network is indifferent as to whether the caller is sitting in Los Angeles or in Manhattan. Why, then, does Verizon's compensation for completing the calls vary so much depending on that apparently irrelevant detail?

The answer is rooted in decades of calcified regulatory tradition. As discussed in chapter 2, regulators have traditionally viewed access charges—the fees that local exchange carriers impose for originating* and terminating long-distance calls—as a critical source of *implicit subsidies* for local telephone incumbents.[13] As a result, the access charges collected by the local carrier at each end of a long-distance call have traditionally exceeded any genuine measure of the costs such carriers incur in handling those calls. That is particularly true of "intrastate" access charges set by state commissions—that is, the charges that long-distance carriers pay local carriers for handling calls that stay within a given state's boundaries. To a lesser extent, it may also be true of "interstate" access charges set by the FCC for long-distance calls that do cross state lines. But the FCC has historically taken more aggressive measures than the states to reduce such charges closer to cost and replace the lost implicit subsidies with explicit funding mechanisms (see chapters 2 and 8).

In 1996, Congress created such an explicit funding mechanism (in section 254), but it took no immediate steps to stem the uneconomic money flow inherent in the access charge regime. To the contrary, in the grandfathering provision of section 251(g), it expressly authorized the FCC to conduct business as usual in its regulation of access charges for long-distance calls.[14] But there was never any analogous subsidy flow—and thus no irrational regulation to protect—in the carrier-to-carrier exchange of local traffic.† Permitting ILECs to charge similarly above-

* *Originating* access charges, which a calling party's LEC charges his long-distance carrier, are relevant only in the now small class of cases where the calling party subscribes to two different companies for "local exchange" and "long-distance" services, respectively. Such charges were economically far more important before about 2003, when the Bell companies won approval to enter the long-distance market. Today, most voice providers, including ILECs, cable VoIP providers, and mobile wireless carriers, sell bundled local and long-distance service, and most consumers no longer deal directly with standalone "long-distance" carriers. *Terminating* access charges, which the *called* party's LEC charges the *calling* party's long-distance carrier, remain important despite these developments because the carrier serving the calling party is very often not the carrier serving the called party.

† Before the 1990s, most of the local calls that crossed different networks were those that linked adjacent rural communities, each of which was served by a different local monopolist. In cases involving a symmetrical exchange of local traffic between adjacent LECs, regulators generally imposed a bill-and-keep regime (i.e., no money changed hands) on the theory that the number of calls sent from monopolist X to monopolist Y would be roughly equivalent to the number of calls sent from monopolist Y to monopolist X. Regulators sometimes devised more nuanced intercarrier compensation arrangements for more complex scenarios involving adjacent Bell companies, "independent" ILECs, and long-distance companies.

cost rates for terminating calls originated by their new local exchange rivals would obviously raise a host of competitive concerns.

In sections 251(b)(5) and 252(d)(2), therefore, Congress prescribed a *reciprocal compensation* regime that permits each terminating carrier to charge each originating carrier only for the "additional costs" of completing a call and no more.[15] Although the statutory language of these provisions seems to anticipate that money will flow from one carrier to another, Congress added that it did not mean "to preclude arrangements that afford the mutual recovery of costs through the offsetting of reciprocal obligations, including arrangements that waive mutual recovery (such as bill-and-keep arrangements)."[16] We return to that beguiling caveat later in this chapter.

In the 1996 *Local Competition Order*, introduced in chapter 2, the FCC interpreted this obscure collection of provisions to require, for the most part, a calling-network-pays regime for "local" calls.[17] Under this approach, the calling party's carrier reimburses the called party's carrier for the latter's costs in (1) *transporting* a call from the point of hand-off between the two carriers to the switch directly serving the called party and then (2) *terminating* the call through that switch en route to the called party.* For purposes of measuring these costs, the FCC chose and the individual states had to apply the same TELRIC cost methodology that governs the leasing of an ILEC's network elements. As discussed in chapter 2, TELRIC is a "forward-looking" cost methodology that analyzes how much it would cost a hypothetical, highly efficient carrier to perform the network functions in question if it built a new network today mostly from scratch. Although the resulting TELRIC-based cost estimates varied wildly from state to state, they typically produced reciprocal compensation rates for CLECs far lower than the access charges that ILECs received for terminating long-distance calls, even though the functions involved are often exactly the same.

This brief synopsis brings us back to the arbitrage scheme described earlier. If you were a long-distance carrier looking to reduce the substantial

* As defined in the FCC's rules, "transport" is "the transmission and any necessary tandem switching . . . from the interconnection point between the two carriers to the terminating carrier's end office switch that directly services the called party, or equivalent facility," and "termination" is "switching . . . at the terminating carrier's end office switch, or equivalent facility, and delivery of such traffic to the called party's premises." 47 C.F.R. § 51.701 (c), (d). For simplicity, we use the word *termination* to encompass both functions except where we explicitly distinguish between them.

terminating access charges you must pay to the called party's local carrier, you might look with some envy on the much lower rates that CLECs are paying that same carrier for the termination of local calls. You might be so envious, in fact, that you would try to disguise your long-distance calls as local calls. Indeed, at the height of this phenomenon, up to 20% of all calls terminated on the PSTN fell within the category of *phantom traffic*—that is, traffic whose origin is unidentified and thus cannot be accurately billed.[18]

Whatever their morality or legality, such arbitrage opportunities inevitably arise whenever regulators treat like services differently. No matter how hard regulators try to close the loopholes, such distinctions induce myriad ways of cheating, and cheating creates not just market distortions, but also significant enforcement costs. In this case, the arbitrary distinction between "local" and "long-distance" calls is what invited the cheating. Our next case study, which embroiled the industry in legal controversy from the late 1990s until 2010, drives home how dysfunctional that local/long distance distinction could become.

C. The ISP reciprocal compensation controversy

In the 1990s, a peculiar type of "call" suddenly became popular: dial-up connections to the Internet over the PSTN.[19] As we saw in chapter 5, such calls—which still exist, particularly in rural areas without broadband[20]—begin like any other telephone call, except that the caller uses a computer modem rather than a telephone handset to dial a local number on the circuit-switched telephone network. The called party that answers is another computer modem, this one belonging to an ISP such as AOL or EarthLink. For the duration of this data "call," the telephone company's circuit switch establishes a fixed connection between the calling party and the ISP, and the ISP translates the analog signals exchanged by the two modems into digital signals and connects the calling party to the websites of her choice.

Although a dial-up ISP may seem at first blush to be something else, it bears key similarities to a traditional stand-alone long-distance provider such as AT&T Corp. or MCI. It enters into a direct relationship with an end user to receive his calls via his local carrier and transport them over a long-distance network—either its own Internet backbone or that of another provider—to a distant called party (the computer server hosting the website). Given the essential similarity between dial-up ISPs and conventional long-distance carriers—both in terms of the network role they play and in their independent relationship with their own fee-paying end users—one would expect that such ISPs would pay the same per minute originating access charges that conventional long-distance carriers have paid ILECs. But, in fact, they never have. Since the 1980s, the FCC has exempted ISPs and other information ("enhanced") service providers from federal or state access charges.* It has entitled them instead to pay the ILEC whatever flat monthly fee is paid by the banks, department stores, and other ordinary, noncarrier customers that purchase local "business lines" of comparable capacity.[21] The FCC adopted this longstanding policy, known as the *ESP access charge exemption*, to encourage the growth of information services by insulating dial-up providers of those services from the obligations imposed on similarly situated non-Internet long-distance carriers.

In the late 1990s, increasing numbers of dial-up ISPs realized that they could avoid paying an ILEC anything at all, including business-line rates, by choosing CLECs as their local service providers. What does it mean for a CLEC to step in as the "local" service provider for an ISP? One curious feature of dial-up Internet access is that it is essentially all one way: an ISP's subscribers place calls to the ISP's modem bank (and from there to the distant websites, with which the subscribers can carry on protracted "conversations"), but the ISP's modem bank originates no calls back to those subscribers. A CLEC serves the ISP by interconnecting with the ILEC's network (the network that the ISP's subscribers are typically calling in from), receiving the incoming Internet-bound calls, routing them through a switch specially designed to terminate such calls to an ISP's modem bank, and then conveying them to that modem bank through high-capacity loops (see figure 7.4).

The task of terminating Internet-bound calls to a dial-up ISP imposes costs on the CLEC, at least in the aggregate. The CLEC is taking those costs off the shoulders of the ILEC, which would otherwise be performing similar functions for the ISPs and charging them business-line rates. Under the traditional reciprocal compensation regime for local calls, the ILEC in this arrangement must cover the total costs of the call all the way to the ISP, which means compensating the CLEC for the latter's share of those costs. As noted, the FCC's rules in the late 1990s provided that the rates the ILEC paid the CLEC for these call-termination functions should be based on TELRIC (unless the parties agree otherwise).[22] *In theory*, this payment system would give no carrier any artificial advantages or

* The distinction between "telecommunications services" and "information services" is discussed in chapter 6. The terms *information service provider* and *enhanced service provider* (ESP) are synonyms for all present purposes; Internet service providers (ISPs) are one subclass within the larger category of information service providers.

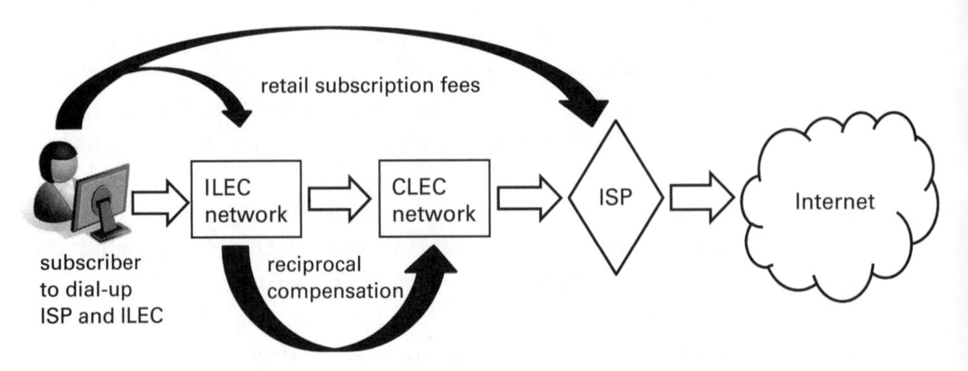

Figure 7.4
ILEC-originated traffic to a dial-up ISP served by a CLEC

incentives if regulators accurately calculated the underlying costs of that capacity, arranged for the CLEC to recover those costs from the ILEC in a rational way, and allowed the ILEC in turn to pass those costs through to the end users that place these ISP-bound calls.

The devil, however, is always in the details. Most state commissions ended up setting the TELRIC-based call-termination rate higher than any genuine measure of what it costs to obtain and maintain a switch to serve an ISP modem bank. CLECs thus had an artificial incentive to specialize in the termination of calls in order to receive, in effect, a regulatory subsidy for each minute of overstated costs for which they could bill the carrier (usually the ILEC) that handed calls off to them. Indeed, as Internet usage increased, the subsidies became so huge that CLECs would compete for them by offering to charge ISPs little or nothing for call-termination services—or even, it was said, offering to pay them money outright for the privilege of serving them. The subsidies produced by such above-cost regulatory rates were both economically inefficient and self-sustaining in that they detached prices from cost without the possibility of a market-based correction.[23]

The state commissions arguably could have alleviated the arbitrage problem created by ISP-bound traffic simply by lowering call-termination rates to levels that would have reflected a CLEC's actual costs of termination and provided for recovery of those costs as they were incurred. But that process would have taken too long, and few ILECs trusted regulators to get those rates "right." Ultimately, in a 2001 order, the FCC supplanted the state role altogether and announced that it would steadily reduce termination rates for ISP-bound traffic to $0.0007 per minute (far lower than prevailing rates at the time) and eventually eliminate them altogether, charting a (still-incomplete) "transition towards a complete bill and keep recovery mechanism."[24]

As noted earlier, under the bill-and-keep approach, each provider—in this case, the ILEC, the CLEC, and the dial-up ISP—looks to its own subscribers rather than to one another to cover its variable costs: the ILEC charges its subscribers, the CLEC charges the ISP, and the ISP charges its own subscribers. Dial-up ISP subscription charges might increase (because CLEC providers would now have to charge ISPs rather than ILECs for termination costs), but by less than if the ISP access charge exemption were eliminated outright (which would mean dramatically increased costs to ISPs).[25] The FCC subjected the ILECs to one proviso, known as the *mirroring rule*. To benefit from the low new rates for ISP-bound traffic, any ILEC had to agree to a similar transition toward bill-and-keep for non-access calls originated by carriers (mostly wireless providers) that *sent* more traffic *to* the ILEC's network than they received from it.[26]

This regulatory initiative devastated the business plans of certain CLECs, whose corporate earnings depended on the continued receipt of above-cost compensation from ILECs for the termination of calls to ISPs. Those CLECs took the FCC to court, where the dispute festered for more than ten years. In 2002, the D.C. Circuit rejected the FCC's legal theory for its 2001 order, but it nonetheless left the FCC's new rules substantively intact even while sending the case back to the Commission to puzzle through the legal issues again on remand.[27] The FCC essentially ignored the court's remand order for half a dozen years until an exasperated D.C. Circuit compelled it to come up with a new legal rationale by November 2008 or face invalidation of its new rules.[28] On the last day of its judicial deadline, the FCC came up with such a rationale, which the D.C. Circuit ultimately affirmed in 2010.[29]

D. Intercarrier compensation and VoIP

Not long after the ISP reciprocal compensation dispute arose, a second battle broke out over which intercarrier compensation rules should govern another technology for connecting the PSTN with the Internet: VoIP.

In the first years of the new millennium, companies such as Vonage and Skype began providing over-the-top "interconnected" VoIP services, which, as discussed in chapter 5, enable subscribers to use any broadband Internet connection to call or receive calls from people who use conventional circuit-switched telephone services on the PSTN. Suppose that you, sitting in Los Angeles, use your Comcast broadband connection to place

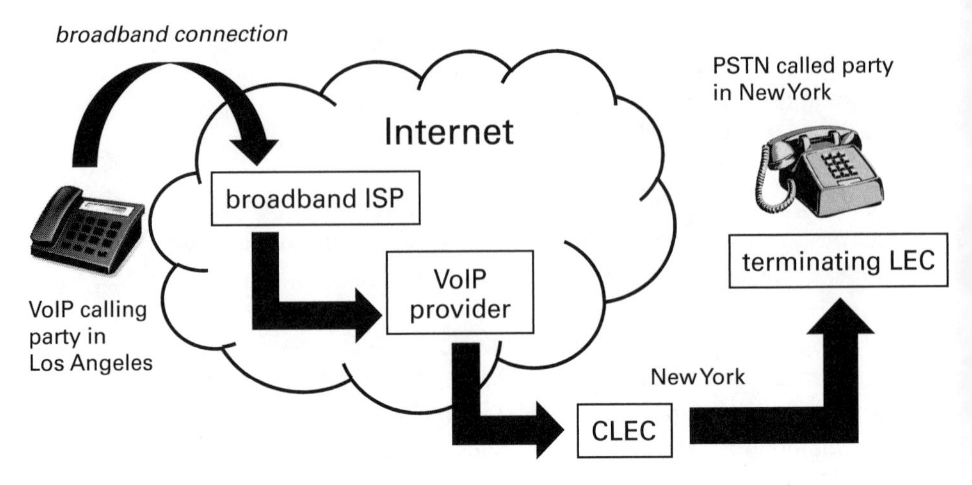

Figure 7.5
Over-the-top VoIP-to-PSTN call

a long-distance VoIP call via your Vonage account to someone on the PSTN in New York. Vonage—serving, in effect, as your long-distance carrier—has never owed access charges to any local provider for the origination of that call on your end, for the call does not pass through the public switched network there. To complete your call, however, Vonage will typically contract with a CLEC partner to convert the call from IP to a circuit-switched format and drop it off at a point of interconnection with the LEC serving the called party (see figure 7.5).

Given that this is a long-distance call, does the VoIP provider (or its CLEC partner) owe terminating access charges to the terminating LEC? Or is that provider, insofar as it is offering an "information service," covered by the ESP access charge exemption, such that it owes the terminating LEC, at most, TELRIC-based reciprocal compensation?[30]

There was never an obvious answer to that question. The access charge exemption traditionally arose as an issue only where, as with dial-up calls to ISPs, the information ("enhanced") service provider needed to rely on the PSTN to establish a connection with one of its own subscribers. VoIP providers turn that traditional relationship between the Internet and the public switched network on its head. Rather than processing "calls" from its subscribers over the PSTN to the Internet, as a dial-up ISP does, an interconnected VoIP provider processes calls from the Internet to the PSTN—and, in particular, to people on the PSTN who are not its subscribers. ILECs argued that the access charge exemption, designed to foster the growth of the fledgling "enhanced services" industry in the 1980s, should be limited to its traditional role in facilitating connections between providers and their own subscribers and that access charges should thus apply on the PSTN end of IP-to-PSTN calls.

VoIP providers and their allies disagreed, contending that the access charge regime is itself an economically irrational relic of a precompetitive age. In their view, so long as a VoIP provider (or its telecommunications partner) sets up a point of interconnection near the ILEC's network, any hand-off of traffic between the ILEC and that provider should be subject to lower reciprocal compensation rates, no matter how far away the VoIP party to the call might be. Each side claimed a long pedigree for its position in FCC access charge doctrine dating back to the early 1980s,[31] and each argued that the other's position would pervert that doctrine.

Many billions of dollars turned on that arcane debate. But the debate was as pointless as it was obscure. Ultimately, there is no good reason to retain a schizophrenic intercarrier compensation regime that gives rise to such questions in the first place. The carriers on the PSTN side of VoIP-to-PSTN calls, much like the ILEC "victims" of the alleged access charge "fraud" discussed in section I.B, are performing the same call-processing functions, no matter how the calls in question are characterized as a legal matter. And identical functions should be compensated identically. At least in the long term, they should not be compensated differently on the basis of distinctions rooted in yesterday's regulatory policy rather than in today's technological or engineering reality.

In its 2011 *USF–ICC Reform Order*, the FCC finally resolved this dispute prospectively by ruling that VoIP-to-PSTN calls would henceforward be subject either to interstate (not intrastate) access charges or to reciprocal compensation rates, depending on whether they would otherwise be classified as "toll" or "local" calls.[32] The FCC reached that resolution only as part of a much broader decision to unify all termination rates by reducing access charges (including those for VoIP-to-PSTN calls) to reciprocal compensation levels over time and then gradually lowering intercarrier charges for all PSTN call-termination functions to zero (for traffic exchanged at a prescribed point on the terminating carrier's network). That zero-rate regulatory outcome, addressed in detail in section II, is called "bill-and-keep."

E. Tariffs, CLEC access charges, and "traffic pumping"

Our next case study explores the regulatory dysfunction that can arise from the use of tariffs to prescribe intercarrier compensation rates. As noted in chapter 2, tariffs are regulatory filings that, when allowed to take effect, give the force of law to whatever rates carriers have posted in those tariffs for their services. Although ILEC access charges have typically exceeded any rigorous measure of the "cost" of terminating calls, those charges are still subject to regulation, which has capped how much long-distance carriers must pay ILECs for terminating their calls. In particular, ILEC access tariffs are subject to regulatory review for "reasonableness," and regulators can reject them if the rates charged are (by some measure) too high.

CLECs can file access tariffs, too, but for a long time the rates they posted were *not* subject to the same scrutiny as ILEC tariffs, and that fact gave rise to economically irrational arbitrage opportunities around the turn of the millennium.

It makes sense, of course, not to cap the *retail* rates posted by the small CLEC providers of "nondominant" services to consumers. If those rates are too high, retail consumers can always switch to some other provider. The same is not true, however, of terminating access charges. Suppose you place a long-distance call to a friend who has subscribed to a CLEC. In general, your carrier *must* deliver that call to the CLEC, even if it has posted very high terminating access charges, because the most basic precepts of Title II regulation force all telecommunications carriers to interconnect with every other and pay whatever rates a terminating carrier has lawfully tariffed.[33]

As a result, many CLECs born in the wake of the 1996 Act began charging radically above-cost fees for the wholesale "service" of routing incoming long-distance calls to their subscribers. And they faced no market response for doing so because their customers were not the ones paying the above-cost fees.* This practice led the FCC in 2001 to curb the interstate access charges that these nondominant CLECs could

* A statutory peculiarity has long exacerbated this lack of market response to above-cost terminating access charges. In 1996, led by senators from remote rural states, Congress improvidently required all carriers to charge nationally averaged rates to every subscriber of a given long-distance service, irrespective of the costs of completing any given subscriber's calls. This retrograde implicit subsidy arrangement, codified in 47 U.S.C. § 254(g), flies in the face of the universal service reform ordered at the local level (see chapter 8) in that it deliberately detaches the prices charged for services from the costs incurred in providing them. The practical effect of prohibiting a carrier from passing terminating access overcharges back to the calling party is that *no* end user, including the calling party, has any real incentive to object to them because the ultimate burden of paying them is diffused among thousands or millions of customers.

impose by tariff, "benchmarking" them to the access charges imposed by the local rate-regulated ILEC (or to proxies for such charges).[34]

The underlying problem in this "CLEC access charge" controversy is sometimes described as a market failure—a manifestation of a *terminating access monopoly* that each LEC is said to hold over all interconnecting carriers for access to its own customers.[35] But the problem was actually a straightforward regulatory failure attributable to Title II rules. Those rules had forced each long-distance carrier not only to interconnect with CLECs to whom its customers placed calls, but also to pay whatever rates those CLECs had unilaterally filed in the form of legally enforceable tariffs.[36] Because no regulator was closely scrutinizing those tariffs before they assumed the force of law, it was hardly surprising that the rates they contained exceeded any realistic measure of cost.

But periodic regulatory scrutiny is not enough to ensure reasonable rate levels in a tariff-based environment. As the ink was drying on the CLEC access charge order, various ILECs and CLECs began exploiting a new arbitrage opportunity known as *access stimulation* or, more colloquially, *traffic pumping*. This scheme typically arose in rural areas where ILECs are subject to rate-of-return regulation (see chapter 2). In those areas, per call costs are high because scale economies are low. Population density and call volumes are a fraction of what they are in urban areas, and the high fixed costs of deploying and operating a network are thus apportioned over many fewer subscribers and calls. In such areas, rate-of-return ILECs have long been permitted to tariff high terminating access charges, keyed to assumptions about these low call volumes. And CLECs operating in those areas have likewise been permitted to tariff terminating access charges benchmarked to the same level as the local ILEC (or to applicable proxies).

The problem is that the rate-of-return proceedings that justify particular access charges are complex affairs that regulators undertake only once in a while; regulators do not continually adjust rates to reflect changes in network costs, subscribership, or call volumes. In its *USF–ICC Reform Order*, the FCC described the predictable arbitrage opportunity that some rural ILECs and CLECs began exploiting to avail themselves of this regulatory lag:

> Access stimulation occurs when a LEC with high switched access rates enters into an arrangement with a provider of high call volume operations such as chat lines, adult entertainment calls, and "free" conference calls. The arrangement inflates or stimulates the access minutes terminated to the LEC, and the LEC then shares a portion of the increased access revenues resulting from the increased

demand with the "free" service provider, or offers some other benefit to the "free" service provider. The shared revenues received by the service provider cover its costs, and it therefore may not need to, and typically does not, assess a separate charge for the service it is offering.[37]

As the Commission added, these "[a]ccess stimulation schemes work because when LECs enter traffic-inflating revenue-sharing agreements, they are currently not required to reduce their access rates to reflect their increased volume of minutes."[38]

These schemes were a win-win-win for the LEC, for the "free" conference call service (or other high-call-volume business), and, in turn, for all of that service's users (e.g., the people availing themselves of "free" conference calls). So who was left holding the bag? All other customers of interconnecting carriers. As the FCC explained, "the wireless and interexchange carriers . . . paying the increased access charges are forced to recover these costs from all their customers, even though many of those customers do not use the services stimulating the access demand."[39] Initially, some of these carriers engaged in self-help, blocking their subscribers' calls to the traffic-pumping LECs. Perceiving a threat to "the ubiquity and reliability of the nation's telecommunications network,"[40] the FCC forbade such self-help measures but promised to investigate the underlying reasonableness of the rates.[41] Ultimately, the Commission acted to fix the problem at its source by requiring any LEC whose accounts bear tell-tale signs of "access stimulation" to "file revised tariffs to account for its increased traffic."[42]

F. Intercarrier compensation for mobile and tandem transit providers

Let's pause here to survey the hodgepodge of mutually inconsistent intercarrier compensation rules we have discussed so far, which the reforms begun by the FCC in 2011 are now gradually rationalizing. When a calling party's ILEC hands a call off to a conventional long-distance carrier, the latter pays the former *originating access charges*. If the ILEC hands the call off to an ISP instead, the latter pays the former the standard *retail rate* for a business line—the rate paid by customers rather than carriers. If the ILEC hands the call off to a CLEC that in turn serves the ISP, then the ILEC pays the CLEC the *specialized low rate for ISP-bound traffic* so long as the ILEC has "opted into" the FCC's grand compromise and charges the same rate to complete incoming local calls from wireless providers and others. If, in its discretion, the ILEC has *not* opted into that compromise, it must pay the CLEC the higher *reciprocal compensation* rate derived from TELRIC. In contrast, when a long-distance carrier hands a call off to the called party's ILEC, it must pay a *terminating access charge*, which may vary radically depending on whether the call crosses state lines. And whether the call crosses state lines or not, this terminating access charge that the long-distance carrier must pay the ILEC will almost certainly be considerably higher than a TELRIC-based reciprocal compensation fee, even though the ILEC is performing essentially the same functions as when it completes incoming calls from CLECs and wireless carriers. Finally, if that long-distance carrier collaborates with a CLEC to disguise the long-distance call as a local call when handing it off to the ILEC—well, that might be a *crime*.

This list of mutually inconsistent compensation rules, moreover, describes only one dimension of the chaos. Mobile and "tandem transit" services add entirely new dimensions of complexity to the picture.

One key issue confronting the FCC around the turn of the millennium concerned whether wireless carriers, like LECs, are entitled to collect access charges when they terminate a long-distance call handled by some other carrier's long-distance network. In 1996, the FCC tentatively suggested that wireless providers should be entitled to collect such charges.[43] The FCC then let the matter fall into regulatory limbo for six years. In 2002, it found that because wireless carriers are subject to mandatory detariffing (see chapter 4), they may not impose terminating access charges by tariff and may obtain them, if at all, only by successfully negotiating for them with long-distance carriers.[44] The practical upshot is that wireless carriers rarely recover terminating access charges.

Recall, though, that wireline CLECs have no such disability, even though there is no clear policy reason for treating them differently from wireless carriers for these purposes. That artificial dichotomy gave rise to its own set of regulatory disputes. A given wireless carrier, recognizing that it probably could not collect access charges directly for the long-distance calls it terminates, would arrange for such calls to be delivered to an intermediate CLEC, which, unlike the wireless carrier, can assess such charges by tariff. Suppose the CLEC in this scenario billed the long-distance carrier for full access charges and did not disclose that the called parties actually subscribed to the wireless carrier rather than the CLEC itself. Once the long-distance carrier paid the bill, the CLEC would pass a portion of its access revenues back to the wireless carrier behind the scenes. But if the long-distance carrier in this arrangement found out that a wireless carrier rather than the CLEC was in fact terminating these calls, it could litigate to recover the portion of the access charges corresponding to the functions performed by the wireless carrier.[45]

Figure 7.6
Tandem transit service

Adding to the confusion is uncertainty about the legal rules applicable to *tandem transit* services.* Suppose you subscribe to one mobile service provider, and you call a friend who subscribes to another. The odds are that the two carriers do not interconnect with each other directly; they rely instead on a third-party network to link them indirectly. More generally, any given wireless carrier, CLEC, or rural ILEC will often choose to avoid the expense of building its own transport facilities out to each of the other networks with which it exchanges calls, including networks operated by other wireless carriers, CLECs, or rural ILECs. Instead, each of these different types of carriers often relies on some other carrier in the middle, which will bridge the calling and called parties' networks in any traffic exchange by routing their calls through an intermediate "tandem switch" and charging per minute fees for the service (see figure 7.6).

The transit provider in the middle was traditionally a large ILEC, with its ubiquitous local network; the originating carrier would drop off the call at the ILEC's tandem switch (see chapter 2), and the ILEC would transport it from there to a point of interconnection with the terminating carrier. More recently, other carriers, such as Level 3 and Inteliquent (Neutral Tandem), have sprung up in many areas to provide tandem transit services in competition with the ILECs.

Critically, a provider of tandem transit services has no independent relationship with either the calling or the called parties and thus cannot

* These tandem transit services should not be confused with the *Internet* transit services offered by backbone providers, which perform analogous but quite distinct functions. Tandem transit services are circuit-switched telecommunications services that make use of PSTN infrastructure to connect two participants in a voice telephone call. As discussed in chapter 5 and later in this chapter, Internet transit services are packet-switched services that connect individual IP networks to much or all of the rest of the Internet by means of unregulated contractual arrangements.

recover its network costs from them. Transit arrangements thus raise critical questions about what the transiting carrier in the middle may charge for its services to the originating and terminating carriers (who *do* serve the calling and called parties) and what, if anything, the terminating carrier may charge the carrier that originated the call. The FCC has resolved a few of those questions, but it has deferred many others, including the all-important issue of whether and when ILECs should be required to offer tandem transit services at regulated rates.[46] The result is more uncertainty and more state-by-state litigation of industry-wide issues that are appropriately subject to a unified national resolution, at least at a high level.

II. The Rise of Bill-and-Keep as the Cornerstone of Intercarrier Compensation Reform

So long as regulators continue enforcing interconnection requirements, the question of who pays what to whom should turn on coherent economic principles applied equally to any exchange of telecommunications traffic. The fee charged for terminating a call should not vary radically, as it traditionally has, on the basis of technologically irrelevant details such as the call's geographic origin or the legacy regulatory classification of the originating carrier.

As we have seen, one basic problem is that most of the distinctions that regulators have historically drawn between carriers and services—such as those between "local" and "long-distance" calls and those between information service providers and long-distance carriers—are unstable and ultimately unprincipled in a competitive world. Such distinctions may have made sense in the wake of the AT&T breakup in 1984, when there was no competition from CLECs, wireless carriers, or VoIP providers. Back then, tidy regulatory fences separated local telephone companies from each other geographically, and the distinction between a local and a long-distance call was relatively straightforward. Consequently, for the decade or so following the AT&T consent decree, the access charge rules created only modest opportunities for economically inefficient gamesmanship. But those days are long gone. In the first years of the new millennium, the FCC understood with increasing urgency that its traditional regulatory regime was a sinking ship—with technology and competition creating new holes faster than regulatory responses could plug the old ones.

But arbitrary regulatory categories are not the only source of the problem. Much of the regulatory dysfunction discussed in the previous section—inflated ISP reciprocal compensation rates, inflated CLEC access charges, and traffic pumping—reflect the danger, inherent in any calling-network-pays regime, that regulators will set termination rates too high. If regulators were able somehow to "get the rates right" so that they perfectly match "cost" (whatever that might mean), these problems would not arise or at least not in the same forms. The main problem is that regulators, no matter how skilled, are neither omniscient nor able to respond immediately to market changes.

These are the reasons why in 2001 the FCC began asking whether it should eliminate the traditional crazy quilt of calling-network-pays rules in favor of a unified bill-and-keep regime, which retains the underlying interconnection obligations but minimizes regulatory mandates for one carrier to pay another fees for calls that cross both carriers' networks.[47] For the ensuing ten years, however, the FCC could not muster the political will needed to undertake such fundamental reform. Although it solicited reform proposals of baroque complexity from various industry coalitions, each of those proposals came under heavy fire from powerful interest groups. The strongest opponents included the rural telephone companies that have historically relied on inflated access charges to supplement the often below-cost rates they charge their end users. Finally, as described in section III, the FCC brought these grand "unification" ambitions to fruition in the *USF–ICC Reform Order* of 2011, charting a course for replacing calling-network-pays with bill-and-keep for the termination of all calls on the PSTN. Before we turn to the specifics of that *Order* and the subsequent legal challenge, we first explore the theoretical foundations of these two opposing philosophies of cost-recovery for calls traversing multiple networks.

We begin with a critical caveat. The regulatory choice between calling-network-pays and bill-and-keep arises in the marketplace for voice telephone services only because the government has imposed interconnection obligations in that context in the first place. In contexts where one network can *refuse* to interconnect with another (because, for example, it dislikes the pricing terms), there is generally no role for regulatory pricing rules to play because those rules are applicable only if the two networks *do* interconnect. But ever since the Kingsbury Commitment of 1913 (see chapter 1), there has been widespread consensus, enshrined in sections 201 and 251 of the Communications Act, that the government should enforce interconnection obligations on common carrier providers of voice telephone service. The concern has always been that, in the absence of such obligations, larger networks would harm smaller ones by denying interconnection on reasonable terms or that the unified national telephone system would fragment, leaving some people unable to call others.

That said, there is one immensely critical setting where the government imposes no interconnection obligations, but where "carriers" nonetheless do interconnect, normally with great efficiency: the marketplace for Internet peering and transit arrangements. At the end of this chapter, we address why U.S. policymakers have generally concluded, sometimes over foreign opposition, that regulation of Internet interconnection arrangements would be both unnecessary and counterproductive. And we discuss the implications of that policy position as "voice" telephone traffic migrates inexorably from the circuit-switched PSTN, with its legacy interconnection obligations, to an all-IP ecosystem, where "voice" packets account for a very small share of data packets generally. For now, however, we address the theoretical contest between calling-network-pays and bill-and-keep as it arises in the one context where interconnection obligations remain a given: the circuit-switched PSTN.

A. The economic challenges of a calling-network-pays regime

Under the calling-network-pays approach, the calling party is deemed to "cause" all the costs of any given call, and the called party's carrier may thus charge the calling party's carrier (a LEC for local calls and a long-distance carrier for long-distance calls) a regulated rate for the costs of completing that call. This was the consensus intercarrier compensation scheme in the United States for many decades. But for historical and policy reasons, regulators devised two very different approaches to measuring those costs, one of which (reciprocal compensation) has applied to "local" traffic, and the other of which (access charges) has applied to "long-distance" traffic.

As we have explained, it is unsustainable over the long term to have two different cost inquiries arising from an artificial regulatory distinction between "local" and "long-distance" calls for what amounts to the same call-termination functions. There have been, to be sure, powerful political reasons for preserving the higher access charges for long-distance service because those have traditionally cross-subsidized low residential local service rates. But the arbitrage opportunities discussed earlier have long eroded this artificial distinction as service providers found more and more ways to avoid paying access charges. Thus, realistic advocates of a calling-network-pays approach have had to concede the necessity over the long term of reconciling these two different schemes and of establishing a single, coherent means of measuring "cost" for purposes of terminating any call.

The main challenge for any calling-network-pays regime is to ensure that regulators get these costs "right" in the sense that the intercarrier rates they set neither overcompensate nor undercompensate the terminating carrier. If regulators set costs and rates too high, they will produce the types of economically inefficient regulatory distortions described earlier, such as the above-cost termination rates that induced a generation of CLECs to specialize in serving ISPs in part to capture a regulation-driven windfall. The problem is that there may be no "right" way, even in theory, to measure call-termination costs. Moreover, even if there were such a way, regulators would face formidable challenges in applying economic theory to political and technological reality.

Several independent factors complicate administration of any calling-network-pays approach. As with the pricing of unbundled network elements, a regulator's first task is to choose a basic cost methodology (such as TELRIC), and the complexities of that threshold inquiry are formidable (see chapter 2). And choosing a particular cost methodology for call-termination services—say, some variant of forward-looking cost—still leaves many basic questions for which there may be no theoretically satisfying answer. First, should a regulator take into account the type of carrier whose forward-looking costs are at issue? There has been much debate between wireless and wireline carriers as well as between ILECs and CLECs about the disparate costs of terminating traffic over different types of networks. Each carrier naturally wants regulators to raise their estimates of its call-termination costs and lower their estimates of other carriers' termination costs. Should regulators compromise by making ILECs and wireless carriers pay each other the same rate for call termination, in which event that rate (as an average) may reflect no carrier's true network costs? Or should different carriers pay different rates that reflect the technology-driven cost differences in their respective networks, in which event regulators will stand accused of artificially favoring some technological choices over others? These questions have always been topics of intractable debate.[48]

Second, how should regulators apportion any given carrier's termination costs among the call-originating carriers that—under the basic premise of the calling-network-pays rule—are said to "cause" those costs? As a thought experiment, assume that a given carrier has installed a switch that does nothing except terminate traffic to a class of customers. Under the calling-network-pays rule, the carrier can expect to recover the entire cost of that switch—not from its own customers, but from the carriers that originate the calls that are then terminated through the switch. We know that the total costs to be recovered are those of the switch, that the period for recovering them is whatever regulators predict to be the useful life of the switch, and that those costs need to be recovered from other carriers (because, by hypothesis, the switch is used only to terminate other carriers' calls). But there is no one economically "correct" way of apportioning payment responsibility among the many different carriers that originate these calls because the short-term marginal costs of actually terminating a call for the benefit of any given carrier, once the network is up and running, are effectively zero. The terminating carrier incurs most of its costs before any call is placed at all—that is, when it orders and installs the switch and transport facilities large enough to accommodate the unusually high call volumes present during "peak load" periods. To cover its high fixed costs, however, the terminating carrier in our example must recover rates of greater than zero from someone. Doing so means that rates will not be paid in close correspondence to how those costs are incurred.[49]

As a practical matter, regulators typically order some variant of per minute pricing because such pricing is the most feasible way to try to allocate responsibility for these costs among the many different carriers that deliver traffic to the facilities whose costs are being recovered. But per minute pricing raises more questions than it answers. If regulators set the same rate for all hours of the day, they may produce what amounts to a cross-subsidy running from those who use the network mainly during off-peak hours to those who use it mainly during peak hours (because many network costs are incurred specifically to ensure enough capacity to handle calls during peak hours). For that reason, regulators sometimes impose higher per minute rates for peak periods than for non-peak periods. But this can be only a partial solution to the cross-subsidy problem because, as the FCC has noted, there may be no "right" mark-up for use of the network during peak periods, no "right" way to define when those peak periods occur, and, more generally, no economically satisfying way to use such premium rates to reflect the up-front, lumpy manner in which carriers incur the fixed costs of switching and transport.[50]

Of course, these pricing questions are an issue not just for intercarrier compensation for call-termination services, but also for retail pricing plans to end users. In competitive markets (such as wireless), carriers offer their subscribers a continuously changing menu of options featuring different combinations of flat and usage-sensitive rates. If the carrier's executives are doing their jobs, these payment plans will produce an income flow that manages to cover the carrier's costs. But such carriers succeed despite the absence of any determinate answer to the problem of recovering their high fixed costs over time. And they succeed because, unlike regulators, they have the flexibility to use several different compensation options at once and to change their customers' menu of options as soon as the need arises.

Finally, even if there *were* a "right" methodology for calculating and apportioning the call-termination costs of the many different carriers in the world, regulators would still confront enormous subjectivity in the task of applying that methodology to produce actual rates. Take the two primary contexts in which regulators have set intercarrier compensation rates: the "reciprocal compensation" regime for local calls and the "access charge" regime for long-distance calls. Throughout much of the 1990s, the rates for interstate access were tied up in protracted litigation while carriers successfully attacked, as arbitrary and capricious, the FCC's "X-factor" formula (which, as discussed in chapter 2, reduced access charges over time by accounting for presumed efficiency improvements in the industry as a whole).[51] As for local calls, reciprocal compensation rates can vary by more than 100% from state to state—presumably not because of any commensurate difference in the underlying "cost," but because state regulators disagree fundamentally about how to measure cost.

Given these multiple layers of theoretical indeterminacy, how can we reliably know when regulators have set intercarrier compensation rates at "correct" levels? We cannot. We can, however, hope for some measure of regulatory rough justice. And we can surmise in hindsight that rates have been set about right if they do not appear to have caused major competitive distortions. We also know in hindsight when the rates have been set incorrectly, as they were in the case of ISP-bound traffic, because then they do cause major distortions. But our foresight can never be so accurate as to ensure that we will avoid such mistakes, so regulatory rate setting of any kind—including reciprocal compensation—necessarily requires constant readjustment.

B. The bill-and-keep alternative

All of the regulatory conundrums just discussed arise wherever regulators choose, as a threshold matter, to entitle carriers that are subject to regulatory interconnection obligations to recover their call-termination costs from *other carriers*. Bill-and-keep, which the FCC has now adopted as its PSTN methodology of the future, represents a radically different threshold choice. Instead of making the calling party's carrier responsible for *all* of the costs of a call, bill-and-keep divides responsibility for those costs between the carrier serving the calling party and the carrier serving the called party. Specifically, it allocates (1) to the calling party's carrier responsibility for the costs of delivering the call to a demarcation point (defined by regulation) between the two networks and (2) to the called party's carrier the responsibility for the costs of transporting the call the rest of the way from that point to the called party.*

Thus, whereas the calling-network-pays rule imposes all "transport and termination" costs on the calling party's carrier, bill-and-keep imposes all "termination" costs and potentially some "transport" costs on the called party's carrier.[52] How will that carrier recover those costs? By building them into the rates it charges its own end users. Of course, that does not necessarily mean that end-user rates will go up, for each carrier is simultaneously freed of the responsibility to cover any costs attributable to the termination of other carriers' calls. Put differently, bill-and-keep addresses the problem of the terminating carrier's cost recovery not by regulating the intercarrier compensation that one carrier must pay another, as calling-network-pays does, but by eliminating intercarrier compensation altogether, at least in all cases where the sending carrier drops off traffic at a specified default point on the terminating carrier's network (the "network edge" concept we describe later in this subsection).

In one special set of cases, bill-and-keep is uncontroversial and indeed dovetails with the calling-network-pays approach. In particular, all agree

* For simplicity, our discussion focuses on the dynamics of bill-and-keep for calls involving only two carriers, but the same basic rules apply to calls involving a third carrier—one that provides transport services in between LECs on each end of a call. Such three-carrier calls fall into two basic categories: those in which the intermediate transport provider has an independent relationship with the calling party and those in which it does not. In both contexts, bill-and-keep requires the terminating carrier to accept calls from the intermediate provider at a designated point without charge. In the first context, the intermediate transport provider (the calling party's long-distance company) incurs the financial responsibility to deliver the calls to that point. In the second context, the originating LEC bears that financial responsibility, and because (by hypothesis) its own facilities do not extend to the terminating carrier's network, it hires the intermediate transport provider as a paid subcontractor to perform that function. In both contexts, the carrier with the financial responsibility recovers its costs from its own end users.

that bill-and-keep is appropriate where two comparable carriers have "balanced" traffic flows—that is, where neither is, with respect to the other, a net originator or terminator of traffic. In those circumstances, their respective liabilities more or less cancel out, and the transaction costs of monitoring all the traffic may exceed any net liability one carrier might be shown to have to the other. Our discussion thus does not focus on whether bill-and-keep is desirable in these circumstances. Instead, we are addressing only the harder case in which the traffic flow between two carriers is "unbalanced" (or "asymmetric"). Such imbalances have historically arisen, for example, in the case of traffic between ILECs and ISP-serving CLECs and between wireline LECs and mobile wireless carriers, which traditionally originated more calls than they terminated.

Proponents of bill-and-keep argue that over the long term, as the telecommunications industry becomes more competitive, bill-and-keep will require far less regulatory intervention in the PSTN than a calling-network-pays approach. To see why this may be so, take the problem of excessive terminating access charges. As we saw earlier, any LEC—even a nondominant CLEC—has incentives to game the regulatory system so that it may charge other carriers significantly more than "cost" for the privilege of completing a call to its customer. Now suppose that the LEC has an obligation to terminate calls originated by another carrier—that is, an interconnection obligation—but no power to charge that other carrier anything at all for performing the task so long as that other carrier delivers the call to the defined point of interconnection. How can the LEC recover these call-termination costs if not from the originating carrier? Simply by passing them on to its own end users. And here is the crucial point: the LEC must compete for those end users against various rivals in the local exchange market. What if the LEC tried to charge those end users the same above-cost rates for call termination that regulators have historically entitled LECs to charge, say, long-distance carriers? *It would lose them as its customers*. The fear of that outcome will cause the LEC, at least in a competitive market, to lower its rates to cost.

This point is a central argument for bill-and-keep. One defining drawback of the calling-network-pays approach is that regulators must continue recalibrating their rules in perpetuity to reflect the changing costs of call termination. That is because, under such an approach, market forces alone are often incapable of producing genuinely cost-based inter-carrier compensation rates no matter how competitive the industry becomes. In contrast, regulators need not remain forever involved in the estimation of a carrier's costs if direct responsibility for paying them is shifted—as bill-and-keep prescribes—from other carriers to the terminating carrier's own end users. After all, a competitive market (by definition) will itself produce end-user rates that reflect the underlying costs of providing service—and it will do so with greater accuracy and far less controversy than any regulatory proceeding could ever do. Also, end users in the aggregate would be no worse off under this approach because (as discussed) each carrier would no longer pass along to its customers the call-termination costs imposed by other carriers. To the contrary, end users may well be better off in the aggregate because the market as a whole would benefit from greater regulatory certainty, lower administrative costs, and fewer competitive distortions.[53]

To be sure, in the near term, regulators would still need to regulate the end-user rates of "dominant" carriers in markets that are not effectively competitive. A carrier whose customers lack alternatives may often get away with charging its customers above-cost rates for any number of services. That, indeed, is the whole rationale for retail rate regulation, as discussed in chapter 2. In the long term, however, the advantage of bill-and-keep is that as competition develops and fewer carriers remain "dominant," rate regulation will become increasingly unnecessary because market forces (acting on end-user rates) could perform the cost-estimation functions traditionally assumed by regulators. Over time, therefore, bill-and-keep may bring an end to much of the PSTN-related regulatory indeterminacy that has plagued the implementation of the calling-network-pays regime for many decades.

Bill-and-keep is nonetheless subject to a number of objections, which run the gamut from the highly theoretical to the highly pragmatic. We address the theoretical objections first, which, as a general rule, are less substantial than the pragmatic ones.

Some critics contend that bill-and-keep defies the economic principle that costs should be allocated to the party that "causes" them because, they say, the costs of a call are attributable to the calling party and her carrier. That objection is somewhat overstated, however, because the called party can also be said to "cause" at least some of the costs of a call.[54] In particular, those costs would never be incurred but for the called party's active cooperation in the form of deciding to be on the network and thus available to receive calls and subsequently answering the telephone when it rings and engaging in conversation rather than hanging up immediately.[55]

The opponents of bill-and-keep further argue that it would create perverse incentives for carriers to specialize in serving customers (such as telemarketers) that mostly originate calls, just as the calling-network-pays regime created perverse incentives for carriers to specialize in serving customers (such as ISPs) that mostly terminate calls. Indeed, the FCC itself expressed that concern in rejecting bill-and-keep for unbalanced traffic in 1996.[56] But this concern appears overstated as well, as the FCC recognized in 2011.[57] The main reason that the traditional calling-network-pays regime favored carriers that specialized in the termination of traffic is that regulators had set termination rates above cost, entitling those carriers to a mini-subsidy whenever they terminated a call. In a bill-and-keep world, each carrier would still incur the considerable costs of originating calls and transporting them to some point on the called party's network, and it would have to recover those costs from end users, at rates those end users are willing to pay, rather than from other carriers at whatever rates a regulator might set. Those end users will be unwilling to pay rates above competitive levels, and thus the market itself will foil an originating carrier's plans to derive supracompetitive revenues for its services.

The deeper controversies about bill-and-keep relate to more pragmatic questions. Recall that bill-and-keep requires regulators to pick a demarcation point between two networks that separates where the first carrier's financial obligations begin and the other's ends. But it is not at all obvious how that point, which the *USF–ICC Reform Order* calls the *network edge*, should be defined. From a competitive standpoint, moreover, much turns on that definition.

One of the earliest proposals—devised in 2000 by Patrick DeGraba, an economist with the FCC at the time—argued for setting the default point of interconnection at the end office used by the terminating carrier to serve the called party.[58] Thus, if Joe, served by Carrier X, places a call to Sue, served by Carrier Y, then Carrier X would bear financial responsibility for delivering the call to the end office that houses the Carrier Y switch that is directly connected to Sue's loop. One concern about this approach is that it could place CLECs at a competitive disadvantage as compared to ILECs. In general, the traditional network of a large ILEC typically features many switches (and thus end offices), and modern CLEC networks tend to feature fewer switches and longer loops. Thus, whereas an ILEC would satisfy its financial obligations under DeGraba's approach by building lines to just a few CLEC end offices, the CLEC could satisfy its corresponding obligations only by building or leasing connections to a much larger number of ILEC end offices. A strong consensus thus arose that where the terminating carrier is a large ILEC with a tandem switch serving several end offices, the sending carrier should be able to satisfy its financial obligations by dropping the call off at the tandem switch. Indeed, the FCC has now built that consensus into its new bill-and-keep regime.[59] Although that rule creates some formal asymmetry between the obligations of CLECs and ILECs (which still bear financial responsibility to deliver calls to CLEC end offices), it does produce rough symmetry in practical result.

Significantly, establishing a "default" point of interconnection does not necessarily require any carrier to build its own facilities out to that point on pain of losing interconnection rights.[60] Carriers remain free to interconnect elsewhere on the terminating carrier's network as well; the default point ("edge") merely prescribes the division of *financial* responsibility for covering the costs of any given call in the absence of a negotiated agreement to the contrary.[61]

In the simplest case, if the first carrier actually does drop the call off at the default point and the second takes it from there, the application of bill-and-keep is straightforward: no money changes hands, and the carriers recover their respective costs from their own end users. But what happens if the sending carrier lacks the transport facilities needed to reach that point? The carrier must hire some other provider to bridge the gap either by leasing fixed transport capacity (e.g., special access links or "entrance facilities") or by selling usage-sensitive transport services billed by the minute (e.g., "switched transport" or tandem transit services). The seller in such scenarios will sometimes, but not always, be the carrier on the other end of a call. Either way, whether that seller is subject to price regulation for these purposes should theoretically depend on whether the relevant market is sufficiently competitive to produce efficiently priced transport services in the absence of regulation.

Achieving that policy outcome, in which rate regulation depends on the market power of the selling carrier, may require the FCC to revisit its position as amicus curiae in the *Talk America* litigation. As noted at the beginning of this chapter, the Commission's brief in that case opined that ILEC-supplied "entrance facilities" designed to bridge CLEC and ILEC networks can be subject to TELRIC-based rate regulation under the interconnection provision of section 251(c)(2) even if they do not meet the "impairment" standard applicable to section 251(c)(3)—that is, even if the ILEC faces substantial competition from other network middlemen. The Supreme Court deferred to that position in upholding a state commission ruling to the same effect, but the FCC would presumably be free to change its mind.

III. The *USF–ICC Reform Order* and the 1996 Act

A. Overview of the *Order*

As we have previewed, the FCC issued its *USF–ICC Reform Order* in late 2011. Weighing in at 740 pages and with more than 2,500 footnotes, the *Order* is one of the most complex and important FCC actions in recent memory, reforming not only intercarrier compensation, but also critical portions of the federal universal service program, which we discuss in the next chapter.

At the most fundamental level, the intercarrier compensation portions of the *Order* chart a course for replacing the traditional hodgepodge of calling-network-pays methodologies with a more unified bill-and-keep approach. The Commission cushioned the financial disruption by drawing the transition out over several years, first reducing intrastate terminating access charges to interstate levels and then gradually reducing all termination rates—for both "access" and "local" traffic—to zero (i.e., bill-and-keep).[62] The Commission gave the (mostly rural) LECs subject to rate-of-return regulation a longer transitional period than the LECs subject to price-cap regulation (see chapter 2), reasoning that the rate-of-return LECs have historically been smaller and more reliant on intercarrier compensation to cover their network costs and are thus more vulnerable to abrupt regulatory changes.

In exchange for these lost intercarrier revenues, the Commission gave LECs both (1) greater flexibility to charge higher rates to their own end users (insofar as retail competition allows) and (2) in some cases rights to additional explicit subsidies from the universal service fund. Here, too, the details of the rules are more solicitous of rate-of-return carriers than of price-cap carriers,[63] although not solicitous enough to avoid triggering a legal challenge by many of the former. In line with the basic rationale for bill-and-keep, the Commission emphasized that although end-user rates for "local" landline services might rise as a result, end-user rates for wireless and long-distance services would fall. Indeed, the Commission projected (somewhat subjectively) that its reforms would eventually "provid[e] over $1.5 billion annually in benefits" for consumers of wireless and long-distance services and that "the average consumer benefits of [its] reforms [will] outweigh any costs by at least 3 to 1."[64]

The *Order* resolved a great many regulatory questions that had paralyzed the FCC in the past, but it deferred a number of key implementation issues, either to future rulemaking proceedings by the FCC itself or to separate proceedings by individual state commissions. For example, the *Order* addressed issues relating to *termination* charges (access and local), but not various corresponding issues relating to *originating* access charges. It left unresolved important details, in various call scenarios, about the location of the "network edge"—which, as discussed, is the notional point in any bill-and-keep regime where the sending carrier's financial obligations end and the terminating carrier's begin. The *Order* discussed but did not resolve key disputes relating to tandem transit and other transport services. And most important of all, it left until another day crucial questions about whether regulation is needed to govern interconnection arrangements between IP networks for the exchange of voice traffic. We address that highly contentious issue in section IV.

B. Legal rationales and litigation

The *Order* promptly triggered one of the most massive lawsuits in the history of the FCC. Several dozen parties—including state commissions, large and small ILECs, major CLECs, and smaller wireless carriers—filed separate appeals in different federal courts on an immense range of intercarrier compensation and universal service issues. A number of parties challenged some aspects of the *Order* but at the same time intervened on the FCC's side to defend other aspects. All of these disparate appeals were ultimately consolidated and assigned by lottery to the U.S. Court of Appeals for the Tenth Circuit, based in Denver, where they remain pending as this edition goes to press.[65]

The FCC's disputed jurisdiction

One key question in that case is whether the FCC has comprehensive legal authority to regulate intercarrier compensation for all categories of PSTN traffic exchanges, both interstate and intrastate. The answer depends on how one interprets several profoundly ambiguous sentences in the 1996 Act.

The FCC has two explicit sources of authority for regulating intercarrier compensation paid to wireline carriers.[66] First, it has long had general authority under section 201 to regulate the terms and conditions of interstate and international services. Second, as discussed in chapter 2, it has more specific authority, under *Iowa Utilities Board*,[67] to issue rules implementing the basic provisions of the 1996 Act, including the "reciprocal compensation" provision of section 251(b)(5). The FCC's discretion to enforce its policy preferences has always been greatest as to interstate traffic within the scope of section 201. Like the other provisions of the original 1934 Act, section 201 places few constraints on the substance of the FCC's rules beyond the general requirements of "reasonableness" and adequate explanation.

What, though, of calls that begin and end within a single state and thus fall outside the FCC's traditional rate-setting authority for "interstate" calls? Many intrastate calls are "local," exchanged between carriers operating in the same local calling areas. Call-termination arrangements for such calls are indisputably subject to section 251(b)(5), which requires "reciprocal compensation arrangements for the transport and termination of telecommunications."[68] And they are thus subject as well to the FCC's general rulemaking jurisdiction, recognized in *Iowa Utilities Board*, to implement all provisions of section 251.

The FCC still confronted a jurisdictional challenge for one critical class of traffic. In 1996, it had construed section 251(b)(5) to cover only "local" intrastate traffic and not long-distance ("access") traffic that begins and ends in remote calling areas *within the same state*.[69] Such *intrastate access* traffic appeared to fall in a no-man's-land between section 201 (*interstate* long distance) and section 251(b)(5) (intrastate *local*)—and thus categorically outside the FCC's jurisdiction.[70] This was a serious problem for any comprehensive intercarrier compensation reform, given that intrastate terminating access charges, approved by 50 state commissions, have historically been subsidy laden and much higher than other termination rates.

In the *USF–ICC Reform Order*, the Commission sought to address this problem by explicitly reinterpreting section 251(b)(5) to encompass not only local traffic, but all transport and termination by a LEC of *any* form of "telecommunications," including both interstate and intrastate access traffic.[71] As the Commission observed, the text of section 251(b)(5) does not expressly distinguish between "local" and "long-distance" calls, and that dichotomy is arguably just a relic of an obsolescent regulatory paradigm in any event.[72]

Critics argued, however, that Congress could not have meant to include "access" (nonlocal) traffic within the scope of section 251(b)(5). First, they said, a provision requiring "reciprocal" compensation for the "transport and termination" of calls appears directed at arrangements between *two* carriers that terminate each other's calls to their respective end users.[73] In 1996, most two-carrier calls were "local," whereas most long-distance calls with different LECs on each end involved an independent third carrier—a long-distance company—as a middleman. The language of section 251(b)(5) seems to have been written without these traditional three-carrier access calls in mind.

Critics also noted that Congress almost certainly did not mean to trigger a multi-billion-dollar flash cut in intrastate access charges—a critical source of universal service subsidies—back when it enacted this provision in 1996. At first blush, this flash cut might seem to be the logical outcome of reading section 251(b)(5) to apply to all "telecommunications," given that the Act permits ILECs to recover at most the "additional costs" of terminating calls subject to that provision.[74] To this argument, however, the FCC had a rejoinder.[75] Section 251(g), noted earlier in this chapter, grandfathers the "access" rules adopted by the FCC itself before 1996 or by the federal court administering the AT&T consent decree—until the FCC explicitly supersedes them with new rules, as it did in the *USF–ICC Reform Order* when it extended section 251(b)(5) to cover access charges. Thus, the FCC suggested, Congress could hardly have feared that a broad interpretation of section 251(b)(5) would immediately trigger an instantaneous flash cut in access charges. Section 251(g) froze them in place, subject to the FCC's ability to phase them out and slowly wean the industry from the implicit subsidies they contained. This argument is strong with respect to interstate access charges, which are plainly covered by section 251(g); it is more debatable with respect to intrastate access charges, which are not as straightforwardly covered by that provision.

All this said, the FCC receives considerable deference from the courts in resolving such statutory ambiguities. The legal question is not what individual members of Congress subjectively thought they were doing in 1996 (if they were thinking about these issues at all). Instead, courts look mainly at whether the words Congress enacted in a statute permit the construction an agency ultimately gives them.[76] And the courts grant the FCC significant leeway when it construes the terms of the 1996 Act, which the Supreme Court has aptly called "a model of ambiguity or indeed even self-contradiction."[77]

The disputed legal basis for bill-and-keep

Apart from these disputes about the breadth of the FCC's jurisdiction vis-à-vis the states, the most important intercarrier compensation issue on appeal from the *USF–ICC Reform Order* concerns the FCC's authority to impose *bill-and-keep* specifically as a universal compensation methodology for all calls on the PSTN. Again, after *Iowa Utilities Board*, there is no question that the FCC has statutory jurisdiction to set intercarrier compensation rules for calls within the scope of section 251(b)(5) as part of its general authority to implement any substantive provision of the 1996 Act, even though the calls themselves are usually intrastate. But even if (as the FCC found) all calls fall within section 251(b)(5), the FCC still faces substantive legal objections to the imposition of bill-and-keep rather than some variant of a calling-network-pays methodology.

The text of section 251(b)(5) itself requires "reciprocal compensation arrangements for the transport and termination of telecommunications." Section 252(d)(2) purports to prescribe substantive rules governing such compensation for all traffic covered by section 251(b)(5), at least "[f]or the purposes of compliance by an incumbent local exchange carrier with section 251(b)(5)." The statutory passages containing those rules, however, are also, to borrow again from the Supreme Court's description, "model[s] of ambiguity or indeed even self-contradiction."[78] Section 252(d)(2)(A) directs regulators (1) to "provide for the mutual and reciprocal recovery by each carrier of costs associated with the transport and termination on each carrier's network facilities of calls that originate on the network facilities of the other carrier" and (2) to "determine such costs on the basis of a reasonable approximation of the additional costs of terminating such calls."[79] But the next subparagraph, section 252(d)(2)(B), provides that the statutory language just discussed "shall not be construed . . . to preclude arrangements that afford the mutual recovery of costs through the offsetting of reciprocal obligations, including arrangements that waive mutual recovery (such as bill-and-keep arrangements)."

This *bill-and-keep savings clause* is amenable to several different interpretations. In 1996, the FCC narrowly interpreted it as confined to the "easy" cases noted earlier, where two carriers' reciprocal obligations are fully offset because the traffic flows from each carrier to the other are balanced.[80] In the *USF–ICC Reform Order*, however, the FCC reversed course and construed the same language to authorize it to order bill-and-keep arrangements (which "waive mutual recovery") even for unbalanced traffic so long as carriers have adequate opportunities for "recovery of costs" from their own subscribers, as they normally do.[81] As with the threshold jurisdictional issues, the FCC has arguments it can invoke in support of its statutory interpretation; the question is whether the Tenth Circuit will view the Commission's conclusion as a reasonable resolution of statutory ambiguity.

One final challenge to the FCC's choice of bill-and-keep warrants brief mention here. Recall that, in essence, bill-and-keep prescribes a rate of zero for all traffic exchanged at the designated "network edge." Some state commissions argue that even though *Iowa Utilities Board* upheld the FCC's authority to prescribe a rate-setting *methodology* under the 1996 Act, it has no authority to prescribe *specific rates* for local and other intrastate traffic. And the zero rate prescribed by bill-and-keep, they say, crosses a jurisdictional line that Congress drew in the 1996 Act between the FCC's authority to issue general rules implementing sections 251 and 252 and the states' authority to set specific rates on the basis of those rules in particular factual circumstances. The FCC and its defenders in their turn argue that the *USF–ICC Reform Order* crosses no such line because bill-and-keep is properly conceptualized as a methodology rather than as a "rate" and because, in any event, the states retain some discretion in applying that methodology by defining the location of the network edge at which the sending carrier's financial responsibilities end.[82]

* * *

Navigating the transition to a unified intercarrier compensation regime is legally treacherous in part because in 1996 Congress did not fully appreciate the long-term unsustainability of the legal distinctions governing this area. As of this writing, the legal fate of the FCC's sweeping intercarrier compensation reforms rests with the Tenth Circuit. If that court finds fault with any major aspect of those reforms, and if the Supreme Court cannot be persuaded to intervene, Congress may need to wade back into this quagmire to facilitate a coherent federal solution.

IV. After the PSTN Sunset: Interconnection Policy in an All-IP World

To this point, we have addressed interconnection and intercarrier compensation issues as they relate to the public switched telephone network, that conceptual aggregation of circuit-switched wireline and wireless networks used primarily to transmit voice telephone calls. Over the next decade, most of those PSTN networks will have completed the transition from circuit-switched technology—often denoted by the acronym TDM, for "time-division multiplexing"*—to packet-switched technology in IP format. The FCC's Technology Advisory Council has even called on

* As noted in chapter 2, TDM is properly conceptualized as a circuit-switched technology because even though a TDM-based physical circuit encompasses a stream of unrelated signals, that circuit reserves dedicated time slots—that is, capacity—for any given call for its entire duration, whether anyone is talking or not.

policymakers to begin planning for an official *PSTN sunset* as early as 2018.[83]

What will it mean for the PSTN to "sunset?" It will *not* mean that people will stop talking on the phone, nor will it necessarily mean that they will stop using ten-digit telephone numbers.[84] Instead, it will mean only that the official U.S. telephone system will no longer consist mainly of TDM-based switches and the legacy SS7 signaling networks that direct traffic between those switches. Like Vonage subscribers today, people may go on using ten-digit telephone numbers, but those numbers will simply be proxies for their devices' IP addresses. Calling and called parties will find each other not by means of traditional PSTN databases, but by means of Internet-based servers and routing tables or some more specialized mechanism for uniting the voice customers of disparate IP networks.

Should these IP networks, to the extent they are exchanging voice traffic, be subject to the same general types of interconnection and inter-carrier compensation rules as their circuit-switched antecedents? At first glance, the answer might seem to be yes. After all, basic policy choices should not turn on the technological details underlying particular services, should they? In fact, the answer is not so clear, as we discuss in subsection IV.A. Then in subsections IV.B and IV.C we turn to broader questions about whether regulatory oversight will be needed for interconnection among IP networks more generally. The debate about these questions, which the FCC has just begun exploring as this edition goes to press, is among the most interesting in modern telecommunications policy.

A. VoIP-to-VoIP interconnection

Every day millions of people exchange voice calls or videochats entirely over the Internet via over-the-top VoIP providers such as Skype and Google. Even though the calling and called parties may connect to the Internet through different ISPs, the calls go through because those ISPs interconnect with each other either directly (through peering) or indirectly (through third-party transit networks). Yet these interconnection arrangements are *entirely unregulated* by any U.S. authority. No governmental agency forces Comcast, as a broadband ISP, to find ways to connect its subscribers to CenturyLink's ISP subscribers for purposes of completing a Skype call. But the call is connected anyway. Why? For the same reason you can reach any website from either ISP's broadband Internet access services. As discussed in the final section of this chapter,

Comcast and CenturyLink each have strong incentives to enable their customers to reach every point on the Internet, which includes all customers served by the other network, VoIP users no less than websites.*

From this perspective, VoIP might potentially become just one application like any other riding on top of broadband IP networks, indistinguishable for most purposes from all other applications exchanged over Internet peering and transit arrangements. If that is the right perspective—which is not yet by any means clear—then when the PSTN fades away and all voice calls are placed and received over broadband IP platforms, there might be no greater need for policymakers to impose physical-layer interconnection obligations for the exchange of voice calls than for the exchange of traffic for any other type of application, such as email, instant messaging, or online gaming. That would reduce to irrelevance much of the voice-centric regulation embodied in Title II of the Communications Act, including the bill-and-keep and other intercarrier compensation rules discussed earlier, all of which presuppose underlying interconnection obligations.† There are, however, several complicating factors, of which the most prominent are telephone numbers and quality of service (QoS).

Telephone numbers in an all-IP world

U.S.-based callers have grown deeply accustomed to using ten-digit numbers, both to identify how they can be reached and to identify the people or businesses they wish to call. There is every indication that people will continue using such numbers, at least for a while, even after "telephone" networks have evolved into IP networks that are at least potentially connected to the broader Internet. The problem is that the calling party's IP network typically does not "know," on the basis of a

* This discussion assumes, of course, that ISPs will observe basic industry norms against *blocking* VoIP traffic over best-effort broadband Internet connections. Such blocking has been rare to nonexistent in the United States since the *Madison River* controversy in 2005 (see chapter 6). But it is more common abroad, where ISPs in some foreign countries may seek to identify and suppress VoIP traffic in order to preserve legacy termination-charge revenues, particularly for international calls.

† Of course, policymakers may independently wish to retain some voice-specific Title II rules that serve important social objectives, such as telecommunications access for the hearing impaired (*see* 47 U.S.C. § 225, 255) and requirements for 911 emergency dialing (see *id.*, at § 251(e)(3); Nuvio Corp. v. FCC, 473 F.3d 302 (D.C. Cir. 2007) (upholding 911 rules for interconnected VoIP providers)).

mere telephone number, how to route VoIP-related packets toward the called party's IP network, and there is today no commonly accessible database that can translate such numbers into IP addresses. In other words, there is not yet any analog, for telephone numbers, to the function that DNS servers perform in translating alphanumeric website names into the IP addresses that Internet routers can understand.

As a result, many calls today between customers using different VoIP providers are not routed from beginning to end over IP networks, like conventional Internet traffic, but must make brief detours through TDM-based tandem switches in the middle.* Those detours into the circuit-switched PSTN are inefficient, and they can limit the higher-layer IP functionality that VoIP applications otherwise support. Indeed, the only real function of these detours is to fill in the call-routing gap left by the use of telephone numbers rather than IP addresses to indicate the called party's network location. Over time, the industry might well work out efficient and wholly IP-based arrangements to connect the customers of different VoIP providers, at least for all such providers that wish to be part of a larger interconnected system of voice services. In the shorter term, these arrangements may include bilateral agreements between two VoIP providers to exchange subscriber information and make their higher-layer VoIP protocols interoperable. In the longer term, if the industry can overcome various collective-action and customer-security concerns, VoIP interconnection arrangements may include a more globally accessible "ENUM"-type database that any VoIP provider can query.[85] To date, these initiatives remain private, and the FCC has shown little inclination to intervene, even though Congress has given it plenary authority to oversee "telecommunications numbering."[86]

Of course, not all VoIP providers will wish to share subscriber information, and their higher-layer voice services will remain closed even if the physical-layer networks on which those services ride are interconnected. Today, for example, Skype has hundreds of millions of subscribers

* The longer-term VoIP-to-VoIP interconnection issues discussed here are conceptually distinct from shorter-term VoIP-to-*PSTN* interconnection issues that arise whenever one party to a call is using a VoIP service and the other is still using a conventional TDM-based telephone service. In the latter context, conversions from IP to TDM format are by definition inevitable, and the main question is who should pay for those conversions: the VoIP provider (and any telecommunications partner it may have) or the interconnecting TDM-based telephone company. As of this writing, the FCC has teed up that question but has not yet resolved it.

worldwide who connect to the Internet through many different ISPs but who can call only one another. Similarly, most instant-messaging platforms are not interoperable either; you cannot use Facebook's chat service to communicate with your friend on Google+. The existence of such closed networks is not currently viewed as a problem because, by virtue of the PSTN, everyone can easily join a voice service that *is* interconnected with other voice services serving almost everyone else. In other words, you need not subscribe to many different services in order to engage in real-time communications with everyone you might wish to call; you need only sign up with one provider interconnected with the PSTN, be it Sprint, AT&T, or Vonage.

That fact is a legacy of the PSTN and its traditional, centrally managed interconnection obligations. But there is no obvious reason to assume that this supernetwork of voice networks will fragment once the PSTN sunsets, even if the government no longer imposes physical-layer interconnection requirements. After all, the Internet itself—the platform on which over-the-top VoIP services ride—has remained open in the absence of interconnection mandates; you can reach any point on the Internet by subscribing to a single ISP. That said, a ubiquitous and open "telephone system" is still considered so fundamental to modern society that even after the PSTN sunsets, policymakers may take steps to preserve it in some form. For example, they might compel the sharing of customer-directory information and impose applications-layer interoperability requirements if they fear that the phone system shows signs of fundamental fragmentation. Or they might take steps to open a closed VoIP service that has accumulated such a large share of subscribers that the whole voice market might tip in favor of that service, much as the PSTN did in favor of the Bell System in the nineteenth century.

Indeed, there is already precedent for such interoperability requirements in the Internet space. When the FCC approved AOL's merger with Time Warner in 2001, it feared that AOL Instant Messenger had accumulated such a large subscriber base that users of instant messaging would feel compelled to choose AOL as their provider. The FCC thus imposed a controversial "interoperability" condition for certain "advanced," broadband-oriented applications of instant messaging.[87] Instant-messaging programs, the FCC reasoned, can be modified to serve as "information platforms" for all sorts of communications applications. Indeed, some people believed that instant messaging would gradually supplant the telephone as the dominant means of person-to-person communication, and they viewed AOL as the twenty-first-century version of the late-nineteenth-century Bell System. As AOL's share of instant-messaging users steadily declined in the early 2000s, however, these concerns began to seem overblown, and the FCC lifted the interoperability requirement.[88] But such concerns reveal regulators' unusual sensitivity to the monopolization threat posed by network effects in the communications industry.

Indirect interconnection and QoS

We asked earlier whether unregulated peering and transit arrangements can handle the exchange of VoIP traffic as easily as they can handle the exchange of traffic relating to any other higher-layer Internet application. In addition to telephone numbers, QoS issues further complicate the answer.

As discussed in chapter 5, when a broadband ISP provides a managed VoIP service to its subscribers over a unified IP platform, it typically prioritizes VoIP packets over ordinary Internet packets or uses other methods to protect those packets from latency and jitter. It is by no means clear whether, over the long term, residential ISPs and ordinary consumers will perceive a continuing need for such "managed VoIP" services and the extra costs they cause. After all, millions of over-the-top VoIP calls (and videochats) are exchanged every day between different ISPs over the Internet without any special packet-handling arrangements at all. Although the quality of these over-the-top calls can vary, it often exceeds the call quality that consumers associate with the cellphones they now use to place most of their calls anyway. For the foreseeable future, however, broadband ISPs will continue offering managed VoIP services on the premise that at least some consumers prefer to pay for the extra reliability that QoS-enhanced packet handling delivers.

When two providers of managed VoIP services exchange calls between their respective subscribers, they typically take steps to ensure QoS in the hand-off. Those steps have traditionally involved interconnecting through a TDM-based tandem switch. If the managed VoIP traffic between two broadband ISPs is voluminous enough, however, the ISPs can bypass such PSTN-in-the-middle arrangements by joining their IP networks directly. Today, some cable VoIP providers reportedly use distinct circuits for the exchange of managed VoIP traffic rather than relying on general-purpose Internet peering and transit arrangements, which, for the most part, are currently used only for best-effort exchanges of IP traffic. In the future, managed VoIP providers may use any number of other techniques to perform QoS-oriented packet exchanges; conceivably, they might exchange managed VoIP traffic over the same Internet peering points used to exchange Internet traffic more generally (see chapter 6). But the industry remains in the very earliest stages of thinking through these issues.

Should the Commission regulate any of these "IP-to-IP interconnection" arrangements for the exchange of voice traffic? In the *USF–ICC Reform Order*, it did not resolve that issue but rather sought additional comment from the industry, and the issue remains pending at press time. As a threshold matter, there is some debate about whether the Commission has *legal* authority to regulate interconnection between VoIP providers. That question is complex and may turn on whether VoIP services are properly classified as "telecommunications services" subject to Title II and its various interconnection obligations (under sections 201 and 251) or instead as Title I "information services" that are subject to no such obligations unless and until the FCC can justify imposing them under its "ancillary" authority. As noted in chapter 8, the FCC has left that threshold classification issue unresolved for many years.

The more interesting issues in this policy debate are about economics rather than law. The parties urging the FCC to regulate IP-to-IP interconnection for the exchange of voice traffic argue that interconnection obligations have always been essential to consumer welfare in a circuit-switched environment; that the shift to IP is a mere technological change that should have little effect on market dynamics; and that interconnection obligations will therefore be just as essential after that technological shift as before it. Opponents of such regulation argue that just as market forces have ensured efficient exchanges of Internet traffic for many years without regulatory intervention, they will also ensure efficient exchanges of all VoIP traffic (managed and over the top), whether those exchanges take place on the Internet (via peering and transit) or through specialized interconnection arrangements.

This debate largely boils down to whether the shift from circuit-switched to IP technologies and other industry-wide changes have reduced the need for the general interconnection mandates at the heart of Title II. No one doubts that such mandates were critical in 1913, when the Justice Department forced AT&T's long-distance arm to interconnect with independent LECs (see chapter 1), and in 1982, when the Department forced AT&T's local exchange affiliates to interconnect with competing long-distance providers such as MCI (see chapter 2). Much of the need for antitrust intervention in those contexts, however, arose from unusual concentrations of market power: throughout the relevant period,

AT&T had the overwhelming majority of local and long-distance customers throughout the United States. No similar aggregations of market power have kept the Internet's constituent IP networks from reaching efficient interconnection agreements in the absence of any regulatory mandate, although antitrust authorities have sometimes intervened to keep mergers from creating undue market concentration (see chapter 5).

Two related features of IP technology may attenuate the need for regulatory intervention to ensure that private actors enter efficient interconnection arrangements in an all-IP environment—whether for voice communications or for any other purpose. The Internet's constituent IP networks are *distributed*, which means that multiple redundant routes are available to get from one point to another on the Internet. And they use *connectionless packet-switched technology*, which means that an entire unit of capacity (the circuit) need not be held open as two Internet users communicate across multiple IP networks; instead, the networks efficiently devote just as much variable capacity as is needed to transmit the users' information. Both of those factors make it particularly efficient for two IP networks to exchange traffic indirectly (through intermediate transit links) rather than directly (though peering), and indirect interconnection has long been extremely common on the Internet as a result.[89]

At least on the public Internet, these technological features of IP networks have given rise to a highly competitive yet unregulated market for IP transit services, with rapidly falling prices per unit of capacity.[90] As noted in chapter 5, the widespread availability of transit and the threat that transit customers will simply peer directly with one another typically keep larger networks from exploiting their greater size to drive anticompetitive bargains with smaller networks. Suppose that Small Network wishes to exchange traffic directly with Large Network through a direct-peering relationship. Large Network might agree to such direct interconnection only if Small Network agrees to pay for the privilege—the so-called *paid-peering* arrangements we discussed in chapters 5 and 6. But Small need not pay the full price that Large demands in order to preserve its customers' ability to communicate with Large's customers. Instead, Small can tell Large that it will hire Backbone Provider to connect the two networks indirectly via transit links. Large may be Backbone Provider's settlement-free peer, in which case Large must accept for free the incoming traffic originated by Small and handed off by Backbone (so long as the extra traffic is not so voluminous as to violate the terms of the peering agreement between Large and Backbone). Or Large may be Backbone's transit customer, in which case Large may end up *paying* Backbone Provider to receive Small's traffic.[91] Either way, if Small hires Backbone Network to carry its traffic to Large, Large will end up with lower revenues than it would have earned had it simply proposed a lower, efficient price for paid peering with Small. And that fact gives Large an incentive to propose and accept that lower price.

In effect, the widespread availability of this transit alternative enables smaller networks to unionize—to pool their bargaining clout through a larger backbone network, which acts as their agent in negotiations with other large networks. So long as the market for transit services remains unconcentrated and competitive, that dynamic can normally be expected to keep transit prices low and give every network, large and small, incentives to peer directly on efficient terms in contexts where direct interconnection makes engineering sense.[92]

Some parties in this debate claim that despite the efficiency of unregulated arrangements for the exchange of *best-effort Internet traffic*, regulation is necessary to ensure efficient interconnection between different providers of *managed VoIP services*. These parties argue that because current Internet transit arrangements come without end-to-end QoS guarantees, they are no substitute for direct interconnection among managed VoIP networks. And these parties thus conclude that, absent regulation, larger managed VoIP providers might extract anticompetitive terms for the provision of end-to-end QoS for calls exchanged between their subscribers and those of smaller managed VoIP providers.

As noted, opponents of such regulation question whether end-to-end QoS is necessary to meet consumer expectations in the first place, given the success of non-QoS-enhanced over-the-top services offered by Skype and Vonage. And they suggest that even if end-to-end QoS is needed for some classes of voice traffic, it would not inevitably require direct interconnection; instead, they hypothesize, the market might work out arrangements for QoS-aware transit arrangements as an alternative. These arrangements, they say, might resemble ordinary IP transit arrangements except that the transit provider might agree to honor the priority instructions of its customer (the calling party's network) and might negotiate with all terminating networks to honor those priority instructions as well. All of these points remain extremely speculative for now, and policymakers are just beginning to puzzle through them.

B. Paid-peering disputes involving "eyeball" ISPs and CDNs

We now move from the specialized policy debates about VoIP-to-VoIP interconnection to more general questions about what role, if any, policymakers should play in overseeing interconnection among the Internet's constituent IP networks. Although unregulated peering and transit agreements have generally produced efficient interconnection arrangements since the privatization of the Internet backbone two decades ago, such arrangements are not permanently immune from market failure. For example, the market for IP transit services might theoretically become less competitive, raising transit prices above efficient levels and weakening the price discipline that indirect interconnection alternatives place on larger networks negotiating with smaller ones for direct interconnection. As noted in chapter 5, these and related concerns about concentration in the transit market led antitrust enforcers in the last years of the twentieth century to constrain the merger-fueled growth of WorldCom's backbone network.[93]

More recently, a debate has flared about whether certain *types* of IP networks—specifically, ISPs serving primarily residential customers (content-receiving "eyeballs")—have such inherent market power that regulators should compel them to enter into settlement-free peering arrangements with the networks delivering high volumes of content to the those customers. This debate assumed prominence during a peering dispute between Level 3 and Comcast beginning in late 2010. Although that dispute has not matured into a public regulatory inquiry, it sheds light on some of the most important policy issues now emerging in telecommunications policy.

Level 3 wears several hats in the Internet space. It is a Tier 1 backbone provider; a CLEC specializing in the delivery of VoIP traffic for wholesale customers such as Vonage; an ISP serving universities and other institutional clients; and a CDN (see chapter 5) that transports data to cache servers near broadband ISPs dispersed throughout the Internet. It was in this CDN role that Level 3 entered into settlement-free peering and other traffic-exchange arrangements with various residential ISPs such as Comcast.

The controversy with Comcast began when Level 3 signed up Netflix as a major CDN customer and began delivering ever-greater volumes of streaming Netflix video directly into Comcast's residential broadband networks. Comcast notified Level 3 that if it wished to increase the capacity of its direct interconnection arrangements to accommodate all this new Netflix traffic, it would have to start paying for traffic volumes exceeding the levels covered by the companies' existing arrangements. Level 3 balked, claiming that Comcast was exploiting its position as the ISP for millions of Internet "eyeballs" to extract anticompetitive fees from interconnecting providers of Internet content. Comcast responded that it was simply applying a longstanding norm reflected in many companies' published peering guidelines: that where there are radical imbalances in traffic flows, the net traffic generator typically pays the net traffic recipient on the theory that the former is imposing greater network costs on the latter. Comcast noted that Level 3 itself had invoked this same peering norm when in 2005 it briefly depeered Cogent until Cogent agreed to pay Level 3 for much of the traffic it was funneling into Level 3's network (see chapter 5).[94] Comcast added that if Level 3 did not wish to pay Comcast for direct interconnection, it was free to choose alternative routes into Comcast's network via transit arrangements that Level 3 could buy from Comcast's other peers or transit providers.[95] In other words, Comcast argued that, for the reasons we discussed in the previous subsection, the widespread availability of indirect interconnection was sufficient to preclude anticompetitive outcomes for direct interconnection.[96]

That indirect-interconnection solution was anathema to Level 3. As a Tier 1 backbone, Level 3 had never bought transit services from other IP networks or otherwise paid for interconnection with them. To buy transit for access to Comcast's network—or, for that matter, to enter into any public paid-peering arrangement with Comcast—would imperil Level 3's Tier 1 standing. In addition, Level 3's allies suggested that QoS concerns could make indirect interconnection an unattractive substitute for direct interconnection, potentially enabling Comcast to charge anticompetitively high fees for direct access to its millions of eyeball customers. In particular, they contended that transit-based routes into Comcast's network were congested and suggested that Comcast would have no incentive to unclog them, given that, by doing so, it would facilitate the delivery of streaming video traffic that competed with its own subscription video services.[97] Comcast responded that those transit links were not generally congested and that the company would have no incentive to alienate its residential customers by degrading the performance of all the diverse Internet content delivered through those links.[98] Ultimately, when it became clear that policymakers had little appetite for intervening, Level 3 and Comcast stopped arguing publicly and resumed private negotiations.

This dispute bears more than a faint resemblance to the net neutrality debate discussed in the previous chapter. Recall that in the *Open Internet Order* the FCC adopted a nondiscrimination principle that generally disfavors any "commercial arrangement" between a broadband ISP and

Internet content provider "to directly or indirectly favor some traffic over other traffic in the broadband Internet access service connection."[99] The Commission justified that rule on the basis of "winding dirt road" concerns—the fear that "if broadband providers can profitably charge edge providers for prioritized access to end users, they will have an incentive to degrade or decline to increase the quality of the service they provide to non-prioritized traffic. . . . [B]roadband providers might withhold or decline to expand capacity in order to 'squeeze' non-prioritized traffic, a strategy that would increase the likelihood of network congestion and confront edge providers with a choice between accepting low-quality transmission or paying fees for prioritized access to end users."[100]

As discussed in chapter 6, there is ample room to question whether the FCC's "winding dirt road" concerns are generally valid in the first place and whether they justify the strong "nondiscrimination" rule the Commission ultimately adopted. Again, however, it is unclear why those concerns, if valid at all, would apply only to *packet-prioritization* arrangements between ISPs and individual *content providers* but not to *paid-peering* arrangements between ISPs and *networks* (such as CDNs) acting as agents for those content providers. In either case, the issue is whether an ISP can and would use its control over its last-mile eyeball network to squeeze content providers (or their CDN agents) for access to the fast lane (packet prioritization or direct interconnection) lest they be relegated to the slow lane.

One might therefore have expected to read an account of paid-peering disputes in the *Open Internet Order* or in a further notice of proposed rulemaking, along with some effort to grapple with whether such disputes either do or do not implicate the Commission's nondiscrimination rule. But there was no such discussion; there was just a footnote, which read, in its entirety: "We do not intend our rules to affect existing arrangements for network interconnection, including existing paid peering arrangements."[101] The terseness of that lone footnote suggests that the Commission had not fully thought through the implications of its own economic logic for imposing net neutrality rules. That task will now be left for future policymakers.

C. Concerns about Internet fragmentation

For decades, market forces have held the Internet together as a unified network of networks, in which any user of one network can communicate with any user of another. But is that outcome inevitable? And what should be the government's role, if any, in keeping the Internet unified?

Let's take a concrete scenario to illustrate the complex economic dynamics involved. What if Level 3, invoking its status as a Tier 1 provider, had simply refused to pay anyone, whether a transit provider or Comcast itself, for access to Comcast's network? Could Comcast's customers have received Netflix traffic *at all?* The answer is yes, for two reasons. First, even though Level 3 is a Tier 1 provider, Comcast is not; it reportedly purchases some transit services from at least one third-party Tier 1 backbone.[102] Level 3 could hand off Comcast-bound Netflix traffic to that third-party backbone, and that third-party backbone could in turn deliver the traffic to Comcast as part of their transit relationship. (This assumes, of course, that the third party backbone would not *itself* try to charge Level 3 for the extra traffic load; if it were to do so, then Level 3's Tier 1 status would remain in jeopardy if it continued serving Netflix.)

What happens, though, if Level 3 has a peering dispute with an eyeball ISP—call it "Network Z"—that is also another Tier 1 backbone? In that case, Level 3 would have no peer of its own that it could hand off Network Z–bound traffic to. That is because peering by definition is a relationship where peers exchange traffic between each other's customers (including transit customers). Would Network Z's customers still be able to watch Netflix movies? The answer again is yes because of a second safeguard built into the Internet's contractual infrastructure. Netflix and most other major content providers today are *multihomed*, which means that they have redundant arrangements with multiple providers to ensure delivery of their content throughout the Internet in case something goes wrong with one provider's network. Thus, if Level 3 were for some reason unable to deliver Netflix's traffic to Network Z, Netflix could turn to some other CDN such as Limelight or Akamai to deliver the traffic instead, or it could build its own CDN (as it has in fact begun to do).[103]

More serious problems can arise for a *single-homed* customer when its solitary Tier 1 network provider reaches a peering impasse with another Tier 1 network. In 2008, Sprint and Cogent—both then Tier 1 providers—reached such an impasse, and Sprint severed the peering links between the two networks. *Forbes* writer Scott Woolley reported that, "in an instant, customers who relied solely on Sprint (like the U.S. federal court system) for Web access could no longer communicate with customers who relied solely on Cogent for their Web connections (like many large law firms), and vice versa"; "major American and Canadian universities lost contact with each other"; "[o]fficials in Maine's state government found they couldn't link up with many town governments"; and "[m]illions of Sprint's wireless broadband customers found themselves cut off from thousands of Web sites."[104] Sprint and Cogent faced an immediate barrage of customer complaints, and they resolved their dispute within days, reconnecting their respective users.

Woolley views this episode as, paradoxically, reason for optimism about the future of unregulated peering and transit arrangements: "The current laissez-faire system has a remarkable ability to encourage privately run networks to voluntarily strike deals that benefit everyone, expanding capacity of the larger Internet while allowing everyone to connect to everyone else. In the rare instances where part of the Net does break down, as in the recent fight between Cogent and Sprint, the market provides overwhelming incentives to repair the breach quickly."[105] In addition, this incident reminded corporate information-technology professionals everywhere that they can use multihoming to shield their companies from the most dire consequence of a peering impasse between Tier 1 networks: complete disconnection from millions of Internet users.

All this said, the stakes here for the Internet as a whole are enormous, and theoretically a major Tier 1 peering impasse might fragment the Internet indefinitely. Were such impasses to arise with greater frequency, policymakers might consider a step they have so far been loath to take: they might hold out the prospect of government intervention, in the form of mandated interconnection, to keep the Internet unified during the pendency of peering negotiations. That would be a serious step. Even the prospect of regulatory intervention might make impasses more likely because it would give the negotiating parties incentives to avoid hard bargaining and concessions in the hope of obtaining a better outcome through the regulatory process. To date, therefore, policymakers have been reticent to involve themselves in the exceptionally complex world of Internet peering and transit arrangements.

Moreover, no U.S. agency has clear legal authority to order two Internet backbone networks to interconnect, even in truly exigent circumstances. As noted, the FCC has never regulated peering and transit arrangements, presumably because it has concluded that peering and transit are not "telecommunications services" subject to Title II and its interconnection obligations. Any new FCC initiative to regulate peering and transit would confront immediate legal challenges with a highly uncertain outcome. And no other federal agency has any clear authority to step into this jurisdictional morass. Against the backdrop of that jurisdictional uncertainty, one of us has proposed that the FCC work with a self-regulatory organization such as the North American Network Operators' Group (NANOG) to bring the major industry stakeholders together to develop and accede to enforceable mechanisms for resolving peering disputes quickly and efficiently if they threaten the wider Internet ecosystem.[106] The challenge, as with direct regulation, would be to structure such mechanisms to avoid creating incentives for disputants to stop negotiating in good faith because they hope to get better outcomes from the dispute-resolution process.

Finally, there is a significant international dimension to any debate about government oversight for peering and transit arrangements. Such arrangements are often global in scope, and they are potentially subject to foreign no less than U.S. regulation. Foreign regulators, moreover, may be less concerned about abstract notions of economic efficiency than about creating money flows from networks carrying (largely U.S.-based) content to networks receiving that content. For years, the International Telecommunication Union, an arm of the United Nations, has been considering various proposals under which individual countries could regulate international peering and transit arrangements in order to compensate for lost voice-termination revenues as VoIP displaces the PSTN. The United States has long opposed such initiatives, known by the generic name *ICAIS* ("International Charging Arrangements for Internet Services").[107] And U.S. policymakers are therefore particularly reluctant to engage in what foreign governments might cite as "Internet regulation" by the United States itself, opening the door to additional foreign interventions.

* * *

As noted, the *USF–ICC Reform Order* contains two main components: the intercarrier compensation reforms discussed in this chapter and sweeping changes to the nation's universal service program. The two components are inextricably linked because, as discussed, many forms of intercarrier compensation have traditionally included substantial cross-subsidies to keep the price of residential service low. The FCC's decision to transition to a bill-and-keep regime thus created a corresponding need to find other, more sustainable subsidy mechanisms to support universal service. We discuss that challenge in the next chapter, along with the equally important challenge, also addressed in the *USF–ICC Reform Order*, of refocusing the universal service program from promoting the availability of conventional telephony services to promoting broadband Internet access services.

8

Universal Service in the Age of Broadband

Most of this book focuses on competition policy—that is, on the rules of the game for markets where consumer demand can support competition among multiple firms, but where a dominant firm may nonetheless harm competition and thus consumers. No treatment of telecommunications policy would be complete, however, without a discussion of *universal service*—a diverse set of initiatives to subsidize communications services in contexts where such services otherwise would not be provided at all, even by a single firm, or would be offered only at "unaffordable" (i.e., much higher than accustomed) rates.

The term *universal service* entered the telecommunications lexicon in the early twentieth century when AT&T President Theodore Vail, a former postal executive, coined the Bell System's six-word slogan: "one system, one policy, universal service."[1] In the words of AT&T's 1910 annual report, Vail envisioned "'a system as universal and as extensive as the highway system of the country which extends from every man's door to every other man's door.'"[2] As Paul Starr notes, "[t]he following year's report endorsed cross-subsidies" to produce that outcome: "'Rates must be so adjusted as to make it possible for everyone to be connected who will add to the value of the system to others.'"[3] And as with the postal system that Vail had helped run, where stamp rates never varied with mailing location, policymakers soon agreed that local telephone rates should be roughly comparable in each area of the country, irrespective of the radical differences in the *costs* of serving each area.[4]

Eighty-six years later Congress reaffirmed that basic policy choice in the Telecommunications Act of 1996. Among the principles codified in new section 254(b), Congress provided that "[q]uality services should be available at . . . affordable rates," and "[c]onsumers in all regions of the Nation, including . . . those in rural, insular, and high cost areas, should have access to telecommunications and information services . . . at rates that are reasonably comparable to rates charged for similar services in urban areas."[5] As in Vail's day, fulfilling these aspirations requires subsidies of one kind or another. As we will see, competitive pressures have forced policymakers to replace the traditional scheme of implicit cross-subsidies, baked invisibly into the retail rates of individual services, with more taxlike explicit subsidies that are reflected in line-item fees on consumer bills.

By 2010, the FCC's system of explicit subsidies, known as the *Universal Service Fund* (USF), had grown to encompass four disparate programs that accounted in the aggregate for $8.7 billion per year in federal disbursements, underwritten by fees assessed on telecommunications providers and passed through to their customers.[6] These programs are (1) the federal *Lifeline* and *Link-Up* programs, which provide need-based subsidies for low-income households; (2) the *High Cost Fund*, a non-need-based program designed to keep telephone rates for customers "affordable" in the mostly rural areas where economies of scale and density are low and per customer costs are therefore high; (3) the *E-rate* program, which funds broadband and other communications services to the nation's schools and libraries; and (4) a similar program for funding such connections to rural health care facilities.[7] In 2010, these four programs accounted, respectively, for $1.2 billion, $4.6 billion, $2.7 billion, and $0.2 billion in annual subsidies.[8]

In this chapter, we focus on the largest of these programs—the High Cost Fund—and the FCC's 2011 decision to overhaul it. Two features of that fund are worth mentioning at the outset. First, until very recently the fund focused anachronistically on subsidizing conventional telephone services rather than on broadband Internet access. Second, for many years the FCC and state regulators often felt bound by principles of "competitive neutrality" to confer equal high-cost subsidies on multiple redundant providers in rural areas where, in the absence of subsidies, economies of scale and density were by hypothesis insufficient to support "affordable" services by even a single provider. As discussed later in this chapter, that so-called *identical-support rule* helped cause the fund to balloon out of proportion to its consumer benefits, and the main victims were ordinary consumers throughout the rest of the country who had to underwrite the extra subsidies through higher fees on their phone bills.

In 2011, the FCC sought to tackle both of these policy concerns in the *USF–ICC Reform Order* we introduced in the previous chapter.[9] Very roughly speaking, the FCC announced that it would eliminate the identical-support rule but maintain the existing size of the high-cost program, using the surplus to subsidize rural broadband deployment and shifting the overall focus of the program from conventional telephony to broadband. We discuss those initiatives in section II of this chapter and then turn in section III to the equally important questions of who pays for these programs and how. Because the details of today's universal service mechanisms are complex and somewhat dry, however, we first review the big picture of universal service policy.

I. Introduction to the Political Dynamics of Universal Service

Policymakers once justified universal service subsidies largely by reference to the *network-externality* concept—a manifestation of the "network-effects" phenomenon discussed in chapter 1. Simply put, the value of a network (or network of networks) to any given user is directly proportional to the number of *other* users who can be reached on it, and no individual user internalizes the full extent of that value in making decisions about whether to join or drop off the network. Ubiquitous subscribership, moreover, benefits not just individual consumers, but society as a whole by enhancing economic development, democratic participation, and public safety.

Universal service subsidies might well have accelerated subscribership levels in the early years of telephone service, but by the final decades of the twentieth century it was no longer clear that such subsidies were still necessary to keep those levels high. Because basic voice service had become so integral to contemporary life, most people of average means would likely have purchased it even if the government had curtailed non-need-based subsidies and if rates had increased commensurately.[10] Certainly one can imagine a policy that restricts subsidies to situations where they are truly necessary for continued access to the network and that eliminates them everywhere else.

That, however, has never been the policy in the United States. The term "universal service" has long encompassed a broad range of subsidy mechanisms, some of which have very little to do, even as a conceptual matter, with keeping people on the network. If, for example, the estimated monthly cost of providing telephone service to a given household in a rural town is $100, the state public utility commission may nonetheless compel the local ILEC, under its *carrier-of-last-resort* obligations (see chapter 2), to provide service to each household at a capped monthly rate of only $25. That is true even if the rural town also happens to be a wealthy resort where raising the price to $100 per subscriber would actually induce very few residents to leave the network.

Why do policymakers keep rates below cost for people who can afford to pay cost-based rates? After all, prices for many goods and services, such as housing or gasoline, often vary tremendously from one place to another, but the government usually perceives no need to equalize them. For example, no one expects homeowners in rural Mississippi or Wyoming to pay a special real estate tax to keep housing rates in Manhattan or San Francisco more "comparable" to those elsewhere. The rationale usually given for the universal service system in telecommunications is that telephone service—like postal delivery and now broadband Internet access—is so fundamental to modern civic life that it should be extended, at comparable rates, to all Americans as a civil right. Whatever the merits of that position, the government's commitment to traditional, non-need-based universal service programs is unlikely to change.

In our example, who pays the extra $75 per month for the $25 telephone service provided to each inhabitant of the remote town where the average cost of service is $100? The short answer is "other consumers," through a complex web of *implicit* and, more recently, *explicit* subsidies. We address each type in turn.

As discussed in chapter 2, the implicit cross-subsidies underlying "affordable" telephone service have taken many forms. First, under the practice known as *geographic rate averaging*, customers in cities often pay roughly the same rates as customers in remote rural locations even though, because of economies of density, the per line cost of installing and maintaining a line in the city is a fraction of the cost of doing the same in the countryside. In other words, some urban consumers pay above-cost rates to enable the local telephone company to serve rural consumers at below-cost rates, though that reality is hidden from them: urban and rural consumers typically perceive only that they are paying roughly the same rate for the same service.

Second, local telephone companies have traditionally charged up to twice as much for a "business line" as for a "residential line," again without any cost-based justification. Third, as discussed in chapter 7, local telephone companies have subjected long-distance companies to tariffed access charges that, particularly on the state level, have exceeded any genuine measure of cost, and the difference has been passed along in the form of inflated per minute rates to heavy users of conventional long-distance services. Finally, *vertical services*, such as call waiting and caller ID, are also priced well above cost. In all of these cases, the people contributing to universal service often have no idea that they are doing so: that is why these types of subsidy are called "implicit."

For many decades, regulators developed a universal service policy dependent on a maze of such implicit cross-subsidies. For two basic reasons, these cross-subsidies were economically problematic even before the age of competition. First, by detaching the rate charged for a service from its underlying cost, implicit cross-subsidies defy basic principles of cost causation, artificially inflating demand for some services and dampening demand for others, such as long distance. Second, such cross-subsidies violate *Ramsey pricing* principles, named after the early twentieth-century Cambridge economist Frank Ramsey.

Roughly speaking, Ramsey pricing holds that when there is no straightforward way to allocate costs (or taxes or fees) among different services, the most efficient solution is to recover them in the form of higher rates for necessary services that consumers would be reluctant to drop (such as basic local service) rather than for more elective services that customers would more easily forgo at the margins (such as long-distance calls or call waiting).[11] The intuition here is that the market will be most efficient if such recovery is structured to distort consumer buying practices as little as possible. Despite its impeccable economic logic, Ramsey pricing is politically unpopular in the universal service context and thus widely ignored there. It would raise prices for basic access to the network—which all subscribers must buy—rather than the elective services typically sold to more affluent consumers. Regulators have almost always followed political pressure to "keep basic rates low" by effectively raising the price of long-distance and other more elective services. Over the decades, that choice has imposed billions of dollars in allocative inefficiencies by diverting social resources from their most productive uses.

The growth of local competition in the late twentieth century added an entirely new dimension of dysfunctionality to implicit cross-subsidies. High "business line" rates, geographic rate averaging, and other implicit subsidies are all unsustainable in a competitive world. Those policies rely on the prevalence of captive customers who have no choice *but* to pay a telephone monopolist the above-cost rates for essential services that subsidize below-cost rates to certain subscribers, such as those living in rural areas. Once competition arises, the erstwhile monopolist—the incumbent LEC—cannot get away with charging downtown business customers rates far above the cost of serving them, for those customers would then switch to competitors who can and do charge much less. And if the ILEC continues imposing above-cost access charges, interconnecting providers will increasingly use the various arbitrage mechanisms discussed in the previous chapter to avoid paying those charges.

In short, policymakers needed to come up with a new type of subsidy scheme designed to function in a competitive environment. This is the avowed purpose of section 254 of the Communications Act, added in 1996.[12] Section 254 envisions a transition from traditional implicit cross-subsidies to a system of "explicit" subsidies that are underwritten by competitively neutral assessments on telecommunications providers generally and are passed through to consumers as line-item fees on telephone bills.

That solution is simply stated; the challenge lies in the implementation. One problem is that regulators, particularly on the state level, need considerable prodding to undertake the transition from implicit to explicit subsidies. Consumers often notice when they must pay taxlike fees, and they look for someone to blame. But they typically do not know when they are simply paying rates that in some abstract economic sense exceed cost. From the short-term perspective of many regulators, the political costs of accelerated universal service reform have outweighed the benefits. And the 1996 Act contained no specific time frame for the elimination of the old implicit subsidies, leaving most regulators content to confront this challenge gradually.

This abiding attachment to implicit cross-subsidies exemplifies what Richard Posner has called *taxation by regulation*—a means of compelling "members of the public to support a service that the market would provide at a reduced level, or not at all," while keeping them largely in the dark about the existence, extent, or purpose of the subsidies they must pay.[13] Of course, universal service programs are more transparent now than they were several decades ago, when virtually all subsidies were implicit and Vail's successors at the Bell System routinely cited "universal service" to compliant regulators as a basis for opposing any competitive threat to its nationwide local and long-distance monopoly (see chapter 2). Even today, however, many regulators, particularly at the state level, still cling nostalgically to nontransparent forms of universal service support.

With this high-level overview, we now dive more deeply into the details of the FCC's implementation of section 254.

II. Federal USF Disbursement Mechanisms

A. The nuts and bolts of the High Cost Fund (circa 2011)

Two somewhat arbitrary distinctions, which date back to the early days of the Bell System monopoly, remain at the heart of universal service policy. First, for purposes of managing universal service subsidy mechanisms, the FCC and the states have divided up responsibility into distinct "interstate" and "intrastate" spheres. Second, regulators in both spheres treat the Bell companies and midsize ILECs, which are subject to price-cap regulation, quite differently for universal service purposes from smaller ILECs, which are typically still subject to rate-of-return regulation.[14] Although both distinctions (interstate/intrastate and price cap/rate of return) are somewhat artificial, they will nonetheless drive universal service policy for the foreseeable future—as will a third arbitrary classification discussed later in this chapter: the distinction between "telecommunications services" and "information services."

We begin with a recap of the interstate/intrastate distinction. Recall from chapter 2 that federal and state regulatory authorities "separate" the costs of ILEC local loops into interstate (federal) and intrastate (state) jurisdictions so that, among other things, courts can evaluate the merits of a takings claim (alleging confiscatory regulation) against either the federal government or a state.[15] Under this system, the FCC, after consulting with the states, splits an ILEC's costs into two arbitrary categories: "interstate" costs, whose recovery the FCC superintends, and "intrastate" costs, whose recovery is the responsibility of the states. For example, under the federal–state separations process, 25% of a Bell company's loop costs have traditionally been allocated to the interstate jurisdiction and 75% to the intrastate jurisdiction.[16]

Once the FCC makes that separation,[17] each jurisdiction (the FCC or the state) must decide how to enable the ILEC to recover the relevant costs. The FCC traditionally enabled ILECs to recover the costs on the interstate side of the ledger through (typically above-cost) per minute access charges assessed on interstate long-distance calls and flat-rated "subscriber line charges" imposed directly on end users.[18] Separately, on the intrastate side of the ledger, the states enabled ILECs to recover their costs through, among other things, inflated intrastate access charges, various toll charges, fees for "vertical features" such as call waiting and caller ID, and monthly local service rates, which were anomalously higher for "business" lines than for "residential" lines.

Many of these charges—on both the federal and the state level—involve implicit cross-subsidies and are priced above any genuine measure of cost. When competition took root and began siphoning off those cross-subsidies, ILECs could not simply drop their high-cost customers or raise their rates to cost. Instead, ILECs were, and in a slowly declining number of states still are, subject to carrier-of-last-resort and related obligations, which compel them to provide service to high-cost customers and all others who request it and to do so at low, often below-cost rates. By eroding implicit cross-subsidies, therefore, competition threatened to leave ILECs holding the bag.

In the 1996 Act, Congress included the highly ambiguous provisions of section 254 to address this concern. Under the prevailing interpretation,[19] section 254 instructs the FCC, after formally consulting with a Federal–State Joint Board,[20] to take steps to keep rates "affordable" and "comparable," both from place to place within each state and from state to state across the country. It also envisions that the FCC will accomplish that goal, in cooperation with the states, by phasing out the unsustainable implicit cross-subsidies that characterized universal service policy before 1996. Again, section 254 directs the FCC to set up an explicit funding mechanism that is underwritten by "equitable and non-discriminatory contribution[s]" by all telecommunications carriers.[21] The FCC responded by setting up the High Cost Fund.

In the immediate aftermath of the 1996 Act, the FCC asked to what extent that fund should bear the burden of replacing *all* traditional subsidy mechanisms, including those that, like geographic rate averaging, have been managed mostly by the states. Conceivably, Congress could have instructed the states, through a process known as *rate rebalancing*, to eliminate implicit cross-subsidies altogether by lowering the above-cost rates charged to business customers, residential customers in densely populated neighborhoods, and others. And Congress then could have made ILECs whole for their sudden revenue shortfall by using the *federal* High Cost Fund to pay them the complete difference between the cost of serving high-cost customers and the retail rates they were allowed to charge them. That, however, would have required a dramatic enlargement of the federal fund by billions of dollars. Congress did not require that radical solution, and the FCC has never seriously considered it.

Instead, the Commission concluded from the outset that explicit federal funding mechanisms should replace (1) the use of cross-subsidies (such as above-cost interstate access charges) on the *interstate* side of the cost ledger and (2) a small portion of traditional cross-subsidies on the *intrastate* side of the cost ledger. The FCC has traditionally addressed those discrete functions in separate proceedings. And, to make matters more complicated still, it has addressed both of them differently depending on whether the geographic territories at issue are served by "price-cap" ILECs (the Bell companies and a few others) or the hundreds of small "rate-of-return" ILECs.*

As to *price-cap* carriers, the FCC's most significant reform efforts on the *interstate* side of the ledger came in 2000 and 2011. First, in June 2000, the FCC adopted the main components of an access charge reform program proposed by a broad-based industry alliance known as the "Coalition for Affordable Local and Long Distance Services" (CALLS). Roughly speaking, the FCC ordered significant reductions in the interstate access charges imposed by the largest ILECs and compensated them for the shortfall by increasing both the size of the federal fund from which they may seek support as well as the flat-rated subscriber line charge they could impose on end users.[22] The *CALLS Order*, as it was called, was a step in the right direction. But the resulting interstate access charges, applicable to *long-distance* calls, were still well above the "reciprocal compensation" rates set for the identical network functions performed in the termination of *local* calls. As discussed in chapter 7, that arbitrary rate discrepancy continued generating arbitrage schemes and ultimately led the FCC to take more decisive measures. In the *USF–ICC Reform Order* of 2011, the Commission announced that it would reduce all access charges to reciprocal compensation levels and then eventually eliminate all termination charges ("access" and "local"), while granting greater flexibility to the price-cap LECs to recover network costs from their end users.

* Through 2011, the FCC focused on a similar but slightly different distinction between "rural" and "nonrural" ILECs, a distinction that turns on the number of lines a given ILEC serves. *See generally* 47 U.S.C. § 153(37) (defining "rural telephone company"). Those terms are a bit confusing because in the aggregate "nonrural" ILECs actually serve many more customers in rural areas than "rural" ILECs do. In the *USF–ICC Reform Order* (Report & Order and Further Notice of Proposed Rulemaking, *Connect America Fund*, 26 FCC Rcd 17663, ¶¶ 129–30 (2011)), the Commission decided to distinguish between ILECs instead on the basis of their method of price regulation: rate of return or price caps (see chapter 2). That change altered the USF status of a small handful of "rural" carriers that had opted into price-cap regulation. For simplicity, we use the current categories in discussing the FCC's pre-2011 regime even though they are slightly imperfect proxies for the prior "rural" and "nonrural" categories.

The FCC has historically taken a much narrower view of the federal fund's role as a replacement for implicit subsidies on the *intrastate* side of the cost ledger for the Bell companies and other price-cap ILECs. As to the areas served by those carriers, the Commission limited its role to helping states with unusually high *average* costs attain rates reasonably comparable to those of other states.[23] Under the Commission's somewhat complicated formula, the federal fund subsidized "76% of the amount that the statewide average cost per line exceeds two standard deviations above the national average cost per line"—in essence, three-quarters of the differential between (1) a state's average costs per line and (2) roughly 135% of the national average cost per line.[24] Only a handful of states had average costs above 135% of the national average and thus qualified for any funding under this approach. The FCC reasoned that unless statewide average costs were unusually high, a state could and should create its own funding mechanisms to equalize rates within its borders so that each subscriber's rates are roughly comparable to those in most other states. This is where matters stood for price-cap carriers as of 2011, when the FCC announced that it would phase out the High Cost Fund in favor of the new broadband funding mechanism—the "Connect America Fund," or "CAF"—discussed below.

Rate-of-return carriers are typically much smaller than price-cap carriers and have traditionally been eligible for more generous universal service support under a different but equally complex scheme. In a nutshell, the FCC has allocated a larger portion of a rate-of-return carrier's loop-cost recovery to the "interstate" jurisdiction to the extent the carrier's costs exceed various federal benchmarks.[25] And the FCC continues to base support levels on those carriers' embedded costs rather than (as with the price-cap carriers) their forward-looking costs, thereby assuring a steady subsidy flow.[26] It has justified this differential treatment on the grounds that the smaller rate-of-return ILECs, unlike the Bell companies and other price-cap ILECs, have "higher operating and equipment costs, which are attributable to lower [average] subscriber density, small exchanges, and a lack of economies of scale," and that these carriers need a degree of "certainty and stability" in confronting the development of competition.[27]

B. Wireless ETCs and the rise and fall of the identical-support rule

The recipients of subsidies under the High Cost Fund have traditionally been ILECs because they are the ones with the carrier-of-last-resort obligations and thus the ones that normally serve high-cost customers. Nonetheless, the FCC sought for many years to promote local competition in high-cost areas by providing "portable" subsidies to any carrier that was willing to serve all customers within a defined geographic area and was designated (usually by the relevant state commission) as an *eligible telecommunications carrier* (ETC).[28] In essence, this *competitive ETC* would step into the ILEC's shoes and receive whatever subsidy the ILEC would also receive for serving the customers at issue.[29] This policy became known as the *identical-support rule.*

This policy ultimately began inflating the size of the High Cost Fund to unsustainable levels. First, because the FCC based support levels for rate-of-return carriers on their actual embedded costs, every line that a competitive ETC took from a rate-of-return ILEC caused the ILEC's scale economies to fall, its per line costs to rise, and thus its per line subsidy levels to increase as well.[30] In effect, this approach held rate-of-return ILECs harmless whenever they lost lines, all at the expense of telecommunications customers nationally. Worse, it simultaneously increased the per customer subsidies available to the competitive ETCs that were winning those lines because the ETCs' own subsidies were keyed to the ILECs' subsidies. The result was a giant feedback loop: federal subsidies encouraged competitive entry into areas that in many cases could not economically support more than a single carrier, which had the effect of increasing those subsidies, which in turn encouraged yet more competitive entry into the same areas. As the FCC understood, this dynamic was unsustainable: "As an incumbent 'loses' lines to a competitive eligible telecommunications carrier, the incumbent must recover its fixed costs from fewer lines, thus increasing its per line costs. With higher per line costs, the incumbent would receive greater per line support, which would also be available to the competitive eligible telecommunications carrier for each of the lines that it serves. Thus, a substantial loss of an incumbent's lines to a competitive eligible telecommunications carrier could result in excessive fund growth."[31]

This dynamic would have been problematic by itself even if the FCC had been correct to assume, as it did in 1997, that competitive ETCs would generally be wireline CLECs that vied with the ILEC for the single landline connection into a customer's home.[32] Under that scenario, CLECs might raise an ILEC's per line costs when winning its customers, but at least the number of subsidized lines would remain the same. What the FCC failed to foresee was that the vast majority of subsidized "lines" served by competitive ETCs—some 98% by 2010—would be mobile wireless services, which consumers in the same high-cost household often bought in addition to rather than instead of subsidized landline services.[33] As the Commission later acknowledged, "Providing the same per-line support amount to competitive ETCs had the consequence of encouraging wireless competitive ETCs to supplement or duplicate existing services while offering little incentive to maintain or expand investment in unserved or underserved areas."[34] Indeed, "many areas ha[d] four or more competitive ETCs providing overlapping service" to the same consumers, "attracting investment that could otherwise be directed elsewhere, including areas that are not currently served."[35] This was bad enough as a matter of distributional equity, but it was even worse from the perspective of the fund's long-term financial integrity. As more and more wireless carriers got in on the game, annual subsidies for competitive ETCs grew from less than $17 million in 2001 to $1.18 *billion* in 2008, all at the expense of the ordinary consumers nationwide who had to pay increasing line-item fees on their telephone bills.

The FCC was initially slow to act on these concerns. In early 2004, a sharply divided Federal–State Joint Board proposed controlling the growth of the High Cost Fund by, among other things, limiting subsidies to one consumer-designated "primary connection" per user.[36] But that recommendation was controversial, not least with the recipients of competitive ETC funding and the states that benefited disproportionately from it. Finally, in 2008, in response to another Joint Board recommendation, the FCC faced up to financial realities and adopted an "interim" cap on high-cost support for competitive ETCs, freezing subsidies at existing levels (subject to some exceptions).[37] Wireless ETCs brought and lost a legal challenge to that cap in the D.C. Circuit.[38] Meanwhile, the Commission extracted commitments from Verizon Wireless and Sprint—two of the largest recipients—to forgo competitive ETC funding in exchange for the Commission's approval of those carriers' separate mergers with other providers.[39]

Finally, in its 2011 *USF–ICC Reform Order*, the FCC dropped the other shoe and announced that it would eliminate the identical-support rule altogether. As it explained, "The interim cap slowed the growth in competitive ETC funding, but it did not address where such funding is directed or whether there are better ways to achieve our goal of advancing mobility in areas where such service would not exist absent universal service support."[40] The Commission had initially proposed to limit funding to a single provider—whether wireline or mobile wireless—in any given geographic area.[41] Given the continued role of wireline ILECs, however, that outcome likely would have cut off most high-cost funding to wireless providers. The proposal therefore drew widespread industry opposition and conflicted with the Commission's own sense that mobile broadband services are worthy subjects of federal subsidies in their own right in high-cost areas where they might otherwise not be provided at all.

The Commission thus established two distinct funds, each of which can provide funding for a single provider in the same area: one fund for wireline providers (the CAF) and a much smaller one for mobile wireless providers (the "CAF Mobility Fund"). Over the long term, the Mobility Fund will allocate $500 million per year to "expand and sustain mobile voice and broadband services in communities in which service would be unavailable absent federal support,"[42] although the Commission deferred the details of that program to further proceedings. For more immediate purposes, the Commission announced that it would conduct "reverse auctions" for the immediate disbursement of $300 million to wireless providers that bid to provide, in particular areas, a defined level of mobile voice and broadband service for the lowest subsidies.[43] These Mobility Fund reverse auctions would give the Commission valuable trial-and-error experience as it geared up for the larger and more consequential reverse auctions it will use to allocate broadband subsidies under its main CAF program. We discuss that program and reverse auctions more generally in the next section.

C. The Connect America Fund

As discussed, the universal service system is designed to ensure that everyone, regardless of wealth, has "affordable" access to whatever communications services are deemed essential to participation in modern society. Until recently, federal policymakers confined the category of such services to voice-grade telephone connections. With the key exception of the E-rate program noted earlier, which funds broadband connections for schools and libraries, policymakers were initially slow to make subsidies available for broadband Internet access, largely because of concerns about the extra burden such subsidies would place on the ordinary consumers who underwrite the USF through line-item fees.[44] It was not until several years into the twenty-first century, when broadband surpassed dial-up as the dominant mode of Internet connectivity, that the FCC began seriously considering new broadband subsidies on the ground that, like telephony, broadband had become an essential service. And it was not until 2011 that the FCC made clear that it would make broadband subsidies the central focus of its universal service program.

That said, there was one historically significant context in which the FCC did fund residential broadband access for many years before 2011. Under the so-called *no-barriers* policy adopted in 2001, the Commission authorized carriers, if they so chose, to use USF funds "to invest in infrastructure capable of providing access to advanced services" in addition to voice services—for example, DSL- or FTTH-based broadband in addition to conventional telephony.[45] The beneficiaries of that policy have been rate-of-return ILECs, which are generally entitled to recover whatever actual costs they incur in upgrading their networks, rather than price-cap ILECs, which by definition are not so entitled because their rates are detached from their actual expenditures.

That difference in regulatory status led to a so-called *rural–rural divide*: consumers living in rural areas served by rate-of-return carriers were more likely to enjoy state-of-the-art broadband access than were consumers living in equally rural areas served by price-cap carriers.[46] As the Commission observed in the *USF–ICC Reform Order*, "More than 83 percent of the approximately 18 million Americans that lack access to residential fixed broadband at or above the Commission's broadband speed benchmark live in areas served by price-cap carriers—Bell Operating Companies and other large and mid-sized carriers."[47]

As part of its 2009 economic stimulus legislation, Congress prompted the FCC to begin closing this rural–rural divide by directing it to produce a comprehensive "National Broadband Plan" to "ensure that all people of the United States have access to broadband capability."[48] The ensuing Plan, issued by the FCC staff in 2010, was a nonbinding set of proposals, but it strongly influenced many of the most important telecommunications initiatives of the FCC under Chairman Julius Genachowski, ranging from spectrum liberalization (see chapter 3) to data-roaming rules (see chapter 4) to intercarrier compensation reform (see chapter 7). With respect to universal service in particular, the Plan deemed broadband "the great infrastructure challenge of the early 21st century," akin to the rural electrification initiatives of the early twentieth century,[49] and it encouraged the Commission to "shift up to $15.5 billion over the next decade from the current High-Cost program to broadband."[50]

In 2011, the FCC responded to these staff proposals by announcing in the *USF–ICC Reform Order* that it would gradually replace the telephony-oriented High Cost Fund with the new broadband-oriented Connect America Fund. In rough outline, the CAF seeks to trigger greater rural broadband deployment by (1) increasing the subsidies for carriers operating in price-cap territories, thereby helping to close the rural–rural divide; (2) conditioning those subsidies on a CAF recipient's provision of broadband services in addition to voice telephony; and (3) making parallel but more modest changes to the universal service rules governing rate-of-return carriers.

The Commission vowed to accomplish all that, and simultaneously create the Mobility Fund, on the same annual budget as the legacy High Cost Fund in 2011: $4.5 billion per year. It derived much of the extra money for broadband-related subsidies in price-cap areas from the savings it would achieve by phasing out support under the identical-support rule in favor of more limited support under the Mobility Fund.* Of course, the Commission could have generated still greater rural broadband deployment had it chosen a larger budget. But striking a quintessentially political balance, it decided to freeze the High Cost Fund's overall size at a firm $4.5 billion in order to "protect consumers and businesses that ultimately pay for the fund through fees on their communications bills."[51]

The details of the CAF are numerous and very complex, and here we discuss only the most important of them, which relate to the areas served by price-cap LECs. As noted, the FCC found that as of 2011 those areas contain the great majority of "unserved" Americans (i.e., those without access to robust broadband service). The FCC announced that in two distinct "phases" the CAF would provide new subsidies to carriers serving those areas. First, in Phase I, the Commission has initially held legacy high-cost funding for price-cap ILECs at existing levels but has offered (in the aggregate) up to $300 million in immediate "incremental support" for those ILECs if and only if they have agreed to meet defined broadband-deployment goals in specified high-cost areas.[52] Among their other obligations, recipients must use the extra money to deploy new broadband services (with actual speeds of 4 Mbps downstream and 1 Mbps upstream) to one new unserved location for every $775 in federal support they accept.[53] Although the FCC acknowledged that the Phase I mechanism contained methodological shortcuts, it explained that this "simplified, interim" approach was needed to boost broadband deployment in the short term while it worked out the more complex details of Phase II funding.[54]

* After the CAF is fully phased in, the $4.5 billion figure will include $1.8 billion for carriers in price-cap areas (a substantial increase over previous support levels); $2 billion for rate-of-return carriers (the same as previous support levels); $500 million for the new Mobility Fund; and $100 million for the "Remote Areas Fund," described later in this chapter. *USF–ICC Reform Order* ¶¶ 18–30.

Phase II then gives price-cap ILECs a *right of first refusal* for broadband funding over the course of five years, after which, the FCC projected, it would open subsidies to competitive bidding. Significantly, whether ILECs elect to receive such funding or not, they will steadily lose their corresponding subsidies under the legacy High Cost Fund.[55] An ILEC that does accept Phase II funding under the right of first refusal must make a *state-level commitment* to offer broadband services (with defined performance characteristics) to all residents in designated high-cost areas throughout its traditional footprint in a state by the end of the five-year funding period.[56] To select the areas eligible for such support, the FCC announced that it would design a new forward-looking cost model "to identify those census blocks where the cost of service is likely to be higher than can be supported through reasonable end-user rates alone," but that it would exclude all areas where an "unsubsidized competitor" (usually a cable company) was already offering a robust broadband service.[57]

For each designated high-cost census block, the Commission offered to provide opting-in ILECs "the difference between the model-determined cost in that census block . . . and the cost benchmark used to identify high-cost areas."[58] Of course, the lower that benchmark is, the more funding the FCC will need to provide, so the choice of a benchmark is a function of the Commission's fixed $1.8 billion annual budget for price-cap carriers. The Commission also accommodated those budget realities by creating a new *Remote Areas Fund*. That fund will subsidize *satellite* broadband services, which are typically slower and more latency prone than terrestrial wireline services, for the very highest-cost areas in the United States (such as isolated ranches). By creating this separate fund, the FCC freed Phase II recipients from any obligation to build out wireline broadband networks to these areas. As the Commission recognized, the per line costs in such areas are so disproportionately high that subsidizing *wireline* broadband connections to them would have quickly consumed the entire budget while benefiting only a relatively small handful of consumers.

What if an ILEC rejects the state-level commitment and thus turns down the offered five-year funding compact altogether under its right of first refusal? In that case, the FCC opens up the provision of Phase II support within the ILEC's footprint to competitive bidding in the form of *reverse auctions*, also known as *procurement auctions*—the same concept we briefly noted earlier in connection with the Mobility Fund. The contestants in a reverse auction compete for government subsidies by bidding down the subsidy amount they are willing to accept in exchange for a promise to provide some defined level of service to a specified number of customers.[59]

Reverse auctions are in some respects similar to conventional auctions and in other respects very different. The most important difference lies in the far greater need to scrutinize the post-auction performance of a reverse-auction winner. In conventional auctions, the auctioneer (here the FCC) can usually rely on market forces to prompt winners to make socially productive use of the auctioned good.[60] In reverse auctions, however, bidders have strong incentives to take the money and run and thus neglect the otherwise undesirable tasks that they won subsidies to perform. Reverse auctions are thus invariably accompanied by deployment deadlines, performance benchmarks (here, for example, relating to broadband speeds), and quite often performance bonds, and the FCC typically imposes substantial penalties for providers that fail to perform as promised.

A reverse auction's design details are critical to its success, and the FCC has left most CAF Phase II details unresolved as this book goes to press. What geographical units can be the subject of competitive bids? Will each auction participant be free to aggregate its bids across different areas, and if so, can they make all-or-nothing package bids? How should the FCC deal with partial geographic overlaps among bidders? Should it assign bidding preferences for small businesses? Should it grant additional bidding credits to providers that commit to provide superior broadband performance, and if so, how should it weight such performance commitments in choosing auction winners? How should it enforce a winning bidder's commitments in the years following an auction? The *USF–ICC Reform Order* sought public comment on these and many other questions about the basic design of the reverse-auction process.[61] By deferring its resolution of those issues, the FCC signaled that it had much work to do before it could actually conduct a workable reverse auction that would produce socially beneficial results.

The complexity of these auction-design challenges explains why Phase I was a necessary interim measure to trigger short-term broadband deployment. That complexity may also explain why the FCC gave price-cap ILECs a right of first refusal for Phase II funding. The more areas there are in which ILECs accept such funding, the fewer areas there will be for which reverse auctions are needed, at least during the initial five-year subsidy term (after which any further subsidies will be allocated by competitive bidding as well). The Commission understood that its reverse-auction mechanism might be fraught with unforeseen complications and prone to dilatory glitches. Thus, to keep the perfect from becoming the enemy of the good, it opted for a simpler means of triggering widespread rural broadband deployment—the right of first refusal—once the Phase I mechanism had run its course.

That said, the FCC's decision to grant ILECs that right of first refusal was controversial. Cable companies and others complained that the FCC had compromised principles of competitive neutrality by singling out a particular class of providers for special regulatory treatment, giving ILECs a unilateral right to take entire geographic areas off the table for competitive bidding, at least for five years. The FCC acknowledged the point but explained that ILECs stood in a special position with respect to the subsidized areas. By definition, those areas exclude census blocks that are already served by cable broadband providers and other "unsubsidized competitors." And in the remaining high-cost areas, ILECs, which have built ubiquitous telephone networks during decades of ILEC-specific carrier-of-last-resort obligations, are likely to have unique "financial and technological capabilities to deliver scalable broadband that will meet [the FCC's] requirements."[62]

The Commission also insulated itself from accusations of pro-ILEC bias by attaching funding conditions that ILEC recipients had strongly opposed. For example, ILECs objected to the obligation to opt into funding on a state-by-state basis rather than on the basis of much smaller geographic units, such as individual wire centers (i.e., the service areas covered by local switches). That requirement for a state-level commitment not only kept participating ILECs from cherry-picking the most desirable service areas but essentially treated them as broadband providers of last resort throughout most of the least desirable service areas in a given state. And rejecting CAF subsidies was hardly an attractive option for ILECs, either. As noted, ILECs will lose legacy subsidies under the High Cost Fund whether they opt into the statewide commitment or not. And the FCC further denied ILEC requests to preempt *state*-level carrier-of-last-resort obligations for voice telephone service.[63] In short, ILECs that decline to participate in the CAF may well be stuck with many of the same universal service obligations as they were previously, but with less money to meet them.

The CAF initiative as a whole was *legally* controversial, too. Section 254(e) provides that "only an eligible telecommunications carrier . . . shall be eligible to receive specific Federal universal service support." Recall from chapter 6 that a "telecommunications carrier" is defined as a provider of "telecommunications services"; that the FCC has classified broadband Internet access as an "information service"; and that, in a highly charged political atmosphere, the FCC resisted calls in 2010 to find that such services include a "telecommunications service" component. How, then, could the FCC legally justify disbursing public funds for the provision of broadband Internet access? A number of parties had argued that section 254 as a whole, although ambiguous, could be reasonably construed to authorize direct FCC funding for information services. For that proposition, they cited section 254(b)(2), which establishes the principle that "[a]ccess to advanced telecommunications and information services should be provided in all regions of the Nation," and section 254(c)(1), which defines "[u]niversal service" as "an evolving level of telecommunications services that the Commission shall establish periodically under this section, taking into account advances in telecommunications and information technologies and services."

The FCC chose a different route to the same outcome. Although it acknowledged that it was refocusing its subsidy mechanisms from circuit-switched telephony (a "telecommunications service") to broadband Internet access (an "information service"), it stressed that it would require every CAF funding recipient to provide *both* voice *and* broadband services. It further noted that for many years the no-barriers policy discussed earlier had uncontroversially permitted rural ILECs to use their high-cost support to deploy facilities capable of providing broadband in addition to voice. And that policy, the Commission concluded, comported with section 254(e), which provides that "[a] carrier that receives such [universal service] support shall use that support only for the provision, maintenance, and upgrading of *facilities* and services for which the support is intended."[64] The Commission concluded: "By referring to 'facilities' and 'services' as distinct items for which federal universal service funds may be used, we believe Congress granted the Commission the flexibility not only to designate the types of telecommunications services for which support would be provided, but also to encourage the deployment of the types of facilities that will best achieve the principles set forth" elsewhere in section 254.[65] In short, the FCC found, it was simply converting the no-barriers policy from a regulatory option to a regulatory mandate, and that shift could not create section 254(e) problems where none existed before.

As with so much else in the *USF–ICC Reform Order*, however, there is a wrinkle in that legal justification. Today, all telco recipients of federal funding do offer "telecommunications services" over their subsidized residential facilities in the form of conventional circuit-switched telephony services. But those telcos will eventually abandon circuit-switched technologies and will offer voice—VoIP—as a higher-layer application on a converged IP platform. At that point, it is unclear whether those providers will remain "telecommunications carriers" because, as noted in chapter 7 (and later in this chapter), the FCC has not yet decided whether VoIP itself is a telecommunications service or an information service. The *Order* essentially defers resolution of that issue until a later date.

To hedge its bets, the Commission alternatively rested its legal authority for broadband subsidies on section 706(b) of the 1996 Act. That provision directs the Commission to "'determine whether advanced telecommunications capability is being deployed to all Americans in a reasonable and timely fashion' and, if the Commission concludes that it is not, to 'take immediate action to accelerate deployment of such capability by removing barriers to infrastructure investment.'"[66] As the Commission noted, "one of the most significant barriers to investment in broadband infrastructure is the lack of a business case for operating a broadband network in high cost areas in the absence of programs that provide additional support."[67] In response, Commissioner Robert McDowell and the FCC's opponents contended that section 706 is simply aspirational and is not an independent source of regulatory authority.[68] All legal challenges to the Commission's authority to fund broadband Internet access remain unresolved as of this writing and pending before the Tenth Circuit on review of the *USF–ICC Reform Order* as a whole.

III. Federal USF Contribution Mechanisms

So far, we have discussed how the money in the federal Universal Service Fund is spent; now we turn to who must pay for all these programs. The short answer is "various segments of the telecommunications industry," which pass their assessments through to consumers in the form of line-item fees. The key controversies involve how to determine which telecommunications providers should bear how much of the burden; how to avoid arbitrage and competitive distortions as technological change blurs regulatory distinctions between categories of providers and services; and how to square any sensible solution to those problems with the pre-broadband-era provisions of the Telecommunications Act of 1996.

Before we get down to details, we pause to consider a threshold policy choice that is responsible for these contribution-related policy challenges in the first place. To *whatever* extent policymakers subsidize communications services—either for poor subscribers or those living in high-cost areas—it is hardly obvious that these subsidies should be underwritten through "contribution obligations" imposed on telecommunications providers, as section 254 requires, rather than through general tax revenues. Industry-specific assessments artificially depress demand for the taxed products or services and are therefore less economically efficient than general taxes.[69] Of course, the telecommunications industry incurs not only the full burden of supporting universal service programs, but also a welter of regular federal, state, and local taxes as well.

Like much of universal service policy, this reliance on industry-specific assessments is largely a function of politics. Politicians are loath to raise income taxes and are only too happy to adopt alternative funding mechanisms that, through their sheer complexity, obscure the extent to which ordinary American voters are indirectly paying taxes by another name. From a politician's perspective, implicit cross-subsidies are most appealing because they are the hardest for individual voters to perceive. But if the growth of competition—which is also popular among voters—makes such cross-subsidies unsustainable, complex universal service assessments are politically preferable to an ordinary tax hike, at least until consumers figure them out and complain. And even when consumers do complain, they may well get lost in the confusing array of fees and not know whether to blame politicians or service providers. Unfortunately, in the process of mandating these industry-specific assessments, Congress has needlessly compelled regulators to answer intractable questions about which types of providers should be required to bear what percentage of the contribution burden. As discussed here, those questions have no fully satisfying, competitively neutral answers.

A. The traditional revenue-based assessment regime

We begin with the main statutory provision governing contribution obligations: section 254(d), enacted as part of the 1996 Act. In 1996, residential broadband Internet access did not exist, many consumers still viewed cellphones as an extravagant novelty, and people relied on circuit-switched wireline connections to place the vast bulk of telephone calls, which could be neatly divided into separate "interstate" and "intrastate" categories on the basis of the fixed physical locations of the calling and called parties. Enacted against that backdrop, section 254(d) requires every carrier that provides "interstate telecommunications services" to contribute to the federal fund in a manner directed by the FCC.[70]

In the wake of the Act's passage, the Commission tested the extent of its authority in this area by declaring that all telecommunications carriers should contribute to the fund on the basis of both their interstate revenues and potentially their intrastate revenues as well.[71] In 1999, the Fifth Circuit rejected the intrastate component of that initiative on jurisdictional grounds. It held that despite the Supreme Court's expansive reading of the FCC's jurisdiction to implement the 1996 Act's competition-related provisions (see chapter 2), the FCC lacked authority to base contribution obligations on the magnitude of a carrier's intrastate revenues—that is, revenues attributable to the provision of telecommunications services to calling and called parties within the same state.[72]

The immediate upshot of this ruling was that the heaviest contribution burden fell largely on standalone long-distance carriers, such as AT&T and MCI, whose traffic disproportionately consisted of interstate and international calls. Today, the lion's share of contributions is still paid by "telephone companies" with major long-distance voice operations, and almost three-quarters is paid by five carriers: AT&T, Verizon (which acquired MCI in 2005), CenturyLink, Sprint, and T-Mobile.[73] The Fifth Circuit's rigid jurisdictional holding was also bad timing. The long-distance market was beginning to implode as the fiber glut of the late 1990s pushed down long-distance rates and as customers began using email and other substitutes for conventional long-distance calling.[74]

In the years that followed, the FCC responded with two types of regulatory patches. First, it continuously raised the "contribution factor" that dictates the percentage of interstate revenues that a carrier must pay into the USF to underwrite all the various federal programs. That percentage rose from around 5.5% in 1998 to around 9% in 2004, when we wrote the first edition of this book, and it now hovers around the midteens, approaching 18% in the first quarter of 2012.[75] Attentive consumers will recognize these fees as ever-increasing line items on their telephone bills.

Second, while maintaining its single-minded focus on "interstate revenues," the FCC began looking to providers other than traditional wireline telcos as major sources of USF contributions. Mobile wireless carriers were one obvious target for these initiatives as consumers increasingly substituted cellphones for landlines when placing long-distance calls. Precisely because these services are mobile, however, it was (and is) difficult to determine from phone records which calls are interstate in the sense that the calling and called parties are in different states. And to complicate matters further, wireless carriers were in the vanguard of selling distance-agnostic buckets of minutes, a practice now shared by many wireline telephone plans as well. As a result, revenues could not be easily assigned to interstate or intrastate categories even if it were possible to determine which individual calls crossed state lines. The Commission has addressed these problems by creating various rebuttable presumptions about the percentage of a given mobile carrier's revenues that are attributable to interstate calls. Over time, it has increased these rebuttable presumptions, known as *safe harbors*, to extract greater contributions from the wireless industry.[76]

Meanwhile, by the first years of the new millennium, consumers also began abandoning traditional circuit-switched long-distance services in favor of "interconnected" over-the-top VoIP services, which, as the FCC defined that term, enabled users to call and be called by subscribers to conventional telephone networks (see chapter 5).[77] In 2006, the Commission extended contribution obligations to these providers for the first time, prompting considerable controversy and a court challenge, which the Commission ultimately won.[78]

Understanding this VoIP contribution dispute requires a brief detour back into the statutory language governing the FCC's contribution authority. Quite apart from the intrastate/interstate distinction, section 254 separately distinguishes between providers of (interstate) "telecommunications *services*"—that is, Title II "telecommunications carriers"—and providers of mere "interstate *telecommunications*." Under section 254(d), the former providers are subject to compulsory contribution obligations; the latter are subject to such obligations only when the FCC, in its discretion, concludes that they should be. Who is in this latter category? One key subclass consists of companies offering "information services" that *include* "telecommunications" as an input but that do not qualify as "telecommunications services" because they provide integrated data-processing functionalities to consumers (see chapter 6).

Many VoIP providers maintain that their services are "information services" rather than "telecommunications services" and are thus exempt not only from automatic contribution obligations, but also from ordinary Title II common carrier regulation. The FCC has avoided resolving that politically fraught issue since it first arose in the early 2000s.[79] In 2006, however, the FCC decided that even if VoIP is not itself a "telecommunications service," VoIP providers nonetheless qualify as "providers of interstate telecommunications" subject to the FCC's discretionary contribution authority.[80] The D.C. Circuit upheld that statutory construction the following year.[81]

From a policy perspective, it was no small step for the FCC to extend contribution obligations to these IP-based providers. In the early years following the 1996 Act, the FCC steadfastly refused to impose contribution obligations on information service providers such as AOL and other ISPs.[82] In a widely cited report to Congress in 1998, the FCC defended this policy in part on the ground that the government could best foster the Internet's growth by exempting information service providers from the various regulatory burdens associated with the conventional telephone industry.[83] And it ultimately maintained that policy even as facilities-based broadband ISPs began supplanting dial-up ISPs as the providers of choice for Internet access.[84]

The Commission treated interconnected VoIP differently for two main reasons. First, as a normative matter, it found that interconnected VoIP providers could hardly complain about contributing to universal service because "much of the appeal of their services to consumers derives from the ability to place calls to and receive calls from the PSTN, which is supported by universal service mechanisms."[85] Second, it invoked "the principle of competitive neutrality" for the proposition that "[a]s the interconnected VoIP service industry continues to grow, and to attract subscribers who previously relied on traditional telephone service, it becomes increasingly inappropriate to exclude interconnected VoIP service providers from universal service contribution obligations."[86]

Indeed, subjecting circuit-switched telephone companies to assessments that their VoIP rivals need not pay might ultimately create a classic death spiral for the USF contribution base. Such asymmetric regulatory burdens would raise conventional telco prices vis-à-vis VoIP prices, thereby artificially accelerating consumer migration to VoIP services. That accelerated migration would in turn reduce the pool of telco revenues from which USF contributions are drawn. And this lower contribution base would then require the FCC to increase the contribution factor—that is, exact from conventional telcos an even larger percentage of their shrinking revenues—simply to maintain the existing size of the fund. And if telcos respond by passing through the higher USF fees to their retail customers, they would further accelerate the loss of those customers and revenues to VoIP providers, starting the vicious cycle anew.

Having subjected interconnected VoIP providers to contribution obligations, the Commission then needed to grapple with the extent of those obligations. That presented another policy challenge. As with mobile services, it is difficult to determine from conventional phone records where exactly an over-the-top VoIP caller is and thus what percentage of VoIP revenues are attributable to interstate (or international) calls. Indeed, the FCC had previously relied on that geographic ambiguity when determining in 2004 that over-the-top VoIP services are indivisibly interstate and thus beyond the scope of state regulatory jurisdiction (see chapter 5). In 2006, it adopted a safe-harbor presumption that most VoIP revenues (64.9%) are attributable to "interstate" calls and thus subject to federal USF assessments.[87] That number is considerably higher than the corresponding number for mobile providers, a disparity that the FCC justified on the ground that interconnected VoIP service, far more than mobile services, "is often marketed as an economical way to make interstate and international calls, as a lower-cost substitute for wireline toll service."[88] As with mobile wireless providers, this figure is rebuttable. A mobile or VoIP provider can try to reduce its contribution obligations by showing that nonassessable intrastate calls account for a greater percentage of its traffic than the default safe harbors suggest, and many providers have submitted "traffic studies" that purport to make that showing.[89]

Even as the FCC patched these holes in its USF contribution base, new ones kept appearing. "Interconnected" VoIP is not the only service that the FCC has never definitively classified as either a "telecommunications service" (automatically subject to contribution obligations) or an "information service" with a "telecommunications" component (exempt from contribution obligations until the FCC rules otherwise). Other legally ambiguous services include SMS text messaging; various high-end communications services offered to enterprise customers; and "one-way" VoIP services (such as SkypeOut) that do not qualify today as "interconnected" because they enable subscribers *either* to call *or* to be called by PSTN users, but not both.

In some cases, as with enterprise services, the result is profound regulatory uncertainty and a competitive race to the bottom. Providers suspecting that they are selling "telecommunications services" face overpowering incentives to conclude otherwise and thus avoid contribution obligations simply because their rivals have done so. As one enterprise-services provider explained to the FCC, it "is not realistic for one or more providers to charge corporate customers 11 to 12 percent more in USF fees on MPLS-enabled [enterprise] services and maintain market share when other providers do not assess their customers for such fees."[90] In other cases, there is less uncertainty about what the existing contribution rules require, but those rules nonetheless threaten to distort competition. For example, today's narrow definition of "interconnected" VoIP services likely influences market behavior to the extent that consumers choose between two-way VoIP services, which are subject to contribution obligations reflected in higher retail rates, and one-way VoIP services, which are not.[91]

Two more factors have further destabilized today's revenue-based regime. The first is the proliferation of service bundles such as mobile voice-and-data plans and landline triple-play offerings. In the FCC's words, "Bundled offerings of telecommunications and information services present two contribution issues concerning how revenues from a bundled offering should be apportioned: (1) how to apportion revenues when the provider does not offer the assessable service (i.e., telecommunications service or interconnected VoIP) in the bundle on a stand-alone basis, and (2) how to apportion revenues when the provider does offer the assessable service on a stand-alone basis, but does not explicitly allocate the discount on a bundled offering to specific services comprising the bundle."[92] As the FCC concedes, its rules give providers "fairly wide latitude" in determining how to conduct this revenue apportionment, in part because there is no economically straightforward formula for attributing either costs or discounts to particular services within a bundle.[93] Here, too, the lack of clear rules has triggered another competitive race to the bottom, as market rivals with narrow margins face powerful incentives to cut costs by erring on the side of overattributing revenues to nonassessable information services.[94]

A second complicating factor arises from the need to avoid inefficient double assessment of revenues associated with a given service. The FCC's rules "require contribution only once along the distribution chain," generally at the retail level, and thus "'[c]arrier's carrier' revenues," collected at the wholesale level, "are not currently assessed."[95] For example, if a retail provider of conventional voice services resells a wholesale carrier's services to the public, the reseller incurs a contribution obligation for the revenues it earns from end users, but the wholesale carrier incurs no contribution obligation for the revenues it earns from the reseller behind the scenes.

The FCC first adopted this "carrier's carrier" rule in 1997. As it acknowledged in 2012, it had not fully anticipated then the "implementation difficulties" that would arise in an increasingly common circumstance: where "a wholesaler sells a service to another firm that incorporates that wholesale telecommunications into a different offering for its retail customers that is *not subject to assessment*," such as "broadband Internet access service."[96] For example, suppose a broadband ISP buys a special-access circuit from a conventional telecommunications carrier to bridge the "middle mile" between the ISP's last-mile network and the wider Internet. Although the broadband ISP has the retail relationship with its end users, it incurs no contribution obligations. In an effort to close that loophole, the Commission clarified that telecommunications carriers can avoid contribution obligations on their wholesale revenues only if their wholesale customers "reasonably would be expected to contribute" to the USF.[97] In practice, however, it is often difficult for carriers to confirm exactly what services their wholesale customers are selling to the public, how those customers classify their retail services for contribution purposes, and whether they are in fact making any required USF payments.[98]

B. Prospects for contribution reform

Reacting to all these regulatory conundrums, the FCC sought public comment in 2002, 2006, and 2008 on successive proposals to overhaul its contribution regime from the ground up.[99] On each occasion, it asked whether it should abandon revenues as the main source of contribution obligations and choose alternative criteria instead, such as a carrier's use of telephone numbers or its provision of "connections to a network." And on each occasion, the Commission's proposals generated much debate but little action. By 2012, continuing industry shifts, combined with the FCC's own restructuring of the USF distribution program, led the Commission to open yet another proceeding on contribution reform.[100] That proceeding remains pending as this edition goes to press, and we outline its basic themes here.

One key impetus behind this renewed reform effort was a growing disconnect between the two sides of the USF regime: disbursements and contributions. As discussed, the pre-2011 federal universal service program focused for decades on promoting affordable voice telephone services. Against that backdrop, it was at least superficially plausible to focus much of the contribution-collection effort on those same services. But it seemed increasingly anomalous to rely disproportionately on voice services as a source of USF contributions once the Commission shifted the focus of USF disbursements in 2011 from affordable telephone service to affordable broadband Internet access. More anomalously still, the FCC has chosen not to assess contribution obligations on broadband

Internet access itself under its discretionary authority, even though it is the very service that in high-cost areas benefits from the new disbursement regime.

More generally, by 2012, the voice-oriented regulatory distinctions governing contribution obligations were fast becoming unhinged not only from the new broadband focus of the USF, but from the technological realities of the emerging telecommunications ecosystem, where voice is increasingly just one application among many on a convergent IP platform. At the applications layer, the traditional contribution regime imposed obligations on some higher-layer applications (interstate voice), but not on others (e.g., streaming video and the Web), even though the latter category accounts for a far greater share of telecommunications traffic overall than voice does. And at the physical layer, the traditional regime imposed obligations on revenues attributed to some transmission functions (those used to provide interstate "telecommunications services" like circuit-switched telephony), but not to others (such as those used to provide broadband Internet access), even though the latter category will eventually account for most of the telecommunications industry.

This mismatch between regulatory distinctions and technological reality creates countless opportunities for arbitrage and competitive distortion. Just as troubling, it threatens the long-term solvency of the USF itself. By 2012, the FCC observed, the broadband revolution and other technological developments had helped cause a multiyear "decline in the contribution base at the same time that the communications market has grown,"[101] largely by supplanting assessable "telecommunications services" with nonassessable "information services." It was against this backdrop that in 2012 the FCC followed up on its *USF–ICC Reform Order* by proposing to reform its contribution regime to match the new broadband-oriented focus of the USF in general.

At the highest level of generality, the FCC sought comment on three alternative methodologies for assessing universal service contributions: interstate revenues, connections, and numbers. The first option—interstate revenues—is of course the same basic methodology the Commission has used all along. But the FCC asked whether certain reforms might rehabilitate that methodology and tailor it to the broadband age. For example, it proposed to expand the contribution base to encompass new service categories on a case-by-case basis, as it did with interconnected VoIP in 2006. In particular, it proposed to use its discretionary authority to impose contribution obligations on certain services that are deemed (or presumed *arguendo*) to be "information services," including broadband Internet access, one-way VoIP, SMS text messaging, and various enterprise communications services.[102]

Alternatively, in lieu of this case-by-case strategy, the Commission proposed a "broader definitional approach" under which "[a]ny interstate information service or interstate telecommunications" would be "assessable if the provider also provides the transmission (wired or wireless) . . . to end users."[103] That approach, the Commission added, would be "intended to include entities that provide transmission capability to their users, whether through their own facilities or through incorporation of services purchased from others, but not to include entities that require their users to 'bring their own' transmission capability in order to use a service."[104]

As the Commission acknowledged, however, this distinction is hardly self-implementing.[105] For example, the Commission had previously found that an interconnected over-the-top VoIP provider such as Vonage "provides telecommunications" to its end users, even though they "bring their own" broadband on *their* end of each call, because, on the *other* end of a PSTN–VoIP call, the VoIP provider must arrange for PSTN transmission behind the scenes and necessarily "provides" that (distant) transmission to its end users whenever it connects a call. Indeed, that was the very rationale on which the Commission subjected Vonage and similar providers to its discretionary contribution authority.[106] Under that logic, however, a broad range of Internet-based "information service" providers, ranging from content delivery networks to cloud computing services, might also be said to "provide telecommunications" to their end users because they, too, must arrange for transmission behind the scenes over some segment of every Internet-based communication. The FCC gave no indication that it wished to include such providers within its "broader definitional approach" or to open a new round of disputes on this set of metaphysical questions. The Commission may ultimately need to retain its traditional case-by-case approach, identifying the specific categories of information services that it wishes to subject to its discretionary contribution authority.

The FCC also proposed to restructure the wholesale–resale distinction discussed earlier, under which wholesale telecommunications carriers today can avoid contribution obligations for "carrier's carrier" revenues if and only if their wholesale customers "reasonably would be expected to contribute" to the USF.[107] Under the proposed alternative, wholesale and retail providers in the same service-distribution chain might be subject to simultaneous contribution obligations, but only to the extent of "the value [each] provider adds to the service."[108] Such value-added assessments are widespread in the general tax regime, but the FCC has never applied them in this industry-specific context, and it asked a broad variety of questions about exactly how such an approach would be implemented and what competitive distortions or other unintended consequences it might introduce.[109]

The Commission next sought comment on a *connections-based* alternative to the current revenues-oriented regime. Although the term "connection" is subject to various definitions, the best-known proposal of this type would tie contribution obligations to a communications provider's physical-layer connections to any "interstate public or private network" (or to any "assessable service"), including the PSTN or the Internet.[110] Supporters of this approach claim that it has two main advantages over a revenues-based system. First, they say, it would "provide a more stable contribution base than a revenue-based system because the number of connections has historically been more stable than end-user interstate telecommunications revenues."[111] Second, they add, a connections-based approach would curb the uncertainties and competitive distortions caused by the current revenue-based regime's emphasis on "differentiat[ing] between revenues from interstate and intrastate jurisdictions and from telecommunications and non-telecommunications services."[112]

That said, a connections-based approach might present formidable methodological challenges of its own. One obvious challenge is that "connections" can take many different forms, and there is no universal standard of measurement for them. For example, how would the FCC count the number of connections provided by, for example, (1) providers of paging services, (2) wireless prepaid services, (3) triple-play fiber-to-the-home services, (4) multiline business plans, (5) machine-to-machine services (such as smart meters and supply chain tracking), and (6) various categories of high-capacity circuits?[113] There is no straightforward solution to such questions, and different industry segments predictably offer mutually antagonistic answers.[114]

Finally, the FCC sought comment on an alternative *numbers-based* approach.[115] As its name suggests, this approach would base contribution obligations largely on the volume of telephone numbers that a provider assigns to its customers; under some variations, this approach would also be supplemented by connections-based assessments for special-access services and private lines. A principal virtue of a numbers-based regime is relative ease of administration. But critics argue that a pure numbers-based approach would irrationally exempt various telecommunications providers from contribution obligations because their services do not involve telephone numbers. On that and other grounds, these critics question the legal and policy merits of basing contributions on the use of telephone numbers as such rather than more directly on the provision of interstate telecommunications.[116]

There is also a deeper concern about long-term reliance on "numbers" as the primary basis for assessing USF contributions. We take it for granted today that most people will indefinitely subscribe to voice services that are assigned ten-digit telephone numbers. But as the circuit-switched PSTN gives way to newer IP-based technologies, it is not inevitable that people will always insist on having such numbers.[117] Some VoIP services today (such as Vonage's) use ten-digit numbers, but others (including some Skype offerings) do not. It is at least conceivable that ten years from now increasing numbers of consumers will begin abandoning voice connections that are assigned ten-digit numbers in favor of voice services that are associated with email addresses or Internet domain names. If so, choosing a numbers-based contribution scheme would simply reintroduce the same type of regulatory dysfunction that the existing revenues-based regime has produced since 1996. That is, the USF would rely for its funding on an obsolescent regulatory premise; it would depend for contributions on an ever-shrinking base of services; and it would inefficiently distort competition by imposing contribution burdens on those services, but not on their market substitutes. Plus ça change, plus c'est la même chose.

* * *

Despite the FCC's renewed focus, there will be no fully satisfying solution to these USF contribution-related policy conundrums. In the end, the FCC must continue to muddle through, patch holes that threaten the funding base, and minimize competitive distortions as best as it can. Again, all these challenges arise from antiquated legal distinctions codified in the 1996 Act and from the choice to recover the costs of universal service from the telecommunications industry in particular rather than from tax revenues in general. These challenges will remain unless and until Congress reenters the fray and rationalizes the basic statutory regime. Depending on the outcome of the pending judicial challenges to the FCC regime described in this chapter, Congress may need to intervene sooner rather than later.

9

Competition in the Delivery of Video Programming

To this point, we have examined the regulation of information platforms that in any given transmission deliver voice and data traffic from one point to another or, at most, to a discrete set of points. Such platforms include wireline and wireless telephone networks and broadband Internet access networks. Of course, the FCC also plays a central role in the regulation of traditional *broadcasting*—that is, the transmission of over-the-air signals by local radio and television stations to their surrounding communities. For several decades after World War II, the FCC's video-distribution policies focused on over-the-air TV. For the past few decades, it has also focused on the complex policy issues raised by the rapid growth of *multichannel video programming distributors* (MVPDs), including cable companies, other wireline video providers (such as Verizon FiOS and AT&T U-verse), and satellite providers such as DirecTV and DISH.

Unlike traditional point-to-point networks, these *video-distribution platforms*—local broadcasters and the various types of MVPDs—have traditionally transmitted the same widely watched TV signals to large numbers of viewers at the same time. In many contexts, this approach is still the most efficient way to disseminate television events to mass audiences. For example, when you watch the Super Bowl on a conventional cable television system, you are "tapping," from a common video stream in the cable running along the street, the same signals that all your neighbors are simultaneously receiving. Given the widespread demand to watch the Super Bowl at the same time and the prodigious bandwidth needed to transmit high-quality images of the action, this arrangement is more efficient than transmitting fully separate and redundant video streams of the Super Bowl to each of the many subscribers that request them.[1]

This chapter discusses the major types of *competition-oriented* rules that Congress and the FCC have imposed on these video-distribution platforms. To their critics, many of these rules are economically unsophisticated and technologically obsolescent, embodying the pre–Chicago School mindset of the 1960s and 1970s and irrationally fixated on propping up traditional local broadcasting against alternative video-distribution platforms.[2] One answer to those critics is that media regulation has always been designed to promote not only traditional competition policy objectives, but also various non-efficiency-related social goals.[3] The latter goals include the preservation of *free over-the-air television*, principally for the benefit of those who cannot afford to subscribe to cable or satellite television services, and the promotion of *localism* and *diversity* in programming content.* In many markets, of course, product diversity, like vigorous price competition, can be an important goal of antitrust policy itself. In the television programming context, however, Congress and the FCC have traditionally sought to generate greater (or more types of) diversity than an efficient market would necessarily produce on its own.[4]

This policy has incited a long-running debate about whether the government *should* take affirmative regulatory steps to promote diversity and localism or whether, as former FCC Chairman Mark Fowler provocatively suggested in 1981, the government should treat television as "just another appliance . . . a toaster with pictures."[5] Because this book primarily concerns competition policy, it is not our purpose to evaluate the highly contested merits of the government's non-efficiency-related objectives in the television world, much less to weigh them against the costs that inevitably accompany regulatory intervention in any market. Instead, we have limited our discussion in this chapter to concise narrative summaries rather than full-blown economic critiques of the federal government's major competition-related policies concerning video programming and distribution.†

* The concept of "diversity" is hardly self-defining. In different contexts, it can mean several distinct things: diversity of sources for programming (source diversity); diversity of types of programs (output or programming diversity); diversity of ownership in terms of numbers of different owners and full representation of different ethnic and racial backgrounds (input diversity); and, of course, diversity of views (viewpoint diversity). *See generally* FCC v. National Citizens Comm. for Broad., 436 U.S. 775, 796–97 (1978) ("[d]iversity and its effects are . . . elusive concepts, not easily defined let alone measured").

† We also do not address wholly non-competition-related areas of TV regulation such as the now defunct "fairness doctrine," restrictions on "indecent" content, obligations to carry certain amounts of educational children's (or other public interest) programming, the various regulations applicable to political advertising, or the federal government's forced migration of broadcast TV technology from analog to digital, which concluded (for full-power stations) in 2009.

After a brief introduction to the structure of the television marketplace, we discuss three major categories of competition-related video regulation. The first consists of regulations that mediate the relationships *among* different video-distribution platforms, entitling or obligating one such platform (e.g., a broadcast station, cable operator, or satellite carrier) to carry the programming of another. These regulations—which encompass *compulsory copyright licenses*, *retransmission-consent* requirements, *must-carry* obligations, and *program-access* rules—are justified as promoting a number of objectives, including greater competition among rival video-distribution platforms. We close that discussion by addressing the nascent debate about whether such rules should extend in some form to online video distributors such as Netflix and Hulu.

The second category consists of rules intended to promote greater programming diversity by protecting video-programming *suppliers* from the putative market power of video-programming *distributors*. These rules include the now-defunct *finsyn* (financial interest and syndication) restrictions once applicable to the broadcast networks; the *program-carriage* rules designed to keep cable operators from discriminating against unaffiliated programmers; and the related *channel occupancy* and *horizontal ownership* limits that the FCC has also sought to impose on cable television systems.

The third category of competition-related policies consists of limits on a given firm's ownership of multiple media outlets—such as newspapers and TV stations—nationally or in particular geographic markets. These policies, too, are intended not just to curb potentially undue market concentration, but also to promote greater programming diversity by ensuring a multiplicity of voices within given communities, particularly on matters of local concern.

I. The Basics of the Video-Distribution Marketplace

A. Introduction to video programmers and distributors

In industry jargon, *linear* television consists of the programming made available for initial home viewing at particular channel locations in designated time slots. Today, the great majority of American households with televisions receive linear programming in one of three ways: (1) by receiving conventional over-the-air broadcasts from television stations, (2) by subscribing to a wireline (cable or telco-based) MVPD service, or (3) by subscribing to a direct-to-home satellite TV service (DirecTV or DISH). As noted in chapter 3, the FCC estimates that only about 10% of television households now rely exclusively on terrestrial broadcasting to receive linear television signals.[6] Of the remaining 90%, roughly two-thirds subscribe to a cable or telco MVPD service, and almost all of the rest to satellite.[7] For simplicity, we refer to all wireline MVPD services, including those offered by traditional telcos, as "cable services," even though one major telco provider (AT&T) has disputed that classification of its video service as a legal matter.[8]

The broadcasters dominated television from their inception in the 1940s until the rapid growth of MVPDs in the 1980s. As a policy matter, broadcasters are characterized as "trustees" (not owners) of their assigned blocks of spectrum and are obligated to air programming "in the public interest."[9] In a typical mid-sized urban market, the FCC has licensed a handful of television stations to broadcast in VHF frequency channels (channels 2–13) and a handful more to broadcast in the UHF channels (channels 14 and up). Each of the most-watched broadcast stations in a given market is typically affiliated with—that is, has obtained the rights to carry the programming distributed by—one of the major television networks, such as CBS, NBC, ABC, Fox, or Univision.

The networks in turn acquire rights to content produced by various content creators. These include the networks' own affiliated studios, which produce shows both for broadcast stations and for nonbroadcast cable channels. A television network typically *owns and operates* a small number of broadcast stations outright, which are known as "O&Os" and are generally located in major metropolitan markets. As noted later in this chapter, however, FCC rules—designed to promote localism and programming diversity—effectively limit the number of stations any one company may own. Each network thus reaches the remainder of U.S. households through *affiliated* stations in which the network companies have no ownership interest. These independently owned affiliates deal with the networks on an arm's length contractual basis and occasionally defect from one network to another, as happened in the mid-1990s when the new Fox network aggressively recruited local affiliates.

The relationship between a television network and its affiliates is mostly, though not entirely, symbiotic.[10] The network contributes national programming for much of an affiliate's broadcast day, including its prime-time schedule. An affiliate typically contracts for exclusive rights in a given geographic market to air the first broadcast of a network's programs. The network is generally free, however, to sell any other station in the same market the *syndication* rights to programs that it owns, including the right to show reruns of previous seasons' episodes in later years. And the affiliate is likewise free to purchase (among its other programming options) syndicated shows that originally aired on other networks. With limited exceptions, however, the affiliate is contractually bound to air its own network's programming during prime time (and often certain other parts of the day). And its revenues tend to increase with the popularity of the network's shows because advertisers will generally pay more to have their commercials run during shows with high ratings. From the network's perspective, the affiliates contribute, in the aggregate, the coast-to-coast potential audience needed to generate the enormous national advertising revenues that in turn underwrite the network's programming expenses. The networks draw substantial revenues from national advertising aired during network programs, and the affiliates draw most of their revenues from their own advertising, aired during (or adjacent to) network programs and during non-network programs, such as the local news.

Policymakers often cite the essential role of local broadcasters—both network affiliates and unaffiliated independent stations—in preserving America's longstanding system of "free" over-the-air television. We place the word "free" in quotation marks because a television broadcast is cost-free to consumers only in the literal sense that they need not pay money out of their pockets to receive broadcast signals over the air. But viewers of commercial TV broadcasts are asked to watch paid advertisements that underwrite the production and distribution costs of the relevant programming. Moreover, from a societal perspective, television broadcasts present substantial opportunity costs: they consume enormous swaths of spectrum that could often be put to more efficient uses (such as mobile voice and data) now that the overwhelming majority of Americans rely on MVPDs rather than on terrestrial over-the-air transmissions to receive television programming. As discussed in chapter 3, Congress acted on this very premise in 2012, authorizing the FCC to hold "incentive auctions" designed to induce broadcasters to cede spectrum voluntarily in exchange for a share of the cash proceeds. But it is not yet clear that these auctions will succeed in inducing many broadcasters to vacate spectrum in large enough numbers and in enough geographic locations to realize the full potential of this spectrum for nonbroadcast uses.

Unlike broadcasters, MVPDs—cable and satellite companies—charge subscription fees. In the early years of cable service, cable providers focused mostly on retransmitting broadcast programming to remote areas where over-the-air signals otherwise might not come through strongly enough to provide clear pictures to viewers. Starting in the

1970s and 1980s, cable providers looked for ways to increase their subscribership and penetrate urban markets by offering new programming that would be available only on their cable platforms. The result was a proliferation of cable-only channels such as HBO, MTV, and ESPN.

The programming on many of these channels was and is produced by major studios affiliated either with the cable companies themselves or with the major television networks. For example, Disney owns ABC Family, the Disney Channel, and most of ESPN Inc., and News Corp. owns the Fox News Channel, several regional sports channels, and FX, among others. Particularly in the largest markets, cable systems are owned by national media companies such as Comcast, Time Warner Cable, and Cox, known as *multiple system operators* (MSOs). As we shall see, Congress directed the FCC to limit the number of markets such MSOs may serve, with the primary goal of protecting the viability of independent programmers.

Cable television service was traditionally viewed as a natural monopoly (see chapter 1), and a single company would operate the sole cable system in a given geographic region, sometimes under a (now banned) exclusive franchise agreement with local authorities.[11] Today, telcos such as Verizon (FiOS) and AT&T (U-verse) offer competing wireline MVPD services to about one-third of American households, most of them in major metropolitan areas.[12] Much smaller cable "overbuilders" such as RCN and WOW! (which recently bought Knology) also provide competing MVPD services in some areas. But the costs of deploying wireline networks capable of carrying high-quality video signals are immense, and in many areas the revenues a new wireline MVPD can earn in a market already occupied by a cable incumbent may be insufficient to cover those costs. As of this writing, therefore, it remains unclear whether "terrestrial" (nonsatellite) MVPD competitors will expand much beyond those densely populated metropolitan areas in which they currently offer service, where revenues and scale economies are greatest.

That brings us to satellite TV providers. Although direct-to-home satellite providers first offered service in the 1980s, it was not until the 1990s that these providers assumed their current position as serious competitors to cable companies. There are two major *direct broadcast satellite* (DBS) firms that beam television programming directly to American consumers: DirecTV and DISH. Because of their nationwide coverage, these two companies are, respectively, the second and third largest U.S. MVPDs by total number of subscribers.[13] In 2002, the FCC effectively blocked a proposed merger of these two companies on the ground that it would subvert MVPD competition.[14]

The satellites operated by these companies occupy highly coveted orbital slots that are both *geostationary*, in that they remain fixed in place above the earth, and located in positions from which the satellites can transmit signals to the entire continental United States. (These are thus known as *full CONUS* slots.) The FCC has assigned separate blocks of spectrum to these providers, each of which uses digital compression technology to transmit hundreds of TV channels. Most subscribers now receive satellite signals via pizza-size dish antennas affixed to their rooftops, whereas the earliest generation of subscribers, most of them in rural areas, relied on much larger "C-band" antennas the size of compact cars.

B. Cable rate deregulation and the "à la carte" debate

Competitive MVPDs are free from retail rate regulation, and, with limited exceptions, that is now generally true of cable incumbents, too. The deregulation of cable rates has been a subject of controversy and congressional vacillation. In the Cable Act of 1992, Congress directed the FCC to impose a scheme of comprehensive retail rate regulation for cable operators, which were then indisputably dominant providers of MVPD services.[15] Four years later, in the Telecommunications Act of 1996, Congress reversed course. It announced that starting in 1999 cable rates would be deregulated for all tiers of cable service other than the stripped-down "basic service tier," which meant in practice that the overwhelming majority of cable subscribers would pay unregulated rates.[16] The rationale for that shift was that although cable incumbents still had commanding market shares, satellite providers and terrestrial overbuilders (including telcos) were poised to unleash price-disciplining competition for MVPD services. That competition was far slower to develop than Congress appeared to expect, and cable rates have increased substantially since then (as, to be sure, have the cable companies' infrastructure investments and programming-acquisition costs). Nonetheless, the basic deregulatory choice adopted in 1996 remains in place today.

The debate about cable rate regulation briefly flared up half a dozen years into the new millennium, when the FCC began considering proposals for *mandatory à la carte pricing*. Apart from the basic service tier, which only a small minority of cable subscribers order, cable operators typically offer only broad service tiers that bundle dozens and sometimes hundreds of cable channels. Any given subscriber may watch only a fraction of those channels, yet he pays for all of them in the sense that his retail cable rates are driven in part by the cable company's own costs of programming acquisition. Consider the simplified example of a subscriber who loves the History Channel, never watches ESPN or most other cable channels, but can watch the History Channel only if he buys a channel bundle that includes ESPN. At least in theory, he must pay more to watch the History Channel than he would if his cable operator sold the History Channel à la carte, unbundled from ESPN. That is not only because he must pay for an extra channel he does not watch, but also because the extra channel in our hypothetical is ESPN, which charges cable operators unusually large fees to carry what many viewers consider must-see sports programming.[17] Indeed, according to one 2011 estimate, cable operators pay ESPN $4.69 per subscriber per month, whereas they pay the average cable channel only 26 cents.[18]

In 2004 and 2006, the FCC's staff, under the direction of two successive FCC chairmen with very different regulatory philosophies and management styles, issued radically different recommendations on proposals to compel à la carte pricing for cable programming packages. In 2004, under Chairman Michael Powell, the staff report recommended against any market intervention. It concluded that mandatory à la carte would increase transaction costs for the industry; would raise subscription fees for consumers who watch more than a small handful of channels; and would disadvantage niche and other lesser-known cable programmers, who would find it harder to reach a critical mass of American households.[19] In 2006, a second report reached strikingly contrary conclusions on the basis of the same underlying data, supporting Chairman Kevin Martin's outspoken advocacy for new à la carte rules.[20]

Martin's à la carte proposal drew support from some consumer groups and from religious conservatives who wished to see more "family tiers." But it sparked widespread opposition from cable operators, cable programmers, and free market advocates. Citing the policy concerns outlined in the 2004 staff report, the large cable programmers opposed any *wholesale* unbundling requirements that would keep them, in their commercial deals with cable operators, from bundling lesser-known cable channels with the highly popular ones that cable operators valued most. These programmers also argued that such compulsory wholesale unbundling would be ineffective anyway in producing the policy outcome that Martin espoused: more *retail* choices offered by *cable operators* at lower prices. For their part, cable operators opposed any regulatory or legislative proposal to unbundle their own retail offerings. As they recognized, retail à la carte rules might resurrect full-blown retail rate regulation because, without such regulation, cable operators could evade the à la carte rules simply by pricing disfavored channel packages at unattractively high levels.

Martin's proposal ultimately collapsed amid widespread political opposition.[21] In 2008, the Democratic members of the FCC's House oversight committee added insult to injury when they criticized Martin in unusually personal terms for (among other things) "manipulat[ing]" the process underlying the 2006 staff report to produce unsound economic conclusions.[22]

II. Regulation of Relationships among Video-Distribution Platforms

Over the years, Congress and the FCC have devoted much time and energy to mediating the relationships among the three conventional video-distribution platforms: broadcasting, cable, and satellite. The government's regulation of these platforms has sought not merely to promote efficient cross-platform competition, but also to protect the particular interests of broadcasters. From the mid-1960s to the late 1970s, for example, the FCC struggled to justify various burdens it had placed on the upstart cable television companies—including an obligation to originate their own local shows and a prohibition on charging extra for premium programming—to protect the incumbent broadcasters' viewership and financial interests. One court later described such regulatory burdens as "hostile to the growth of the cable industry, as the FCC sought to protect, in the name of localism and program diversity, the position of the existing broadcasters, and particularly, the struggling UHF stations."[23]

Today, Congress and the FCC are less obviously protectionist in their television policies. Nonetheless, much of the regulation in this area reflects an abiding solicitude for the nation's local television broadcasters. Critics view this approach as symptomatic of the public choice pressures discussed throughout this book; proponents view it as an important part of a continuing effort to support local broadcasting as a basic institution of American democracy.

A. Compulsory copyright licenses and retransmission consent

In the early years of cable television, cable operators, known then as "community antenna television" providers, received the signals of broadcasters off the public airwaves and retransmitted them over wires to households that would otherwise receive poor or no reception. Local broadcasters viewed such retransmission of their signals with mixed feelings. On the one hand, so long as the signals were retransmitted in their entirety to the same local communities, the broadcasters and their commercial sponsors benefited from the successful efforts of cable systems to increase the number of "eyeballs" watching the broadcasters' programming and, in particular, the paid advertising accompanying that programming. On the other hand, the broadcasters feared, among other things, that cable companies would also pipe signals from distant broadcast stations into local broadcasting markets and would thereby dilute the share of the local viewership that any given local broadcaster could promise advertisers (a concern we discuss in more detail later in this section). The broadcasters thus challenged the cable operators' right to retransmit the copyrighted programming contained in the signals.

In the late 1960s and early 1970s, the courts ruled in favor of the cable operators in the ensuing copyright litigation, reasoning that cable systems, as passive transmitters, were more like "viewers" of the original programming than like infringing "performers" of it.[24] Congress then stepped in and struck a compromise in the Copyright Act of 1976: It deemed the cable systems "performers" for copyright purposes but granted them a blanket statutory license to retransmit broadcast programming.[25] That *compulsory license* freed cable operators from and eliminated the transaction costs of individualized negotiations between them and the many thousands of applicable copyright holders.[26]

The price of this compulsory license is determined by a statutory formula implemented by the Library of Congress.[27] The result, as that institution has described it, is an arrangement that is "technical, complex, and, many would say, antiquated."[28] As critics of the existing system observe, moreover, the compulsory license covers only broadcast programming, not programming that appears solely on cable channels, yet the transaction costs of individualized copyright negotiations have hardly kept cable channels from succeeding. The critics cite this success as evidence that the "transaction cost" rationale for the existing, broadcast-oriented compulsory license is hollow. In their view, the compulsory license is simply a regulatory mechanism for lowering prices below the levels that market negotiations would produce.[29] Despite recent reform advocacy, however, Congress appears unlikely to make fundamental changes anytime soon.

The compulsory license also addresses only *copyright* restrictions on the retransmission of programming content. In the Cable Act of 1992,

Congress granted broadcasters a separate and distinct property right in their broadcast *signals* and forbade any MVPD from retransmitting any such signal without "the express authority of the originating station," subject to a few enumerated exceptions.[30] In practical effect, this *retransmission-consent* provision requires cable companies and other MVPDs to give something of value to major, network-affiliated broadcast stations in exchange for the right to carry their "must-see" programming. Such compensation can take the form of cash payments or agreements to carry new or less prominent cable channels affiliated with the broadcast networks or both.[31] Some network–affiliate agreements specify that affiliates will share a portion of their MVPD cash payments with their networks, and networks are potentially interested parties in retransmission-consent negotiations involving their affiliates.[32] Of course, networks are also the main stakeholders in negotiations involving the stations they own and operate in major metropolitan markets.

Retransmission-consent negotiations occasionally degenerate into high-profile games of chicken. For example, in March 2010, negotiations between Disney and Cablevision broke down over the latter's right to retransmit Disney's ABC stations in the New York area. When the two companies' existing contract expired, about 3 million Cablevision households lost the ability to watch the Academy Awards ceremony until the signal was restored in mid-broadcast, after a new agreement was reached. Several months later, in October 2010, News Corp. and Cablevision fought their way into a longer impasse over the terms on which Cablevision could retransmit the signals of two Fox stations, also in the New York area. Until that impasse was broken two weeks later, Cablevision subscribers could not use their cable connections to view the National League baseball championship series or the first two games of the World Series.[33] Of course, viewers theoretically can watch the withheld programming by switching to another MVPD or, if they live within the station's broadcast footprint, by attaching antennas to their televisions and receiving the signals over the public airwaves. Nonetheless, consumers find it disruptive to switch providers each time retransmission-consent negotiations veer toward impasse, and many consumers cannot receive high-quality signals over the air.

In 2010, a group of cable and satellite companies petitioned the FCC to take regulatory measures that, in the name of industry stability, would blunt the broadcasters' bargaining power in various ways.[34] For example, they asked the FCC to rule that retransmission-consent impasses would henceforth be subject to binding arbitration and to grant cable and satellite operators interim rights to continue retransmitting broadcast programming once an existing retransmission agreement expires. These and other steps, the petitioners argued, were necessary to protect consumers. They claimed more generally that the retransmission-consent right was "artificial," that it subverted free market principles, and that it had enabled broadcasters to extract excessive fees from cable and satellite providers, which those MVPDs then passed on to consumers in the form of higher subscription fees. The broadcasters responded that the retransmission-consent right was no more artificial than any other intellectual property right and that it was a necessary counterweight to the compulsory copyright license, which MVPDs understandably wished to preserve. In Disney's words, "by giving broadcasters the right to bargain over the value of their programming, Congress meant to place them in a position closer to that occupied by nonbroadcast content providers like Discovery and HBO," whose programming is not subject to compulsory copyright licensing.[35]

In a 2011 notice of proposed rulemaking, the FCC essentially punted this policy dispute to Congress. As it noted, "Section 325(b) of the [Communications] Act expressly prohibits the retransmission of a broadcast signal without the broadcaster's consent," and the Commission thus lacks "authority to adopt either interim carriage mechanisms or mandatory binding dispute resolution procedures applicable to retransmission-consent negotiations."[36] The FCC nonetheless sought comment on a variety of procedural issues, such as how to interpret and enforce the parties' statutory duty to pursue retransmission-consent negotiations in "good faith."[37] That proceeding remains pending at press time.

As part of the same proceeding, the MVPDs had also asked the FCC to eliminate its longstanding *network-exclusivity rules*. To understand how these rules operate, suppose that a cable company wishes to carry programming from a given broadcast network, Network Z, but does not wish to pay the retransmission-consent fee required by Network Z's local broadcast affiliate. Suppose further that the cable company can threaten to obtain the same programming from any *distant* broadcast affiliate of Network Z. If that threat were credible, it would largely undermine the bargaining power of the local affiliate, which can still withhold *local* programming but—critically—not must-see *network* programming. From the local affiliate's perspective, ceding its carriage slot on local cable systems to the distant affiliate would be catastrophic: its local advertising revenues would quickly evaporate as the local cable subscribers, who constitute most of the affiliate's viewers, begin watching the distant station, with its own set of local commercials. To avoid that outcome, local stations typically bargain with networks for exclusive rights to broadcast network programming in a defined geographic area. This arrangement benefits the network, too, in that it keeps multiple affiliates from bargaining down the value of the network's programming in negotiations with MVPDs.

The FCC's *network-nonduplication rule* essentially provides that when exclusive rights to broadcast network programming appear in network–affiliate agreements, a station can generally seek FCC intervention if a cable operator enters into a contractually prohibited arrangement to carry duplicative network programming from a distant affiliate of the same network.[38] A similar regulation, known as the *syndicated-exclusivity rule*, enables stations to assert contractual exclusivity rights with respect to syndicated programming.[39] Cable companies argue that these rules artificially shield local broadcasters in retransmission-consent disputes from the price-disciplining effects of competition from other network affiliates. Local broadcasters and broadcast networks respond that the rules are hardly artificial because they give local stations no exclusivity rights beyond what they have privately bargained for and merely provide an additional forum—the FCC—for the enforcement of private contractual rights so that aggrieved stations need not sue in court. In its 2011 notice, the FCC sought comment on this controversy as well.[40]

B. Must carry

To this point, we have addressed the right of cable operators to carry broadcast programming they wish to carry; now we turn to their obligation to carry broadcast programming they do *not* wish to carry. If left to their own devices, cable companies might not carry, even for free, some of the least-watched stations, which are generally unaffiliated with the major networks.[41] Instead, cable operators might select different programming channels from other sources, particularly if they can sell local advertising on those channels. That market-based outcome would arguably spell financial doom for many of these smaller television stations. Because most Americans watch television by means of cable, any broadcast station not carried by cable would have few viewers, would be greatly limited in what it could charge advertisers for commercial airtime, and would thus lack the financial resources needed to acquire or produce better programming. In 1992, to preserve these stations from that threat, Congress entitled any broadcaster that does not wish to engage in retransmission-consent negotiations to rely instead on a statutory right to have its programming carried on the local cable system, albeit without compensation.[42]

This *must-carry* provision has been justified primarily as a necessary means of preserving a vibrant system of "free" and locally based over-the-air television. In 1997, by a vote of five to four, the Supreme Court relied mostly on that rationale to reject a First Amendment challenge to this regime.[43] The must-carry rules nonetheless remain controversial as a policy matter. They are undeniably overbroad in that they have benefited not just (or even primarily) stations with unusually meritorious "local" programming, but also stations with little or no local programming. The latter have included, for example, home-shopping channels, which obtained broadcasting licenses in part to guarantee themselves a free slot on the local cable line-ups.[44] More generally, as former FCC Commissioner Glen Robinson has observed, the must-carry regime and similar rules designed to protect "local" programming "assume[] that there is some plausible case to be made that the market will not assure local service sufficient to provide what the public needs in addition to what they want."[45] For now, however, that assumption remains engrained in U.S. television regulation.

C. Satellite retransmission of broadcast signals

The same traditional concerns that underlie the cable must-carry rules—localism and the sustainability of non-subscription-based television—also govern the relationship between local broadcasters and the major satellite television providers. Because of their centralized national transmissions and limited spectrum, such providers historically found it infeasible to carry simultaneously the signals of many different local affiliates of the same broadcast network (although, as we shall discuss, technological advances have eased the problem). The satellite companies thus long sought to retransmit the signals only of select network affiliates to everyone in the continental United States. This arrangement understandably concerned other local network affiliates because it enabled viewers to watch the same network programming with commercials meant for distant communities and thus placed the non-carried affiliates' local advertising revenues in jeopardy.

Similar concerns do not arise, of course, in remote rural areas where no terrestrial broadcasting signals can reach prospective viewers in the first place. Balancing the interests of local broadcasters against those of television viewers in remote areas (for whom satellite television was and still often is the only means of receiving television signals), Congress enacted the Satellite Home Viewer Act of 1988 (SHVA, pronounced "SHIH vuh"). This legislation granted satellite providers a limited copyright license to retransmit the programming, including the coveted network programming, of a distant broadcast station, but only to households that would otherwise be "unserved" by (i.e., unable to receive clear signals from) the relevant local network affiliate.[46] Congress delegated to the FCC much of the inquiry into whether particular areas should be deemed "unserved" for this purpose, and the FCC issued a set of highly specific technical criteria.

Satellite providers nonetheless followed their own expansive interpretation of which households were "unserved," relying not on the FCC's objective criteria, but instead on viewers' subjective judgments about whether they received clear television signals from their local broadcasters. The broadcasters went to court and ultimately obtained injunctions banning this practice.[47] By the time the broadcasters prevailed, however, many consumers in technically "served" households had already grown accustomed to receiving network programming over their satellite systems—a politically charged concern that prompted Congress to intervene again in 1999.

Throughout the 1990s, it became increasingly obvious that a satellite provider's inability to retransmit network and local programming to all of its subscribers placed it at a clear competitive disadvantage to the cable companies. The challenge was technological as well as legal. Satellite providers had little hope of persuading Congress to let them undercut the market position of local network affiliates by beaming to satellite subscribers the signals of the same network's distant affiliate. As noted earlier, however, satellite providers initially lacked the transmission capacity to beam to each of their millions of subscribers nationwide the feed of that subscriber's local network affiliate—so-called *local-into-local* retransmission. In contrast, locally operated cable systems faced no analogous technological constraints and could readily retransmit the signals of a network affiliate to the affiliate's local viewers. Of course, satellite subscribers who lived in areas "served" by local broadcasters—and who were thus generally ineligible to receive network programming via satellite—could attach rabbit ears to their televisions, flick a switch, and receive such programming from those local broadcasters directly. But the picture quality would generally be worse than on the channels carried either on satellite systems or on cable TV systems. Moreover, market research revealed that "viewers want to be able to receive all of the television channels they watch from a single source" and that, in the words of one industry official, “most people who walk into a satellite dealer’s showroom turn around and walk out because they can’t get their local TV channels through [satellite].”[48]

By the late 1990s, satellite providers had made some of the *technological* advances they would need in order to retransmit local-into-local programming in direct competition with the cable companies. First, progress in digital compression technology enabled them to squeeze more channel capacity from their allotted spectrum, and they could therefore beam a greater number of retransmitted broadcast signals than previously. (Although the same signals might be beamed nationwide, the scrambler in a given subscriber’s set-top box would restrict her to viewing only local, not distant, broadcast channels.) Second, satellite providers began developing “spot beam” technologies that could target particular signals to particular parts of the country, thereby reusing their assigned spectrum efficiently.[49]

Despite these technological advances, however, the satellite companies still lacked a statutory copyright license to transmit network programming into areas adequately served by the signals of local broadcasters. This was a propitious time for the satellite companies to seek a legislative fix for this problem. As noted, Congress had all but pulled the plug in 1996 on the brief federal experiment with cable rate regulation. The rationale for this deregulatory step was that free market competition—principally from satellite providers and potentially from telcos and cable overbuilders—would keep the incumbent cable companies’ rates in check. By the late 1990s, such competition was still slow to develop, and cable subscription fees had steadily increased during the intervening years. Amid a chorus of complaints, Congress concluded that more was needed to generate cross-platform competition in the MVPD market.

To that end, Congress passed the Satellite Home Viewer Improvement Act of 1999. SHVIA (pronounced “shuh VEE uh”) grants satellite providers a conditional copyright license to make local-into-local retransmissions to all households in a given market, while holding those providers subject to the same retransmission-consent obligation as cable companies (and retaining SHVA’s general prohibition on the transmission of a *distant* network affiliate’s signals into a *local* network affiliate’s market).[50] This compulsory license is subject to a significant condition, which took effect in 2002. If a satellite provider invokes its right to retransmit one or more local broadcasting signals in a given geographic market, such as those of the network affiliates, it incurs an obligation (subject to several exceptions) to carry the signals of *all* other broadcasting stations in the same geographic market, including the less-viewed stations.[51]

This *carry-one, carry-all* rule is the satellite industry's counterpart to the must-carry rule in the cable industry. The satellite version is in one sense more voluntary than the cable version because satellite carriers remain free to avoid the must-carry obligation altogether simply by declining to retransmit any local broadcaster's signals in a given market. For that reason and relying on the Supreme Court's prior decision to uphold the cable must-carry rules, the courts rejected the satellite industry's First Amendment challenge to SHVIA.[52]

D. The program-access rules

Our discussion so far has focused mostly on the rights and obligations of cable and satellite providers to carry *broadcast* programming. Under the *program-access* requirements, cable companies are themselves required in many contexts to make their *own* programming (including the programming of their corporate affiliates) available on reasonable terms to rival MVPDs, such as telcos and satellite providers.[53] The concern underlying these requirements is that a cable incumbent, left to its own devices, might withhold affiliated programming from its MVPD rivals in the hope that the programming is so indispensable to the television experience of many viewers that they will forgo the rivals' service in favor of the cable incumbent's service.[54] This rule is often described as a ban on exclusive self-dealing by cable companies even though it sweeps more broadly than that.[55]

Two related statutory provisions underlie the main program-access rules, and their precise language is central to program-access disputes that arose many years after their initial enactment in 1992. First, section 628(b) of the Communications Act broadly prohibits "unfair" acts by vertically integrated cable companies that have the "purpose or effect" of "hinder[ing] significantly" the ability of rival MVPDs to provide competitive video services.[56] Second, section 628(c)(2), entitled "[m]inimum contents of regulations," instructs the FCC to ban more specific forms of conduct. These include unjustified "discrimination" in the sale of programming and, until recently, also included "exclusive contracts" designed to keep MVPD rivals from offering the cable channels of a cable-affiliated programmer.[57]

Sections 628(b) and 628(c)(2) differ in several key respects. First, section 628(b) applies to a broader range of conduct but gives the FCC greater flexibility to determine what conduct should be prohibited.

Second, section 628(c)(2) applies only to cable programming that is distributed to local cable systems from geostationary satellites ("satellite cable programming"), as almost all cable programming was in 1992. It does not extend to programming that is delivered to local cable systems via terrestrial fiber-optic facilities. This curious statutory feature, which rival MVPDs later disparaged as the *terrestrial loophole*, does not apply to section 628(b)'s more general prohibition on "unfair acts" that "hinder" the ability of rival MVPDs to compete. Third, whereas section 628(b) is permanent, Congress specified that the "exclusive contracts" ban in section 628(c)(2) would automatically expire in 2002 unless the FCC determined that "such prohibition continue[d] to be necessary to preserve and protect competition and diversity in the distribution of video programming."[58] The FCC twice made that determination and extended the prohibition for five-year terms—once in 2002 and once in 2007.[59] In October 2012, the FCC denied a third five-year extension, allowed the flat ban on exclusive contracts to expire, and announced that it would continue to review such exclusive contracts on a case-by-case basis under section 628(b).[60]

The program-access rules might seem to resemble the traditional responses to monopoly leveraging concerns in the telecommunications industry, such as the AT&T divestiture (see chapter 2) or the *Computer Inquiry* rules (see chapter 6). In fact, however, they address quite different market dynamics. Rather than protecting suppliers of rival *applications* (programming) from a large firm's perceived strength in the *platform* (MVPD) market, these rules are designed to protect rival *platform* providers (such as telco-affiliated MVPDs and satellite providers) from a large, vertically integrated firm's perceived strength in the *applications* market—here, the market for "must-see" TV channels.

The rationale behind the program-access rules is controversial. On the one hand, as one former FCC commissioner has observed, "A marketplace that pressures competitors to produce new original programming fosters diversity and competition; it certainly does not harm it."[61] For example, as these critics observe, DirecTV competed with cable incumbents in part by developing original programming of its own, such as the highly successful *NFL Sunday Ticket*. On the other hand, defenders of the rules claim that program-sharing rights are necessary to lower the barriers to entry into the MVPD market. On this view, without a regulatory jump-start in the programming market, prospective new entrants may be unable to develop a sufficient customer base quickly enough to justify the massive investments needed to enter the distribution market in competition with cable incumbents.

These contrasting viewpoints were on full display when in 2010 a divided panel of the D.C. Circuit affirmed the FCC's second five-year extension of the section 628(c)(2) exclusive-contracts ban. The cable incumbents had challenged that extension on both statutory and First Amendment grounds, arguing that the ban was more harmful than ever because it suppressed incentives to produce innovative programming and no longer served any substantial purpose, given the enormous competitive inroads new entrants had made in the MVPD market since the ban was first imposed in 1992. The majority appeared sympathetic to these points in principle but deferred to the FCC's expert judgment, even while suggesting that it would be less deferential to future extensions:

> While cable no longer controls 95 percent of the MVPD market, as it did in 1992, cable still controls two thirds of the market nationally. In designated market areas in which a single cable company controls a clustered region, market penetration of competitive MVPDs is even lower than nationwide rates. . . . [But w]e anticipate that cable's dominance in the MVPD market will have diminished still more by the time the Commission next reviews the prohibition, and expect that at that time the Commission will weigh heavily Congress's intention that the exclusive contract prohibition will eventually sunset.[62]

In dissent, Judge Brett Kavanaugh condemned the exclusive-contract ban as an unconstitutional anachronism that, without any countervailing regulatory purpose, burdens incentives to create original programming. In his view, "the justification [for that ban]—counteracting the bottleneck monopoly power of cable operators—has collapsed" with the growth of MVPD and programming competition.[63]

There is, however, somewhat greater consensus for market intervention with respect to one narrow but unusually important class of TV content: regional sports programming.[64] By the first decade of the new millennium, cable incumbents in several metropolitan areas had begun using exclusive contracts to keep MVPD rivals from showing hometown sports events, either at all or in high definition. For example, Cablevision owned not only cable systems in the New York area, but also a large attributable interest in Madison Square Garden, Inc. (MSG), a regional sports network. In turn, MSG owned the exclusive rights to license live broadcasts of games involving various professional sports teams in the New York area, and in some cases it obtained those rights because it also owned the teams themselves, including the Knicks (basketball) and

Rangers (hockey). Controversy arose when MSG categorically refused to license the high-definition feed of these sporting events to Verizon and AT&T. Those companies, which had entered the MVPD market in separate parts of Cablevision's geographic footprint, filed complaints with the FCC.

If Cablevision had received this affiliated sports programming by means of satellite signals, MSG's categorical refusal to deal with these MVPD rivals would have violated the exclusive-contracts ban in section 628(c)(2), then still in full force. But section 628(c)(2) was inapplicable because Cablevision, like a number of cable companies, had instead arranged to receive the signals for its regional sports networks by means of terrestrial fiber-optic transmission.[65] The FCC was nonetheless concerned that such refusals to deal would impair MVPD competition because, it found, a significant percentage of potential subscribers would rule out competing MVPD services that excluded high-quality coverage of local sports teams. And such programming is uniquely nonreplicable: there is typically only one home team for any given professional sports league, and that team typically grants only one network the rights to televise its games.[66]

To address these concerns, the FCC invoked the more general prohibition in section 628(b), which, unlike section 628(c)(2), contains no exemption for terrestrially delivered programming. Under the new regime initiated in 2010, MVPD entrants aggrieved by exclusive contracts or discrimination regarding cable-affiliated programming may file complaints with the FCC alleging that the conduct in question significantly hinders MVPD competition. The Commission simultaneously adopted a rebuttable presumption that such conduct will meet that standard if it relates to regional sports programming, which the Commission deemed "very likely to be both non-replicable and highly valued by consumers."[67]

The cable incumbents took the FCC to court, arguing that section 628(b) grants the FCC no authority to compel access to terrestrially delivered programming outside the scope of section 628(c)(2). The statutory language is in fact oblique on that point. Although section 628(b) is not explicitly limited to programming that cable incumbents obtain via satellite, it is nonetheless limited to conduct by those incumbents that significantly hinders a *competing MVPD* from providing *satellite-delivered* programming.[68] The cable incumbents cited that qualification as evidence that Congress meant to ban exclusive contracts with respect to *their* programming—that is, the programming that the competitor wants—only insofar as that programming is satellite delivered. In rejecting that argument, the FCC ruled that competing MVPDs, to trigger section 628(b)'s protections, need show only that an incumbent's conduct hinders them from offering their customers video-programming *packages* that include at least *some* programming that the competitors receive via satellite, even if that is not the programming that the incumbent has withheld. In 2011, the D.C. Circuit found the statute ambiguous, upheld the FCC's key interpretative decisions as reasonable, and generally upheld the substance of the FCC's new regime.[69]

E. Online video distributors

With the widespread deployment of ever-faster broadband Internet services, a new breed of video-programming distributor began appearing in the first decade of the new millennium: Netflix, Hulu, iTunes, and other online aggregators of television content that consumers can stream on demand. These *online video distributors* (OVDs) do not themselves provide physical transmission of video programming all the way to viewers; instead, their viewers must independently purchase a broadband connection from another provider.

As of this writing, there is no clear consensus that OVD services strongly compete with, rather than supplement, conventional MVPD services. Some consumers have begun "cutting the cord" by canceling their conventional MVPD services and relying on OVD services in combination with over-the-air reception of conventional broadcast signals.[70] As the FCC found in 2011, however, such "cord-cutting is relatively infrequent," and "most consumers today do not see OVD service as a substitute for their MVPD service, but as an additional method of viewing programming."[71] Reaffirming that conclusion in mid-2012, the FCC observed that "online viewership is still dwarfed by . . . traditional distributors" and noted new evidence that, "surprisingly, the more alternative platforms consumers use, the more they tend to spend on traditional television subscription services."[72] Nonetheless, consumer behavior is evolving rapidly, and OVDs will likely play an increasingly central role in the marketplace for video distribution in the years to come.

OVDs, copyright, and the definition of "MVPDs"

Like conventional MVPDs, an OVD must obtain rights to a wide variety of programming in order to attract viewers. But unlike conventional MVPDs, to which a typical consumer subscribes one at a time, online video is "multihomed" in the sense that consumers can make use of several OVDs during the same day, clicking through different OVD websites or apps until they find the programming they want. That dynamic reduces the pressure on any given OVD to carry the same comprehensive suite of must-see programming that consumers have come to expect of conventional MVPDs. Nonetheless, access to popular programming is typically the greatest challenge that OVDs face, and a key obstacle to that access arises from copyright law.

As discussed, Congress enacted the "compulsory copyright" in 1976 to spare cable operators the need to negotiate individually with each of the thousands of copyright holders who have intellectual property interests in broadcast TV programming. But that compulsory copyright can be claimed only by "cable systems," as that term is defined in the Copyright Act.[73] As of this writing, the leading judicial precedent has read that term to exclude OVD services.[74] As a result, the typical over-the-top video service must negotiate individual deals with copyright holders in order to stream content, which effectively means that even if they pay considerable sums for the privilege, they can stream only some of the programming that appears on broadcast television.*

This and other intellectual property barriers may make it more difficult for over-the-top video providers to compete with conventional MVPDs. In 2012, the FCC cited two cases in point. First, under a deal with The CW (a second-tier broadcast network), Netflix has reportedly paid "$600,000 an episode for established shows like *Gossip Girl*."[75] That, the FCC noted, was "a significant cost for what will amount to a small part of Netflix's overall content library."[76] The FCC also found that some well-capitalized would-be entrants had reconsidered the OVD business model in the face of copyright complications. For example, Microsoft "put its plans to start an online subscription service for television shows and movies on hold after determining that constant licensing costs would be too high for the company's envisioned business model."[77]

* We say "typical" because some over-the-top providers have devised ingenious ways to offer broadcast programming over the Internet while maintaining plausible arguments that they are not themselves engaged in activities that might constitute copyright infringement and thus need no copyright license in the first place. *See, e.g.*, American Broadcasting Companies, Inc. v. Aereo, Inc., 874 F. Supp. 2d 373 (S.D.N.Y. 2012) (refusing to enjoin an OVD service that receives local broadcast signals over the air via antennas notionally assigned to individual subscribers and that streams those signals to the subscribers over the Internet if they certify that they are viewing the signals in the relevant station's local market), *aff'd sub nom.* WNET v. Aereo, No. 12-2786 (2d Cir. Apr. 1, 2013).

The FCC concluded: "Given the costs faced by established companies, it is even more difficult for new entrants with less capital to enter into the many high-priced content deals required to build an adequate content library."[78]

OVDs are at a further disadvantage when compared to conventional MVPDs because, to date, they have also been unable to avail themselves of the program-access rules, which confer rights of nondiscriminatory treatment only on "multichannel video programming distributors."[79] Under the applicable statutory definition, "MVPD" means "a person such as, but not limited to, a cable operator, . . . a direct broadcast satellite service, or a television receive-only satellite program distributor, who makes available for purchase, by subscribers or customers, multiple *channels* of *video programming*."[80] In a separate provision whose precise meaning and relevance are subject to debate, "channel" is defined as "a portion of the electromagnetic frequency spectrum which is used in a cable system and which is capable of delivering a television channel."[81]

When Congress enacted these provisions in the 1980s and 1990s, it likely envisioned that new entrants in the video-distribution business would arrange for the provision of complete transmission paths to end users as part of their service, as cable operators and direct broadcast satellite providers have done. It certainly did not foresee the rise of streaming video over the Internet, where viewers arrange for their own transmission paths by purchasing separate broadband services. The question now is whether the definition of "MVPD" and the regulatory rights (and obligations) that classification bestows can be stretched to include OVDs as well.[82] To date, the FCC has left that question unresolved, but its staff has proceeded as though the answer is no.[83] In 2012, the Commission sought broader public comment on the scope of its statutory authority to treat OVDs as MVPDs, and that inquiry remains pending as this edition goes to press.[84]

OVDs and the Comcast–NBC Universal merger conditions

The statutory nuances we have discussed deterred neither the FCC nor the Department of Justice from addressing the same basic concerns when they jointly reviewed the application filed by Comcast—the nation's largest MVPD and broadband ISP—to acquire a controlling interest in NBC Universal (NBCU), a major broadcast network owner and programming producer. As the FCC explained when conditionally approving this transaction in 2011, the combination of these two companies "effectuate[d] an unprecedented aggregation of video programming content with control over the means by which video programming is distributed to American viewers offline and, increasingly, online as well."[85] And both the FCC (in the form of administratively enforceable merger conditions) and the Department of Justice (in the form of an antitrust consent decree) thus imposed "targeted, transaction-related" OVD protections "to mitigate the potential harms the proposed combination might otherwise cause."[86] These protections fall into two main categories: (1) *net neutrality–type* requirements, which prohibit Comcast from harming the performance of competing OVD services (or favoring its own such services) delivered via Comcast's high-speed Internet access service, and (2) *program-access-type* requirements, which compel Comcast to provide affiliated programming to OVDs in certain circumstances.

Among the requirements in the first category, the Department of Justice consent decree provides that "Comcast shall not prioritize Defendants' Video Programming or other content over other Persons' Video Programming or other content."[87] It further provides that "[i]f Comcast offers consumers Internet Access Service under a package that includes caps, tiers, metering, or other usage-based pricing, it shall not measure, count, or otherwise treat Defendants' affiliated network traffic differently from unaffiliated network traffic."[88] Comcast separately committed to the FCC, among other things, that it would comply with the net neutrality rules the FCC had adopted in 2010 even if the reviewing court ultimately invalidates them.[89] And to ensure that Comcast will offer sufficient bandwidth for best-effort Internet traffic at attractive enough prices to support OVD services, the FCC extracted commitments from Comcast (1) to offer stand-alone broadband Internet access at $49.95 or less for three years and at downstream speeds no less than 6 Mbps and (2) to offer, in its upgraded DOCSIS 3.0 markets, a service that typically provides 12 Mbps in downstream bandwidth (subject to potential adjustments).[90] In each case, however, the FCC gave Comcast considerable discretion to impose data caps and usage-based pricing.

As noted in chapter 6, Comcast's net neutrality commitments received new scrutiny in early 2012 when the company introduced a new method of transmitting video programming to its MVPD customers. As Comcast told its subscribers, the data caps applicable (at that time) to its broadband Internet access service would not apply to the IP-based video-on-demand services that Comcast had begun streaming to users of its proprietary Xfinity Xbox 360 application. Netflix and certain consumer groups claimed that this practice unfairly discriminated against Internet-based OVD services, which remained subject to the cap, and that Comcast had thereby violated open Internet principles and the Comcast–NBCU merger conditions.[91] Comcast responded that although it was streaming the Xbox 360 packets over the same IP platform used for Internet access, the Xbox 360 app was delivered as part of Comcast's regular MVPD service rather than its "Internet access service." Among the distinctions Comcast cited was the control it maintained over the Xfinity Xbox packets from source to destination.[92] As a result, Comcast contended, net neutrality requirements were no more applicable to this practice than to AT&T's longstanding use of its IP-based U-verse platform to deliver AT&T's voice and video services separately from AT&T's Internet access service. As discussed in chapter 6, this debate vividly illustrates the conceptual instability at the heart of the FCC's net neutrality regime and in particular its radically disparate rules for "broadband Internet access" and "specialized IP services."

The FCC and Department of Justice separately required Comcast–NBCU to comply with a variety of stringent program-access rules that extend far beyond the industry-wide obligations of other cable operators. These rules create, among other things, two novel mechanisms under which an OVD may obtain programming from Comcast–NBCU on terms analogous to those the OVD itself or Comcast–NBCU has reached in deals with specified third parties.[93]

First, and roughly speaking, a requesting OVD may generally obtain Comcast–NBCU's programming line-up as a package on terms that are "economically equivalent" to those in Comcast–NBCU's contracts with conventional MVPDs, but only if the OVD obtains a narrow majority (55%) of its overall programming from other content sources, including at least one "peer" of NBCUniversal.[94] Second, an OVD that has struck a program-acquisition deal with such a peer is generally entitled, with some exceptions, to obtain "comparable" programming from Comcast–NBCU, this time on terms "economically equivalent" to those that the *OVD* has struck with the *peer*.[95]

The details of these program-access conditions are open ended, prone to radical interpretive disagreement, and subject to binding arbitration. Both program-access mechanisms also raise a variety of practical challenges. For example, because both turn on the terms of independent deals involving third parties (other MVPDs such as DirecTV or other content sources such as Disney), complex disputes have arisen about how to protect the legitimate confidentiality interests of those third parties. As we complete work on this edition, the FCC remains in the early stages of addressing these implementation challenges.

III. Efforts to Protect Programming Competition and Diversity through Vertical Regulation of Broadcast Networks and MVPDs

A. Vertical integration concerns and the finsyn case study

The program-access rules reflect a longstanding suspicion of vertical integration by large media companies in the separate markets for the production and distribution of television programming. From a strictly antitrust perspective, this suspicion, when expressed in broad-brush terms, is sometimes difficult to substantiate.[96] Nonetheless, before the rise of the Chicago School of antitrust economics in the final quarter of the twentieth century (see chapter 6), the received wisdom was that all vertical integration, at least by large media companies, should be viewed with immense skepticism—such that the Hollywood studios, for example, could not be trusted with ownership interests in movie theaters.[97]

Carried too far, such suspicion not only exaggerates the anticompetitive potential of vertical integration but also ignores the efficiencies that such integration permits. In the 1970s, an influential academic movement known as *New Institutional Economics* drove this point home.[98] Building on the insights of Ronald Coase, this scholarship showed that in the absence of market failure, ordinary market forces will generally induce a firm to integrate vertically when, and only when, doing so is economically efficient.[99] For example, a firm in a particular market will typically contract out for complementary goods or services when the transaction costs of doing so are low; conversely, when those transaction costs are high, it will vertically integrate rather than outsource. This scholarship explains how the strategic use of insourcing or outsourcing by firms throughout the economy enables them to operate more efficiently and maximizes overall consumer welfare.

In the media context, when a single integrated firm both produces and distributes a given program, it economizes on transaction costs by streamlining the program-development process, cutting out middlemen, and eliminating the concern that independent producers might extort supracompetitive prices during program-renewal periods for the rights to continue airing long-running popular shows. At the same time, the major broadcast networks still sometimes contract out for programming to independent firms when that course is more efficient.[100] Of course, any distributor might irrationally favor its own production affiliates because of managerial politics or corporate hubris, even when doing so results in fewer viewers, lower distribution revenues, and perhaps lower earnings for the corporate family as a whole.[101] Over time, however, competition in the video-distribution market, along with the prospect that concerned corporate boards (or hostile corporate takeovers) will replace incompetent management, may well limit the extent and duration of such inefficient favoritism.

As noted earlier in this chapter, however, traditional television regulation rests less on pure antitrust-type concerns than on the more nebulous goals of localism and, of particular relevance here, programming "diversity"—in this context, the promotion of the "widest possible dissemination of information from diverse and antagonistic sources."[102] The "financial interest and syndication" (*finsyn*) rules, first adopted in 1970 and significantly revised in 1991, present the lead case study in the FCC's ill-fated efforts to promote such diversity through restrictions on vertical integration.

With complex exceptions irrelevant to our discussion, the finsyn rules largely precluded each of the major broadcast networks (then ABC, CBS, and NBC) from, among other things, "syndicating" television programs—for example, licensing individual stations to broadcast them after the network had initially aired them. Given the centrality of syndication revenues at the time, the effect of that and the other finsyn restrictions was to sideline the three broadcast networks from most of the market for program production. The FCC justified these restrictions as necessary to promote greater diversity in television programming and to protect independent studios from the dominance of the networks by making it easier for such studios to earn the syndication revenues needed to subsidize their production operations.

By the 1990s, however, the growth of cable had diminished any "bottleneck" the broadcast networks may have once enjoyed in the market for programming distribution. And as Judge Richard Posner observed for the Seventh Circuit in 1992, the finsyn rules "appear[ed] to harm rather than to help" their supposed beneficiaries—non-network producers, in particular those independent of major studios.[103] Posner explained that the restriction on the sale of syndication rights made it harder for such producers to market their programs to the networks, "a class of buyers that may be the high bidders for them. . . . Since syndication is the riskiest component of a producer's property right—for its value depends on the distinctly low-probability event that the program will be a smash hit on network television—restricting its sale [by keeping it from the networks] bears most heavily on the smallest, the weakest, the newest, the most experimental producers, for they are likely to be the ones least able to bear risk."[104]

That, Posner speculated, might well have been an animating purpose of these rules. "It becomes understandable why the existing producers support the [finsyn] rules: the rules protect these producers against new competition both from the networks . . . and from new producers. The ranks of the outside producers of prime-time programming have been thinned under the regime of financial interest and syndication rules. The survivors are the beneficiaries of the thinning. They do not want the forest restored to its pristine density. They consent to have their own right to sell syndication rights curtailed as the price of a like restriction on their potential competitors, on whom it is likely to bear more heavily."[105]

In the past, the Supreme Court had given considerable deference to the FCC's "predictive judgment" about the types of structural regulation needed to ensure programming diversity.[106] Posner, however, brushed off the FCC's invocation of that institutional expertise. He explained that "while the word diversity appears with incantatory frequency [in the FCC's order], it is never defined," and, in all events, the FCC had made "no attempt to explain" how finsyn would increase diversity in the marketplace of ideas.[107] "Stripped of verbiage," he concluded, "the [FCC's] opinion, like a Persian cat with its fur shaved, is alarmingly pale and thin."[108]

In response to the Seventh Circuit's decision, the FCC ultimately abandoned the finsyn rules altogether and thereby dismantled what amounted to an artificial barrier to entry into both the network and programming markets. The result has been not only significant vertical integration—Disney bought ABC, and Universal merged with NBC, for example—but also far greater consumer choice among network-programming lineups. Fox, which had previously provided only about ten hours of programming per week (and thus remained exempt from the finsyn rules) became a full-fledged major network. Warner Brothers and Paramount created entirely new networks—the WB and UPN, which, with CBS's involvement, have since merged to form The CW. Indeed, new broadcast networks flourished in the years following finsyn's abolition, and several of them, including The CW, have made their mark on popular culture by targeting their broadcast programming to previously underserved audiences. This is not to say, of course, that such vertical integration always makes business sense, and sometimes it does not. For example, programming giant Viacom decided in 2005 to spin off CBS, which it had acquired just six years earlier, amid skepticism that the two companies were achieving genuine synergies.[109]

Over the two decades following finsyn's repeal, some popular support has remained for vertical restrictions on the major media companies. A number of commentators, including media mogul Ted Turner, have proposed reinstituting the finsyn rules or some other type of limitation on vertical integration.[110] They claim, in essence, that a major media company is less innovative than smaller companies and typically refuses to deal with them. In 2004, FCC Commissioner Michael Copps likewise championed "some sort of set-aside, like 25–35% of prime time hours, for independent creators and producers."[111] None of these proposals, however, is likely to be enacted. The FCC in particular has expressed little interest in devising bold new ways of intervening in the market for conventional television programming.[112]

B. Cable operators, horizontal ownership restrictions, and the program-carriage rules

If there are counterparts to finsyn for the cable operators, they have lain in two sets of FCC rules: the (now vacated) *cable ownership* rules and the so-called *program-carriage* rules.

Cable ownership restrictions

In the 1992 Cable Act, Congress ordered the FCC to develop rules designed in different ways to limit any undue influence by large cable companies in the market for video programming.[113] The FCC responded by imposing, as Congress directed, (1) a *horizontal* restriction on the number of subscribers a cable company or affiliated companies may reach nationwide and (2) a separate *vertical* restriction on the number of channel slots on any given cable system that can be occupied by programmers in which the system operator has an attributable interest.[114] Although the FCC's statutory responsibility to adopt both horizontal and vertical restrictions has never been in doubt, the cable industry has successfully challenged the specifics of the FCC's rules in the D.C. Circuit.

The vertical restriction was never particularly onerous. The FCC limited the number of affiliated channels to 40% of a system's channel capacity—but only up to 75 channels, beyond which the operator could include as much affiliated programming as it wished. As the FCC noted on appeal, no cable operator had ever actually complained that this rule seriously constrained its business decisions. Nonetheless, the D.C. Circuit invalidated the rule in 2001 on the ground that, in the face of First Amendment concerns, the FCC had offered no analytically sound basis for concluding that a 40% limit was appropriate, particularly for cable operators subject to competition.[115] This issue, which technically remains before the FCC on remand more than a decade later,[116] is likely to have ever-diminishing significance as the deployment of digital technology expands the channel capacity of all major cable systems.

The FCC's horizontal rules were more commercially significant to the cable industry and thus more controversial, for they imposed clear limits on how large a cable operator could grow on a national level. The *monopsony* rationale for such limits requires a brief explanation.* Despite some inroads made by telcos and cable overbuilders, there is often only one cable system operating in any given market. Thus, even if a single cable company—say, Comcast—purchased all of the cable companies in the country, most consumers would suffer no immediate loss of video-delivery alternatives: in any given community, consumers would still be able to choose among Comcast, the two major satellite providers, all terrestrial broadcast stations within range, and online streaming-video services such as Hulu and Netflix. Nonetheless, Comcast's ability to expand horizontally by buying additional cable systems might raise a different type of competitive risk: that Comcast might suppress competition in the *programming* market by favoring affiliated programmers in its purchasing decisions and disfavoring others. The principal concern is that an independent creator of programming cannot hope to recover its production expenses unless its shows reach a critical number of viewers and may thus be doomed by a cable company's decision not to carry those shows if that company owns cable systems that in the aggregate account for a critical mass of the potential national audience.

The horizontal cable ownership rules were designed to preclude cable companies from growing large enough to exercise such monopsony power, principally through collusive favoritism among cable system operators for certain programming suppliers. In 1999, the FCC limited any single company (and all others in which the company had "an attributable interest") to ownership of cable systems that in the aggregate reached 30% of the MVPD market nationwide.[117] In 2001, the D.C. Circuit invalidated the 30% limit on two basic grounds. First, the court found that the limit rested on unsupported assumptions about the potential for collusion among separate cable companies in program-purchasing decisions. Second, it held that the FCC's analysis understated the extent to which satellite-based MVPD competition would force cable companies

* The term *monopsony* describes a firm's dominance as a *purchaser* (as opposed to a seller) in a given market.

to purchase attractive programming, regardless of the originating source, lest they lose their customers to these alternative video platform providers.[118]

In 2008, the FCC tried to devise a new justification for the same 30% horizontal ownership limit.[119] This time it dispensed with any collusion rationale but adopted both (1) a more aggressive assumption about the number of viewers a cable programming network must reach in order to attain "minimum viable scale" (i.e., critical mass) and (2) a more pessimistic assumption about the percentage of viewers the cable programming network actually will reach (the "penetration rate") once a cable operator agrees to carry it, a function of the cable service tier on which it is carried. The FCC acknowledged that, as before, it was largely ignoring the role of inter-MVPD competition in disciplining a cable operator's refusal to carry programming that viewers would find attractive, but it offered several speculative reasons why it believed that such discipline would be minimal. In 2011, the D.C. Circuit summarily rejected all of these rationales and vacated the ownership limit on the ground that the FCC "either cannot or will not fully incorporate the competitive impact" of satellite- and telco-based MVPD competition—which, as the court noted, had grown significantly since the previous round.[120]

One aspect of the D.C. Circuit's treatment of the cable ownership rules warrants further attention, for it reveals an important ideological fault line in the regulation of the television media. In addition to the FCC's fairly tenuous antitrust-like arguments in support of its 30% horizontal limit, the Commission had also invoked its expertise to determine what steps were needed to ensure robust programming diversity. In its 2001 decision, the court rejected this logic and expressed doubt, particularly in light of First Amendment concerns, that the FCC could adopt aggressive horizontal ownership limits solely to promote greater programming diversity in the absence of any showing of a genuine market failure in a rigorous economic sense.[121]

Although the relevant statutes (and affected media outlets) are different, this holding stands in at least philosophical tension with the Third Circuit's subsequent decision in the broadcasting context (discussed in the next section). In the Third Circuit's view, the FCC not only can but perhaps must consider diversity values in addition to standard antitrust concerns in developing horizontal ownership limitations.[122] This subtle difference in perspective is likely to remain one of the key debates in media policy: the extent to which ownership restrictions remain necessary to promote greater programming diversity and localism than would be produced by an efficient and increasingly competitive video-distribution market superintended by generalist antitrust authorities.[123]

The program-carriage rules

In addition to cable ownership restrictions, Congress has sought to protect independent cable programmers more directly by creating a so-called "program-carriage" mandate and directing the FCC to administer it. Enacted in 1992, section 616 of the Communications Act prohibits an MVPD from, among other things, "unreasonably restrain[ing] the ability of an unaffiliated video programming vendor to compete fairly by discriminating in video programming distribution on the basis of affiliation or nonaffiliation of vendors in the selection, terms, or conditions for carriage of video programming."[124] This provision entitles aggrieved programmers (such as the Tennis Channel, WealthTV, or the Mid-Atlantic Sports Network) to file complaints with the FCC alleging that MVPDs have "discriminated" against them and in favor of their affiliated channels.

Until recently, this complaint process was little used and had never resulted in any FCC finding of actual discrimination.[125] In 2012, however, the FCC broke new ground by ruling that Comcast had violated the program-carriage rules by discriminating against the Tennis Channel. Comcast challenged the FCC's carriage remedy as a violation of its First Amendment rights, and the D.C. Circuit issued a stay order in litigation that remains pending as this edition heads to the printer.[126] Meanwhile, in 2011, the FCC also added some new programmer protections to its program-carriage rules and sought further comment on additional measures that would benefit independent programmers.[127] The cable industry responded by challenging the larger program-carriage regime on First Amendment grounds, and that litigation also remains pending as this edition goes to press.[128]

Finally, quite apart from prescriptive regulation, the federal government sometimes uses the merger-review process to extract program-carriage commitments from vertically integrated MVPDs to deal evenhandedly with unaffiliated producers of programming. For example, in 1994, the FTC imposed important conditions on the merger of Time Warner and Turner Broadcasting, including one that required Time Warner Cable to carry a rival news network so that favoritism for Turner's CNN would not limit the entry of upstarts such as the Fox News Network.[129] And in 2011, the FCC conditioned its approval of the Comcast–NBCU merger on far-reaching commitments from Comcast "to add at least ten new independently owned and operated programming services" to its cable lineup over eight years, to submit to programming nondiscrimination requirements more stringent than the general program-carriage rules, and to place independent news and business channels in the same channel "neighborhood" as Comcast's own affiliated news and business channels.[130]

IV. Restrictions on Ownership of Television Broadcast Stations

We now turn to a final set of issues that has generated unusually intense political attention and some of the most prodigious letter-writing campaigns in the communications policy world: horizontal restrictions on the ownership of television broadcast stations.

Until the final quarter of the twentieth century, many American cities had only three major television stations; few homes subscribed to cable television; and (as is still the case) many cities had only one major newspaper. To promote greater viewpoint diversity and to protect competition in advertising and other markets, longstanding FCC rules precluded a single firm from owning more than one television station in the same local market or more than seven television stations nationwide.[131] In the mid-1970s, the FCC also precluded any single firm from owning a full-service radio or TV station and a daily newspaper in the same community.[132] In 1978, the Supreme Court rejected a First Amendment challenge to this rule, reasoning that most Americans got their news from television and newspapers and that diverse ownership is necessary to ensure diverse viewpoints.[133]

In the 1980s, the FCC loosened a number of these ownership restrictions. In 1984, under the direction of Chairman Mark Fowler, a staunch advocate of media deregulation, the FCC concluded that there should be no restrictions on the number of local broadcasting stations that a single corporate family could own nationwide.[134] Fowler's approach likely would have dramatically increased the number of stations owned and operated by the broadcast networks and decreased the number of network affiliates owned by independent media companies. Amid the ensuing uproar, Congress stepped in and authorized only the more modest step of increasing the number of television stations a single firm could own nationwide, with the proviso that any given firm's stations could reach no more than a combined total of 25% of the national viewing audience.[135] In 1988, the FCC also took the more incremental step of providing for waivers of the television–radio cross-ownership restriction in particular circumstances.[136] But the FCC retained its flat ban on common ownership of two television stations or of a television station and newspaper in the same market.

In the Telecommunications Act of 1996, Congress set the stage for easing broadcast ownership restrictions more comprehensively. Among other measures, it relaxed the horizontal limit on ownership of TV stations nationwide by increasing from 25% to 35% the percentage of the national television audience that a single firm's stations may reach in the aggregate.[137] Congress also directed the FCC to open periodic inquiries into whether it should further loosen that and other broadcast ownership restrictions.[138] In response, the FCC revised its rules to permit a single firm, for the first time, to own two television stations in a large market, subject to the restriction (among others) that at least one of the two stations must fall outside the top four in viewership in that market.[139] Otherwise, however, the FCC decided to leave almost all of the existing rules in place, suggesting that it needed time to observe the effect of the changes Congress had already ordered.[140]

In *Fox Television Stations, Inc. v. FCC*,[141] the D.C. Circuit in 2002 rejected the FCC's "go slow" approach as too cautious, reasoning that the 1996 Act required the FCC to lift ownership restrictions promptly unless it affirmatively found that they should be retained. In particular, the court concluded that the 1996 Act's mandate for regular reviews of the broadcast ownership rules was closer to "Farragut's order at the battle of Mobile Bay ('Damn the torpedoes! Full speed ahead.') than to the wait-and-see attitude" adopted by the FCC.[142] As a consequence, it directed the FCC to complete its inquiry into whether the 35% horizontal ownership cap should be further relaxed. And in a striking display of judicial assertiveness, it ordered the FCC simply to abolish, as unnecessary for either competition or programming diversity, the Commission's 22-year-old ban on common ownership of a broadcast station and a cable system in the same community.

On remand, the FCC opened a new inquiry into the future of its broadcast ownership restrictions. The proceeding, which has now persisted for ten years and has generated two lengthy court of appeals decisions, has been one of the most contentious in the FCC's recent history. Advocates of deregulation argued from the outset that the existing restrictions are both inefficient in that they artificially constrain economies of scale and scope and are unnecessary for viewpoint diversity, given the explosion of new media outlets such as cable systems and the Internet.[143] Those opposing further deregulation argued that there is increasing public unease with growing media concentration in the United States, observed (among other things) that neither the Internet nor cable systems feature much independent local news reporting, and concluded that permitting a single firm to own both a major television station and a major newspaper in a single community might leave the inhabitants without adequate diversity in the coverage of local politics and other matters of local interest.[144]

The FCC's *2003 Media Ownership Order*, decided by a three–two vote, greatly relaxed a number of its horizontal broadcast ownership restrictions.[145] First, the FCC responded to the *Fox* decision by loosening once more the national ownership restrictions on local television broadcast stations. Specifically, it raised from 35% to 45% the total percentage of the national viewing audience that can be reached by a group of broadcast stations under common ownership (by, for example, a broadcast network).[146] Many criticized this decision on the ground that independently owned network affiliates are more responsive to their local communities' needs than are the stations owned and operated by the networks themselves. In reality, however, many such affiliates are themselves controlled by giant media companies such as Sinclair and Hearst-Argyle, each of which owns or operates dozens of stations across the nation. Such companies may have no keener understanding of the particular communities served by their stations than do the networks. And the networks have similar (though not necessarily identical) economic incentives as these independent media companies to increase audience share in the communities they serve by airing programs of local interest. Also, in contrast to the cable ownership limits, no national *broadcast* ownership limit can be persuasively justified as a necessary safeguard against monopsony power in the market for broadcast television programming. Unlike cable companies, which exercise significant discretion in what types of programming to carry in scores of channel slots, any broadcast station affiliated with one of the major networks—whether it is owned by that network or not—almost invariably airs whatever programming the network has designated for prime-time viewing.

When the issue is viewed in the cold light of economic analysis, there is thus no obvious reason why raising the national ownership cap from 35% to 45% should have alarmed anyone. Indeed, from a pure efficiency perspective, the real debate is whether such caps are needed at all. As we have discussed, however, media-consolidation issues present an array of non-economic concerns that elude formal economic analysis. Much like Chairman Fowler's proposed elimination of a national cap in the 1980s, this new relaxation of that cap triggered such a political outcry that Congress immediately stepped back into the fray, setting a new statutory limit at 39% of the national audience instead of the FCC's 45% cap.[147] On the surface, this legislation was a rebuke to the FCC. In reality, the 39% figure was gerrymandered to remain as low as possible while permitting the two largest broadcast station owners—Viacom (which then owned CBS) and News Corp. (owner of Fox)—to hold on to all of their existing stations nationwide (including those authorized by previous waivers).[148]

The other set of controversies sparked by the *2003 Media Ownership Order* concerned the FCC's restrictions on common ownership of multiple media outlets *within a given locality*. In a nutshell, the FCC replaced its existing restrictions of this type with new ones that varied with the number of total media outlets in the community at issue. First, the FCC abolished all restrictions on a firm's cross-ownership of a television station with a newspaper or radio station in the largest markets (i.e., those with nine or more television stations), retained the preexisting restrictions for the smallest markets (those with three or fewer stations), and adopted a middle-ground approach for all remaining markets.[149] Second, it loosened its limits on common ownership of multiple TV stations in a single community, allowing a firm to own up to three stations in the largest markets and two in most others, so long as no more than one of the jointly owned stations ranks in the top four by viewership in the relevant market.[150] The FCC claimed that these new rules addressed the diversity concerns served by the former ownership restrictions but with "more precision and with greater deference to First Amendment interests."[151] It further found that transformations in the media marketplace over the past 20 years, including the growth of the Internet, provided viewers with an unprecedented diversity of media outlets and justified a commensurate relaxation of the various ownership rules.

The FCC's analysis of diversity in particular markets was based on a modified version of the Herfindahl–Hirschmann Index, a standard economic measure of market concentration used by the Department of Justice and the FTC in assessing proposed mergers. The controversy surrounding the FCC's "diversity index," as it was called, lay in the details. Critics challenged the index's reliance on the Internet as an important source of independent local news, observing that the news-related content on the Internet tends to be either national or, if local, derivative of traditional media outlets (e.g., webpages duplicating print newspaper articles). The critics further assailed the FCC's across-the-board assumption that any two television stations (or any two newspapers) contribute equally to diversity of programming in local markets. As these critics pointed out, the resulting regime would anomalously allow the most popular television station in Tallahassee, Florida—the CBS affiliate with a 59% share for local news—to be owned by the same firm as the city's major newspaper (with five times the circulation of its competitor).[152]

Many parties sought review of the *2003 Media Ownership Order* in different courts. The appeal was assigned by lottery to the Third Circuit in Philadelphia rather than back to the D.C. Circuit. In June 2004, in *Prometheus Radio Project v. FCC* ("*Prometheus I*"), a split panel of the Third Circuit invalidated much of the order.[153] Although the majority upheld the FCC's abolition of the flat ban on television–newspaper cross-ownership, it agreed with the criticisms of the FCC's diversity index, including the objections that it overweights the Internet as a source of local content and irrationally assigns equal weight to all media outlets within a given category. The court thus remanded the matter to the Commission to conduct a more reasoned analysis and to adopt a new set of cross-ownership rules for particular markets.[154] The majority further upheld in principle the Commission's decision to relax restrictions on ownership of multiple TV stations in a single market, agreeing that the efficiencies gained from consolidation can "translate[] into improved local news and public interest programming."[155] Here again, however, the majority invalidated the specifics of the FCC's new rules on the ground that they were tainted by the methodological flaws of the agency's diversity index.[156] The Third Circuit remanded all of these issues to the FCC for further analysis and "retained jurisdiction" over the case, which controversially required any subsequent appeal to go back to the same court rather than to the D.C. Circuit, which has been more sympathetic to media deregulation.

Several years later the FCC issued its *2008 Media Ownership Order*, which responded to the Third Circuit's 2004 decision by adjusting the broadcast ownership rules in various ways.[157] Among other things, the Commission reaffirmed its earlier decision to abolish the flat newspaper–broadcast cross-ownership ban but adopted a new set of presumptions that vary by market. In the top-20 TV markets, the presumptions would generally permit common ownership of a newspaper and one broadcast station (TV or radio) if the combination does not involve one of the top four TV stations and leaves at least eight "major media voices" in the affected area.[158] Outside of the top-20 markets, the presumptions would generally bar all newspaper–broadcaster combinations except in special circumstances, such as a failing newspaper or broadcaster. The FCC separately abandoned its 2003 initiative to relax restrictions on the number of television stations a single company may own in the same market. It reasoned that despite the growth of alternative media outlets, the more stringent pre-2003 restrictions remain "necessary in the public interest to protect competition for viewers and in local television advertising markets."[159]

Another massive set of appeals followed. In its 2011 decision in *Prometheus II*, the Third Circuit upheld most of the FCC's rules, including the restrictions on ownership of multiple TV stations.[160] The court nonetheless invalidated the FCC's new approach to newspaper–broadcasting combinations—not because it found fault with the FCC's substantive rationale, but because it concluded that, as procedural matter, the FCC had given the public inadequate notice of its intended course of action.[161] As a result, the FCC's original cross-ownership ban sprang back into force pending further action on remand.[162]

A number of media companies promptly asked the Supreme Court to intervene, arguing that the Third Circuit had applied an impermissibly lenient standard of review in upholding government restrictions on their First Amendment rights to communicate with the public. They further urged the Supreme Court to overrule several of its own dated precedents that had relied on claims of spectrum "scarcity" (see chapter 3) to justify a lenient standard of review for broadcast regulation.[163] Since the 1990s, courts and many academic commentators have argued that this broadcast-specific, "scarcity"-based standard of review, known informally as the *Red Lion* doctrine, is anachronistic, but the Supreme Court has always found ways to avoid fundamentally revisiting it.[164] The Court took the same course here, denying review of *Prometheus II* in June 2012.

10

The Future of Telecommunications Competition Policy

In 1977, President Carter selected Alfred Kahn, the noted regulatory economist, to head the now defunct Civil Aeronautics Board. The Board was responsible for comprehensive regulation of the commercial airline industry: it awarded routes to airlines, limited carriers' entry into and exit from particular markets, and regulated passenger fares.[1] This system, which had been in place since the New Deal, was ready for a complete overhaul because few informed observers believed that any part of the airline industry was a "natural monopoly" in need of such pervasive economic oversight by the government.[2] And Kahn concluded that such oversight was not just unnecessary, but affirmatively harmful in that it produced flights with few passengers, barred potential competitors from entering the market, and created significant consumer welfare losses.[3] A year later, armed with Kahn's criticisms, policy leadership, and broad political support,[4] Congress passed the Airline Deregulation Act of 1978.[5] This legislation banished all of the classic command-and-control tools previously used by the Board—it ended price regulation, tariffs, and limits on market entry and exit. After a period of transition, the Board shut its doors for good in 1985.

When Congress passed the Telecommunications Act of 1996, some hoped that it would "deregulate" the telecommunications industry in much the same way that the 1978 legislation deregulated the airline industry. After all, the 1996 Act advertised itself as the "most deregulatory [law] in history."[6] Anyone who believed that characterization at the time, however, was quickly disillusioned. The Act was not at all deregulatory in the straightforward sense of "tending to abolish regulation." To the contrary, it added an entirely new *dimension* to pre-1996 regulation by creating a broad new set of wholesale requirements, including the unbundling obligations discussed in chapter 2, to the existing edifice of retail regulation.[7] And its new rules for interconnection, intercarrier compensation, and universal service spawned massive FCC regulatory initiatives and fundamental legal disputes that the courts are still sorting out seventeen years later.

Some of the regulatory tumult has been inevitable, but many observers have expressed dismay that it has taken so long to complete the shift from legacy monopoly-style regulation to a more market-oriented regulatory model based on convergence and the layered nature of modern communications. Professor Kahn, who died in 2010 at the age of 93, assigned culpability to the FCC. In the late 1990s, he claimed that the FCC was "micromanag[ing] the process of deregulation itself" and in the process was crossing the line "between regulatory interventions establishing the conditions under which competition may be relied on to determine the outcome and interventions intended, whether consciously or unconsciously, to *dictate* that outcome."[8] Peter Huber, a noted regulatory commentator and lawyer, likewise contends that the FCC is culturally incapable of letting go of the industry's reins, now or ever. That concern led Huber in 1997 to propose abolishing the FCC, much like the Civil Aeronautics Board before it, and letting disinterested, non-bureaucratic antitrust courts remedy any anticompetitive practices in the industry.[9]

Occasional calls for the FCC's abolition have come not just from libertarians such as Huber, but also from liberal activists on the other end of the political spectrum. For example, Lawrence Lessig argued in 2008 that the FCC had become a "junior varsity Congress," staffed with political hacks beholden to corporate interests; that "[y]ou can't fix DNA"; and that "President Obama should get Congress to shut down the FCC and similar vestigial regulators, which put stability and special interests above the public good." In the FCC's place, Lessig would create a new, non-industry-specific agency with jurisdiction over communications policy and intellectual property rights, "a staff absolutely barred from industry ties," and a generalized mandate to "protect innovation."[10]

In this final chapter, we explain why we disagree with such proposals and why the FCC will indefinitely remain the least problematic institution to oversee telecommunications policy, at least for most issues within its jurisdiction. In that respect, the FCC's continued role recalls Winston Churchill's observation about democracy itself: "Many forms of Government have been tried, and will be tried in this world of sin and woe. No one pretends that democracy is perfect or all-wise. Indeed, it has been said that democracy is the worst form of Government except all those other forms that have been tried from time to time."[11] Continued reliance on the FCC is likewise the worst way to superintend the communications industry—except for the alternatives.

We wish to make one point clear from the outset. This book focuses on *competition* policy in the telecommunications industry, and this chapter addresses the institutional dimensions of competition policy. Of course, the FCC is responsible for a great many issues beyond competition policy. These include truth-in-billing and other consumer-protection issues, rights of telecommunications access for people with disabilities, E-911 emergency calling, and various universal service initiatives beyond those discussed in chapter 8. In speculating about the future of competition policy, we do not mean to address how the FCC or other agencies, including the Federal Trade Commission (FTC), should manage these non-competition-oriented regulatory concerns. Those concerns are important, but they lie outside the scope of this book.

I. First Principles of Institutional Reform

Before we can analyze comparative institutional competence or propose meaningful steps for reform, we must first identify our objectives. At bottom, any institutional arrangement for managing competition policy should promote five basic values, which we label expertise, determinacy, transparency, neutrality, and humility. The first two of these values are straightforward; the last three less so.

We begin with *expertise*, which is perhaps the most self-evident of these five values. Expertise simply means that primary decisionmaking authority should be committed to institutions that understand or can easily learn the esoteric technology of telecommunications, the structure of the industry, and the complex economic and regulatory issues that have defined this field for many years.

We next move to *determinacy*. A regulatory regime is determinate if its governing institutions—Congress, regulatory agencies, and the courts—work together smoothly and expeditiously enough that the industry knows as quickly as possible what the ground rules for competition policy will be and can predict with reasonable precision how those rules will be applied. The more determinate these ground rules are, the more comfortable investors will be in placing bets on the future of this industry, and the more likely it is that innovators will obtain financing to put their ideas to work for the public good. Particularly in a dynamic industry such as telecommunications, it is often "more important," as Justice Louis Brandeis once put it, "that the applicable rule of law be settled than that it be settled right."[12] This intuition is an application of the Coase theorem (discussed in chapter 3), which holds that parties can generally order their behavior efficiently within a given legal framework if, but only if, transaction costs are low and that framework is reasonably clear and stable.

Transparency, our third value, is a commitment to openness in the process of decisionmaking. The concept is more complex than it may first appear. It often is used to denote full disclosure of the inputs into the decisionmaking process: if a party makes an argument to a decisionmaker, it should have to tell its adversaries what it said and to whom. But such transparency of inputs does not itself produce public knowledge about the evolving deliberations of the decisionmakers themselves as they work through complex and often shifting sets of issues. Indeed, sometimes—as in the case of formal judicial proceedings—the inputs are completely transparent, yet the deliberations of the decisionmaking body (such as a jury or panel of judges) are a black box. Both forms of transparency—of inputs and of deliberations—are important whenever the questions relevant to the outcome of a proceeding can be expected to shift unpredictably as technology evolves and as decisionmakers climb the learning curve on particularly complex issues. That, more often than not, is true of the critical policy issues discussed in this book.

The fourth value that an ideal institutional arrangement should promote is analytically rigorous *neutrality* in the resolution of controversial policy issues. This concept is subtle. By "neutrality," we mean that whoever writes the rules for competition policy should think of problems by reference to first principles about how to maximize consumer welfare—principles that, for the most part, are informed by antitrust analysis. And when considering proposals for regulatory reform, policymakers should begin by asking, "Why have we done it this way?" rather than "Why should we incur the political costs of changing our longstanding policy on this issue?"

Our fifth value, and admittedly the vaguest, is *humility*. We use the term to describe the attitude of a policymaker who, with every important decision, remembers the many times in which other policymakers have been flatly wrong in their predictions of how the telecommunications market would take shape and in their assessments of the regulatory measures needed to enhance consumer welfare within that evolving market. Humility also reminds policymakers that over the long term the unintended, undesired consequences of regulation can dwarf the intended, desired outcomes. That fact is not a reason for doing nothing when action is needed to correct genuine market failures. But it is a reason for policymakers to respect the market's ability to enhance consumer welfare and, as they evaluate the predicted benefits of their own regulatory involvement, to give due regard to the unpredictable course of technological and economic change.

Having defined our criteria for institutional success, we now examine the relationships among the principal actors in this field—Congress, agencies, and the courts.

II. Judging Congress

In the United States, the ultimate source of competition policy for the telecommunications industry (or any other) is Congress, which enacts the legislation applied by the courts, the FCC, NTIA, the federal antitrust agencies, and, in many contexts, the state public utility commissions. Ideally, Congress would enact rules of conduct that are substantively sound, unambiguous, and yet articulated at a high enough level of generality that courts and agencies can apply them sensibly over time, despite unpredictable changes in technology and industry structure. In key legislation, however, Congress has fallen short of that ideal, with predictable consequences for the principle of determinacy.

You may have been struck by the number of times we have concluded our discussion of important legal disputes in this field—from net neutrality to intercarrier compensation reform to broadband subsidies—by noting that they remain pending as this edition heads to press. Why are people still arguing about the meaning of core statutory provisions enacted nearly two decades ago? Why couldn't Congress simply have spoken more clearly and avoided all this investment-deterring uncertainty? The problem is twofold: first, Congress has institutional incentives to speak ambiguously on politically divisive topics; and, second, technology evolves more quickly than Congress can legislate.

For an example of the first phenomenon, consider the fate of the FCC's rules governing how much a new entrant must pay local telephone incumbents to lease their network elements. As discussed in chapter 2, the FCC predicated those rules in particular—and its own pricing jurisdiction in general—on a particular interpretation of sections 251 and 252, added to the Communications Act by the Telecommunications Act of 1996. Those rules lingered in a state of suspended animation for six years after they were first adopted, having been vacated twice by the Eighth Circuit and reinstated twice by the Supreme Court, the second time in 2002. Congress could have avoided those six years of investment-chilling uncertainty both by making it explicit from the outset that, contrary to the Eighth Circuit's initial reading of the 1996 Act, the FCC has the authority to issue pricing rules for network elements and by giving the Commission greater guidance on the appropriate content of those rules.

We should not be surprised, however, that Congress spoke as vaguely as it did. The 1996 Act was effectively written by warring interest groups that believed they were playing a zero-sum game. On the one side were the legacy long-distance companies, which wanted the FCC to grant them low-priced access to local facilities. On the other side were the local telephone monopolists, which wanted *state commissions* to set any pricing rules because the monopolists believed that the states would set higher prices. In these circumstances, it is always easier for a legislator to vote for an ambiguous provision—thereby punting hard issues to regulators and the courts—than to vote for a provision that decides important controversies clearly and directly. The latter course, although better for industry stability and the public interest, is sure to alienate one powerful interest group or another. Ambiguity, by contrast, enables politicians to waffle about what they really meant when they cast their votes and to blame someone else if their votes are later construed by a court or agency against the interests of a particular constituency.

The 1996 Act's origins in interest-group politics explain why so much of it seems written as though for a law school moot court: the opposing positions coexist in such perfect equipoise that no side enjoys any advantage over the others, and each side can litigate decisions with which it disagrees in a state of almost perfect interpretive competition.[13] On the other hand, we do not want to overstate Congress' irresponsibility in enacting such a deliberately ambiguous statute, for the reality is that no significant legislation would have been enacted in the first place if it had addressed such highly consequential issues too clearly.

Sometimes the statutory language produces prolonged uncertainty not so much because Congress means to be ambiguous as because Congress simply fails to see around the technological bend. The current telecommunications marketplace bears little resemblance to the one Congress envisioned when it enacted the 1996 Act, mostly because Congress did not anticipate the full consequences of the Internet. Although it anticipated some degree of convergence between different technology platforms, it focused almost exclusively on the prospect that cable companies would offer circuit-switched telephony and that telephone companies would offer video programming. It did not anticipate that voice, video, and data would someday be reduced to streams of bits running on top of a single and universal logical-layer platform, the Internet protocol, which itself can be used on virtually any physical-layer transmission medium.[14] As a result, Congress never straightforwardly specified whether broadband Internet access is a Title I "information service" or a Title II "telecommunications service"; indeed, for a few years around the turn of the millennium, some parties even insisted that it was a Title VI "cable service."[15] As discussed in chapters 6 and 8, this disputed characterization of broadband Internet access services has produced massive regulatory uncertainty across a range of issues, from net neutrality to federal broadband subsidies.

III. The Antitrust Alternative

As IP technology engulfs the telecommunications world, it becomes increasingly unsustainable to attach profound legal significance to the legacy service definitions contained in the Communications Act ("information service," "telecommunications service," "cable service," "MVPD service," etc.). Policymakers will ultimately need to complete the transition toward a more functional model of regulation that takes greater account of the Internet's established layers: physical, logical, applications, and content.[16] Under this approach, much of telecommunications-related competition policy would collapse into an antitrust-oriented objective of keeping any genuinely dominant provider on one layer from limiting competitive entry into markets on that layer or exerting undue influence on adjacent markets at other layers.

If those substantive antitrust-oriented goals become the central stated purpose of telecommunications policy, would it make sense to shut down the FCC and commit the future of telecommunications competition policy to federal antitrust agencies (the Department of Justice and the FTC) and ultimately antitrust courts? Generalist courts, largely immune from interest-group politics and from entrenched assumptions about how the world must work, are arguably better equipped than specialized agencies to resolve competition policy issues on their economic merits. Perhaps the purest endorsement of courts over agencies as guardians of telecommunications competition policy appears in the scholarship of Peter Huber. In his view, the FCC and its "army of federal employees hanging around indefinitely to meddle and mess up" should be abolished altogether in favor of minimalist, case-by-case antitrust enforcement.[17]

Of course, antitrust courts are only as neutral as the judges who sit on them, and not all judges are exemplars of neutrality. A district judge selected to decide a critical antitrust case might do quite a bit worse than Judge Harold Greene, who administered the twelve-year antitrust regime spawned by the AT&T consent decree (see chapter 2) and in that capacity made a number of controversial policy judgments about the trajectory of competition within the industry.[18] The risks of vesting such enormous power in a single generalist district court judge are compounded by forum-shopping opportunities that sometimes enable an industry faction to choose, for the resolution of critical industry-wide controversies, whatever court it considers unusually sympathetic to its cause. In one well-known example, two Bell companies brought a constitutional challenge to section 271 (see chapter 2) in the remote Wichita Falls Division of a federal district court in Texas, where their desired judge, Joe Kendall, obliged them by invalidating that provision and several others on New Year's Eve in 1997.[19] Although Kendall's decision was eventually reversed, it symbolizes the dangers of letting individual judges play too significant a role in shaping the future of this uniquely volatile industry.

Huber is nonetheless correct that the FCC, like any entrenched bureaucracy, has developed a self-sustaining bias in favor of keeping itself important by intervening heavily in the industry it regulates. That bias indulges rent-seeking behavior and invites overregulation, with all of its attendant inefficiencies. Whenever an industry requires broad oversight, however, similar bureaucracies tend to arise spontaneously no matter what institution is formally charged with conducting the oversight. For example, the AT&T consent decree regime produced its own small "army of federal employees" in the Department of Justice who devoted a dozen years of their careers to the zealous enforcement of the decree's manifold restrictions. That group of lawyers and economists was not necessarily less disposed to government intervention in the telecommunications market than the FCC is today. Ultimately, Huber's *institutional* preference for antitrust courts over regulatory agencies could eliminate such bureaucracies only when he and his fellow libertarians are granted their *substantive* wish for a negligible government role in the oversight of telecommunications competition.

The more basic problem with relying on antitrust courts to superintend the telecommunications industry is that the judicial process is deficient in the areas of determinacy and expertise. Consider determinacy first. Companies with market power are better off knowing now, not at the end of a multiyear antitrust suit, whether the aggressive business strategy they are contemplating will be deemed anticompetitive. Likewise, the absence of a regulatory agency to develop and enforce preset rules would make it more difficult for industry pioneers to compete with a dominant firm, particularly if they lack the money or endurance to prosecute an antitrust suit. Of course, the need for such rules varies with the market power of the main players; as an industry becomes more competitive, it becomes increasingly appropriate to replace prophylactic rules with after-the-fact enforcement of basic competition norms.

Generalist courts also lack the technical expertise needed to make fully informed judgments about the market consequences of any substantive remedies they order. As Frank Easterbrook observes, "Judges are the regulators with the broadest portfolios, and thus are the least competent."[20] In theory, this shortcoming could be alleviated either by relegating telecommunications competition issues to specialized courts (much as Congress has assigned all patent law appeals to the Federal Circuit) or by permitting judges to retain experts who can explain the industry to them. But these measures present considerable challenges of their own.[21] Indeed, ever careful to preserve the appearance of judicial self-sufficiency, the courts have sometimes expressed outright hostility to the use of retained experts to help resolve technically complex litigation on the merits, as illustrated by the D.C. Circuit's 1998 order barring Judge Thomas Penfield Jackson from using Lawrence Lessig as a special master in one phase of the Microsoft antitrust case.[22] Just as important, generalist judges lack both the resources and the technical proficiency to resolve the thousands of day-to-day disputes, on pricing and other issues, that must be decided under any local competition regime that involves interconnection obligations and even minimal leasing rules (whether in the form of "unbundled network elements" or price-regulated special-access services).[23] Aware of these limitations, the judiciary has generally shown great solicitude for the greater expertise of regulatory agencies within the scope of their substantive authority.[24]

These institutional concerns form the backdrop to the Supreme Court's 2004 decision in *Verizon Communications, Inc. v. Law Offices of Curtis V. Trinko LLP*,[25] which concluded that antitrust courts are generally inappropriate forums for the ongoing management of telecommunications competition policy, at least so long as the industry remains subject to pervasive regulation. Our discussion of *Trinko* first requires a brief review of the historical intersection between antitrust law and prescriptive telecommunications regulation.

Before 1996, as we have noted, much of telecommunications competition policy was managed by a single judge: Harold Greene. In 1982, AT&T acquiesced in the consent decree that ultimately spun off the Bell companies only after Greene rejected the company's argument that the FCC's comprehensive oversight of the industry precluded any role for antitrust enforcement. Greene found that argument unpersuasive because, as demonstrated by years of regulatory indecision, "the Commission is not and never has been capable of effective enforcement of the laws governing AT&T's behavior."[26] When Congress called for the termination of the consent decree in 1996, it directed the FCC and its state counterparts to implement wireline competition provisions that, as discussed in chapter 2, go far beyond the FCC's traditional mandate to require interconnection among carriers on just and reasonable terms. At the same time, however, Congress included an antitrust "savings clause" providing that "nothing in the Act or in the amendments made by this Act shall be construed to modify, impair or supercede the applicability of any antitrust laws."[27]

Like so many other provisions of the 1996 Act, this one led to widespread disagreement. Some argued that despite the savings clause, antitrust courts should generally defer to regulators in deciding whether particular conduct is genuinely anticompetitive and, if so, what sorts of enforcement mechanisms would be appropriate for addressing it.[28] But others contended that the savings clause preserves antitrust remedies as a backstop for the protection of competitors whenever prescriptive regulation proves ineffective in keeping incumbents from exploiting any market power they might have.[29] Advocates of this position further maintained that, as in the AT&T antitrust litigation, defendants should have to bear the burden of proving that, as a factual matter, regulatory mechanisms are sufficiently effective to make antitrust intervention unnecessary. The major theories of liability expounded by these proponents of continued antitrust enforcement included the controversial *essential facilities doctrine*, under which some lower courts have forced monopolists to cooperate with their rivals' market-entry plans by selling those rivals access to bottleneck facilities.[30]

In *Trinko*, the Supreme Court resolved this debate with a resounding victory for telecommunications antitrust defendants in general and the Bell companies in particular. The Court began by explaining that although the antitrust savings clause "preserves claims that satisfy existing antitrust standards, it does not create new claims that go beyond existing antitrust standards."[31] Then, in setting forth those existing standards, the Court sharply limited the circumstances in which courts may impose antitrust remedies—under the essential facilities doctrine or any other—for a monopolist's refusal to help rivals compete with it.[32] Such remedies often do more harm than good, the Court reasoned, "because of the uncertain virtue of forced sharing and the difficulty of identifying and remedying anticompetitive conduct by a single firm," and because "[m]istaken inferences and the resulting false condemnations are especially costly" in that "they chill the very conduct the antitrust laws are designed to protect" by "lessen[ing] the incentive for the monopolist, the rival, or both to invest in . . . economically beneficial facilities."[33]

Moving from substantive to institutional concerns, the Court added that "[e]nforced sharing also requires antitrust courts to act as central planners, identifying the proper price, quantity, and other terms of dealing—a role for which they are ill-suited."[34] The Court further found that "[t]he 1996 Act's extensive provision for access" to an incumbent's facilities and services on regulated terms makes it as unnecessary as it is potentially harmful "to impose a judicial doctrine of forced access."[35] The Court echoed this holding in its 2007 *Credit Suisse* decision, when it held that the securities laws, together with comprehensive regulation by the Securities and Exchange Commission, implicitly bar application of the antitrust laws to certain types of underwriting practices.[36] Although interpretations vary, these two decisions suggest that as prescriptive regulation of a field waxes, antitrust enforcement must wane. In effect, the 1996 Act, together with the *Trinko* and *Credit Suisse* cases, has turned the pre-1996 regulatory scheme upside down. Whereas the Department of Justice once displaced the FCC in the field of telecommunications competition, the FCC's current ascendancy has sharply curtailed the role of traditional antitrust enforcement.

Although both of us generally support this institutional arrangement, we disagree about the role of antitrust institutions in overseeing net neutrality disputes in particular. For that class of disputes, one of us would divest the FCC of any regulatory authority it may have (thereby nullifying the *Trinko* presumption against antitrust enforcement) and entrust the oversight role instead to a non-industry-specific competition authority such as the FTC.[37] The other of us favors FCC jurisdiction over net neutrality disputes while proposing an alternative model of "coregulation" under which the FCC would rely more heavily on industry *self-regulatory organizations* (SROs, discussed in the next section) to generate consensus on disputed network-management practices and would also emphasize after-the-fact adjudication.[38] But this disagreement is fairly narrow: we agree that the FCC should maintain jurisdiction over most of its traditional areas of regulatory responsibility, including spectrum policy, PSTN interconnection, and the price and other terms of unbundled network elements, special access, and other telecommunications services. As the *Trinko* court noted, such responsibilities are ordinarily far more easily entrusted to regulatory agencies with rate-setting authority over "common carriers" than to generalist antitrust courts focused instead on specific complaints of anticompetitive behavior.

IV. The FCC in Transition

Whatever the right institutional arrangement might have been as an original matter, specialized regulatory agencies, led by the FCC, will almost certainly play the dominant role in setting telecommunications competition policy for the foreseeable future. We now turn to the FCC's performance in that role and the prospects for improving upon it, using as our frame the five institutional values we have identified: expertise, determinacy, transparency, neutrality, and humility.

A. Expertise

Unlike the antitrust agencies—the Department of Justice and the FTC—the FCC focuses all of its regulatory energy on a single industry: telecommunications. That fact is simultaneously the FCC's greatest strength and greatest weakness as an institution. On the one hand, the Commission's fixation on this single industry raises concerns about its neutrality. It makes the Commission susceptible to capture by various interest groups and gives it structural incentives to err on the side of intervening in its designated marketplace simply to keep itself relevant. On the other hand, the Commission's narrow focus on this single industry also gives it unmatched insights into that industry's peculiar challenges.

Even so, the FCC cannot be expected to understand the industry's complex and rapidly evolving engineering challenges as well as the industry's own practicing engineers do. That fact poses challenges for the FCC in a number of areas. For example, the FCC's net neutrality rules anticipate that it will conduct case-by-case inquiries into the "reasonableness" of particular network-management practices. Because that set of engineering issues is arcane and fluid, one of us has encouraged the FCC to involve SROs more closely in the process to give the engineering community's perspective.[39] One SRO in particular—the Broadband Internet Technical Advisory Group (BITAG)—has begun playing an important role in that area, although it does not yet have any official relationship with the FCC or any other federal agency.[40] The FCC should likewise look increasingly to other SROs, ranging from the Internet Engineering

Task Force (IETF) and the North American Network Operators' Group (NANOG), to build consensus more generally on best practices for Internet network operators.[41]

B. Determinacy

In theory, Congress delegates legislative rulemaking authority to administrative agencies not just because they are expert in their designated fields, but also because they, unlike Congress, have the institutional agility needed to adjust the rules promptly to accommodate changes in market conditions. In reality, the FCC has long tended to string out its decisions on important matters, sowing regulatory uncertainty and ignoring the D.C. Circuit's exasperated admonition to the Commission more than 60 years ago that "[a]gency inaction can be as harmful as wrong action."[42]

Congress has sometimes addressed this problem by giving the FCC strict statutory deadlines for the resolution of particularly time-sensitive issues. For example, the FCC completed its work on the initial implementation of sections 251 and 252 within the specified 180-day period after passage of the 1996 Act,[43] and, at least on paper, met the separate 90-day deadline under section 271 for deciding each Bell application to enter the long-distance market in a particular state.[44] Sometimes, however, the FCC manages to elude even congressionally mandated deadlines. For example, Congress directed the Commission to resolve within fifteen months any petition seeking forbearance from particular regulatory obligations, and it even provided that such petitions shall be "deemed granted" if the FCC misses that deadline.[45] In practice, however, the FCC sometimes evades the deadline by creating new procedural reasons for refusing to address a forbearance petition on the merits, and even if a reviewing court later invalidates those reasons, the Commission can then let the reinstated petition lie dormant on remand, potentially for years.[46] And a great many of the Commission's most significant rulemaking proceedings have dragged on for many years because they are subject to no statutory deadline at all. In 2011, House Republicans began considering legislation that would require "the Commission to establish shot clocks for each type of proceeding it oversees," similar to the informal shot clocks the Commission currently uses to review merger applications.[47] But that legislative initiative appears unlikely to bear fruit anytime soon.

The FCC's delays in resolving important industry controversies stem in part from the elaborate behind-the-scenes deal making needed to reach consensus among the Commission's five members. Like cabinet officials, those members are appointed by the president and confirmed by Congress; unlike cabinet officials, however, no more than three of them may belong to the same political party, and, once confirmed, they may be removed during their five-year terms only for cause.[48] The result is that members other than the chairman, even those who belong to the chairman's (and president's) party, may worry more about pleasing their separate constituencies within Congress or the industry than about pleasing the White House. This is a recipe for internecine intrigue and deliberative inefficiency.

Some have cited these concerns as a reason to place the FCC more firmly within the Executive Branch, eliminate the current five-member structure, and vest plenary authority in a single decisionmaker at the top of the FCC's organizational chart, much as Congress has organized the Food and Drug Administration and the Environmental Protection Agency.[49] That proposal has much to commend it. Given the unprecedented pace of change that the Internet has brought to the telecommunications industry, the dilatory costs of the FCC's multimember structure may now outweigh whatever benefits it was once thought to present in the form of internal checks and balances. Any such proposal, however, would face severe political obstacles in Congress, where legislators have exploited the FCC's instability at the top and its partial detachment from the White House as bases for exerting more direct influence over it than over more traditional Executive Branch agencies.

In all events, any reform of the FCC's own processes could serve as no more than a first step in bringing greater regulatory determinacy to this industry. Equally in need of reform are the FCC's relationships with other institutional players: with the courts that review the FCC's policy choices, with the state commissions that implement those choices, and with sister agencies on the federal level—the Justice Department and the FTC—that share the FCC's responsibility to review proposed mergers between telecommunications companies. We address each of these institutional relationships in turn.

Relations with reviewing courts

Any final FCC order is subject to judicial review in a federal court of appeals.[50] The availability of such review contributes to the indeterminacy of telecommunications regulation, particularly when undertaken by activist generalist courts that consider themselves equally equipped as specialist agencies to understand the complexities of this industry.

Under the doctrine of judicial deference formalized in *Chevron U.S.A. v. National Resources Defense Council*,[51] a reviewing court may not act as a policymaker in its own right. Instead, it may serve only as a backstop against agency action that is patently irrational, inconsistent with a clear statutory (or constitutional) mandate, or inadequately justified in the written document that accompanies the agency's order.[52] A court that finds fault with an agency's decision is expected to remand the matter back to the agency itself for further deliberation within broad bounds.[53] The traditional reason given for such deference is that agencies have greater topical expertise than judges and are subject to continuing congressional oversight as a check.[54] But an equally important rationale, particularly in the telecommunications field, is the value of regulatory determinacy.

Every time a court invalidates an FCC rule, it injects legal indeterminacy into the industry that may distort economically efficient behavior for many years. Examples include the Eighth Circuit's invalidation of key aspects of the FCC's initial rules implementing the 1996 Act;[55] the Eleventh Circuit's invalidation of FCC rules implementing access by Internet and wireless providers to electric utility infrastructure;[56] and the Ninth Circuit's invalidation of the FCC's conclusion that cable modem services are a unified "information service" without a "telecommunications service" component.[57] In each case, judicial invalidation of considered FCC policy choices threw the industry into years of investment uncertainty, even though, each time, the FCC eventually persuaded the resource-constrained Supreme Court to intervene and reinstate the Commission's rules.

This is no way to run a major sector of the economy. If, alternatively, the courts stand down when faced with close questions about the lawfulness of agency decisions, the worst that can happen is that we as a society will get what we allow our politically accountable institutions—Congress and its administrative delegates—to give us. We are not suggesting that courts should play no role in reviewing the FCC's decisions. Sometimes the FCC does act irrationally, and sometimes it ignores clear statutory directives. Such abuses of delegated authority warrant judicial intervention. Reviewing courts should nonetheless pick their fights and remedies carefully, generally deferring to the FCC's greater expertise and, just as important, to the industry's need for regulatory predictability.

Relations with the states

As discussed in chapter 2, the 1996 Act partially displaced the traditional model of dual jurisdiction, which divides the *subject matter* of wireline telecommunications regulation into mutually exclusive federal and state spheres, with a new model of cooperative federalism in which the FCC and the states often work together in complementary roles on the same wireline competition issues. For Congress, this new model was the only feasible choice for regulating telecommunications competition. The need to hammer out the innumerable details of carrier-to-carrier relations under the 1996 Act presents an immense bureaucratic challenge. Congress was not about to create a series of FCC branch offices or increase the FCC's staff many times over, and the state commissions provided a ready source of labor for the task at hand.[58]

Delegating such responsibility to the states, however, unavoidably gives all of them substantial discretion, as sovereign actors, in deciding how federal law will be implemented. This carries both benefits and costs. As one of us has argued, the states' discretion in competition matters allows them to tailor regulatory approaches to local conditions, encourage public participation in the policymaking process, experiment with acceptable alternatives, and compete with one another to develop an optimal scheme of regulation.[59] As the other of us has noted, however, these benefits can come at a high cost in the form of delay and confusion as well as massive lawyering and lobbying expenses.[60] The more complicated and multidimensional the regulatory scheme is, the more investment-chilling indeterminacy there will be about the rules of the road.

The debates about state participation in telecommunications policy are only just beginning, as demonstrated by the pending Tenth Circuit challenge to the FCC's new intercarrier compensation regime, which preempts a broad new swath of traditional state-level regulation (see chapter 7). For as long as there is telecommunications regulation in the United States, there will be controversy about the states' proper role in its implementation. Nonetheless, as the entire industry gradually coalesces around the Internet protocol, traditional state public utility regulation will gradually give way to two of the Internet's most cherished characteristics: its federally enforced freedom from state regulation and its tendency to efface political boundaries of all kinds.

Relations with coordinate merger-review authorities

The FCC shares decisionmaking authority not just vertically with the states, but also horizontally with other federal agencies. As discussed in chapter 3, for example, the FCC's efforts to reform spectrum policy are inevitably complicated by NTIA's independent jurisdiction over spectrum allocated to the government itself. Here we discuss the similar bureaucratic challenges caused by the FCC's sharing of merger-review authority with the Justice Department and occasionally the FTC.[61]

The Justice Department normally conducts its own merger reviews under section 7 of the Clayton Act, which requires an inquiry into whether the effect of a proposed merger "may be substantially to lessen competition, or to tend to create a monopoly."[62] This standard makes it much more difficult for two companies to merge if they *already* compete with one another than if they only *might* do so in the future.[63] In 2011, for example, the Justice Department moved to block the proposed merger of AT&T and T-Mobile on the ground that those two companies are close competitors in wireless markets (see chapter 4). In contrast, the Department repeatedly permitted massive mergers between different Bell companies—reducing their numbers from seven in 1996 to three today—because, as former antitrust chief Joel Klein explained, these geographically separated local telephone carriers had generally not "invaded [one] another's territory."[64]

Persuading the Justice Department to clear a telecommunications merger, however, is only half the task. The merging parties must also persuade the FCC to approve the transfer of lines or wireless licenses from the merging parties to the merged entity. In deciding whether to approve such transfers, the FCC is officially unconstrained by the Clayton Act standard or, for that matter, by any standard more determinate than whether, in the FCC's view, the merger would advance the public interest.[65]

The FCC has exploited this statutory freedom to great effect. For example, unlike the Justice Department, the Commission may inquire broadly into whether a merger would foreclose merely potential competition between the merging parties.[66] The procedures for obtaining clearance are also decidedly different. Whereas the Justice Department must seek judicial intervention to block a proposed merger,[67] the FCC may unilaterally quash any merger simply by declining to approve the necessary license transfers. Similarly, whereas the Justice Department bears the burden of proving that a merger would violate the Clayton Act standard, the FCC maintains that the burden falls on the merging parties to prove that their combination would affirmatively advance the public interest.[68] In practice, the FCC often teams up with the Justice Department to review a merger and will withhold approval until the Department gives the green light. The threat of FCC nonapproval effectively liberates the Justice Department from any need to go to court to block a merger and grants both agencies enormous leverage to extract concessions from the merging parties.

How does the FCC exercise that leverage? After meeting informally with the FCC's members and staff, merger parties typically end up making a range of "voluntary" commitments to persuade the FCC that, in some amorphous sense, the public interest benefits of the merger will outweigh the harms.[69] These commitments sometimes bear only a tenuous relationship to any concerns raised specifically by the merger itself, and they create ongoing obligations that in many cases the FCC could not have lawfully imposed on the parties through its ordinary rulemaking authority.[70] But because the parties have agreed to them "voluntarily," such obligations are effectively immune from judicial scrutiny.

Many have questioned the legitimacy of this merger-review process.[71] In 1999, then-Commissioner Michael Powell expressed discomfort with the open-endedness of the Commission's inquiry, which "places harms on one side of a scale and then collects and places any hodgepodge of conditions—no matter how ill-suited to remedying the identified infirmities—on the other side of the scale."[72] As he explained, "the process of obtaining 'voluntary' conditions inevitably involves bilateral negotiations with the parties that leave the integrity of the Commission's process vulnerable to criticism," particularly when "we pursue conditions that do not go simply to the harms occasioned by the merger, but reach further into the [more general] rights and concerns of other parties."[73] Although the FCC later purported to curb this latter practice,[74] there is still no real check on the Commission's power to extract broad promises only loosely designed to alleviate any merger-specific concerns. Other commentators have likewise condemned these aspects of the FCC's merger-review process as well as its speculative inquiries into potential, rather than actual, competition between the merging parties.[75]

Whatever the merit of these criticisms, there can be no doubt that the FCC's independent review of any merger adds considerable uncertainty to the merger-approval process. To compound the problem, many states view mergers as opportunities for imposing their own wish lists on the merging parties as preconditions for obtaining any necessary state-level approvals.[76] It is debatable whether the public interest demands these additional, largely unchecked layers of regulatory intervention beyond the basic inquiries already conducted by the Justice Department or the FTC, although those agencies have sometimes welcomed the FCC's unique remedial oversight.[77] Also, Congress itself is capable of prescribing more specific and stringent standards for particular categories of mergers thought to present special competitive concerns, as it has done for proposed combinations of cable and telephone companies operating in the same region.[78]

To be sure, the Justice Department and the FTC review only the strictly economic costs and benefits of mergers, and mergers between media companies may raise various *non*economic concerns about society's need for a diversity of voices on television and radio (see chapter 9). Whatever the merit of those concerns, however, they can play no role when the merging parties are, for the most part, providers of transmission services rather than content. In that case, the inefficiencies of redundant merger review by the FCC may outweigh whatever benefits such redundancy is thought to promote. To address those concerns, one of us has proposed a rationalized regime for dual-agency review of telecommunications mergers, under which the FCC would defer to the relevant antitrust agency on competition policy issues, the antitrust agency would entrust the enforcement of post-merger conduct remedies to the FCC, and any such remedies would be subject to judicial review.[79]

C. Transparency

In our democratic system, different governmental institutions follow radically different rules governing the transparency of their decisionmaking processes. On one end of the spectrum stands Congress, which routinely hears the views of competing interests behind closed doors before producing its legislative sausage. There is of course much debate about the integrity of that process, but reform proposals tend to focus on money: shining a spotlight on who is giving what to whom and regulating when and how much money may be spent. No one seriously proposes to make members of Congress record and publicize the substance of their countless discussions with the public about proposed legislation. On the other end of the spectrum lies the judiciary. With rare exceptions, all communications a party makes with a judicial decisionmaker (a jury, judge, or panel of judges) must appear either in written briefs or in open court in the presence of all other parties.

The transparency norms of administrative agencies fall somewhere between these two extremes, and each agency has its own institutional traditions about precisely how transparent to make its decisionmaking process. The FCC tends to use a more informal approach than many agencies, at least for rulemaking proceedings of industry-wide significance. In such proceedings, the FCC typically solicits formal "comments" (advocacy papers) from interested parties on a specified timetable but then allows any party to supplement its comments through so-called *ex parte presentations*. These presentations can take the form either of substantive "ex parte letters" filed with the Commission and publicly accessible on its website or, more informally, of in-person meetings or telephone calls with commissioners or FCC staff. Under the Commission's *permit-but-disclose* rules,[80] parties generally must memorialize such meetings or calls with an ex parte letter that tells the public who met with whom and, at some level of generality, what they said. The FCC has tried in recent years to require parties to be more detailed and explicit in these follow-up letters,[81] but many parties find ways to keep the letters cryptic, particularly if the underlying discussions involve politically charged horse-trading.[82]

Not all FCC proceedings are this informal; enforcement matters and some adjudications are subject to much more rigorous restrictions on informal presentations to the agency.[83] But the main FCC proceedings discussed in this book—from the net neutrality rulemaking to the AT&T/T-Mobile merger review—were subject to these relaxed permit-but-disclose rules and thus a flurry of post-comment ex parte presentations. Given that FCC proceedings commonly last many months or years, the Commission often relies disproportionately on these presentations, which are often filed very late in a proceeding, and which therefore, unlike the formal comments filed at the very outset of the proceeding, address the questions the FCC ultimately signals are most relevant.

FCC observers have very different views on the merits of this procedural informality. One of us has called on the FCC—in the interests of transparency, procedural fairness, and public accountability—to rely far less on ex parte presentations in its decisionmaking and far more on formal comments, rigorous information-gathering hearings before administrative law judges, and public workshops.[84] The other one of us finds the ex parte process less problematic so long as it is accompanied by genuinely full disclosure to the public about the communications of interested parties with the Commission's decisionmakers.

Each of us agrees, moreover, that the FCC should not only require parties to disclose on the record what they have told the agency, but also keep the public reasonably informed about the evolution of its own deliberations. So far the Commission has accomplished this latter goal informally, telling individual parties in ex parte meetings which way various commissioners and staff members have begun to lean and what new questions they are asking. Alternatively, the FCC might try to accomplish the same objective more formally and publicly by relying more on "further notices of proposed rulemaking" that disclose how its inclinations have evolved and that seek new rounds of formal comment on the merits of those new inclinations. In theory, this latter approach would be far more transparent and fairer to interested parties who cannot afford to hire FCC experts to represent their interests. In practice, however, a highly formalized approach, unaccompanied by informal give-and-take with individual parties, may present challenges in implementation, at least in highly complex and fluid proceedings.

The integrity of the Commission's proceedings will ultimately depend as much on the integrity of its personnel as on the details of its procedural rules, and any agency confronted with fast-changing policy choices will require some play in the joints of its decisionmaking process. Indeed, rigid restrictions designed to produce greater agency transparency can sometimes defeat the very objectives for which they were designed. A case in point is a post-Watergate-era statute optimistically entitled the "Government in the Sunshine Act." The Sunshine Act generally prohibits the FCC's chairman and commissioners from discussing substantive matters as a group outside of formal public hearings.[85] Among its other unintended consequences, this restriction increases the importance of each commissioner's "legal advisors," who are subject to no such bar, and transforms meetings among the FCC's actual members into prebaked press events devoid of any genuine deliberation. Few knowledgeable people today believe that these Sunshine Act restrictions make sense in their present form, but it would be politically naive to expect Congress to repeal a statute entitled the "Government in the Sunshine Act" anytime soon.

D. Neutrality

As anyone who has watched the FCC in action is aware, the Commission often seems more adroit at jury rigging intellectually sloppy deals to appease industry factions in the short term than at making the analytically sound but politically difficult policy choices needed to promote long-term economic efficiency. Judge Richard Posner memorably described one FCC regulatory scheme (the "finsyn rules," discussed in chapter 9) as a set of "unprincipled compromises of Rube Goldberg complexity among contending interest groups viewed merely as clamoring suppliants who have somehow to be conciliated."[86] Such dealmaking is particularly common in regulatory areas where the FCC's decisions have immediate and quantifiable effects on consumer bills or on the bottom lines of the regulated parties. In chapters 7 and 8, for example, we documented the FCC's historical preference for short-term patches over long-term solutions in the fields of intercarrier compensation and universal service.

There is no straightforward institutional reform that would force the FCC to stand up to political pressures and chart a course of analytically rigorous neutrality. One obvious priority is to staff the FCC with principled leaders who have demonstrated as much of an appetite for making sound policy decisions on the merits as for appeasing political constituencies. And so long as the person at the top meets that description, vesting ultimate authority in one decisionmaker rather than five would reduce much of the horse-trading that not only delays resolution of important issues but compromises the analytical integrity of the FCC's decisions when they are finally issued.

Another impediment to the FCC's neutrality comes from the formal arrangement of its staff into "bureaus" and "offices" corresponding to the obsolete regulatory categories drawn by the Communications Act of 1934 and thus to arbitrarily defined industry segments. This organizational structure invites parochialism and occasionally outright protectionism. Spectrum policy reform, for example, has occasionally been distorted by the differing perspectives of the Wireless Bureau, which takes special care to protect the incumbent mobile providers; the Office of Engineering and Technology, which looks after unlicensed uses; the Media Bureau, which regulates the broadcast industry; and the International Bureau, which oversees the spectrum used by satellite providers. A similarly arbitrary division of authority has also complicated the evolution of broadband policy, although the problem is less acute now than it was at the dawn of the broadband era. For example, the Wireline Competition Bureau, which until 2002 was known as the "Common Carrier Bureau," tends to focus on the broadband services offered by the telephone companies it has historically regulated, whereas the Media Bureau is more likely to view broadband issues through the lens of the cable companies that it regulates. Replacing this legacy structure with a regulatory orientation more in tune with industry realities would alleviate these institutional concerns.

E. Humility

The FCC was created during the New Deal for two basic missions: management of the radio spectrum and traditional command-and-control regulation of telephone monopolies. In the wake of the technological upheavals of the past twenty years, the FCC must now respond by redefining its own role in an industry characterized, however imperfectly, by competition, convergence, and a commensurate decrease in the need for competition-related regulatory oversight.[87] To make that transition, the FCC must embrace the elusive virtue of regulatory humility.

In Alfred Kahn's words, the FCC's basic challenge is to reorient its efforts toward "establishing the conditions under which competition may be relied on to determine the outcome" and away from policies "intended, whether consciously or unconsciously, to *dictate* that outcome."[88] In practice, this means that in an increasing number of regulatory areas the FCC should focus more on back-end enforcement of basic competition norms, remedying only clear acts of anticompetitive conduct instead of developing front-end prophylactic safeguards designed to anticipate all possible scenarios. Of course, at least for the foreseeable future, there will be innumerable areas—such as PSTN interconnection and intercarrier compensation—in which preset rules will be essential to industry stability. And the FCC thus will need to develop and superintend such rules for years to come. It is also conceivable—though not very likely—that the most dystopian predictions about the future of this industry will come true: that cable incumbents will become monopolists in the residential broadband market, free from the discipline of cross-platform competition (see chapter 1). In that event, there will be calls for the FCC to reverse course and reinstate some appropriately tailored form of public utility regulation in local markets subject to such monopolization. Barring that outcome, however, as competition relieves the need for comprehensively prescriptive regulation, the Commission should embrace the more neutral, adjudicative approach of an enforcement agency such as the FTC.

This institutional reorientation will force the FCC to check its traditional instinct to "plan in advance of foreseeable events, instead of waiting to react to them."[89] The basic problem with such preemptive intervention is that, as the history of regulation has shown, policymakers are often wrong both in their predictions of how the market will develop and in their judgments of what regulatory measures will best promote consumer welfare. As discussed in chapters 3 and 4, nowhere has such bureaucratic miscalculation harmed the public interest more than in the FCC's assumption of the "wise man" role in dictating how the airwaves should be used.

Of course, when particular economic conditions demonstrably lead to market failure, regulators should intervene sooner rather than later. Our point is that regulators should not blithely assume that a market will fail if it has not already done so, nor should they proceed on the assumption that because they have regulatory authority, they should exercise it somehow or another. Regulatory humility means knowing when one's judgments as a policymaker cannot do better and might do much worse than the collective judgments of competing firms and millions of self-interested consumers operating in a genuinely free market.

* * *

In retrospect, it should not be surprising that an institution initially designed to regulate monopolies in perpetuity is poorly designed to intervene in the market just enough to promote competition and then, as appropriate, stand out of the way.[90] That new mission, however, is the FCC's most important assignment. The ultimate end game in telecommunications regulation—which, to be sure, will take many years to reach—should be a deregulatory environment in which market forces rather than FCC officials dictate the most productive uses of the radio spectrum, create cross-platform competition in the last mile, and devise efficient solutions to interconnection. Of course, even in that world, regulators will play an important vestigial role in managing social welfare priorities such as universal service and 911 emergency dialing, and they may need to exercise some continuing oversight of basic interconnection arrangements, particularly as long as the PSTN remains in place. For the most part, however, the FCC, like Alfred Kahn's Civil Aeronautics Board, should define success as creating the conditions necessary for phasing out its legacy regulatory functions.

Notes

Chapter 1

1. *See The Underwater Web: Cabling the Seas*, Smithsonian Institution Libraries http://www.sil.si.edu/Exhibitions/Underwater-web/uw-optic-02.htm; James Gleick, The Information: A History, a Theory, a Flood 129–36 (Pantheon, 2011).

2. One renowned study of these early years concludes that before AT&T's Bell System cemented its lock on most major markets, the competition for market dominance prompted rival telephone companies to build out infrastructure throughout population centers as quickly as possible, with quite significant consumer benefits. *See* Milton Mueller, Universal Service: Competition, Interconnection, and Monopoly in the Making of the American Telephone System (AEI Press, 1997). That may well be so, but there is broad consensus that such competition, left to its own devices, was likely in the end to produce a single dominant provider in any given market.

3. *See generally* Gerald W. Brock, Telecommunication Policy for the Information Age 65–66 (Harv. Univ. Press, 1994).

4. *See* United States v. Microsoft Corp., 84 F. Supp. 2d 9, 18-23 (D.D.C. 1999), *aff'd in relevant part*, 253 F.3d 34, 55–56 (D.C. Cir. 2001).

5. For a classic exposition of such network-effects phenomena throughout the economy, see Carl Shapiro & Hal Varian, Information Rules: A Strategic Guide to the Network Economy 173–225 (Harv. Bus. School Press, 1998).

6. The relationship between monopoly and innovation is complex and is the subject of extensive academic analysis. *See, e.g.*, Jonathan B. Baker, *Beyond Schumpeter vs. Arrow: How Antitrust Fosters Innovation*, 74 Antitrust L.J. 575 (2007); Michael L. Katz & Howard A. Shelanski, *Mergers and Innovation*, 74 Antitrust L.J. 1 (2007); Richard Gilbert, *Looking for Mr. Schumpeter: Where Are We in the Competition–Innovation Debate?* 6 Innovation Policy and the Economy 159 (2006).

7. Joseph A. Schumpeter, Capitalism, Socialism, and Democracy 81–90 (Harper & Bros., 2d ed. 1947). For a discussion of the Schumpeterian perspective, see Philip J. Weiser, *The Internet, Innovation, and Intellectual Property Policy*, 103 Colum. L. Rev. 534, 576–583 (2003).

8. *See* Howard A. Shelanski & J. Gregory Sidak, *Antitrust Divestiture in Network Industries*, 68 U. Chi. L. Rev. 1, 10–11 (2001); Richard Schmalensee, *Antitrust Issues in Schumpeterian Industries*, 90 Am. Econ. Rev. 192, 194 (2000); Clayton M. Christensen, The Innovator's Dilemma (Harv. Bus. School Press, 1997) (discussing related concept of "disruptive technology"). Richard Posner aptly sums up the Schumpeterian view of the digital economy: "The gale of creative destruction that Schumpeter described, in which a sequence of temporary monopolies operates to maximize innovation that confers social benefits far in excess of the social costs of the short-lived monopoly prices that the process also gives rise to, may be the reality of the new economy." Richard A. Posner, *Antitrust in the New Economy*, 68 Antitrust L.J. 925, 930 (2001); *see also* J. Bradford DeLong, *Creative Destruction's Reconstruction: Joseph Schumpeter Revisited*, Chron. Higher Educ., Dec. 7, 2007, at B8.

9. William Baumol explains the distinction between fixed and sunk costs:

There is considerable confusion in the literature about two pertinent concepts, fixed costs and sunk costs, which are really very different. . . . [F]*ixed* costs are costs that must be incurred in a lump in order for any output at all to be provided, and they do not vary when the magnitude of output changes. These costs are not variable either in the short or the long run. Any cost that is not fixed is defined to be *variable*. A *sunk* cost, however, is a cost that cannot be avoided for some limited period of time, but after that period it becomes *avoidable* or *escapable*. A cost that is fixed may or may not be sunk, and a cost that is sunk may not be fixed. For example, one cannot operate an airline between, say, New York and Milwaukee without investing in at least one airplane, an outlay whose amount does not vary with number of passengers until capacity is reached. Thus, this cost is fixed, and does not become variable even in the long run, because one cannot run an airline on the route with zero airplanes. In contrast, this cost is not sunk because, if traffic between New York and Milwaukee declines drastically, the plane can be shifted to serve another route. (William J. Baumol, *Predation and the Logic of the Average Variable Cost Test*, 39 J.L. & Econ. 49, 57–58 n.13 (1996).)

10. 2 Alfred E. Kahn, The Economics of Regulation: Principles and Institutions 119 (MIT Press, 1988).

11. *See generally* W. Kip Viscusi et al., Economics of Regulation and Antitrust 337–60 (MIT Press, 3d ed. 2000).

12. Omega Satellite Prods. Co. v. City of Indianapolis, 694 F.2d 119, 126 (7th Cir. 1982).

13. *See, e.g.*, Warren G. Lavey, *The Public Policies That Changed the Telephone Industry into Regulated Monopolies: Lessons from Around 1915*, 39 Fed. Comm. L.J. 171 (1987).

14. *See* Daniel A. Farber & Philip P. Frickey, *The Jurisprudence of Public Choice*, 65 Texas L. Rev. 873 (1987); Dennis C. Mueller, Public Choice (Cambridge Univ. Press, 3d ed. 2003).

15. For a classic explanation of this point, see Richard Posner, *Taxation by Regulation*, 2 Bell J. Econ. & Mgmt. Sci. 22 (1971).

16. To some extent, this was because regulators became more self-consciously familiar with public choice theory's accounts of the many respects in which regulation can help individual industry participants but harm the consumers it is designed to serve. *See generally* Joseph D. Kearney & Thomas W. Merrill, *The Great Transformation of Regulated Industries Law*, 98 COLUM. L. REV. 1323, 1384, 1397 (1998).

17. *See, e.g.*, Ex Parte Submission of the United States Department of Justice, *Economic Issues in Broadband Competition, a National Broadband Plan for Our Future*, GN Docket No. 09-51, at 15 (FCC filed Jan. 4, 2010), http://www.justice.gov/atr/public/comments/253393.htm (*"DoJ Broadband Plan Comments"*) (competition policy "starts from the presumption that in highly concentrated markets consumers can be significantly harmed when the number of strong competitors declines from four to three, or three to two").

18. *See, e.g.*, Susan P. Crawford, *The Communications Crisis in America*, 5 HARVARD L. & POL'Y REV. 245, 248, 261 (2011) ("Given the tremendous economies of scale and cost advantages of the cable industry, being a wireline phone company is not a great business these days . . . . The emergence of a de facto cable monopoly in high-speed wired Internet access in most of the country cannot stay a secret"); Craig Moffett et al., *U.S. Cable and U.S. Telecommunications: Broadband End Game?* Bernstein Research, at 1, 7 (2010) ("[C]able's advantaged infrastructure will win the broadband wars. . . . Cable's share of 2Q 2010 net broadband additions rose steeply, to 91.4%, versus 67% in the prior quarter and a mere 41% in the year-ago quarter").

19. Jim Barthold, *Bye-bye FiOS? Verizon Winding Down Its FTTH Rollout*, FierceCable (Mar. 29, 2010), http://www.fiercecable.com/story/bye-bye-fios-verizon-winding-down-its-ftth-rollout/2010-03-29 ("It turns out those cable guys who predicted that FiOS was too expensive, even for a deep-pocketed telco like Verizon, were right all along").

20. *See* Memorandum Opinion & Order and Declaratory Ruling, *Applications of Cellco Partnership d/b/a Verizon Wireless and SpectrumCo LLC and Cox TMI, LLC for Consent to Assign AWS-1 Licenses*, WT Docket No. 12-4, FCC No. 12-95, ¶¶ 143–58 (Aug. 23, 2012) (describing remedial measures).

21. *See, e.g.*, Susan Crawford, *Verizon to the Cable Industry: Let's Be Friends*, BLOOMBERG (Mar. 20, 2012), http://www.bloomberg.com/news/2012-03-20/verizon-to-the-cable-industry-let-s-be-friends.html.

22. *See* Marisa Plumb, *Copper at the Speed of Fiber?* IEEE SPECTRUM (Oct. 2011), http://spectrum.ieee.org/telecom/internet/copper-at-the-speed-of-fiber (noting that with new "vectoring" techniques, telcos can "push broadband speeds over 100 Mbps" using "legacy telephone access networks"); Alcatel-Lucent, *Get to Fast, Faster* (Strategic White Paper), at 2 (2011), http://www.alcatel-lucent.com/vdsl2-vectoring/ (click on link under "Resources") (projecting that with VDSL2 vectoring technology, "copper can easily meet the bandwidth demand curve" projected through 2020).

23. *See* RICHARD A. POSNER, ANTITRUST LAW 223–29 (Univ. of Chicago Press, 2d ed. 2001); Herbert Hovenkamp, *Antitrust Policy after Chicago*, 84 MICH. L.

Rev. 213, 255–83 (1985). For comprehensive treatments of this issue as it applies to information industries, see Joseph Farrell & Philip J. Weiser, *Modularity, Vertical Integration, and Open Access Policies: Towards a Convergence of Antitrust and Regulation in the Internet Age*, 17 Harv. J.L. & Tech. 85, 104 (2003), and Christopher S. Yoo, *Vertical Integration and Media Regulation in the New Economy*, 19 Yale. J. on Reg. 171 (2002).

24. *See* United States v. Microsoft Corp., 253 F.3d 34 (D.C. Cir. 2001). Indeed, the reported ambition of Netscape chief technologist Marc Andreessen was to "reduce Windows to a set of poorly debugged device drivers." Ken Auletta, World War 3.0: Microsoft and Its Enemies 82 (Random House, 2001). Netscape CEO James Barksdale sounded a more nuanced tone at Microsoft's antitrust trial, acknowledging only that Netscape believed it could "substitute for some of the characteristics" of Windows. *Id.*; *cf.* Thomas Hazlett, *US v Microsoft: Who Really Won*, FT.com, Jan. 28, 2008 (expressing skepticism about the government's Netscape-related theory of the case).

25. United States v. Southwestern Cable Co., 392 U.S. 157 (1968).

26. *See* 47 U.S.C. § 332.

27. *See* Stephen Blumberg & Julian Luke, *Wireless Substitution: Early Release of Estimates from the National Health Interview Survey, January—June 2011*, Centers for Disease Control and Prevention (Dec. 21, 2011), http://www.cdc.gov/nchs/data/nhis/earlyrelease/wireless201112.htm; Memorandum Opinion and Order, *Verizon Communications Inc. and MCI, Inc. Applications for Approval for Transfer of Control*, 20 FCC Rcd 18433 ¶91 (2005) (citing 2004 data).

28. Indus. Anal. & Tech. Div., Wireline Competition Bur., *Local Telephone Competition: Status as of June 30, 2010*, at 4 (FCC Mar. 2011).

29. FCC, *Connecting America: The National Broadband Plan* (2010), http://download.broadband.gov/plan/national-broadband-plan.pdf ("*National Broadband Plan*"); Report & Order and Further Notice of Proposed Rulemaking, *Connect America Fund*, 26 FCC Rcd 17663 (2011).

30. *See* National Cable & Telecommunications Ass'n v. Brand X Internet Servs., 545 U.S. 967 (2005) (affirming reasonableness of FCC determination).

31. Comcast Corp. v. FCC, 600 F.3d 642, 659 (D.C. Cir. 2010).

32. *National Broadband Plan* at 37.

33. *Id.* at 38.

34. Remarks of FCC Chairman Julius Genachowski, CTIA Wireless 2011 Convention, at 9 (Mar. 22, 2011), http://hraunfoss.fcc.gov/edocs_public/attachmatch/DOC-305309A1.pdf.

35. *See generally DoJ Broadband Plan Comments* at 15 ("[C]onsumers can enjoy substantial benefits when the number of strong competitors rises from two to three, or three to four, especially if the additional competitor offers products based on a new and distinct technology. Developments in both the MVPD and the wireless markets over the past 15 years underscore this point.").

Chapter 2

1. *See* Alcatel-Lucent, *Get to Fast, Faster* (Strategic White Paper), at 2 (2011), http://www.alcatel-lucent.com/vdsl2-vectoring/ (click on link under "Resources"); *see also* Marisa Plumb, *Copper at the Speed of Fiber?* IEEE Spectrum (Oct. 2011), http://spectrum.ieee.org/telecom/internet/copper-at-the-speed-of-fiber ("Vectoring is a technique that reduces copper-wire interference simultaneously for multiple customers in order to push broadband speeds *over 100 Mbps*. While running fiber all the way to the home can produce speeds of several hundred megabits per second, VDSL2 vectoring can use legacy telephone access networks") (emphasis added); Michelle Amodio, *The New Evolution of Copper: Getting Faster, Not Pricier*, TMCnet (Sept. 23, 2011), http://next-generation-communications.tmcnet.com/topics/high-leverage-network/articles/221522-new-evolution-copper-getting-faster-not-pricier.htm ("Alcatel-Lucent is taking existing copper access networks and making them better and faster—not pricier. . . . Using new DSL technologies like VDSL2 Vectoring and VDSL2 Bonding, providers can deliver 100Mbps and beyond per subscriber").

2. This figure is expressed as a 4 followed by 16 zeros. In mathematical terms, if N is the number of subscribers, the number of lines needed to link each with every other equals $N(N - 1)/2$.

3. In contrast, *connection-oriented* packet-switching technologies such as asynchronous transfer mode (ATM) set up *virtual circuits* by allocating packets associated with a particular communication to preassigned slots within a packet flow. For simplicity and at the risk of some imprecision, we use the term *packet switched* to refer to connectionless technologies except where otherwise indicated. Significantly, all of the technologies discussed here—connection-oriented and connectionless packet switching as well as all modern circuit switches—are digital rather than analog; the differences among them lie in how a switching device processes digital packets.

4. *See generally* Jean-Jacques Laffont & Jean Tirole, Competition in Telecommunications 29–35 (MIT Press, 2000).

5. *See, e.g.*, Paul Starr, The Creation of the Media 205–212 (Basic Books, 2004).

6. Verizon Communications Inc. v. FCC, 535 U.S. 467, 477 (2002) (citations omitted). In 1876, the Supreme Court upheld this system of regulation as consistent with the constitutional protection of property rights, including the protection against "takings" of property without just compensation, on the theory that the regulated entities were "affected with a public interest" and that "when private property is devoted to a public use, it is subject to public regulation." Munn v. Illinois, 94 U.S. 113, 130 (1876) (internal quotation marks omitted).

7. Duquesne Light Co. v. Barasch, 488 U.S. 299, 312 (1989).

8. *Verizon Communications*, 535 U.S. 467.

9. *See generally* United States Telecom Ass'n v. FCC, 188 F.3d 521 (D.C. Cir. 1999).

10. *See, e.g.*, W. Kip Viscusi et al.., Economics of Regulation and Antitrust 369–70 (MIT Press, 3d ed. 2000); National Rural Telecom. Ass'n v. FCC, 988 F.2d 174 (D.C. Cir. 1993).

11. 47 U.S.C. § 152(b); *see* Louisiana Pub. Serv. Comm'n v. FCC, 476 U.S. 355 (1986).

12. 282 U.S. 133 (1930).

13. *See* Gerald W. Brock, Telecommunication Policy for the Information Age 68, 190–91 (Harv. Univ. Press, 1994) (discussing the "Ozark Plan" and subsequent separations regime).

14. *See generally* Texas Office of Pub. Util. Counsel v. FCC, 265 F.3d 313 (5th Cir. 2001); National Ass'n of Reg. Util. Comm'rs v. FCC, 737 F.2d 1095 (D.C. Cir. 1984).

15. Brock, Telecommunication Policy at 139–45, 177–89.

16. Under a longstanding FCC rule, a special-access circuit is deemed interstate in character and subject to regulation by the FCC if 10% of the traffic it carries is interstate. *See* Decision & Order, *MTS and WATS Market Structure, Amendment of Part 36 of the Commission's Rules and Establishment of a Joint Board*, 4 FCC Rcd 5660 (1989); *see generally* Qwest Corp. v. Scott, 380 F.3d 367 (8th Cir. 2004).

17. *See generally* Illinois Bell Tel. Co. v. FCC, 966 F.2d 1478 (D.C. Cir. 1992).

18. *See, e.g.*, American Tel. & Tel. Co. v. Central Office Tel., Inc., 524 U.S. 214 (1998).

19. *See* MCI WorldCom, Inc. v. FCC, 209 F.3d 760, 763–65 (D.C. Cir. 2000).

20. *See id.*

21. 2 Alfred E. Kahn, The Economics of Regulation: Principles and Institutions 119 (MIT Press, 1989).

22. Steve Coll, The Deal of the Century: The Breakup of AT&T 105 (Atheneum, 1986).

23. *See* Note, *Competition in the Telephone Equipment Industry: Beyond Telerent*, 86 Yale L.J. 538, 552–53 (1977) (describing states' fears that competition in equipment markets would threaten a valued source of cross-subsidies).

24. *Hush-A-Phone Corp.*, 20 F.C.C. 391, 420 (1955).

25. *See* Hush-A-Phone Corp. v. United States, 238 F.2d 266, 269 (D.C. Cir. 1956).

26. Decision, *Use of the Carterfone Device in Message Toll Tel. Serv.*, 13 F.C.C.2d 420 (1968).

27. *See* 47 C.F.R. § 68.1 *et seq*. These and similar FCC rules, discussed in chapter 3, explain the certifications of compliance you may have noticed on many consumer electronic devices, particularly those that use either the telephone network or wireless spectrum.

28. *See, e.g.*, Litton Systems, Inc. v. Am. Tel. & Tel. Co., 700 F.2d 785 (2d Cir. 1983).

29. 47 U.S.C. § 273. See the discussion of the section 271 process later in this chapter.

30. Coll, Deal of the Century, at 47–52; *see also* Brock, Telecommunication Policy at 124–35.

31. *See* MCI Telecomm. Corp. v. FCC, 580 F.2d 590, 597 (D.C. Cir. 1978) (reaffirming an "expansive view of the scope of the interconnection obligations of AT&T," which the court attributed to prior FCC policies, in order "to allow carriers such as MCI to enter the market and compete with AT&T"); *see also* Coll, Deal of the Century at 83–91; Brock, Telecommunication Policy, at 135–39.

32. *See* Report & Order, *Resale and Shared Use of Common Carrier Domestic Public Switched Network Services*, 83 F.C.C.2d 167 (1980); Report & Order, *Resale and Shared Use of Common Carrier Services and Facilities*, 60 F.C.C.2d 261 (1976), *aff'd*, AT&T v. FCC, 572 F.2d 17 (2d Cir. 1978).

33. Coll, Deal of the Century, at 52.

34. *See* United States v. American Tel. & Tel. Co., 552 F. Supp. 131 (D.D.C. 1982), *aff'd sub nom.* Maryland v. United States, 460 U.S. 1001 (1983); *see also* MCI Communications Corp. v. American Tel. & Tel. Co., 708 F.2d 1081 (7th Cir. 1983).

35. *See* United States v. Western Elec. Co., 993 F.2d 1572, 1580–81 (D.C. Cir. 1993); California v. FCC, 39 F.3d 919, 926–27 (9th Cir. 1994).

36. *See* Southwestern Bell Tel. Co. v. FCC, 153 F.3d 523, 548 (8th Cir. 1998) (dismissing "price squeeze" concerns); Supplemental Order Clarification, *Implementation of the Local Competition Provisions of the Telecommunications Act of 1996*, 15 FCC Rcd 9587, ¶¶ 19–20 (2000) (same), *aff'd*, Competitive Telecommunications Ass'n v. FCC, 309 F.3d 8 (D.C. Cir. 2002).

37. *See* Marius Schwartz, *The Economic Logic for Conditioning Bell Entry into Long Distance on the Prior Opening of Local Markets*, 18 J. Reg. Econ. 247, 286 (2000).

38. Access charges are keyed to particular links within the incumbent's network. To the extent that a long-distance carrier arranges to circumvent one of those links, it avoids the obligation to pay the associated access charge to the incumbent. *See* Declaratory Ruling, *Implementation of the Local Competition Provisions in the Telecommunications Act of 1996*, 14 FCC Rcd 3689, ¶ 9 (1999) ("When two carriers jointly provide interstate access (e.g., by delivering a call to an interexchange carrier (IXC)), the carriers will share access revenues received from the interstate service provider"), *vacated and remanded on other grounds*, Bell Atl. Tel. Cos. v. FCC, 206 F.3d 1 (D.C. Cir. 2000); *see also* Memorandum Opinion and Order, *Waiver of Access Billing Requirements and Investigation of Permanent Modifications*, 2 FCC Rcd 4518, ¶ 2 (1987).

39. *See* Report & Order and Notice of Proposed Rulemaking, *Expanded Interconnection with Local Telephone Company Facilities, Amendment of Part 69 Allocation of General Support Facility Costs*, 7 FCC Rcd 7369 (1992); Second Report & Order and Third Notice of Proposed Rulemaking, *Expanded*

Interconnection with Local Telephone Company Facilities, Amendment of Part 36 of the Commission's Rules and Establishment of a Joint Board, 8 FCC Rcd 7374 (1993). The D.C. Circuit invalidated the "physical collocation" requirements of these orders in 1994 (*see* Bell Atl. Tel. Cos. v. FCC, 24 F.3d 1441 (D.C. Cir. 1994)), but that ruling was itself superseded by the 1996 Act, as discussed later in this chapter.

40. *See* Verizon Tel. Cos. v. FCC, 292 F.3d 903, 906–07 (D.C. Cir. 2002) (discussed later in this chapter).

41. *See* Department of Justice, *AG Unveils Plan to Allow Ameritech in Long Distance Market*, press release (Apr. 3, 1995), http://www.justice.gov/opa/pr/Pre_96/April95/186.txt.html.

42. Pub. L. No. 104-104, 110 Stat. 56 (1996) (codified as amended in scattered sections of 47 U.S.C.).

43. H.R. Rep. No. 104-458, at 113 (1996) (Conf. Rep.), *reprinted in* 1996 U.S.C.C.A.N. 124. As we explain in chapter 10, not everyone agrees that the 1996 Act, at least as implemented, is "deregulatory." *See, e.g.*, Alfred E. Kahn, Letting Go: Deregulating the Process of Deregulation (MSU Public Utilities, 1998).

44. 47 U.S.C. § 253. Although the principle is clear, its application is often not. For example, courts have disagreed about the relationship among the subsections of section 253 and the showing a carrier must make to demonstrate that a municipal ordinance "ha[s] the effect of prohibiting [its] ability" to offer service. *See, e.g.*, Level 3 Communications, L.L.C. v. St. Louis, 477 F.3d 528 (8th Cir. 2007) (noting conflicts).

45. 1996 Act, Pub. L. 104-104, § 302(b)(1) (repealing 47 U.S.C. § 533(b)); *see* Chesapeake & Potomac Tel. Co. of Va. v. United States, 42 F.3d 181 (4th Cir. 1994) (invalidating section 533(b) as First Amendment violation), *vacated as moot*, 516 U.S. 415 (1996).

46. There was no corresponding bar on the Bell companies' provision of long-distance service as part of a *wireless* plan because the Bells had never monopolized the wireless market. *See* 47 U.S.C. § 271(b)(3), (g)(3).

47. 47 U.S.C. § 160(a) (codifying new section 10 of the Communications Act). As noted in chapter 4, this provision builds on a narrower forbearance provision that Congress had added in 1993 to free mobile wireless carriers from undue regulation. The new, broader authority in section 10 extends to all requirements under Title II of the Communications Act; it is unclear whether it extends to requirements arising from elsewhere in the Act.

48. *See, e.g.*, MCI WorldCom, 209 F.3d at 762–63; Earthlink v. FCC, 462 F.3d 1 (D.C. Cir. 2006).

49. 47 U.S.C. § 153(53) (defining "telecommunications service"); *see also* 47 U.S.C. § 153(50) (defining "telecommunications").

50. *See, e.g.*, Virgin Is. Tel. Corp. v. FCC, 198 F.3d 921 (D.C. Cir. 1999); *see also* 47 U.S.C. § 153(51) ("a telecommunications carrier shall be treated as a common carrier under this [Act] only to the extent that it is engaged in providing telecommunications services").

51. *See, e.g.*, Iowa v. FCC, 218 F.3d 756, 759 (D.C. Cir. 2000); National Ass'n of Reg. Util. Comm'rs v. FCC, 533 F.2d 601, 608–09 (D.C. Cir. 1976) ("*NARUC II*").

52. FCC v. Midwest Video Corp., 440 U.S. 689, 701 (1979); *see, e.g.*, *Virgin Is. Tel.*, 198 F.3d at 925.

53. *See generally* National Ass'n of Reg. Util. Comm'rs v. FCC, 525 F.2d 630 (D.C. Cir. 1976) ("*NARUC I*"); Declaratory Ruling, *NORLIGHT Request for a Declaratory Ruling*, 2 FCC Rcd 132, ¶¶ 19–21 (1987).

54. National Cable & Telecommunications Ass'n v. Brand X Internet Servs., 545 U.S. 967 (2005).

55. *See* 47 U.S.C. § 153(24) (defining "information service" as "the offering of a capability for generating, acquiring, storing, transforming, processing, retrieving, utilizing, or making available information via telecommunications").

56. 47 U.S.C. § 153(32); *see also* 47 U.S.C. § 153(20), (54).

57. 47 U.S.C. § 153(32), (51).

58. *See* 47 U.S.C. § 251(h) (defining "incumbent local exchange carrier").

59. Technically, the Regional Bell Operating Companies are the holding companies created as a result of the consent decree, each of which inherited a number of formally separate BOCs. *See generally* 47 U.S.C. § 153(5) (listing the "Bell operating companies"). Today, however, when people speak of BOCs or Bell companies, they are generally referring to the holding companies.

60. *See* 47 U.S.C. § 251(f)(2).

61. This LEC-specific provision is distinct from the main *pole attachment* provision of section 224, which grants cable operators and telecommunications carriers various rate and other protections when they negotiate with "utilities," including electric power companies, about the terms of access to the utilities' poles and conduits. *See* 47 U.S.C. § 224. In 2011, the FCC issued a major order under section 224, concluding that ILECs can claim certain protections under that provision and taking steps to ensure that providers of "telecommunications services" can obtain roughly the same favorable rate levels as pure providers of conventional cable television services. *See, e.g.*, Report & Order, *Implementation of Section 224 of the Act*, 26 FCC Rcd 5240 (2011), *aff'd*, American Elec. Power Serv. Corp. et al. v. FCC et al., No. 11-1146 (D.C. Cir., Feb. 26, 2013). *See generally* National Cable & Telecommunications Ass'n v. Gulf Power Co., 534 U.S. 327 (2002) (upholding favorable rates for Internet services).

62. *See* Verizon Tel. Cos. v. FCC, 292 F.3d 903, 906–07 (D.C. Cir. 2002).

63. The statutory reference to "unbundled" access therefore does *not* mean that an ILEC may unilaterally disconnect requested facilities from one another simply for the purpose of inconveniencing the competitor, a point the Supreme Court confirmed in 1999. AT&T Corp. v. Iowa Utils. Bd., 525 U.S. 366, 394–95 (1999).

64. 47 U.S.C. § 153(35) (defining "network element"); *see also Iowa Utils. Bd.*, 525 U.S. at 386–87.

65. The same provision imposes a somewhat higher hurdle (the so-called "necessary" standard) to the imposition of unbundling obligations for "proprietary"

elements—that is, those that cannot be shared without presenting a theoretical risk of compromising the ILEC's business secrets. Very few elements fall into this category, and the FCC's analysis under section 251(d)(2) almost always involves application of the "impairment" standard.

66. *See* United States Telecom Ass'n v. FCC, 359 F.3d 554, 572, 579–82 (D.C. Cir. 2004) ("*USTA II*").

67. *See* 47 C.F.R. § 51.505. In 2003, the FCC opened a far-reaching inquiry into proposed methodological reforms for TELRIC. *See* Notice of Proposed Rulemaking, *Review of the Commission's Rules Regarding the Pricing of Unbundled Network Elements and the Resale of Service by Incumbent Local Exchange Carriers*, 18 FCC Rcd 20265 (2003). That proceeding promptly lost much of its commercial significance when, in a 2004 judicial decision discussed later in this chapter, the D.C. Circuit limited the network elements that are subject to leasing obligations in the first place. The TELRIC reform proceeding remains pending but ignored even as this second edition goes to press nine years later.

68. 535 U.S. 467 (2002).

69. Iowa Utils. Bd. v. FCC, 120 F.3d 753, 800 (8th Cir. 1997). For a description of the traditional dual-jurisdiction regime, see *Louisiana PSC*, 476 U.S. at 355.

70. 525 U.S. 366 (1999).

71. *Id.* at 378. In an oft-quoted footnote, the majority then dismissed the dissent's view that close questions about the FCC's preemptive jurisdiction should be resolved on the basis of federalism concerns in general and section 2(b) in particular:

[T]he question . . . is not whether the Federal Government has taken the regulation of local telecommunications competition away from the States. With regard to the matters addressed by the 1996 Act, it unquestionably has. The question is whether the state commissions' participation in the administration of the new *federal* regime is to be guided by federal-agency regulations. If there is any "presumption" applicable to this question, it should arise from the fact that a federal program administered by 50 independent state agencies is surpassing strange. The appeals by both Justice [Clarence] THOMAS and Justice [Stephen] BREYER to what might loosely be called "States' rights" are most peculiar, since there is no doubt, even under their view, that if the federal courts believe a state commission is not regulating in accordance with federal policy they may bring it to heel. This is, at bottom, a debate not about whether the States will be allowed to do their own thing, but about whether it will be the FCC or the federal courts that draw the lines to which they must hew. To be sure, the FCC's lines can be even more restrictive than those drawn by the courts—but it is hard to spark a passionate "States' rights" debate over that detail. (*Id.* at 378–79 n.6 (paragraph break omitted).)

72. For a discussion of the different models, see Philip J. Weiser, *Federal Common Law, Cooperative Federalism, and the Enforcement of the Telecom Act*, 76 N.Y.U. L. REV. 1692 (2001).

73. At least in principle, the traditional dual-jurisdiction framework still applies to "intrastate" matters *outside* the scope of the 1996 Act. *Iowa Utils. Bd.*, 525

U.S. at 381–82 n.8 ("After the 1996 Act, § 152(b) may have less practical effect. But that is because Congress, by extending the Communications Act into local competition, has removed a significant area from the States' exclusive control. Insofar as Congress has remained silent, however, § 152(b) continues to function"). Even that framework, however, is subject to the so-called "impossibility exception" set forth in footnote 4 of *Louisiana PSC*, under which the FCC may preempt state laws that frustrate federal objectives relating to matters that are inseverably both interstate and intrastate in character. *See Louisiana PSC*, 476 U.S. at 376 n.4; Memorandum Opinion and Order, *Vonage Holdings Corp. Petition for Declaratory Ruling Concerning an Order of the Minn. Pub. Utils. Comm'n*, 19 FCC Rcd 22404 (2004), *aff'd* Minnesota Pub. Utils. Comm'n v. FCC, 483 F.3d 570 (8th Cir. 2007).

74. *See* 47 U.S.C. § 252(e)(2)(A).

75. *See* 47 U.S.C. § 252(b), (c). Many of the important regulatory decisions under sections 251 and 252 are made not in the context of a specific arbitration proceeding between two carriers, but in so-called generic rulemaking proceedings that address the obligations of incumbents to *all* competitors. For example, state commissions do not necessarily decide in specific arbitration proceedings the rates that a given competitor must pay the incumbent for leasing the latter's network facilities. Instead, state commissions may hold lengthy "cost proceedings" in which they decide, among other things, the rates that any carrier must pay for access to the incumbent's loops, switches, and transport facilities. The state commissions then include those rates as terms in any arbitrated interconnection agreement.

76. *See* 47 U.S.C. § 252(e)(6). In 2002, the Supreme Court rejected Eleventh Amendment challenges to the exercise of federal court jurisdiction against named state commissioners under the *Ex parte Young* doctrine. Verizon Md. Inc. v. Pub. Serv. Comm'n of Md., 535 U.S. 635, 645–48 (2002).

77. First Report & Order, *Implementation of the Local Competition Provisions of the Telecommunications Act of 1996*, 11 FCC Rcd 15499 (1996) ("*Local Competition Order*").

78. 47 U.S.C. § 251(d)(2).

79. 47 U.S.C. §§ 251(c)(4), 252(d)(3); *see also* Iowa Utils. Bd. v. FCC, 219 F.3d 744, 754–56 (8th Cir. 2000) (finding that ILECs "avoid" only a limited portion of their retail-specific costs when they lose retail customers to resellers), *rev'd in other respects*, Verizon Communications Inc. v. FCC, 535 U.S. 467 (2002).

80. In addition to the difference in pricing methodologies, section 251(c)(4) is expressly limited to local services that the incumbent "provides at retail to subscribers who are not telecommunications carriers." By definition, that category excludes carrier-to-carrier "access services"—the task of connecting long-distance companies to their customers for the origination and termination of long-distance calls. Thus, if a long-distance carrier chose the section 251(c)(4) resale option as a means of providing a complete package of local and long-distance services, it would need to purchase both the ILEC's local retail service at the avoided-cost discount and "access" from the ILEC for the origination and

termination of long-distance calls, traditionally at rates above the cost of that service as measured by TELRIC. In contrast, if the carrier chose the UNE-P option instead, it could lease, at TELRIC, the underlying facilities without service-related restrictions and use them to provide both local exchange services and access services. Thus, unlike the resale option, the UNE-P option both (1) removed any need to pay access charges to the ILEC for outgoing long-distance calls and (2) entitled the CLEC to collect access charges of its own for incoming long-distance calls delivered by unaffiliated carriers. Because access charges traditionally exceeded any rigorous measure of "cost," that factor tended to make the UNE-P option more attractive than the resale alternative.

81. *Iowa Utils. Bd.*, 525 U.S. at 389–90.

82. Third Report & Order, *Implementation of the Local Competition Provisions of the Telecommunications Act of 1996*, 15 FCC Rcd 3696 (1999) ("*UNE Remand Order*").

83. United States Telecom Ass'n v. FCC, 290 F.3d 415 (D.C. Cir. 2002) ("*USTA I*").

84. *Id.* at 429.

85. *Id.* at 424.

86. *Id.* at 427 (emphasis added).

87. *Id.* (emphasis added).

88. Report & Order and Order on Remand and Further Notice of Proposed Rulemaking, *Review of the Section 251 Unbundling Obligations of Incumbent Local Exchange Carriers*, 18 FCC Rcd 16978 (2003) ("*Triennial Review Order*").

89. United States Telecom Ass'n v. FCC, 359 F.3d 554 (D.C. Cir. 2004). Even the choice of the D.C. Circuit as the forum for this review was fiercely contested. The various review petitions (including the CLECs') were assigned by lottery to the Eighth Circuit, where the CLECs hoped to obtain a more sympathetic hearing, but the ILECs persuaded that court to transfer the case back to the D.C. Circuit on the ground that the case was substantially related to *USTA I*.

90. *Id.* at 565.

91. *Id.* at 566.

92. *Id.* at 565.

93. *Id.* at 565–66.

94. *Id.* at 566.

95. *See* Philip J. Weiser, *Chevron, Cooperative Federalism, and Telecommunications Reform*, 52 Vand. L. Rev. 1 (1999).

96. *USTA II*, 359 F.3d at 569.

97. *Id.* at 570.

98. On remand, the FCC formally eliminated rights to mass market unbundled switching the next year. *See* Order on Remand, *Unbundled Access to Network Elements*, 20 FCC Rcd 2533, ¶¶ 199–228 (2005), *aff'd*, Covad Communications Co. v. FCC, 450 F.3d 528 (D.C. Cir. 2006).

99. *See* Memorandum Opinion & Order, *Verizon Communications Inc. and MCI, Inc. Applications for Approval for Transfer of Control*, 20 FCC Rcd 18433 (2005); Memorandum Opinion & Order, *SBC Communications Inc. & AT&T Corp. Applications for Approval for Transfer of Control*, 20 FCC Rcd 18290 (2005).

100. See Memorandum Opinion & Order, *AT&T Inc. and BellSouth Corp. Application for Transfer of Control*, 22 FCC Rcd 5662, app. F (2007) (*"AT&T–BellSouth Merger Order"*).

101. The obligations discussed in this chapter likewise relate to the facilities that ILECs must lease to *telecommunications carriers*, not to the services that ILECs were required to provide to unaffiliated *information service providers* (including ISPs) under the pre–1996 Act FCC regulations known as the *Computer Inquiry* rules (discussed in chapter 6). The facilities-leasing rights discussed here are circumscribed not just by section 251(d)(2), but also by the clause of section 251(c)(3) restricting the class of lessors to "telecommunications carrier[s] for the provision of . . . telecommunications service[s]." Because ISPs are generally deemed to fall outside this class, they benefit from any section 251 leasing rights only indirectly by purchasing telecommunications services from CLECs that exercise those rights directly.

102. *See* 47 U.S.C. § 251(h) (defining ILECs essentially as legacy telephone monopolists).

103. State regulators often (though not invariably) construed the applicable cost-allocation principles to preclude the ILEC from charging anything to the CLEC that leased the high-frequency portion of the loop, reasoning that ILECs were already recovering loop costs as a whole through their retail rates for voice services.

104. *USTA I*, 290 F.3d at 428–30.

105. *Id.* at 429.

106. *Triennial Review Order* ¶¶ 255–69.

107. 18 FCC Rcd at 17505 (separate statement of Chairman Michael K. Powell, approving in part and dissenting in part); *see also id.* at 17521–22 (separate statement of Commissioner Kathleen Q. Abernathy, approving in part and dissenting in part). In a similar vein, Alfred Kahn explained that the sunk nature of the copper loop infrastructure "would seem to [present] the archetypal case for mandatory sharing—a heritage of [the ILECs'] franchised monopolies, the sharing of which would therefore not seem to involve any discouragement of future risk-taking investment." Alfred E. Kahn, *Regulatory Politics as Usual*, Pol'y Matters (Mar. 1, 2003), http://www.aei.org/article/regulatory-politics-as-usual.

108. *USTA II*, 359 F.3d at 584–85.

109. *See Triennial Review Order* ¶¶ 285–97.

110. *See* Order on Reconsideration, *Review of the Section 251 Unbundling Obligations of Incumbent Local Exchange Carriers*, 19 FCC Rcd 20293 (2004); *see also* Order on Reconsideration, *Review of the Section 251 Unbundling*

Obligations of Incumbent Local Exchange Carriers, 19 FCC Rcd 15856 (2004) (extending same rules to fiber loops serving multitenant residential buildings).

111. In more technical terms, the *Triennial Review Order* provides that CLECs cannot invoke section 251 unbundling rights to use the fiber portion of an incumbent's hybrid loop for the transmission of packetized data from an end user all the way to the central office. Instead, if they want to provide DSL services of their own to customers served by hybrid loops, they will have to build their own fiber links to the remote terminal and install their own electronic equipment there alongside the ILEC's—which can sometimes be a costly proposition subject to various logistical obstacles. At the same time, the Commission required ILECs, as before, to continue providing a direct *circuit-switched* pathway from end users all the way to the central office over the ILEC's legacy TDM electronics also located at the remote terminal. *See Triennial Review Order* ¶¶ 288–89. But ILECs have no obligation to build TDM capability into networks that have never had it.

112. *Id.* ¶ 286.

113. *See id.* ¶ 291.

114. Pub. L. 104-104, Title VII, § 706(a), Feb. 8, 1996, 110 Stat. 153 (codified at 47 U.S.C. § 1302).

115. *Triennial Review Order* ¶ 288.

116. *Id.* ¶¶ 295, 272.

117. *Id.* ¶ 292.

118. *See*, *e.g.*, *USTA II*, 359 F.3d at 581, 583.

119. *Id.* at 580.

120. *Id.* at 582 (emphasis added).

121. *Id.* at 585. The FCC's decision to eliminate these leasing obligations under section 251 did not of itself eliminate the Bell companies' vaguer obligations to lease similar functionalities as a continuing condition under section 271. In 2004, the Commission separately eliminated the latter obligations. *See* Memorandum Opinion and Order, *Petition for Forbearance of the Verizon Telephone Companies Pursuant to 47 U.S.C. § 160(c)*, 19 FCC Rcd 21496 (2004), *aff'd*, Earthlink, Inc. v. FCC, 462 F.3d 1 (D.C. Cir. 2006).

122. *See*, *e.g.*, Memorandum Opinion & Order, *Petition of Qwest Communications Int'l Inc. for Forbearance from Enforcement of the Commission's Dominant Carrier Rules as They Apply after Section 272 Sunsets*, 22 FCC Rcd 5207 (2007).

123. *See generally* 47 U.S.C. § 160 (discussed earlier in the chapter).

124. Memorandum Opinion and Order, *Petition of Qwest Corp. for Forbearance in the Omaha Metropolitan Statistical Area*, 20 FCC Rcd 19415 (2005), *review dismissed in part and denied in part*, Qwest Corp. v. FCC, 482 F.3d 471 (D.C. Cir. 2007); Memorandum Opinion & Order, *Petition of ACS of Anchorage for Forbearance from Sections 251(c)(3) and 252(d)(1)*, 22 FCC Rcd 1958 (2007),

appeals dismissed, Covad Communications Group v. FCC, Nos. 07-70898 et al. (9th Cir. 2007).

125. Memorandum Opinion and Order, *Petition of Qwest Corp. for Forbearance Pursuant to 47 U.S.C. §160(c) in the Phoenix, Arizona Metropolitan Statistical Area*, 25 FCC Rcd 8622 (2010) ("*Phoenix Forbearance Order*"), *aff'd*, Qwest Corp. v. FCC, 689 F.3d 1214 (2012); Memorandum Opinion and Order, *Petitions of the Verizon Tel. Cos. for Forbearance in the Boston, New York, Philadelphia, Pittsburgh, Providence and Virginia Beach MSAs*, 22 FCC Rcd 21293 (2007), *remanded*, Verizon Tel. Cos. v. FCC, 570 F.3d 294 (D.C. Cir. 2009); Order, *Petitions of Qwest Corp. for Forbearance in the Denver, Minneapolis–St. Paul, Phoenix, and Seattle MSAs*, 23 FCC Rcd 11729 (2008), *remanded*, Qwest Corp. v. FCC, No. 08-1257 (D.C. Cir. Aug. 5, 2009).

126. *Phoenix Forbearance Order*, 25 FCC Rcd 8622.

127. *Id.* ¶¶ 55–61.

128. *Id.* ¶ 59.

129. *Id.* ¶ 60.

130. Order on Remand, *Unbundled Access to Network Elements*, 20 FCC Rcd 2533, ¶ 36 (2005), *aff'd*, Covad Communications Co. v. FCC, 450 F.3d 528 (D.C. Cir. 2006).

131. *See Covad*, 450 F.3d at 535–36 (summarizing relevant FCC regulations).

132. Ad Hoc Telecommunications Users Comm. v. FCC, 572 F.3d 903, 904 (D.C. Cir. 2009).

133. United States Telecom Ass'n v. FCC, 188 F.3d 521 (D.C. Cir. 1999).

134. First Report & Order, *Price Cap Performance Review for Local Exchange Carriers*, 10 FCC Rcd 8961, ¶ 64 (1995).

135. Fifth Report & Order and Further Notice of Proposed Rulemaking, *Access Charge Reform; Price Cap Performance Review for Local Exchange Carriers*, 14 FCC Rcd 14221 (1999) ("*Pricing Flexibility Order*").

136. For a concise overview of the competitive triggers, see Report & Order, *Special Access for Price Cap Local Exchange Carriers*, WC Docket No. 05-25, FCC No. 12-92, at ¶ 30 n. 93 (Aug. 22, 2012) ("*2012 Special Access Order*").

137. *See* WorldCom, Inc. v. FCC, 238 F.3d 449 (D.C. Cir. 2001).

138. *E.g.*, Memorandum Opinion & Order, *Petition of AT&T for Forbearance under 47 U.S.C. § 160(c) from Title II and Computer Inquiry Rules with Respect to Its Broadband Services*, 22 FCC Rcd 18705 (2007), *aff'd*, *Ad Hoc*, 572 F.3d 903. Verizon had previously won *complete* deregulation for its services in this category—not just relief from dominant-carrier regulation—because of a statutory peculiarity. Under the forbearance provision, if the FCC does not formally deny a forbearance petition within fifteen months of its filing, the petition is "deemed granted." 47 U.S.C. § 160(c). In Verizon's case, because the five-member FCC was down one member at the fifteen-month deadline, it failed to reach a majority vote denying Verizon's forbearance petition, which sought very

expansive relief, and the petition was thus deemed granted by operation of law. *See* Sprint Nextel Corp. v. FCC, 508 F.3d 1129 (D.C. Cir. 2007).

139. Order & Notice of Proposed Rulemaking, *Petition for Rulemaking to Reform Regulation of Incumbent Local Exchange Carrier Rates for Interstate Special Access Services*, 20 FCC Rcd 1994 (2005).

140. In the early years of this debate, CLECs and ILECs disputed the precise evidentiary significance of a 2006 report issued by the U.S. Government Accountability Office (GAO), which oversees the federal government's procurement of special-access services. *See* U.S. GAO, *FCC Needs to Improve Its Ability to Monitor and Determine the Extent of Competition in Dedicated Access Services* (Nov. 2006), http://www.gao.gov/products/GAO-07-80. CLECs cited the GAO findings for the proposition that ILEC special-access services were noncompetitive because prices generally rose rather than fell after ILECs won pricing flexibility. ILECs responded that the CLECs were improperly focused on tariffed "rack rates" rather than on actual negotiated rates and that even if prices did rise after pricing flexibility was granted, it was only because pre-flexibility price caps were set below cost. CLECs also argued that ILEC books reflected astronomically high rates of return for special-access services, whereas ILECs responded that such accounting measures of profitability are economically meaningless.

141. For examples of how antitrust law analyzes product discounts, see Cascade Health Solutions v. PeaceHealth, 515 F.3d 883 (9th Cir. 2008), and LePage's Inc. v. 3M, 324 F.3d 141 (3d Cir. 2003) (en banc). *See also* Brooke Group Ltd. v. Brown & Williamson Tobacco Corp., 509 U.S. 209 (1993). The analysis varies depending on whether the discount is deemed to encompass multiple "products" or only one. *See* 3A Phillip E. Areeda & Herbert Hovenkamp, Antitrust Law, ¶ 749a at 309–310 (Aspen, 3d ed. 2008) ("[W]hen a discount is offered on a single product, such as a quantity or a market share discount, the discount is ordinarily lawful if the price after all discounts are taken into account exceeds the defendant's marginal cost or average variable cost"). Of course, regulatory agencies have far greater flexibility than courts to fashion industry-specific discount rules that protect competition without unduly precluding dominant firms from offering low-priced services.

142. As noted in chapter 4, Sprint relied on this wireless backhaul concern in 2011 as a basis for opposing the AT&T/T-Mobile merger. But neither the Justice Department nor the FCC focused on this concern in rejecting the proposed merger, in part because the merger opponents never clearly explained how the merger itself could exacerbate the concern.

143. Apart from industry-wide rulemaking proceedings, the FCC has used its merger-review authority to impose ad hoc price regulation on the largest ILECs. For example, in early 2007 the FCC conditioned its approval of the AT&T–BellSouth merger on the combined company's acquiescence in more stringent rate regulation of its special-access services. *See AT&T–BellSouth Merger Order*, Appx. F (listing "commitments"). But those conditions were temporary (and have since expired), and they were never justified by any rigorous economic analysis; they were simply concessions to AT&T's regulatory adversaries.

144. *See 2012 Special Access Order*.

Chapter 3

1. *See* Stephen Blumberg & Julian Luke, *Wireless Substitution: Early Release of Estimates from the National Health Interview Survey, January–June 2011*, Centers for Disease Control and Prevention (Dec. 21, 2011), http://www.cdc.gov/nchs/data/nhis/earlyrelease/wireless201112.htm.

2. Fourteenth Report, *Annual Assessment of the Status of Competition in the Market for the Delivery of Video Programming*, MB Docket No. 07-269, FCC No. 12-81, at ¶ 211 (July 20, 2012) ("After a steady decline over the last few years, the percentage of television households relying exclusively on over-the-air broadcast service (as opposed to access to broadcast stations via an MVPD) has remained stable since 2010," at approximately "9.6 percent (10.97 million households) at the end of 2011"); *see also* FCC, *Connecting America: The National Broadband Plan*, at 89 (2010), http://download.broadband.gov/plan/national-broadband-plan.pdf ("*National Broadband Plan*"). Other surveys have placed the percentage somewhat higher, in the mid-teens. *See*, *e.g.*, Press Release, *Over-the-Air TV Homes Now Include 46 Million Consumers*, Knowledge Networks (June 6, 2011), http://www.knowledgenetworks.com/news/releases/2011/060611_ota.html.

3. Nicholas Negroponte, *Wireless Revisited*, WIRED (Aug. 1997), http://www.wired.com/wired/archive/5.08/negroponte_pr.html.

4. Barack Obama, *Presidential Memorandum: Unleashing the Wireless Broadband Revolution*, the White House (June 28, 2010), http://www.whitehouse.gov/the-press-office/presidential-memorandum-unleashing-wireless-broadband-revolution ("*2010 Presidential Memorandum*").

5. Remarks of FCC Chairman Julius Genachowski, CTIA Wireless 2011, at 5 (Mar. 22, 2011), http://hraunfoss.fcc.gov/edocs_public/attachmatch/DOC-305309A1.pdf.

6. Some commentators—such as Marty Cooper, credited with inventing the first cellphone—suggest that mobile wireless providers do not need additional dedicated spectrum and can cope with escalating bandwidth demands simply by using their existing spectrum more efficiently. *See* Brian X. Chen, *Q.&A.: Martin Cooper, Father of the Cell Phone, on Spectrum Sharing*, N.Y. TIMES, May 31, 2012, http://bits.blogs.nytimes.com/2012/05/31/qa-marty-cooper-spectrum-sharing/. Although this is a minority view, even the most ardent advocates of spectrum reallocation agree that freeing up more spectrum for mobile broadband uses can be only part of the solution and that more efficient use of spectrum will need to play an essential role as well.

7. We are indebted to Kevin Werbach for bringing this reported quote to our attention, although, as he points out, the attribution to Einstein is quite possibly apocryphal. *See* Kevin Werbach, *Supercommons: Toward a Unified Theory of Wireless Communication*, 82 TEX. L. REV. 863, 882, and n. 97 (2004).

8. Thomas W. Hazlett, *The Wireless Craze, the Unlimited Bandwidth Myth, the Spectrum Auction Faux Pas, and the Punchline to Ronald Coase's "Big Joke": An Essay on Airwave Allocation Policy*, 14 HARV. J.L. & TECH. 335, 338 (2001).

9. This form of interference is the most obvious and intuitive but is not the only one. For example, *intermodulation interference* can arise when one signal transmitted from one location combines with another signal transmitted from a different location to produce a third signal that causes in-band interference.

10. NBC v. United States, 319 U.S. 190, 216 (1944). The Supreme Court reaffirmed this conclusion in its more famous decision in Red Lion Broad. Co. v. FCC, 395 U.S. 367 (1969). In that case, the Court relied on the same scarcity rationale to uphold the FCC's now defunct "fairness doctrine," which obligated broadcasters to give equal time to opposing viewpoints on matters of public interest. Although *Red Lion* technically remains good law, the Court has suggested skepticism about whether the scarcity rationale still justifies an exception to otherwise applicable First Amendment restrictions on content-based regulation. *See, e.g.*, Turner Broad. Sys., Inc. v. FCC, 512 U.S. 622, 637 (1994) ("the rationale for applying a less rigorous standard of First Amendment scrutiny to broadcast regulation, whatever its validity in the cases elaborating it, does not apply in the context of cable regulation"); *see also id.* at 638 n.5 (citing criticisms of *Red Lion*); FCC v. Fox Television Stations, Inc., 556 U.S. 502, 530–35 (2009) (Thomas, J., concurring) (calling for *Red Lion* to be revisited); Christopher S. Yoo, *The Rise and Demise of the Technology-Specific Approach to the First Amendment*, 91 Geo. L.J. 245 (2003) (comprehensively critiquing *Red Lion*).

11. *See* Philip J. Weiser & Dale N. Hatfield, *Policing the Spectrum Commons*, 74 Fordham L. Rev. 663 (2005).

12. *See generally* Philip J. Weiser & Dale Hatfield, *Spectrum Policy Reform and the Next Frontier of Property Rights*, 15 Geo. Mason L. Rev. 549 (2008).

13. Ronald Coase, *The Federal Communications Commission*, 2 J. Law & Econ. 1, 2 (1959) (quoting S. Rep. No. 659, 61st Cong., 2d Sess. 4 (1910)).

14. Lawrence Lessig, The Future of Ideas: The Fate of the Commons in a Connected World 73 (Basic Books, 2001).

15. United States v. Zenith Radio Corp., 12 F.2d 614 (N.D. Ill. 1926) (no authority to impose restrictions on license); Hoover v. Intercity Radio Co., 286 F. 1003 (D.C. Cir. 1923) (no discretion to deny licenses).

16. *See* Radio Act of 1927, 44 Stat. 1162.

17. 47 U.S.C. § 301; *see also* Note, *Federal Control of Radio Broadcasting*, 39 Yale L.J. 245, 250 (1929) (the premise that the "'the government owns the ether' . . . was an idée fixe in the debates of Congress" over the Radio Act of 1927).

18. FCC Frequency Allocations and Radio Treaty Matters; General Rules and Regulations, 47 C.F.R. § 2.1 (definitions of "interference" and "harmful interference").

19. U.S. Government Accountability Office (GAO), *Spectrum Management: NTIA Planning and Processes Need Strengthening to Promote the Efficient Use of Spectrum by Federal Agencies*, GAO-11-352, at 2 (2011), http://www.gao.gov/assets/320/318264.pdf (figures as of December 2009); *see also* U.S. GAO, *Telecommunications: Comprehensive Review of U.S. Spectrum Management with Broad Stakeholder Involvement Is Needed*, GAO-03-277, at 11 (2003), http://

www.gao.gov/assets/240/237138.pdf (finding that in 2003 the U.S. government controlled outright nearly 14% of all allocated spectrum bands and shared 56% of all bands with other users).

20. For a critique of U.S. preparations for international spectrum policy proceedings, see *U.S. Preparation for the World Radio Conferences: Too Little, Too Late: Hearing before the Subcommittee on National Security, Emerging Threats, and International Relations*, 108th Cong. 108–80 (2004), http://www.gpo.gov/fdsys/pkg/CHRG-108hhrg95268/pdf/CHRG-108hhrg95268.pdf.

21. Robert Caro, The Years of Lyndon Johnson: Means of Ascent 89–105 (Knopf, 1990).

22. Thomas G. Krattenmaker & Lucas A. Powe Jr., Regulating Broadcast Programming 148 (MIT Press, 1994).

23. *See* Ashbacker Radio Corp. v. FCC, 326 U.S. 327 (1945).

24. Second Report & Order, *An Inquiry Relative to the Future Use of the Frequency Band 806–960 MHz*, 46 F.C.C.2d 752 (1974); Report & Order, *An Inquiry into the Use of the Bands 825–845 MHz and 870–890 MHz for Cellular Communications Systems*, 86 F.C.C.2d 469 (1981).

25. 47 U.S.C. § 309(i); *see* Report & Order, *Amendment of the Commission's Rules to Allow the Selection from among Mutually Exclusive Competing Cellular Applications Using Random Selection or Lotteries Instead of Comparative Hearings*, 98 F.C.C.2d 175 (1984).

26. *See* Hazlett, *Wireless Craze*, at 399.

27. Sylvia Nasar, A Beautiful Mind 374–78 (Simon & Schuster, 1998).

28. For an excellent discussion of such issues, see John McMillian, Reinventing the Bazaar: A Natural History of Markets 80–85 (W. W. Norton, 2003); *see also* National Academy of Sciences, *Beyond Discovery: The Bidding Game* (Mar. 2003), http://www.beyonddiscovery.org/includes/DBFile.asp?ID=1118.

29. *See* 47 U.S.C 309(j)(1) & (2) (setting forth requirement and delineating exceptions); Report & Order, *Implementation of Sections 309(j) and 337 of the Communications Act of 1934 as Amended*, 15 FCC Rcd 22709 (2000).

30. Paul Klemperer, *How Not to Run Auctions: The European 3G Telecom Auctions*, 46 Eur. Econ. Rev. 829, 837–39 (2002).

31. FCC v. NextWave Personal Communications, Inc., 537 U.S. 293 (2003). Before the Supreme Court ruled, the FCC sought to broker a sale of the licenses that would resolve the litigation and enable the spectrum to be used promptly. It proposed that the licenses be transferred to those firms that had bid the highest for them at a second auction (in an amount exceeding $15 billion), with the proceeds split between NextWave and the U.S. Treasury. But Congress, whose legislative intervention was needed to preserve this compromise from legal challenge, was unwilling to sign off on any such deal. In the eyes of many, such as Senator John McCain (who led the Senate Commerce Committee), any deal that left NextWave with close to $6 billion constituted an unthinkable spectrum

"giveaway." The result, however, was that the spectrum remained idle for longer than it otherwise would have—a significant loss for consumer welfare.

32. Coase, *The Federal Communications Commission*, at 14. This critique followed an earlier piece by a University of Chicago law student, Leo Herzel, who likewise advocated a market-based system. *See* Leo Herzel, *"Public Interest" and the Market in Color Television Regulation*, 18 U. CHI. L. REV. 802 (1951). Many economists and legal scholars have since advanced versions of this argument. *See* Ellen P. Goodman, *Spectrum Rights in the Telecosm to Come*, 41 SAN DIEGO L. REV. 269, 271 n. 3 (2004) (listing property-rights advocates).

33. For an early analysis of those complexities, see Arthur S. De Vany et al., *A Property System for Market Allocation of Electromagnetic Spectrum: A Legal–Economic–Engineering Study*, 21 STAN. L. REV. 1499 (1969). For a more recent analysis, see Weiser & Hatfield, *Spectrum Policy Reform*.

34. Hazlett, *The Wireless Craze*, at 405–53.

35. Douglas W. Webbink, *Frequency Spectrum Deregulation Alternatives*, FCC WORKING PAPER 10 (Oct. 1980), http://www.fcc.gov/Bureaus/OPP/working_papers/oppwp2.pdf.

36. Report & Order, *Formulation of Policies Relating to the Broadcast Renewal Applicant, Stemming from the Comparative Hearing Process*, 66 F.C.C.2d 419, 432 n. 18 (1977) (Separate Statement of Commissioners Benjamin L. Hooks and Joseph R. Fogarty).

37. Hazlett, *The Wireless Craze*, at 427, quoting O. CASEY CORR, MONEY FROM THIN AIR 235 (Crown, 2000) (internal quotations omitted).

38. Hazlett, *The Wireless Craze*, at 427 (emphasis deleted).

39. As Thomas Hazlett put it, "Bringing radio spectrum out of an unproductive employment should not be such a tricky business. . . . Entrepreneurs should have to make their mark innovating in the marketplace, inventing technologies or marketing 'killer apps,' not out-foxing competing lawyers. The countless other businesses that have flunked this test—most of them unknown and deterred from the start—constitute economic carnage without offsetting social advantage." *Id.* at 428.

40. 47 C.F.R. § 24.3.

41. *See, e.g.*, Public Notice, *Nonbroadcast and General Action Report No. 1142*, 12 F.C.C.2d 559, 560 (1963) (a.k.a. *Intermountain Microwave Decision*).

42. Report & Order, *Promoting Efficient Use of Spectrum through Elimination of Barriers to the Development of Secondary Markets*, 18 FCC Rcd 20604 (2003) ("*Secondary Markets Order*").

43. *See* Second Report & Order, *Promoting Efficient Use of Spectrum through Elimination of Barriers to the Development of Secondary Markets*, 19 FCC Rcd 17503 (2004).

44. In particular, Commissioner Michael Copps dissented because he found such flexibility inconsistent with section 310(d) of the Communications Act, which provides that "[n]o . . . station license, *or any rights thereunder*, shall be transferred, assigned, or disposed of in any manner . . . except upon application to

the Commission and upon finding by the Commission that the public interest, convenience, and necessity will be served thereby." 47 U.S.C. § 310(d) (emphasis added). Copps argued: "[T]oday we allow licensees to transfer a significant right—the right to control the spectrum on a day-to-day basis—without applying to the Commission and without the requirement of any Commission public interest finding. How can this be legal under Section 310(d)?" *Secondary Markets Order*, 19 FCC Rcd at 20797.

45. For a detailed summary of these physical properties and their policy consequences, see Weiser & Hatfield, *Spectrum Policy Reform*, at 575–80.

46. *2010 Presidential Memorandum*; *National Broadband Plan* at xii.

47. *National Broadband Plan* at 84.

48. *Id.*

49. *Id.*

50. *Id.* at 85.

51. *See, e.g.*, U.S. GAO, *Better Knowledge Needed to Take Advantage of Technologies That May Improve Spectrum Efficiency*, GAO-04-666 (2004), http://www.gao.gov/assets/250/242645.pdf.

52. *See, e.g.*, U.S. Department of Commerce, *Plan and Timetable to Make Available 500 Megahertz of Spectrum for Wireless Broadband* (Oct. 2010), http://www.ntia.doc.gov/files/ntia/publications/tenyearplan_11152010.pdf; U.S. Department of Commerce, *Spectrum Policy for the Twenty First Century—the President's Spectrum Policy Initiative* (June 2004), http://www.ntia.doc.gov/files/ntia/publications/spct_pol_part_1_rl.pdf; The White House, *Fact Sheet on Spectrum Management*, press release (June 5, 2003), http://georgewbush-whitehouse.archives.gov/news/releases/2003/06/20030605-5.html; U.S. GAO, *History and Current Issues Related to Radio Spectrum Management*, GAO-02-814T (2002), http://www.gao.gov/assets/110/109399.pdf.

53. U.S. Department of Commerce, *An Assessment of the Viability of Accommodating Wireless Broadband in the 1755–1850 MHz Band* (Mar. 2012), http://www.ntia.doc.gov/files/ntia/publications/ntia_1755_1850_mhz_report_march2012.pdf.

54. *See* Phil Goldstein, *CTIA Embraces Spectrum Sharing, but Sees It as Second-Best Option*, FierceWireless (May 31, 2012), http://www.fiercewireless.com/story/ctia-embraces-spectrum-sharing-sees-it-second-best-option/2012-05-31. Despite these misgivings, one major U.S. mobile provider, T-Mobile, sought and obtained approval from the FCC in mid-2012 to begin spectrum-sharing tests in the 1755–1780 MHz band. *See* Phil Goldstein, *FCC Allows T-Mobile to Test Spectrum Sharing in 1755–1780 MHz Band*, FierceWireless (Aug. 15, 2012), http://www.fiercewireless.com/story/fcc-allows-t-mobile-test-spectrum-sharing-1755-1780-mhz-band/2012-08-15.

55. PCAST, *Report to the President: Realizing the Full Potential of Government-Held Spectrum to Spur Economic Growth*, at vi (2012), http://www.whitehouse.gov/sites/default/files/microsites/ostp/pcast_spectrum_report_final_july_20_2012.pdf.

56. *Id.* at vi–vii.

57. *CTIA–The Wireless Association® Statement on PCAST Government Spectrum Report*, press release (July 20, 2012), http://www.prnewswire.com/news-releases/ctia-the-wireless-association-statement-on-pcast-government-spectrum-report-163217616.html.

58. Middle Class Tax Relief and Job Creation Act of 2012, Pub. L. No. 112-96, Tit. VI, §§ 6403 et seq., 126 Stat. 225 et seq. (Feb. 22, 2012) ("2012 Act") (codified at 47 U.S.C. § 1452 et seq.). This legislation followed the rough outlines of a proposal contained in the *National Broadband Plan* (at 88–93), which in turn built on years of scholarship on this topic. *See, e.g.*, Philip J. Weiser, *The Untapped Promise of Wireless Spectrum*, Hamilton Project, Brookings Inst. (July 2008), http://www.brookings.edu/research/papers/2008/07/wireless-weiser; Evan Kwerel & John Williams, *A Proposal for a Rapid Transition to Market Allocation of Spectrum*, OPP Working Paper Series No. 38, at iv (FCC 2002), http://hraunfoss.fcc.gov/edocs_public/attachmatch/DOC-228552A1.pdf. In October 2012, shortly before we completed work on this second edition, the FCC proposed a rough roadmap for implementing this incentive-auction legislation. *See* Notice of Proposed Rulemaking, *Expanding the Economic and Innovation Opportunities of Spectrum through Incentive Auctions*, GN Docket No. 12-268, FCC No. 12-118 (Oct. 2, 2012) ("*Incentive Auction NPRM*"). Given the complexity of the subject matter, however, the proposals inevitably raised as many questions as they sought to answer.

59. For example, under one approach, winning bidders would receive exactly what they bid (if the auction otherwise succeeds). "Another mechanism, known as 'threshold' pricing, would pay a winning bidder the highest amount it could have bid and still have had its bid accepted. . . . Threshold pricing gives bidders an incentive to bid its station's value regardless of the bids submitted by others: if it bids an inflated value, it may forfeit the opportunity to be bought out at a price at least as high as the station's value, and if it bids an understated value, it may relinquish its rights at a price below the station's value." *Incentive Auction NPRM* ¶ 51.

60. *See id.* ¶ 16 ("Those broadcasters that are able to take advantage of . . . opportunities offered by an evolving marketplace have every prospect of continuing successfully to provide the public the benefits of free over-the-air television. For those that cannot, Congress's mandate to conduct a broadcast television spectrum incentive auction creates alternative opportunities. Broadcasters struggling financially and interested in exiting the business entirely, but unable to find a buyer for their facilities, may be able to obtain compensation in an amount acceptable to them by participating in the reverse auction. Their exit from the business would reduce the overall number of broadcast television stations competing for the same limited pool of advertising revenue. Broadcasters that wish to remain in the business also have an opportunity to strengthen their finances through the cash infusion resulting from a winning reverse auction bid to channel share or to move from a UHF to a VHF channel") (footnote omitted); *see also* Oral Testimony of Coleman Bazelon (Brattle Group), U.S. House of Representatives, Committee on Energy and Commerce Subcommittee on Communication

and Technology, at 1–3 (Apr. 12, 2011) ("Bazelon 2011 testimony") ("Exactly how much more radio spectrum is needed for wireless broadband is uncertain. Given this uncertainty, policymakers should apply a principle of spectrum reallocation—based on current allocations, if a higher valued use exists, spectrum should be reallocated from the lower valued use to the higher valued use. . . . One of the key advantages of incentive auctions is that they are designed with [this] Principle of Spectrum Reallocation in mind. That is, by design, they will not reallocate spectrum from a higher valued use to a lower value use").

61. 2012 Act, § 6403(c)(2) (47 U.S.C.§ 1452(c)(2)).

62. *See generally* Lawrence H. Summers, *Technological Opportunities, Job Creation, and Economic Growth*, Remarks at the New America Foundation on the President's Spectrum Initiative (June 28, 2010), http://www.whitehouse .gov/administration/eop/nec/speeches/technological-opportunities-job-creation -economic-growth.

63. *See, e.g.*, *Secondary Markets Order* at 20797 (statement of Commissioner Copps: "Generally we limit our actions to commercial telecommunications providers that paid for their spectrum licenses at auction. Allowing leasing by companies that have already compensated the public for the use of spectrum is both significantly different and far more defensible than allowing companies that were given their spectrum rights for free to lease it and reap windfall profits"); *see also* Norman Ornstein & Michael Calabrese, *A Private Windfall for Public Property*, Wash. Post, Aug. 12, 2003, at A13, http://www.aei.org/article/ economics/a-private-windfall-for-public-property/. Congress incorporated a similar intuition in the Communications Act, directing the FCC to design its auction system to meet, among various other goals, "recovery for the public of a portion of the value of the public spectrum resource made available for commercial use and avoidance of unjust enrichment through the methods employed to award uses of that resource." 47 U.S.C. § 309(j)(3)(C).

64. For a thoughtful critique of the view that fairness considerations are irrelevant to spectrum policy, see Ellen Goodman, *Spectrum Equity*, 4 J. Telecom & High Tech. L. 193 (2005). *See also* Thomas W. Hazlett et al., *What Really Matters in Spectrum Allocation Design*, 10 Nw. J. Tech. & Intell. Prop. 93, 93 (2012) ("[F]unds generated without the use of taxes do not cause tax-distorting social losses. Each tax dollar raised, for instance, is expected to cost society about $0.33 in deadweight loss. Auction dollars, as pure transfers, cost less") (citing Paul Klemperer, *What Really Matters in Auction Design*, 16 J. Econ. Persp. 169, 179 (Winter 2002)).

65. Eli Noam, *Spectrum Auctions: Yesterday's Heresy, Today's Orthodoxy, Tomorrow's Anachronism*, 41 J.L. & Econ. 765–66 (1998). Judge Stephen Williams has similarly explained that "spectrum is government property only in the special sense that it simply has not been allocated to any real 'owner' in any way." Time Warner Entm't Co. v. FCC, 105 F.3d 723, 727 (D.C. Cir. 1997) (dissenting from denial of hearing en banc).

66. *See, e.g.*, Robert M. Entman, Challenging the Theology of Spectrum: Policy Reformation Ahead 20–21 (Aspen Inst., 2004).

67. Nextel's wireless rivals proposed that any brokered deal involve a relocation of Nextel's spectrum to the 2.1 GHz band instead because, as the *Wall Street Journal* reported, the 1.9 GHz band "is considered by some to be more valuable because existing cellphone equipment would need only slight alterations to work on it. Standard cellphone service isn't currently offered over [2.1 GHz]." Jesse Drucker & Anne Marie Squeo, *Interference Call: Nextel's Maneuver for Wireless Rights Has Rivals Fuming*, WALL ST. J., Apr. 19, 2004, at A1.

68. Report & Order, *Improving Public Safety Communications in the 800 MHz Band*, 19 FCC Rcd 14969 (2004).

69. *Id.* (Separate Statement of Chairman Michael Powell), http://hraunfoss.fcc.gov/edocs_public/attachmatch/DOC-249414A2.pdf.

70. Letter from William Barr, General Counsel, Verizon, to Chairman Michael K. Powell, at 1 (June 28, 2004) (emphasis added), http://apps.fcc.gov/ecfs//retrieve.cgi?native_or_pdf=pdf&id_document=6516282241. Lest there be any confusion about the implication of this message, Verizon added: "[T]hese proscriptions . . . hold [an] official accountable in his personal, individual capacity" and in this case give rise to "a substantial probability of criminality." *Id.* at 6.

71. 47 U.S.C. § 309(j).

72. 47 U.S.C. § 765f.

73. Report & Order, *Flexibility for Delivery of Communications by Mobile Satellite Service Providers in the 2 GHz Band, the L-Band, and the 1.6/2.4 GHz Bands*, 18 FCC Rcd 1962 (2003), *petition for review dismissed sub nom.* Cellco Partnership v. FCC, No. 03-1191 (D.C. Cir. July 14, 2005).

74. *See, e.g.*, Order & Authorization, *LightSquared Subsidiary LLC*, 26 FCC Rcd 566 (2011).

75. *See* Order, *New DBSD Satellite Serv. & TerreStar Licensee Inc. Request for Rule Waivers*, IB Docket Nos. 11-149 et al., DA No. 12-332 (IB Mar. 2, 2012).

76. *See* Report & Order, *Service Rules for Advanced Wireless Services in the 2000-2020 MHz and 2180-2200 MHz Bands*, WT Docket Nos. 12-70 et al., FCC No. 12-151 (Dec. 17, 2012) ("*AWS-4 Order*"). The FCC imposed stringent in-band power and out-of-band emissions limits on DISH's operations in the 2000-2005 MHz band in order to protect future operations in the adjacent "H Block" spectrum at 1995-2000 MHz, and it likewise concluded that DISH must accept some interference from those future H Block operations. *See id.* ¶¶ 56-156. DISH objected that the FCC could not reasonably impose those disabilities on DISH in order to achieve more speculative benefits for the still-unassigned H Block, which the FCC had not yet even decided to set for auction. The FCC responded that its rules struck the best balance between competing uses of these adjacent spectrum blocks. *See id.* ¶ 71. It also noted that its decision to protect the H Block from interference (and thereby increase its value) would likely benefit the public safety community because the 2012 spectrum legislation "directs that the proceeds from the auction of licenses in the H Block, including 1995-2000 MHz, be deposited into the Public Safety Trust Fund, which will be used to fund" a next-generation public safety network. *Id.*; *see* 2012 Act § 6401.

77. *AWS-4 Order* ¶¶ 195–99.

78. *Id.* ¶¶ 200–04.

79. *Id.* ¶¶ 195–96 (citing build-out requirements in the 700 MHz and WCS bands); *see, e.g.*, Second Report & Order, *Service Rules for the 698–746, 747–762, and 777–792 MHz Bands*, 22 FCC Rcd 15289, ¶ 153 (2007) (imposing build-out requirements for new 700 MHz licensees).

80. *See, e.g.*, Weiser, *Untapped Promise*, at 14–15.

81. Fresno Mobile Radio, Inc. v. FCC, 165 F.3d 965, 969 (D.C. Cir. 1999) (citations omitted).

82. Report & Order, *Flexibility for Delivery of Communications by Mobile-Satellite Service Providers in the 2 GHz Band, the L-Band, and the 1.6/2.4 GHz Band*, 18 FCC Rcd 1962, ¶ 39 (2003) (footnotes and internal quotations omitted).

83. *Id.*

84. *See generally* Second Memorandum & Order, *Unlicensed Operation in the TV Broadcast Bands*, 23 FCC Rcd 16807 (2010) ("*2010 White Spaces Order*").

85. *See id.*; Second Report & Order, *Unlicensed Operation in the TV Broadcast Bands*, 23 FCC Rcd 16807 (2008); Notice of Proposed Rulemaking, *Facilitating Opportunities for Flexible, Efficient, & Reliable Spectrum Use Employing Cognitive Radio Technologies*, 18 FCC Rcd 26859, ¶ 36 (2003).

86. *See 2010 White Spaces Order.*

87. In addition to the purer commons approach discussed in the text, the FCC has also established a modified commons approach for patches of spectrum that it allocates for common use by classes of defined users. Familiar examples are the frequencies set aside for pilot-to-ground communications, amateur radio, and CB radio, a trucker's medium that briefly developed a faddish popularity with the general public in the mid-1970s. The FCC requires some of these users to obtain licenses, as in the case of amateur radio, but it sometimes dispenses with licensing requirements altogether, as in the case of CB radio, where the maximum allowable transmission power level is lower. But the common denominator is that users must share the spectrum in all of these bands and find ways to cooperate to avoid interference. See, e.g., Kenneth R. Carter et al., *Unlicensed and Unshackled: A Joint OSP–OET White Paper on Unlicensed Devices and Their Regulatory Issues*, OSP Working Paper No. 39 (FCC 2003), http://hraunfoss.fcc.gov/edocs_public/attachmatch/DOC-234741A1.pdf.

88. *See generally* Kenneth R. Carter, *Unlicensed to Kill: A Brief History of the Part 15 Rules*, 11 INFO 8 (2009). Although our discussion of unlicensed spectrum focuses on "intentional radiators" (i.e., devices that are intended to emit radio-frequency radiation), there are also "unintentional" radiators (such as personal computers) that generate internal radiation, which is sometimes emitted into the outside world, and "incidental" radiators (such as electric motors), which are not designed to transmit radio-frequency radiation at all but nevertheless may do so through their operation. The FCC has authority to address the radiation emitted by these devices as well.

89. *See* Weiser & Hatfield, *Policing the Spectrum Commons*, at 671–72.

90. Under its current Part 15 regime, the FCC regulates access to unlicensed spectrum through ex ante rules that require all equipment designed for use in such bands to be certified as compliant with certain technical standards (e.g., maximum power limits and listen-before-talk protocols). *See, e.g.*, 47 C.F.R. § 15.209 (2002) (power-level limitations); 47 C.F.R. §15.321 (2002) (imposing listen-before-talk etiquette). To expedite the process of certifying such equipment, the FCC has authorized private organizations known as Telecommunication Certification Bodies to perform this function. In some cases, where there are notable reasons for concern, the FCC has gone even further to devise special protective rules to govern specific unlicensed uses, as in its decision to make radar detectors comply with specified emissions levels to avoid interference with licensed uses. Notice of Proposed Rulemaking, *Revision of Parts 2 and 15 of the Commission's Rules to Permit Unlicensed National Information Infrastructure (U-NII) Devices in the 5 GHz Band*, 18 FCC Rcd 11581 (2003). Finally, to police compliance with all relevant requirements (special ones and otherwise), the FCC's Enforcement Bureau identifies, investigates, and addresses violations; as a practical matter, however, most such issues come to the FCC's attention through complaints by competitors. *See, e.g.*, Transp. Intelligence, Inc. v. FCC, 336 F.3d 1058, 1060 (D.C. Cir. 2003).

91. First Report & Order, *Revision of Part 15 of the Rules Regarding the Operation of Radio Frequency Devices without an Individual License*, 4 FCC Rcd 3493, ¶ 130 (1989).

92. *See* Claude E. Shannon, *A Mathematical Theory of Communications*, 27 Bell Sys. Tech. J. 623 (1948) (continuing discussion of issue initiated in 27 Bell Sys. Tech. J. 379 (1948)); James Gleick, The Information: A History, a Theory, a Flood 215–31 (Pantheon, 2011) (describing Shannon's insights).

93. *See* Charles Jackson, Raymond Pickholtz, & Dale Hatfield, *Spread Spectrum Is Good—but It Does Not Obsolete NBC v. U.S.!* 58 Fed. Comm. L.J. 245, 250, 257 (2006) (surveying legal scholarship and concluding that, contrary to its basic premise, "spread spectrum does not eliminate interference; rather, it changes the nature of interference" while "provid[ing] efficient distributed access to a range of frequencies").

94. First Report & Order, *Revision of Part 15 of the Commission's Rules Regarding Ultra-wideband Transmission Systems*, 17 FCC Rcd 7435 (2002).

95. Yochai Benkler, *Some Economics of Wireless Communications*, 16 Harv. J.L. & Tech. 25, 36–37 (2002).

96. *See* Yochai Benkler, *Overcoming Agoraphobia: Building the Commons of the Digitally Networked Environment*, 11 Harv. J.L. & Tech. 287 (1998).

97. Garrett Hardin, *The Tragedy of the Commons*, 162 Sci. 1243, 1244–45 (1968).

98. Stuart Minor Benjamin, *Spectrum Abundance and the Choice between Private and Public Control*, 78 N.Y.U. L. Rev. 2007, 2031 (2003).

99. *Id.* at 2032; *see also* Douglas Sicker et al., *Examining the Wireless Commons* (Aug. 15, 2006), http://papers.ssrn.com/sol3/papers.cfm?abstract_id=2103824.

100. *See* Weiser & Hatfield, *Policing the Spectrum Commons*, at 664–65.

101. Hazlett, *Wireless Craze*, at 495–509; James B. Speta, *A Vision of Internet Openness by Government Fiat*, 96 NW. L. REV. 1553, 1572 (2002) ("Where there is a single licensee either operating its own service or acting as a bandwidth manager, that licensee can mandate the use of equipment or protocols that fully utilize the spectrum"); Durga P. Satapathy & Jon M. Peha, *Etiquette Modification for Unlicensed Spectrum: Approach and Impact*, 1 PROC. IEEE VEHIC. TECHNOLOGIES CONF. 272, 273 (1998), http://repository.cmu.edu/cgi/viewcontent.cgi?article=1005&context=epp ("It has been shown, for both fluid flow and bursty traffic, that while greed may benefit a device even with contention for spectrum, it always degrades performance of other devices, forcing them to also resort to greed, in order to regain their performance"). Critics of the commons model also argue that to the extent the commons approach is efficient, spectrum licensees can establish "private commons," as the FCC has authorized. *See* Second Report & Order, *Promoting Efficient Use of Spectrum through Elimination of Barriers to the Development of Secondary Markets*, WT Docket 00-230, ¶¶ 91–99 (July 8, 2004), http://hraunfoss.fcc.gov/edocs_public/attachmatch/FCC-04-167A1.pdf. Whether market forces will often produce this outcome, however, is still unclear.

102. *See* Gerald R. Faulhaber & David Farber, *Spectrum Management: Property Rights, Markets, and the Commons*, AEI–Brookings Joint Ctr. for Regulatory Studies, Working Paper No. 02-12 (2003), http://assets.wharton.upenn.edu/~faulhabe/SPECTRUM_MANAGEMENTv51.pdf; *accord* Benjamin, *Spectrum Abundance*, at 2097–101.

103. 2012 Act, § 6407 (codified at 47 U.S.C. § 1454). The FCC asserts some discretion to make band-plan choices that would incrementally increase the spectrum available for unlicensed uses. *See Incentive Auction NPRM* ¶¶ 175–76.

104. *See, e.g.*, Coleman Bazelon, *Licensed or Unlicensed: The Economic Considerations in Incremental Spectrum Allocations*, IEEE COMMUNICATIONS MAGAZINE 110 (March 2009).

105. For a fuller discussion of these issues, see Weiser & Hatfield, *Spectrum Policy Reform*, from which some of the material in this section is taken. For an earlier (and prescient) discussion of the challenges inherent in enforcing disputes under either a property or commons-based system of spectrum management, see Goodman, *Spectrum Rights*, at 398–403.

106. *See, e.g.*, R. Paul Margie, *Can You Hear Me Now? Getting Better Reception from the FCC's Spectrum Policy*, 2003 STAN. TECH. L. REV. 5.

107. *See* Stephen Labaton, *F.C.C. Heads for Showdown with Congress over Radio Plan*, N.Y. TIMES, Mar. 27, 2000, at C1, http://www.nytimes.com/2000/03/27/business/fcc-heads-for-showdown-with-congress-over-radio-plan.html?src=pm; Pub. L. No. 106-553 § 632, 114 Stat. 2762, 2762A-111 (2000).

108. *See 2010 White Spaces Order*.

109. Coase, *The Federal Communications Commission*, at 30.

110. *See National Broadband Plan* at 82 (noting that many incumbents "have inflexible licenses that limit the spectrum to specific uses" and observing that, as a result, "[t]hese licensees do not incur opportunity costs for use of their spectrum; therefore, they are not apt to receive market signals about new uses with potentially higher value than current uses. The result can be inadequate consideration of alternative uses, artificial constraints on spectrum supply and a generally inefficient allocation of spectrum resources").

111. Hazlett, *Wireless Craze*, at 432–33.

112. Some have suggested that generalist courts can manage such disputes, but the sheer complexity of those disputes would pose substantial challenges for that approach. *See* Weiser & Hatfield, *Spectrum Policy Reform.*

113. Order & Authorization, *LightSquared Subsidiary LLC*, 26 FCC Rcd 566 (2011).

114. *See generally id.*; Public Notice, *International Bureau Establishes Pleading Cycle for LightSquared Petition for Declaratory Ruling*, 27 FCC Rcd 463 (2012).

115. *See generally* Harold Feld, *My Insanely Long Field Guide to Lightsquared v. the GPS Guys*, Wetmachine (June 14, 2011), http://tales-of-the-sausage-factory.wetmachine.com/my-insanely-long-field-guide-to-lightsquared-v-the-gps-guys/. An analogous controversy arose when satellite radio providers Sirius and XM (which have now merged) used "terrestrial repeaters" to amplify their signals, thereby causing interference with traditional FM radio broadcasts. Sirius and XM agreed to various limits on these terrestrial transmissions. *See, e.g.*, Order, *Sirius Satellite Radio Inc.*, 23 FCC Rcd 12301 (2008).

116. *See* Public Notice, *International Bureau Invites Comment on NTIA Letter Regarding LightSquared Conditional Waiver*, 27 FCC Rcd 1596 (2012).

117. *See* Report & Order, *Qualcomm Inc. Petition for Declaratory Ruling*, 21 FCC Rcd 11683 (2006). This Qualcomm proceeding is discussed in greater detail in Weiser & Hatfield, *Spectrum Policy Reform*, at 566–67.

118. Weiser & Hatfield, *Spectrum Policy Reform*, at 591–603.

119. For another example of the FCC's use of this inputs-oriented approach, see Order and Authorization, *GlobalStar LLC: Application for Modification of License for Operation of Ancillary Terrestrial Component Facilities*, 23 FCC Rcd 15975 (2008) (authorizing a licensee, after some delay, to expand its transmission technologies beyond cdma2000 to WiMAX and other air interface standards).

120. Weiser & Hatfield, *Spectrum Policy Reform*, at 594 (quoting FCC Spectrum Policy Task Force, *Report of the Spectrum Rights and Responsibilities Working Group* 29 (2002), http://www.fcc.gov/sptf/files/SRRWGFinalReport.pdf ("*Working Group Report*")); *see* also *Working Group Report* at 27 ("Parameters based on [outputs] provide licensees with greater flexibility in determining their system architecture to meet customer density, geographic location and scope, and cost considerations, while maintaining what should be the Commission's most basic regulatory concern: the extent to which they impact the service of other licensees and operations").

121. *See generally* Weiser, *Untapped Promise*, at 26–28.

122. *See generally* Bazelon 2011 testimony.

123. For one such proposal, which outlines a system of "harm claim thresholds," see Testimony of Pierre de Vries, "The Role of Receivers in a Spectrum Scarce World," Hearing Before the House Energy and Commerce Committee (Nov. 29, 2012), http://energycommerce.house.gov/sites/republicans.energycommerce.house.gov/files/Hearings/CT/20121129/HHRG-112-IF16-WState-deVriesP-20121129.pdf.

124. For example, two FCC staffers recently published a detailed analysis of how auction mechanisms can be used to allocate interference rights efficiently. *See* Mark M. Bykowsky & William W. Sharkey, *Using a Market to Obtain the Efficient Allocation of Signal Interference Rights*, FCC Staff Working Paper No. 4 (June 2012), http://www.fcc.gov/working-papers/using-market-obtain-efficient-allocation-interference-rights.

125. FCC, *Spectrum Policy Task Force Report*, ET Docket No. 02-135, at 39 (Nov. 15, 2002); *see also* Goodman, *Spectrum Rights*, at 389 n. 384.

126. Notice of Proposed Rulemaking, *Establishment of an Interference Temperature Metric to Quantify and Manage Interference and to Expand Available Unlicensed Operation in Certain Fixed, Mobile and Satellite Frequency Bands*, 18 FCC Rcd 25309, ¶¶ 29–51 (2003).

127. Order, *Establishment of an Interference Temperature Metric to Quantify and Manage Interference and to Expand Available Unlicensed Operation*, 22 FCC Rcd 8938 (2007).

128. *See generally* Margie, *Can You Hear Me Now?*

Chapter 4

1. Robert F. Roche & Liz Dale, *CTIA's Wireless Industry Indices, Semi-annual Data Survey Results: A Comprehensive Report from CTIA Analyzing the U.S. Wireless Industry*, at 1–2 (Nov. 2011).

2. *See* Fifteenth Report, *Implementation of Section 6002(b) of the Omnibus Budget Reconciliation Act of 1993*, 26 FCC Rcd 9664, ¶ 45, table 5 (2011) ("*Fifteenth Wireless Report*") (estimating that approximately 90 percent of Americans live in census blocks covered by at least five facilities-based mobile service providers); *id.* ¶ 191 (pricing statistics).

3. *See generally* Nat'l Cable & Telecomm. Ass'n v. Gulf Power Co., 534 U.S. 327, 339–42 (2002) (discussing statutory right to use utility poles for this purpose).

4. 47 U.S.C. § 332(c)(7)(B); *see generally* Timothy J. Tryniecki, *Cellular Tower Siting Jurisprudence under the Telecommunications Act of 1996—the First Five Years*, 37 Real Prop. Prob. & Tr. J. 271 (2002).

5. PrimeCo Pers. Communications v. City of Mequon, 352 F.3d 1147, 1149 (7th Cir. 2003) (Posner, J.) (citations omitted). The FCC's role in setting general rules for tower-siting disputes is the topic of pending litigation as this second edition

goes to press. *See* City of Arlington v. FCC, 668 F.3d 229 (5th Cir. 2012), *cert. granted in part*, Nos. 11-1545 et al. (Oct. 5, 2012).

6. *See* Report & Order, *Gen. Mobile Radio Serv.*, 13 F.C.C. 1190, 1212 (1949); Fourth Report & Order, *Amendment of Section 3.606 of the Commission's Rules and Regulations*, 41 F.C.C. 131 (1951).

7. Jeffrey H. Rohlfs et al., *Estimate of Loss to the United States Caused by the FCC's Delay in Licensing Cellular Communications*, National Economic Research Associates (Nov. 8, 1991).

8. *See generally* Illinois Bell Tel. Co., 63 F.C.C.2d 655 (1977), *aff'd sub nom*, Rogers Radio Commun. Sys., Inc. v. FCC, 593 F.2d 1225 (D.C. Cir. 1978).

9. *See* Report & Order, *An Inquiry into the Use of Bands 825–845 and 870 and 890 MHz for Cellular Communications Systems*, 86 F.C.C.2d 469 (1981).

10. *See* chapter 3 of this volume; *see generally* Philip Palmer McGuigan et al., *Cellular Mobile Radio Telecommunications: Regulating an Emerging Industry*, 1983 BYU L. Rev. 305.

11. United States v. Western Elec. Corp., 578 F. Supp. 643, 648–49 (D.D.C. 1983).

12. For a succinct telling of the rise of the cellular industry, see Stephanie N. Mehta, *Cellular Evolution*, Fortune (Aug. 23, 2004).

13. 47 C.F.R. § 24.202 (defining the terms *major trading areas* and *basic trading areas*, which are based on Rand McNally concepts).

14. *See generally* Second Report & Order, *Service Rules for the 698–746, 747–762, and 777–792 MHz Bands*, 22 FCC Rcd 15289 (2007); Public Notice, *Auction of Advanced Wireless Services Licenses Rescheduled for August 9, 2006*, 21 FCC Rcd 5598 (2006).

15. *Fifteenth Wireless Report* ¶ 296. The FCC has noted that advanced "capacity enhancement technologies such as multiple-input and multiple-output (MIMO) may perform better at higher frequencies." *Id.* MIMO is a form of "smart antenna" technology that increases bandwidth through the use of multiple antennas within transmitters and receivers. MIMO tends to be easier to use at higher frequencies (i.e., shorter wavelengths): the shorter a wavelength is, the smaller the corresponding antennas will be, and it is somewhat more feasible to deploy multiple short antennas in an array than multiple long ones.

16. In 2012, while seeking regulatory approval for a major spectrum acquisition from cable companies (discussed later in this chapter), Verizon Wireless announced that once that deal was approved, it would sell off 700 MHz Lower A and B Block spectrum, potentially (at least in part) to carriers that have little low-band spectrum. *See* Phil Goldstein, *Verizon: 64 Companies Interested in Our 700 MHz Spectrum*, FierceWireless (Aug. 20, 2012), http://www.fiercewireless.com/story/verizon-64-companies-interested-our-700-mhz-spectrum/2012-08-20. Verizon will retain its 700 MHz Upper C Block spectrum for its LTE network.

17. Middle Class Tax Relief and Job Creation Act of 2012, Pub. L. 112-96, Tit VI, § 6404, 126 Stat. 230 (Feb. 22, 2012) ("2012 Act") (codified at 47 U.S.C.

§ 309(j)(17)(A) and (B)). In September 2012, the FCC responded by opening a new inquiry into proposed limitations on the spectrum holdings a carrier may possess in given markets. *See* Notice of Proposed Rulemaking, *Policies Regarding Mobile Spectrum Holdings*, WT Docket No. 12-269, FCC No. 12-119 (Sept. 29, 2012) ("*Spectrum Aggregation NPRM*").

18. 47 U.S.C. § 332(d)(1)-(2).

19. 47 U.S.C. § 332(d)(3).

20. As discussed in chapter 2, Congress authorized the FCC in 1996 to "forbear from applying any regulation or any provision of this [Act] to a telecommunications carrier or telecommunications service" in a wide range of circumstances. 47 U.S.C. § 160.

21. 47 U.S.C. § 332(c)(1).

22. Second Report & Order, *Implementation of Sections 3(n) and 332 of the Communications Act; Regulatory Treatment of Mobile Services*, 9 FCC Rcd 1411, ¶¶ 124–219 (1994). As discussed later in this chapter, section 332(c)(3)(A) preserves a limited regulatory role for the states in this area but, with very limited exceptions, forbids them to regulate rates or terms of entry.

23. Orloff v. FCC, 352 F.3d 415, 420 (D.C. Cir. 2003) (emphasis omitted), *aff'g* Orloff v. Vodafone AirTouch Licenses LLC d/b/a Verizon Wireless, 17 FCC Rcd 8987 (2002).

24. *See* Final Decision, *Amendment of Section 64.702 of the Commission's Rules and Regulations (Second Computer Inquiry)*, 77 F.C.C.2d 384, ¶ 149, *modified on recon.*, 84 F.C.C.2d 50 (1980), *further modified*, 84 F.C.C.2d 50, *aff'd sub. nom.*, Computer and Communications Indus. Ass'n v. F.C.C., 693 F.2d 198 (D.C. 1982), *cert. denied*, 461 U.S. 938 (1983). For representative discussions of the complex antitrust considerations presented by the bundling practices of dominant firms, see Cascade Health Solutions v. PeaceHealth, 515 F.3d 883 (9th Cir. 2008); LePage's Inc. v. 3M, 324 F.3d 141 (3d Cir. 2003) (en banc); 3A PHILIP E. AREEDA & HERBERT HOVENKAMP, ANTITRUST LAW, ¶ 749a (Aspen, 3d ed., 2008).

25. *See, e.g.*, Yannis Bakos & Erik Brynjolfsson, *Bundling Information Goods: Pricing, Profits, and Efficiency*, 45 MGMT. SCI. 1613, 1619 (1999) ("Bundling can create significant economies of scope even in the absence of technological economies in production, distribution, or consumption").

26. *See* Report & Order, *Bundling of Cellular Customer Premises Equipment and Cellular Service*, 7 FCC Rcd 4028 (1992).

27. *See* FCC, *FCC Survey Confirms Consumers Experience Mobile Bill Shock and Confusion and Early Termination Fees*, news release (May 26, 2010), http://hraunfoss.fcc.gov/edocs_public/attachmatch/DOC-298415A1.pdf (expressing concern about industry practices); *see also* Remarks of FCC Chairman Julius Genachowski at the Bill Shock Event, Brookings Inst. (Oct. 27, 2011) ("Genachowski Bill Shock Remarks"), http://www.fcc.gov/document/chairman-genachowski-remarks-bill-shock-event.

28. *See, e.g.*, In re Cellphone Termination Fee Cases, 193 Cal. App. 4th 298, 122 Cal. Rptr. 3d 726 (Cal. App. 2011) (rejecting preemption challenge).

29. *See* Report & Order, *Regulatory Policies Concerning Resale and Shared Use of Common Carrier Services and Facilities*, 60 F.C.C.2d 261 (1976), *aff'd*, AT&T v. FCC, 572 F.2d 17 (2d Cir. 1978); *see also* Report & Order, *Resale and Shared Use of Common Carrier Domestic Public Switched Network Services*, 83 F.C.C.2d 167 (1980).

30. *See* First Report & Order, *Interconnection and Resale Obligations Pertaining to Commercial Mobile Radio Servs.*, 11 FCC Rcd 18455 (1996), *aff'd* Cellnet Commun., Inc. v. FCC, 149 F.3d 429 (6th Cir. 1998); *see also* Fourth Report & Order, *Interconnection and Resale Obligations Pertaining to Commercial Mobile Radio Servs.*, 15 FCC Rcd 13523, ¶ 1 (2000) ("deny[ing] requests for mandatory interconnection between resellers' switches and CMRS providers' networks").

31. For a discussion of the complex economics of price discrimination, see 8 Phillip Areeda & Herbert Hovenkamp, Antitrust Law 226, ¶ 1616f (Aspen, 2d ed. 2002) ("[P]rohibition [of price discrimination] would worsen consumer welfare if—as I tend to believe—prohibiting the restraint would lead the manufacturer . . . to maintain the higher wholesale price while adopting more costly or less effective strategies for reaching low-price customers, or even . . . to abandon his efforts to reach those customers altogether. I therefore regard price discrimination in aid of deep market penetration as a benefit"); William J. Baumol, *Predation and the Logic of the Average Variable Cost Test*, 39 J.L. & Econ. 49, 65–67 & n. 17 (1996) (noting circumstances in which economic efficiency requires the use of differential pricing); In re Brand Name Prescription Drugs Antitrust Litig., 288 F.3d 1028, 1032 (7th Cir. 2002) ("Since price discrimination is not (in general) unlawful, neither are efforts to prevent arbitrage"). Curiously, a quite different rule applies in the wireline context, where the 1996 Act requires even the least competitively significant CLEC to make its services available for resale. *See* 47 U.S.C. § 251(b)(1).

32. As additional examples of this phenomenon, the FCC also now permits all mobile telephony providers to "disaggregate" their licenses (i.e., divide up their allotted spectrum) or "partition" them (i.e., divide up a license's geographic area) and, after jumping through some bureaucratic hoops, sell the reconfigured licenses to other providers. And the FCC's facilitation in the *Secondary Markets* proceeding (see chapter 3) of spectrum-leasing arrangements has also helped ensure more efficient use of this spectrum.

33. *See* Third Report & Order, *Implementation of Section 3(n) and 332 of the Communications Act*, 9 FCC Rcd 7988, ¶ 16 (1994).

34. Report & Order, *2000 Biennial Regulatory Review, Spectrum Aggregation Limits for Commercial Mobile Radio Services*, 16 FCC Rcd 22668, ¶ 2, 6 (2001); *see also* Notice of Proposed Rulemaking, *Service Rules for Advanced Wireless Services in the 2155–2175 MHz Band*, 22 FCC Rcd 17035, ¶ 101 (2007) (noting that the Commission had eliminated the spectrum cap because it "found that the cap, by setting an *a priori* limit on spectrum aggregation without looking at the particular circumstances of specific proposed transactions, was unnecessarily inflexible and could be preventing beneficial arrangements that promote efficiency without undermining competition").

35. 2012 Act § 6404 (codified at 47 U.S.C. § 309(j)(17)(A) and (B)); *see Spectrum Aggregation NPRM.*

36. 47 U.S.C. § 332(c)(3)(A).

37. *Id.*

38. 47 U.S.C. § 332(c)(3)(A)(i) & (ii). As the FCC has observed, this statutory language is deeply anomalous. If literally construed, the "such market conditions exist" qualifier at the beginning of section 332(c)(3)(A)(ii) would make that provision superfluous because all market conditions satisfying that provision would by definition also satisfy section 332(c)(3)(A)(i), which itself encompasses those same market conditions. In an effort to resolve this statutory anomaly, the FCC interpreted that qualifier as a presumption-shifting device, signaling that "concerns about anticompetitive conditions in the market for CMRS will be given greater weight where a state can show that such service is the sole means of obtaining telephone exchange service in a substantial portion of the state." Report & Order and Order on Reconsideration, *Petition of Arizona Corporation Commission, to Extend State Authority over Rate and Entry Regulation of All Commercial Mobile Radio Services*, 10 FCC Rcd 7824 ¶ 66 (1995).

39. *See* Connecticut Dep't of Pub. Util. Control v. FCC, 78 F.3d 842 (2d Cir. 1996) (upholding FCC denial of state petition). In addition, section 332(c)(3)(B) also provided in 1993 for the temporary grandfathering of existing state rate regulation, but such grandfathering ended long ago.

40. *See, e.g., In re Cellphone Termination Fee Cases*, 122 Cal. Rptr. 3d 726 (rejecting preemption challenge to ETFs and summarizing case law through 2011); Memorandum Opinion and Order, *Wireless Consumers Alliance, Inc.*, 15 FCC Rcd 17021, ¶¶ 28–36 (2000) (discussing case law); Leonard J. Kennedy & Heather A. Purcell, *Wandering along the Road to Competition and Convergence—the Changing CMRS Roadmap*, 56 FED. COMM. L.J. 489, 500–511 (2004) (same).

41. *Wireless Consumers Alliance* ¶ 39.

42. The California Public Utilities Commission adopted one of the most prominent of these initiatives but later rescinded it after a change in commission membership. To some extent, these initiatives are filling a regulatory void left by the statutory exclusion of the FTC from regulating "common carriers." 15 U.S.C. § 45(a)(2); *see also* FTC v. Miller, 549 F.2d 452 (7th Cir. 1977) (applying exception).

43. *See* CTIA Consumer Code for Wireless Service, http://files.ctia.org/pdf/The_Code.pdf.

44. *See* Genachowski Bill Shock Remarks.

45. 2 ALFRED E. KAHN, THE ECONOMICS OF REGULATION: PRINCIPLES AND INSTITUTIONS 119 (MIT Press, 1989).

46. The FCC ordered such interconnection in the form of requiring telephone companies to provide access to wireless networks as *customers* rather than as fellow *carriers*. The distinction between a customer's access to a network and a co-carrier's access to a network helps drive critical issues such as (1) where

interconnection may occur; (2) whether interconnection is comparable to what the carrier gives its own affiliates; and (3) what price the carrier may charge for interconnection. Under the prevailing rule that emerged in the common law of common carriage, a carrier could refuse to interconnect with a rival, making the regulation of interconnection—and the resolution of the associated issues—a matter for the regulatory authorities. *See, e.g.*, Pac. Tel. & Tel. Co. v. Anderson, 196 F. 699, 703 (D.C. Wash. 1912); *see also* James B. Speta, *A Common Carrier Approach to Internet Interconnection*, 54 Fed. Comm. L.J. 225, 258 (2002) (discussing the issue); Report & Order, *Cellular Communications Systems*, 86 F.C.C.2d 469, ¶ 56 (1981) (in mandating interconnection for cellular providers, noting that "[a] cellular system operator is a common carrier and not merely a customer" and requiring that interconnection arrangements should be designed to minimize "unnecessary duplication of switching facilities").

47. Report & Order, *Cellular Communications Systems*, 86 F.C.C.2d 469, ¶¶ 53–57 (1981). The Commission specifically provided that the "particular arrangements involved in interconnection of a given cellular system should be negotiated among the carriers involved," but it mandated that interconnection should be provided upon "reasonable demand" and at "terms no less favorable than those offered to the cellular systems of affiliated entities or independent telephone companies." *Id.* ¶¶ 56–57.

48. *See, e.g.*, Third Report & Order, *Interconnection and Resale Obligations Pertaining to Commercial Mobile Radio Services*, 15 FCC Rcd 15975, ¶ 24 (2000) (terminating consideration of "automatic roaming" rules); Notice of Proposed Rulemaking, *Automatic and Manual Roaming Obligations Pertaining to Commercial Mobile Radio Services*, 15 FCC Rcd 21628, ¶ 32 (2000) (tentatively concluding that any roaming rules should sunset).

49. Report & Order, *Reexamination of Roaming Obligations of Commercial Mobile Radio Service Providers*, 22 FCC Rcd 15817 (2007). In 2010, the FCC further required carriers to offer roaming not only to carriers without transmission facilities in a given area, but also to carriers *with* facilities in that area if they need extra network capacity or functionality—an arrangement known as *home-on-home* roaming. *See* Order on Reconsideration, *Reexamination of Roaming Obligations of Commercial Mobile Radio Service Providers and Other Providers of Mobile Data Services*, 25 FCC Rcd 4181 (2010) (eliminating home-on-home exception adopted in 2007 order).

50. Second Report & Order, *Reexamination of Roaming Obligations of Commercial Mobile Radio Service Providers and Other Providers of Mobile Data Services*, 26 FCC Rcd 5411 (2011) ("*Data Roaming Order*"), *aff'd*, Cellco Partnership (d/b/a Verizon Wireless) v. FCC, 700 F.3d 534 (D.C. Cir. 2012).

51. In particular, sections 332(d)(1) and (2) make interconnectedness with the "public switched network" a criterion of "commercial mobile services"; section 332(d)(3) provides that any mobile wireless service that does not qualify as such a service or its functional equivalent is a "private mobile service" instead; and section 332(c)(2) provides that a provider of private mobile services "shall not, insofar as such [provider] is so engaged, be treated as a common

carrier for any purpose" under the Communications Act. 47 U.S.C. § 332(c)(2), (d)(1)–(3).

52. *See Data Roaming Order* ¶ 68.

53. Cellco Partnership (d/b/a Verizon Wireless) v. FCC, 700 F.3d 534 (D.C. Cir. 2012).

54. Report & Order, *Year 2000 Biennial Regulatory Review—Amendment of Part 22 of the Commission's Rules to Modify or Eliminate Outdated Rules Affecting the Cellular Radiotelephone Service and Other Commercial Mobile Radio Services*, 17 FCC Rcd 18401, 18406 ¶ 8 (2002).

55. *Spread Betting: How Code-Division Multiple Access (CDMA) Technology Emerged as the World Standard for Mobile Phones*, THE ECONOMIST TECH. Q. (June 19, 2003), http://www.economist.com/node/1841059?story_id=1841059.

56. For a discussion of the two approaches, see Neil Gandal et al., *Standards in Wireless Telephone Networks*, 27 TELECOMM. POL'Y. 325 (2003).

57. For a broader treatment of standards-setting issues in the telecommunications field, see chapter 12 of this book's first edition. For recent judicial decisions addressing high-profile allegations that particular firms gained anticompetitive intellectual property advantages by manipulating the standards-setting process, see Rambus Inc. v. FTC, 522 F.3d 456 (D.C. Cir. 2008), and Broadcom Corp. v. Qualcomm Inc., 501 F.3d 297 (3rd Cir. 2007).

58. *See* Petition for Rulemaking Regarding the Need for 700 MHz Mobile Equipment to Be Capable of Operating on All Paired Commercial 700 MHz Frequency Blocks (FCC filed Sept. 29, 2009) (assigned to RM No. 11592), http://apps.fcc.gov/ecfs/document/view?id=7020040176.

59. *See* Public Notice, *Wireless Telecommunications Bureau Seeks Comment on Petition for Rulemaking Regarding 700 MHz Band Mobile Equipment Design and Procurement Practices*, DA-1072, RM-11592 (FCC Feb. 18, 2010), http://hraunfoss.fcc.gov/edocs_public/attachmatch/DA-10-278A1.pdf ("*2010 Interoperability PN*").

60. *See, e.g.*, AT&T Comments, WT Docket No. 12-69 (filed June 1, 2012), http://apps.fcc.gov/ecfs/document/view?id=7021921411.

61. *See 2010 Interoperability PN*.

62. *See* Notice of Proposed Rulemaking, *Promoting Interoperability in the 700 MHz Commercial Spectrum*, WT Docket No. 12-69, FCC No. 12-31 (Mar. 21, 2012).

63. Of course, no consumer has a property right to any particular number. *See* In the matter of Starnet, Inc., 355 F.3d 634, 637 (7th Cir. 2004).

64. For a concise history of the FCC's wireless-to-wireless number portability rules, see Cellular Telecomm. & Internet Ass'n v. FCC, 330 F.3d 502, 504–07 (D.C. Cir. 2003) ("*CTIA*").

65. For competing views of the need for governmental intervention in the market to counteract switching costs and the associated phenomenon of *path dependence*, compare Joseph Farrell & Garth Saloner, *Installed Base and

Compatibility: Innovation, Product Preannouncements, and Predation, 76 AM. ECON. REV. 940 (1986), and Michael Katz & Carl Shapiro, *Technology Adoption in the Presence of Network Externalities*, 92 J. POL. ECON. 822 (1986), with STAN J. LIEBOWITZ & STEPHEN E. MARGOLIS, WINNERS, LOSERS, AND MICROSOFT (Independent Institute, 1999).

66. Memorandum Opinion and Order, *Verizon Wireless's Petition for Partial Forbearance from the Commercial Mobile Radio Services Number Portability Obligation*, 17 FCC Rcd. 14972, ¶ 18 (2002) (voicing fear that, absent portability rules, consumers "will find themselves forced to stay with carriers with whom they may be dissatisfied because the cost of giving up their wireless phone number in order to move to another carrier is too high").

67. *CTIA*.

68. *See* Memorandum Opinion & Order, *Carrier Requests for Clarification of Wireless–Wireless Porting Issues*, 18 FCC Rcd 20972 (2003).

69. Memorandum Opinion and Order, *Telephone Number Portability—CTIA Petitions for Declaratory Ruling on Wireline–Wireless Porting Issues*, 18 FCC Rcd 23697 (2003). The D.C. Circuit found that the FCC had adopted these rules without taking all required procedural steps. *See* U.S. Telecom Ass'n v. FCC, 400 F.3d 29 (D.C. Cir. 2005). But the rules ultimately went into effect anyway.

70. *See* Memorandum Opinion & Order, *Applications of AT&T Wireless Services, Inc. and Cingular Wireless Corp. for Consent to Transfer Control of Licenses and Authorizations*, 19 FCC Rcd 21522 (2004).

71. *See* Memorandum Opinion & Order, *Applications of Nextel Communications, Inc. and Sprint Corp. for Consent to Transfer Control of Licenses and Authorizations*, 20 FCC Rcd 13967 (2005).

72. U.S. Department of Justice and FTC, *Horizontal Merger Guidelines*, § 10 (Aug. 19, 2010), http://www.justice.gov/atr/public/guidelines/hmg-2010.pdf. The relationship between industry concentration and innovation is complex and is the topic of considerable scholarship. *See, e.g.*, Jonathan B. Baker, *Beyond Schumpeter vs. Arrow: How Antitrust Fosters Innovation*, 74 ANTITRUST L.J. 575 (2007); Michael L. Katz & Howard A. Shelanski, *Mergers and Innovation*, 74 ANTITRUST L.J. 1 (2007); Richard Gilbert, *Looking for Mr. Schumpeter: Where Are We in the Competition–Innovation Debate?* 6 INNOVATION POLICY AND THE ECONOMY 159 (2006).

73. U.S. Department of Justice and FTC, *Horizontal Merger Guidelines*, § 10.

74. U.S. Government Accountability Office (GAO), *Telecommunications: Enhanced Data Collection Could Help FCC Better Monitor Competition in the Wireless Industry*, GAO-10-779, at 24 (July 2010), http://www.gao.gov/assets/310/308167.pdf. Of course, much of that price drop is likely due to greater scale and technological advances as well.

75. *See* Fourteenth Report, *Implementation of Section 6002(b) of the Omnibus Budget Reconciliation Act of 1993*, 25 FCC Rcd 11407, ¶¶ 364–67 (2010).

76. For a summary of the merging parties' core economic analyses, see Decl. of Dennis W. Carlton, Allan Shampine, and Hal Sider, attached to AT&T/T-Mobile

Pub. Int. Statement, WT Docket No. 11-65 (Apr. 21, 2011), http://apps.fcc.gov/ecfs/document/view?id=7021240428. For the primary merger opponents' response, see Joint Declaration of Steven C. Salop, Stanley M. Besen, Stephen D. Kletter, Serge X. Moresi, and John R. Woodbury, Charles River Associates, attached to Sprint Pet. to Deny, WT Docket No. 11-65 (May 31, 2011), http://apps.fcc.gov/ecfs/document/view?id=7021675883.

77. *See* Complaint, United States v. AT&T Inc. et al., No. 1:11-cv-01560, 2011 WL 3823252 (D.D.C. Aug. 31, 2011), http://www.justice.gov/atr/cases/f274600/274613.pdf. Sprint inserted those input-market arguments into the litigation by subsequently filing its own complaint.

78. *See* Staff Analysis and Findings, WT Docket No. 11-65 (Nov. 29, 2011), http://hraunfoss.fcc.gov/edocs_public/attachmatch/DA-11-1955A2.pdf. The Staff Report noted the merger opponents' input-market arguments as well but did not fully analyze them.

79. *Id.* ¶ 17.

80. *Id.*

81. *Id.* ¶ 138.

82. As noted in chapter 1, Verizon Wireless also entered into a cross-marketing agreement with those same cable companies. In 2012, the Justice Department and the FCC approved those agreements, subject to various conditions designed to preserve competition between the cable companies and Verizon Wireless's majority shareholder, Verizon Communications, which provides fixed-line broadband and video services in competition with those cable companies. *See* Memorandum Opinion & Order and Declaratory Ruling, *Applications of Cellco Partnership d/b/a Verizon Wireless and SpectrumCo LLC and Cox TMI, LLC for Consent to Assign AWS-1 Licenses*, WT Docket No. 12-4, FCC No. 12-95, ¶¶ 143–58 (Aug. 23, 2012) ("*VZW–SpectrumCo Order*") (describing remedial measures).

83. *See, e.g.*, Harold Feld, *Verizon/SpectrumCo: Spectrum Gap v. Spectrum Crunch, Why Competition Is Actually Worse Off If Verizon Swaps AWS for 700 MHz (Part III)*, WetMachine (Apr. 23, 2012), http://tales-of-the-sausage-factory.wetmachine.com/verizonspectrumco-spectrum-gap-v-spectrum-crunch-why-competition-is-actually-worse-off-if-verizon-swaps-aws-for-700-mhz-part-iii/. Whether AT&T and Verizon actually enjoy a spectrum advantage over other providers is itself a controversial issue, which depends in part on how one counts spectrum in various bands. By one measure, Sprint and its wholesale partner Clearwire have far more spectrum than any other provider (*see Fifteenth Wireless Report* ¶ 303), although Sprint is quick to observe that Clearwire's spectrum is subject to various encumbrances and occupies a high-frequency band, with its associated propagation challenges. For a general discussion of spectrum holdings and their competitive significance, see *id.* at ¶¶ 286–307.

84. *VZW–SpectrumCo Order* ¶ 6 ("The transactions will result in an expeditious transfer of valuable spectrum into the hands of multiple national service providers that will put it to use in providing the latest generation mobile broadband services well in advance of the Commission's current deadlines").

Chapter 5

1. Of course, this fidelity to the original might be sacrificed if the senders and recipients use "lossy" compression technologies (such as JPEG) that can subtly degrade a document with each new copy.

2. As related to technology products, modularity involves "breaking up a complex system into discrete pieces—which can then communicate with one another only through standardized interfaces within a standardized architecture—[to] eliminate what would otherwise be an unmanageable spaghetti tangle of systemic interconnections." Richard N. Langlois, *Modularity in Technology and Organization*, 49 J. Econ. Behav. & Org. 19, 19 (2002). "When a design becomes 'truly modular,' the options embedded in the design are simultaneously multiplied and decentralized. The multiplication occurs because changes in one module become independent of changes in other modules. Decentralization follows because, as long as designers adhere to the design rules, they are free to innovate (apply the modular operators) without reference to the original architects or any central planners of the design." Carliss Y. Baldwin & Kim B. Clark, Design Rules, Vol. 1: The Power of Modularity 14 (MIT Press, 2000).

3. For a description of this standard, see T. Socolofsky & C. Kale, *A TCP/IP Tutorial,* RFC No. 1180, at 2-8 (1991), http://www.rfc-editor.org/rfc/rfc1180.txt.

4. *See* Resolution of the Federal Networking Council (Oct. 24, 1995) ("'Internet' refers to the global information system that—(i) is logically linked together by a globally unique address space based on the Internet Protocol (IP) or its subsequent extensions/follow-ons; (ii) is able to support communications using the Transmission Control Protocol/Internet Protocol (TCP/IP) suite or its subsequent extensions/follow-ons, and/or other IP-compatible protocols; and (iii) provides, uses or makes accessible, either publicly or privately, high level services layered on the communications and related infrastructure described herein"), *quoted in* Barry Leiner et al., *Brief History of the Internet*, Internet Society, http://www.internetsociety.org/internet/what-internet/history-internet/brief-history-internet.

5. Since 1998, ICANN has operated under contracts with the U.S. Department of Commerce to perform the Internet-governance functions previously overseen by the U.S. government itself in the early years of the Internet's evolution. Some foreign governments, such as those of Russia, India, and some developing nations, have periodically criticized the U.S. government for maintaining, through ICANN, a preeminent role for the United States in Internet governance. And there have been occasional international initiatives to move greater authority to the ITU, a United Nations agency. *See, e.g.*, World Summit on the Information Society (WSIS) Executive Secretariat, *Compilation of Comments Received on the Chair's Paper (DT/10), Chapter Three: Internet Governance*, Document WSIS-II/PC-3/DT/14 (Rev.2)-E, (Sept. 29, 2005), http://www.itu.int/wsis/docs2/pc3/working/dt14rev2.pdf; Declan McCullagh, *Perspective: Power Grab Could Split the Net*, CNET News (Oct. 3, 2005), http://news.cnet.com/2010-1071_3-5886556.html. To date, however, the U.S. government and other Internet stakeholders have successfully opposed those initiatives. *See generally* Organization for Economic

Cooperation and Development (OECD), *Communiqué on Principles for Internet Policy-Making*, OECD High Level Meeting on the Internet Economy, at 3 (June 28–29, 2011), http://www.oecd.org/internet/innovation/48289796.pdf ("As a decentralised network of networks, the Internet has achieved global interconnection without the development of any international regulatory regime. The development of such a formal regulatory regime could risk undermining its growth"). For early and critical academic commentaries on ICANN's role in Internet governance, see A. Michael Froomkin, *Wrong Turn in Cyberspace: Using ICANN to Route Around the APA and the Constitution*, 50 Duke L.J. 17 (2000), and Jonathan Weinberg, *ICANN and the Problem of Legitimacy*, 50 Duke L.J. 187 (2000).

6. John Naughton, A Brief History of the Future 102 (Overlook Press, 2001).

7. Jerome H. Saltzer et al., *End-to-End Arguments in System Design*, 2 ACM Transactions on Computer Systems 277 (1984), *reprinted in* Innovations in Internetworking 195 (Artech House, Craig Partridge ed., 1988). For an alternative account that denies the centrality of any end-to-end principle in the Internet's origins, see Richard Bennett, *End-to-End Arguments, Internet Innovation, and the Net Neutrality Debate*, Information Technology and Innovation Foundation (Sept. 2009), http://www.itif.org/files/2009-designed-for-change.pdf.

8. The posting of the original specifications for SMTP can still be found at http://www.ietf.org/rfc/rfc0821.txt. For an explanation of its invention, see Janet Abbate, Inventing the Internet 108–09 (MIT Press, 1999). Other (mostly proprietary) versions of email existed before SMTP, but they failed to reach critical mass.

9. *See* Tim Berners-Lee, Weaving the Web: The Original Design and Ultimate Destiny of the World Wide Web (Harper San Francisco, 1999).

10. This was the first major initial public offering of the Internet age. Ironically, the University of Illinois turned around and sold the Mosaic rights to Microsoft, which used Mosaic as the basis for the Internet Explorer browser that would soon eclipse Navigator and lead to the landmark Microsoft antitrust litigation, as discussed in chapter 1. For a telling of Netscape's rise and fall, see Michael Lewis, The New New Thing: A Silicon Valley Story (W. W. Norton, 1999).

11. Naughton, A Brief History, at 246–47.

12. In some cases, dictionaries can play an unofficial gatekeeping role in a language's development, as the *Oxford English Dictionary* does for English. The French have taken the idea one step further, founding the Académie Française in 1635 and maintaining it ever since as the official organization for overseeing the development of their language. *See* http://www.academie-francaise.fr.

13. Paul Hoffman, *The Tao of IETF: A Novice's Guide to the Internet Engineering Task Force*, IETF, at 1 (August 2001), http://www.ietf.org/tao.html. For other descriptions of the IETF, see Scott Bradner, *The Internet Engineering Task Force*, *in* Open Sources: Voices from the Open Source Revolution 47, 47–52 (O'Reilly, Chris DiBona et al. eds., 1999); Committee on the Internet in the Evolving Information Infrastructure, The Internet's Coming of Age

124, 134–35 (National Academies Press, 2001); A. Michael Froomkin, *Habermas@discourse.net: Toward a Critical Theory of Cyberspace*, 116 HARV. L. REV. 749, 796–817 (2003).

14. *See Session Initiation Protocol*, IETF, http://www.ietf.org/dyn/wg/charter/sip-charter.html.

15. Classification of a service as "interconnected VoIP" subjects the service provider to various regulatory duties, including obligations to connect 911 calls to the appropriate authorities and to contribute to the Commission's universal service fund. *See generally* Notice of Proposed Rulemaking and Third Report & Order, *Amending the Definition of Interconnected VoIP Service in Section 9.3 of the Commission's Rules*, 26 FCC Rcd 10074 (2011). The details of this classification are complex and contested. For example, in 2011 the Commission opened a proceeding to address whether it should expand the definition to include not only VoIP services that allow subscribers to call *and* be called by PSTN subscribers, but also, for example, services that allow subscribers to make outbound calls to the PSTN but *not* receive inbound calls from it. *See id.*; *see also* First Report & Order, *IP-Enabled Services; E911 Requirements for IP-Enabled Service Providers*, 20 FCC Rcd 10245, 10246 (2005), *aff'd sub nom.* Nuvio Corp. v. FCC, 473 F.3d 302 (D.C. Cir. 2007).

16. Memorandum Opinion, *Vonage Holdings Corp. Petition for Declaratory Ruling Concerning an Order of the Minn. Pub. Utils. Comm'n*, 19 FCC Rcd 22404 (2004), *aff'd*, Minnesota Pub. Utils. Comm'n v. FCC, 483 F.3d 570 (8th Cir. 2007). As noted in chapter 8, the FCC later entitled nomadic VoIP providers to reduce their universal service contribution obligations by conducting traffic studies identifying the percentage of their calls that are interstate or international. But that right comes with an important caveat: "[A]n interconnected VoIP provider with the capability to track the jurisdictional confines of customer calls would no longer qualify for the preemptive effects of our *Vonage Order* and would be subject to state regulation. This is because the central rationale justifying preemption set forth in the *Vonage Order* would no longer be applicable to such an interconnected VoIP provider." Report & Order and Further Notice of Proposed Rulemaking, *Universal Service Contribution Methodology*, 21 FCC Rcd 7518 ¶ 56 (2006), *aff'd in part and vacated in part on other grounds*, Vonage Holdings Corp. v. FCC, 489 F.3d 1232 (D.C. Cir. 2007). The FCC has also concluded that "state universal service fund contribution rules for nomadic interconnected VoIP are not preempted if they are consistent with the Commission's contribution rules for interconnected VoIP providers and the state does not enforce intrastate universal service assessments with respect to revenues associated with nomadic interconnected VoIP services provided in another state." Declaratory Ruling, *Universal Service Contribution Methodology*, 25 F.C.C.R. 15651, ¶ 1 (2010).

17. *See Minnesota Pub. Utils. Comm'n* v. FCC, 483 F.3d at 582–83 ("The [FCC's] order only suggests the FCC, if faced with the precise issue, would preempt fixed VoIP services. Nonetheless, the order does not purport to actually do so and until that day comes it is only a mere prediction. . . . Indeed, as we noted, the FCC has since indicated VoIP providers who can track the geographic

end-points of their calls do not qualify for the preemptive effects of the Vonage order. . . . As a consequence, [the] contention that state regulation of fixed VoIP services should not be preempted remains an open issue").

18. This traditional military-oriented account of the Internet's physical-layer origins is subject to debate. *Compare* KATIE HAFNER & MATTHEW LYON, WHERE WIZARDS STAY UP LATE: THE ORIGINS OF THE INTERNET 10 (Simon & Schuster, 1996), *with* Roy Rosenzweig, *Wizards, Bureaucrats, Warriors, and Hackers: Writing the History of the Internet*, 103 AMER. HIST. REV. 1530, 1532–33 (1998) (arguing that Hafner and Lyon's telling of the Internet's history downplays Baran's role and its military origins).

19. *See* HAFNER & LYON, WHERE WIZARDS STAY UP LATE, at 10.

20. Various "online" services, such as CompuServe and Prodigy, did achieve some commercial success in the 1980s and early 1990s, but they were not originally genuine ISPs. Instead, an online service sold subscribers access to "walled gardens": proprietary content (news, stock quotes, and the like), sometimes with the ability to exchange emails with other subscribers to the same service, but without full-blown access to the Internet.

21. For a discussion of the government's role in developing the Internet, see Edward L. Rubin, *Computer Languages as Networks and Power Structures: Governing the Development of XML*, 53 SMU L. REV. 1447, 1449–52 (2000); ABBATE, INVENTING THE INTERNET, at 54–60.

22. *See* Scientific and Advanced Technology Act of 1992, Pub. L. No. 102-476, § 4, 106 Stat. 2300 (codified at 42 U.S.C. §1862(g)); ABBATE, INVENTING THE INTERNET, at 196–200 (explaining the change in government policy).

23. Robert E. Kahn & Vinton G. Cerf, *What Is the Internet (and What Makes It Work)*, at 5 (1999), http://www.cnri.reston.va.us/what_is_internet.html; *see also* Notice of Proposed Rulemaking, *Appropriate Framework for Broadband Access to the Internet over Wireline Facilities*, 17 FCC Rcd 3019, ¶ 10 (2002) (noting that after a period of government support "the Internet entered a commercial phase characterized by more widespread network interconnection, an explosion of applications and access to a growing universe of websites utilizing common, interoperable protocols").

24. President William J. Clinton & Vice President Albert Gore Jr., *A Framework for Global Electronic Commerce* 4 (1997), http://clinton4.nara.gov/WH/New/Commerce/read.html (first principle of Internet policy).

25. *See* Jim Hu, *Study: Broadband Leaps Past Dial-Up*, CNET News (Aug. 18, 2004), http://news.cnet.com/2102-1034_3-5314922.html.

26. *See* Craig Moffett et al., *U.S. Cable and U.S. Telecommunications: Broadband End Game?* Bernstein Research, at 1, 7 (2010) ("[C]able's advantaged infrastructure will win the broadband wars. . . . Cable's share of 2Q 2010 net broadband additions rose steeply, to 91.4%, versus 67% in the prior quarter and a mere 41% in the year-ago quarter."); *see also* Susan P. Crawford, *The Communications Crisis in America*, 5 HARV. L. & POL'Y REV. 245, 248, 261 (2011) ("Given the tremendous economies of scale and cost advantages of the cable industry, being a wireline phone company is not a great business these days . . . .

The emergence of a *de facto* cable monopoly in high-speed wired Internet access in most of the country cannot stay a secret").

27. *See* Marisa Plumb, *Copper at the Speed of Fiber?* IEEE Spectrum (Oct. 2011), http://spectrum.ieee.org/telecom/internet/copper-at-the-speed-of-fiber (noting that, with new "vectoring" techniques, telcos can "push broadband speeds over 100 Mbps" using "legacy telephone access networks"); Alcatel-Lucent, *Get to Fast, Faster* (Strategic White Paper), at 2 (2011), http://www.alcatel-lucent.com/wireline/copper-access.html (projecting that, with VDSL2 vectoring technology, "copper can easily meet the bandwidth demand curve" projected through 2020).

28. *See, e.g.*, Peyman Faratin, David Clark, et al., *The Growing Complexity of Internet Interconnection*, 72 Communications & Strategies 51, 59–60 (4Q 2008); David Clark, William Lehr & Steven Bauer, *Interconnection in the Internet: The Policy Challenge* (Aug. 9, 2011), http://people.csail.mit.edu/wlehr/Lehr-Papers_files/clark%20lehr%20bauer%20TRPC2011%20%20Interconnection%20RC2.pdf.

29. *See Level 3 and Cogent Reach Agreement on Equitable Peering Terms*, PR Newswire (Oct. 28, 2005), http://www.prnewswire.com/news-releases/level-3-and-cogent-reach-agreement-on-equitable-peering-terms-55637437.html ("The modified peering arrangement allows for the continued exchange of traffic between the two companies' networks, and includes commitments from each party with respect to the characteristics and volume of traffic to be exchanged. Under the terms of the agreement, the companies have agreed to the settlement-free exchange of traffic subject to specific payments if certain obligations are not met").

30. *See Level 3 Issues Statement Concerning Internet Peering and Cogent Communications*, PR Newswire (Oct. 7, 2005), http://www.prnewswire.com/news-releases/level-3-issues-statement-concerning-internet-peering-and-cogent-communications-55014572.html.

31. *See* chapter 7 in this volume; Scott Woolley, *The Day the Web Went Dead*, Forbes (Dec. 2, 2008), http://www.forbes.com/2008/12/01/cogent-sprint-regulation-tech-enter-cz_sw_1202cogent.html?feed=rss_news.

32. Michael Kende, *The Digital Handshake: Connecting Internet Backbones*, FCC OPP Working Paper No. 32, at 7 (2000), http://transition.fcc.gov/Bureaus/OPP/working_papers/oppwp32.pdf. Although Kende discusses these relationships as being among Internet "backbone" networks, the same peering and transit relationships can apply to any pair of IP networks, including CDNs and conventional ISPs as well.

33. *See* Constance K. Robinson, *Network Effects in Telecommunications Mergers; MCI Worldcom Merger: Protecting the Future of the Internet*, Practicing Law Inst. of San Francisco (Aug. 23, 1999), http://www.usdoj.gov/atr/public/speeches/3889.htm.

34. *See* Complaint, United States v. WorldCom, No. 00-CV-1526 (D.D.C. filed June 26, 2000), http://www.justice.gov/atr/cases/f5000/5051.pdf.

35. *See* Memorandum Opinion & Order, *AT&T Inc. and BellSouth Corp. Application for Transfer of Control*, 22 FCC Rcd 5662, Appx. F (2007) ("For a period

of three years after the Merger Closing Date, AT&T/BellSouth will maintain at least as many discrete settlement-free peering arrangements for Internet backbone services with domestic operating entities within the United States as they did on the Merger Closing Date").

36. *See generally* Faratin & Clark et al., *Growing Complexity*; Christopher S. Yoo, *Innovations in the Internet's Architecture That Challenge the Status Quo*, 8 J. TELECOMM. & HIGH TECH. L. 79 (2010).

37. Akamai White Paper, *Why Performance Matters*, at 1 (2002).

38. *See* Yoo, *Innovations*, at 95.

39. *See, e.g.*, DrPeering International, *Internet Transit Prices—Historical and Projected* (Aug. 2010), http://drpeering.net/white-papers/Internet-Transit-Pricing-Historical-And-Projected.php.

Chapter 6

1. Report & Order, *Preserving the Open Internet*, 25 FCC Rcd 17905 (2010) ("*Open Internet Order*"), *pets. for review pending*, Verizon v. FCC, No. 11-1355 & consolidated cases (D.C. Cir.).

2. For a concise summary of the *Computer Inquiries*, see Robert Cannon, *Where ISPs and Telephone Companies Compete: A Guide to the Computer Inquiries*, 9 COMMLAW CONSPECTUS 49 (2001).

3. *See* Tentative Decision & Further Notice of Inquiry & Rulemaking, *Amendment of Section 64.702 of the Commission's Rules and Regulations* ("*Computer II*"), 72 F.C.C.2d 358 (1979), 77 F.C.C.2d 384 (1980) ("*Computer II Final Decision*"), 84 F.C.C.2d 50, 88 F.C.C.2d 512 (1981), *aff'd sub nom.* Computer and Communications Indus. Ass'n v. FCC, 693 F.2d 198 (D.C. Cir. 1982) ("*CCIA*").

4. *Computer II Final Decision* ¶¶ 93, 96.

5. 47 C.F.R. § 64.702(a); *see Computer II Final Decision* ¶ 120.

6. *See* First Report & Order, *Implementation of the Non-Accounting Safeguards of Sections 271 and 272 of the Communications Act of 1934*, 11 FCC Rcd 21905, ¶ 99 (1996) ("*Non-Accounting Safeguards Order*"), *modified*, 12 FCC Rcd 2297 (1997), 12 FCC Rcd 8653 (1997), *aff'd*, Bell Atl. Tel. Cos. v. FCC, 131 F.3d 1044 (1997). The AT&T consent decree is discussed in chapter 2.

7. *See* 47 U.S.C. § 153(24), (50), (53); *Non-Accounting Safeguards Order* ¶ 102; Report to Congress, *Federal–State Joint Board on Universal Service*, 13 FCC Rcd 11501, ¶¶ 29–33 (1998) ("*1998 Report to Congress*"); *see also* chapter 2 of this volume (discussing statutory definitions). The *1998 Report to Congress* is sometimes known as the *Stevens Report*, after Senator Ted Stevens of Alaska, who sponsored the legislation requesting it.

8. *See Computer II Final Decision* ¶ 231; *see generally* Further Notice of Proposed Rulemaking, *Policy & Rules Governing the Interstate, Interexchange Marketplace*, 13 FCC Rcd 21531, ¶ 33 (1998). This *Computer II* unbundling rule should not be confused with the *facilities* unbundling obligations of the 1996

Act. The former rule deals with obligations to provide transmission *services*; the 1996 Act facilities-unbundling requirements deal with obligations to lease *facilities* (or capacity on such facilities). At the margins, the two concepts merge because leasing out capacity on network facilities is often functionally equivalent to providing a transmission service over those facilities. Also, as noted in chapter 2, the *Computer II* rules separately imposed "unbundling" obligations on telephone companies to sell "customer premises equipment," such as telephones and computer modems, separately from telecommunications services. *See CCIA*.

9. *See generally* Joseph Farrell & Philip J. Weiser, *Modularity, Vertical Integration, and Open Access Policies: Toward a Convergence of Antitrust and Regulation in the Internet Age*, 17 HARV. J.L. & TECH. 85 (2003).

10. *See* California v. FCC, 905 F.2d 1217, 1231–32 (9th Cir. 1990).

11. *See generally* California v. FCC, 39 F.3d 919 (9th Cir. 1994).

12. *See id.* at 931–33.

13. As an AOL spokesperson exclaimed in August 1999, "I'm in awe of the magnitude and money involved. . . . But at its core this is really about an important principle. This involves a lot more than whose garage the Mercedes will be parked in, as the saying goes about big money cases. It involves who will control the Internet." Stephen Labaton, *Fight for Internet Access Creates Unusual Alliances*, N.Y. TIMES (Aug. 13, 1999), http://www.nytimes.com/1999/08/13/us/fight-for-internet-access-creates-unusual-alliances.html?pagewanted=all&src=pm.

14. AT&T Corp. v. City of Portland, 216 F.3d 871, 875 (9th Cir. 2000) ("*Portland*").

15. *Id.* at 878–80.

16. Agreement Containing Consent Orders; Decision & Order, *America Online, Inc. and Time Warner, Inc.*, FTC Docket No. C-3989, 2000 WL 1843019 (proposed Dec. 14, 2000).

17. *Portland*, 216 F.3d at 876. The FCC had filed an oddly inconclusive friend-of-the-court brief with the Ninth Circuit in the *Portland* case that touched on key legal issues, but it offered no clear guidance on any of them. Nor did it ask the court to refer the matter to it under the well-established doctrine of "primary jurisdiction." United States v. W. Pac. R.R. Co., 352 U.S. 59 (1956). The Commission seemed more anxious to avoid this political hot potato than to help establish a coherent body of law on a nationwide basis. At the time, FCC Chairman William Kennard justified the policy of deliberate nondecision as an effort to "do no harm." William E. Kennard, *How to End the World Wide Wait*, WALL ST. J., Aug. 24, 1999, at A18. As he further explained, he was less focused on regulating broadband than on spurring its deployment, explaining: "[W]e don't have a duopoly in broadband. We don't even have a monopoly in broadband. We have a 'no-opoly.' The bottom line is that, most Americans don't even have broadband." William E. Kennard, Remarks before the Federal Communications Bar Association (N. Cal. Chapter), *The Unregulation of the Internet: Laying a Competitive Course for the Future* (July 20, 1999), http://transition.fcc.gov/Speeches/Kennard/spwek924.html. Critics later noted that "had the FCC chosen to implement its 'hands off' policy through formal regulatory action, rather than

through oblique pronouncements, it might have avoided the series of conflicting judicial open access decisions that eventually threatened the agency's ability to set broadband policy on a national basis." Barbara S. Esbin & Gary S. Lutzker, *Poles, Holes, and Cable Open Access: Where the Global Information Superhighway Meets the Local Right-of-Way*, 10 COMMLAW CONSPECTUS 23, 55 (2001); *see also* Nat'l Cable & Telecomm. Ass'n, Inc. v. Gulf Power Co., 534 U.S. 327, 348 (2002) (Thomas, J., concurring in part and dissenting in part) (the FCC's responsibility to implement its regulatory statute "does not permit [it] to avoid this question").

18. *Portland*, 216 F.3d at 878. Although the legal debate ultimately devolved to a binary contest between the "information service" and "telecommunications service" characterizations, early advocates of multiple ISP access argued that cable modem service should be treated as a kind of "cable service" subject to regulation under Title VI. Because traditional cable services had long been subject to expansive regulation by local franchising authorities, these advocates hoped that if cable modem service were treated as just another cable service, it too could be subject to broad regulation by these local authorities, who were the first to impose broadband open access requirements. This argument soon foundered on an uncooperative statutory definition of "cable service," however, and little is heard of it today. Specifically, Title VI defines "cable service" as "(A) the one-way transmission to subscribers of (i) video programming, or (ii) other programming service, and (B) subscriber interaction, if any, which is required for the selection or use of such video programming or other programming service." (47 U.S.C. § 522(6)). In turn, "video programming" means "programming provided by, or generally considered comparable to programming provided by, a television broadcast station" (47 U.S.C. § 522(20)), and "other programming service" means "information that a cable operator makes available to all subscribers generally" (47 U.S.C. § 522(14)). As the Ninth Circuit held in *Portland*, cable modem service provides a range of "interactive and individual" services, such as Web browsing and email, that collectively defy this statutory definition, which appears designed more narrowly to encompass conventional television programming plus a rudimentary "interactive TV" service whose content is supplied by the cable company itself. *See* 216 F.3d at 873. Interestingly, the opposing parties in the Ninth Circuit case actually agreed (although the Ninth Circuit ultimately did not) that cable modem service is a Title VI "cable service" but disagreed about the implications of that characterization for the legality of open-access regulations.

19. Declaratory Ruling & Notice of Proposed Rulemaking, *Inquiry Concerning High-Speed Access to the Internet over Cable and Other Facilities*, 17 FCC Rcd 4798 (2002) ("*Cable Broadband Order*"). The subsequent judicial history of this order is discussed later in the chapter.

20. *Id.* ¶ 38; *see also 1998 Report to Congress* ¶ 88.

21. *See 1998 Report to Congress* ¶ 13; Report & Order, *Federal–State Joint Board on Universal Service*, 12 FCC Rcd 8776 ¶¶ 788–89 (1997).

22. 47 U.S.C. §§ 153(50), (53).

23. *Id.* § 153(50); *see id.* § 153(24) (definition of "information service").

24. *Cable Broadband Order* ¶ 43. Although the term "wireline" today can be used to denote both telephone and cable companies, the Commission specified that "by 'wireline,' [it] refer[red] to services provided over the infrastructure of traditional telephone networks." *Id.* ¶ 43 n.169.

25. *Id.*

26. Brand X Internet Servs. v. FCC, 345 F.3d 1120 (9th Cir. 2003).

27. National Cable & Telecomm. Ass'n v. Brand X Internet Servs., 545 U.S. 967 (2005) ("*Brand X*").

28. *See* Notice of Proposed Rulemaking, *Appropriate Framework for Broadband Access to the Internet over Wireline Facilities*, 17 FCC Rcd 3019, ¶¶ 17–25 (2002).

29. *See* Report & Order, *Appropriate Framework for Broadband Access to the Internet over Wireline Facilities*, 20 FCC Rcd 14853, ¶¶ 19, 44 (2005) ("*Wireline Broadband Order*"), *aff'd*, Time Warner Telecom, Inc. v. FCC, 507 F.3d 205 (3d Cir. 2007).

30. Passages in this section are adapted from Jonathan E. Nuechterlein, *Antitrust Oversight of an Antitrust Dispute: An Institutional Perspective on the Net Neutrality Debate*, 7 J. TELECOMM. & HIGH TECH. L. 19 (2009).

31. Tim Wu, *Network Neutrality, Broadband Discrimination*, 2 J. TELECOMM. & HIGH TECH. L. 141, 149 (2003).

32. *See id.*; Ex Parte Letter of Professors Tim Wu and Lawrence Lessig, *Appropriate Regulatory Treatment for Broadband Access to the Internet over Cable Facilities*, CS Dkt. 02-52 (FCC filed Aug. 22, 2003).

33. J. H. Saltzer, D. P. Reed, & D. D. Clark, *End-to-End Arguments in System Design*, 2 ACM Transactions in Computer Systems 277 (Nov. 1984), http://web.mit.edu/Saltzer/www/publications/endtoend/endtoend.pdf.

34. *Open Internet Order* ¶ 54.

35. Philip J. Weiser, *The Next Frontier for Net Neutrality*, 60 ADMIN. L. REV. 273 (2008).

36. *Id.* at 290 (quoting LOUIS D. BRANDEIS, OTHER PEOPLE'S MONEY AND HOW THE BANKERS USE IT 62 (Nat'l Home Lib. Found., 1933)).

37. Michael K. Powell, *Preserving Internet Freedom: Guiding Principles for the Industry*, at 5 (Feb. 8, 2004), http://hraunfoss.fcc.gov/edocs_public/attachmatch/DOC-243556A1.pdf.

38. Policy Statement, *Appropriate Framework for Broadband Access to the Internet over Wireline Facilities*, 20 FCC Rcd 14986, ¶¶ 4–5 & n.15 (2005) ("*FCC Broadband Policy Statement*").

39. *Id.*

40. *See* Order, *Madison River Communications LLC*, 20 FCC Rcd 4295 (2005).

41. *E.g.*, Memorandum Opinion & Order, *SBC Communications Inc. and AT&T Corp. Applications for Approval of Transfer of Control*, 20 FCC Rcd 18290 (2005). Subsequently, in connection with approving the AT&T–BellSouth merger, the FCC extracted from the combined company a further commitment

not to enter into certain arrangements with Internet content, applications, or service providers for two years. *See* Memorandum Opinion & Order, *AT&T Inc. and BellSouth Corporation Application for Transfer of Control*, 22 FCC Rcd 5662, app. F, at 5814–15 (2007).

42. *See* Notice of Inquiry, *Broadband Industry Practices*, 22 FCC Rcd 7894 (2007).

43. The anti-blocking issue is somewhat more complicated with respect to wireless broadband platforms, given the more extreme scarcity of network bandwidth (i.e., licensed spectrum). *See generally* Robert W. Hahn et al., *The Economics of "Wireless Net Neutrality,"* AEI–Brookings Joint Center Working Paper No. RP07-10 (Apr. 2007), http://papers.ssrn.com/sol3/papers.cfm?abstract_id=983111. As discussed later in this chapter, the FCC recognized that distinction in its *Open Internet Order* by subjecting wireless broadband networks to more lenient regulatory obligations.

44. *See, e.g.*, Jacqui Cheng, *Evidence Mounts That Comcast Is Targeting BitTorrent Traffic*, Ars Technica (Oct. 19, 2007), http://arstechnica.com/uncategorized/2007/10/evidence-mounts-that-comcast-is-targeting-bittorrent-traffic/.

45. *See* Memorandum & Order, *Formal Complaint of Free Press and Public Knowledge against Comcast Corporation for Secretly Degrading Peer-to-Peer Applications*, 23 FCC Rcd 13028 ¶¶ 47–48 (2008) ("*Comcast Order*"). The Commission also asserted that "Comcast . . . has not been fully forthcoming about its own practices." *Id.* ¶ 31; *see also id.* ¶ 9 n. 31 (alleging that Comcast's initial responses to complaints "raise troubling questions about Comcast's candor during this proceeding"). As a remedy, the Commission ordered Comcast to follow through on its commitment to implement "protocol-agnostic network management technique[s]," *id.* at ¶ 54, an objective that Comcast worked cooperatively with the Internet Engineering Task Force to meet. The Commission stopped short of imposing any monetary sanctions on Comcast, but only because it acknowledged that it was announcing new standards of conduct for the first time in this adjudicative proceeding. *Id.* ¶ 34.

46. Comcast Corp. v. FCC, 600 F.3d 642 (D.C. Cir. 2010).

47. *Open Internet Order* ¶¶ 63, 66.

48. *Id.* ¶ 99 (emphasis added).

49. *Id.* (emphasis added).

50. *Id.* ¶ 100.

51. *Id.* ¶¶ 94–95.

52. *Id.* ¶ 82.

53. *Id.*

54. *Id.* ¶ 85.

55. *Id.* ¶ 83.

56. *See, e.g.*, Cisco Visual Networking Index: Global Mobile Data Traffic Forecast Update, 2010–2015 (Feb. 1, 2011), http://www.cisco.com/en/US/solutions/collateral/ns341/ns525/ns537/ns705/ns827/white_paper_c11-520862.html.

57. *See* Philip J. Weiser, *The Future of Internet Regulation*, 43 U.C. Davis L. Rev. 529 (2009) (describing the benefits and challenges presented by such a "co-regulation" regime).

58. Broadband Internet Technical Advisory Group, *BITAG History*, http://www.bitag.org/bitag_organization.php?action=history.

59. "While BITAG is intended to derive consensus resolutions of technical issues with public policy implications, it has no formal relationship with any government agency. It is not formed under or governed by the U.S. Federal Advisory Commission Act; therefore, it is continuing to explore exactly how it should relate to federal agencies." Joe Waz & Phil Weiser, *Internet Governance: The Role of Multistakeholder Organizations*, Silicon Flatirons Roundtable Series on Entrepreneurship, Innovation, and Public Policy, at 8 (2012) (footnote omitted), http://siliconflatirons.com/documents/publications/report/InternetGovernanceRoleofMSHOrgs.pdf.

60. *See, e.g.*, *Comcast Order* ¶ 5 ("Peer-to-peer applications, including those relying on BitTorrent, have become a competitive threat to cable operators such as Comcast because Internet users have the opportunity to view high-quality video with BitTorrent that they might otherwise watch (and pay for) on cable television. Such video distribution poses a particular competitive threat to Comcast's video-on-demand ('VOD') service"); *id.* ¶ 50 ("To the extent . . . that providers choose to utilize practices that are not application or content neutral, the risk to the open nature of the Internet is particularly acute and the danger of network management practices being used to further anticompetitive ends is strong").

61. *Open Internet Order* ¶ 87; *see id.* ¶ 73.

62. *Id.* ¶ 91.

63. *Id.* ¶ 72.

64. *See, e.g.*, Marvin Ammori, *Time Warner Goes Back to the Future*, Save the Internet (Jan. 25, 2008) (arguing that metered pricing "raises Net Neutrality issues" because Time Warner is unlikely "to apply its new high-bandwidth surcharges to its own product," and "favoring its own content over other channels or programs like BitTorrent would be discriminatory"); *see also* Fred von Lohmann, *Time Warner Puts a Meter on the Internet*, Electronic Frontier Foundation (Jan. 22, 2008), https://www.eff.org/deeplinks/2008/01/time-warners-puts-meter-internet (expressing concern that metered pricing "could be used as a cover for price increases on existing customers (bad)" and insisting that "the pricing for 'overages' should bear some relation to costs"); Press Release, Free Press, *Time Warner Metering Exposes America's Bigger Broadband Problems* (Jan. 17, 2008), http://www.freepress.net/release/328 (quoting policy director Ben Scott: "telling consumers they must choose between blocking and metered pricing is a worrying development").

65. *See, e.g.*, David Goldman, *AT&T Starts Capping Broadband*, CNNMoney (May 9, 2011), http://money.cnn.com/2011/05/03/technology/att_broadband_caps/index.htm. As discussed below, usage tiers have sparked controversy when broadband providers have applied them to "Internet" traffic (such as over-the-top

video services) but not "specialized" video services (such as managed IPTV services).

66. *Open Internet Order* ¶ 76. For representative academic commentary on this set of issues, see Tim Wu & Christopher Yoo, *Keeping the Internet Neutral? Tim Wu and Christopher Yoo Debate*, 59 Fed. Comm. L.J. 575 (2007); Robert D. Atkinson & Philip J. Weiser, *A "Third Way" on Network Neutrality*, Information Tech. & Innovation Foundation (May 3, 2006), http://www.itif.org/files/netneutrality.pdf; Tim Wu, *Why Have a Telecommunications Law? Anti-discrimination Norms in Communications*, 5 J. Telecomm. & High Tech. L. 15 (2006); J. Gregory Sidak, *A Consumer-Welfare Approach to Network Neutrality Regulation of the Internet*, 2 J. Competition L. & Econ. 349 (2006); Robert Hahn & Scott Wallsten, *The Economics of Net Neutrality*, AEI–Brookings Joint Center, Working Paper No. RP06-13 (Apr. 2006), http://ssrn.com/abstract=943757; Benjamin E. Hermalin & Michael L. Katz, *The Economics of Product-Line Restrictions with an Application to the Network Neutrality Debate*, Competition Policy Center (July 28, 2006), http://escholarship.org/uc/item/81r3b7xs; Christopher S. Yoo, *Beyond Network Neutrality*, 19 Harv. J.L. & Tech. 1 (2005).

67. *See Open Internet Order* ¶ 67.

68. *See*, *e.g.*, C. Scott Hemphill, *Network Neutrality and the False Promise of Zero-Price Regulation*, 25 Yale J. Reg. 135 (2008) ("*False Promise*").

69. *See*, *e.g.*, Testimony of Timothy Wu, *Hearing before the Task Force on Telecom and Antitrust on the Comm. on the Judiciary*, H. Rep., 109th Cong., 2d Sess., at 7 (Apr. 25, 2006), http://commdocs.house.gov/committees/judiciary/hju27225.000/hju27225_0.HTM#52 (asserting that "[t]he best proposals for network neutrality rules . . . leave open legitimate network services that the Bells and Cable operators want to provide, such as offering cable television services and voice services along with a neutral internet offering").

70. *See* Comments of AT&T Inc., GN Docket No. 09-51, at 51–53 (FCC filed Jan. 14, 2010) (discussing AT&T's methods for distinguishing different service classes over same networks).

71. Lawrence Lessig, The Future of Ideas: The Fate of the Commons in a Connected World 176 (Vintage, 2001) (quoting Charles Platt, *The Future Will Be Fast but Not Free*, Wired (May 2001)) (emphasis omitted).

72. *See*, *e.g.*, Quality of Service Working Group, *Inter-provider Quality of Service, White Paper Draft 1.1*, MIT Communications Futures Program (Nov. 17, 2006), http://cfp.mit.edu/publications/CFP_Papers/Interprovider%20QoS%20MIT_CFP_WP_9_14_06.pdf; *About InterStream*, http://interstream.com/about.

73. *See*, *e.g.*, Tejas Patel, *Telcos Team Up to Interconnect Their Business Video Communities*, RCR Wireless (Feb. 1, 2012), http://www.rcrwireless.com/india/20120201/carriers/telcos-team-up-to-interconnect-their-business-video-communities/; Press Release, *AT&T and Orange Business Services Expand Telepresence Community* (Feb. 2, 2012), http://www.att.com/gen/press-room?pid=22346&cdvn=news&newsarticleid=33819; Paul Taylor, *AT&T and BT in*

Telepresence Exchange, FT.com (Dec. 1, 2010), http://www.ft.com/cms/s/0/6adeaf40-fcb1-11df-bfdd-00144feab49a.html.

74. This flat ban on commercial agreements was a central feature of the most prominent net neutrality bills proposed in Congress. *See, e.g.*, S. 215, 110th Cong. § 12(a)(4)(C), (5) (2007); H.R. 5273, 109th Cong. § 4(a)(6), (7) (2006); H.R. 5417, 109th Cong. § 3 (2006).

75. *See* 47 U.S.C. §§ 202, 211; *see generally* Report & Order, *Competition in the Interstate Interexchange Marketplace*, 6 FCC Rcd 5880 (1991). For an example of this type of proposal, see John Windhausen Jr., *Good Fences Make Bad Broadband: Preserving an Open Internet through Net Neutrality*, Public Knowledge, at 40–45 (Feb. 6, 2006), http://www.publicknowledge.org/pdf/pk-net-neutrality-whitep-20060206.pdf.

76. *Open Internet Order* ¶ 68.

77. One apt example comes from the natural gas industry, where pipelines are "bottleneck" utilities subject to common carrier regulation. As the Federal Energy Regulatory Commission (FERC) has explained, "A large percentage of a pipeline's capacity may be reserved at any given time for firm sales and firm transportation. However, customers that have reserved or 'booked' pipeline capacity and thus have first claim on its use may not always use the entire amount they have reserved. Traditionally, pipelines have taken advantage of that unused (but 'booked') capacity by offering a sales or transportation service that is subject to being terminated or 'interrupted' by the prior claim of firm sales or transportation customers. Although this interruptible service is inferior to and less valuable than firm service, its offering seeks to maximize utilization of idle pipeline capacity and therefore is in the public interest and must be encouraged by ratemaking." *Regulation of Natural Gas Pipelines after Partial Wellhead Decontrol*, 50 Fed. Reg. 42408, 42435 (FERC Oct. 18, 1985); *see generally* Tennessee Gas Pipeline Co. v. FERC, 972 F.2d 376, 379 (D.C. Cir. 1992); Associated Gas Distributors v. FERC, 824 F.2d 981, 1013 (D.C. Cir. 1987) ("Firm sales contracts give the customer the right to demand, and obligate the pipeline at all times to stand ready to deliver, a certain quantity of gas per day"); Complex Consol. Edison Co. v. FERC, 165 F.3d 992, 998 n.12 (D.C. Cir. 1999) (interruptible service "provides gas on a 'when available' basis and may be interrupted after notice to the subscriber").

78. *Open Internet Order* ¶ 76.

79. As Alfred Kahn had explained several years earlier, net neutrality advocates were sometimes "guilty of using the term 'discrimination' sloppily, to embrace mere *differences* in price for different qualities of service." Statement of Alfred E. Kahn, Robert Julius Thorne Professor of Political Economy, Emeritus, Cornell University, before the FTC Workshop on Broadband Connectivity Competition Policy, at 4 (delivered Feb. 13, 2007), http://www.ftc.gov/opp/workshops/broadband/presentations/kahn.pdf (Feb. 21, 2007 rev.) (some emphasis omitted).

80. *Open Internet Order* ¶¶ 112–14.

81. *Id.* ¶ 112.

82. *Id.*

83. *See* Atkinson & Weiser, *A "Third Way."*

84. United States v. Comcast Corp., No. 1:11-cv-00106, 2011 WL 5402137, § V.G.5 (D.D.C. Sept. 1, 2011) ("*Comcast–NBCU Final Judgment*"); Memorandum Opinion & Order, *Applications of Comcast Corp., General Elec. Corp., and NBC Universal, Inc. for Consent to Assign Licenses and Transfer Control of Licenses*, 26 FCC Rcd 4238, Appx. A § IV.E.3 (2011) ("*Comcast–NBCU Order*").

85. Netflix Inc., Quarterly Report (Form 8-K), at 8 (Apr. 25, 2011), http://files.shareholder.com/downloads/NFLX/1872575600x0xS1193125-11-107751/1065280/filing.pdf ("AT&T recently imposed caps of 250 gigabytes on U-verse fiber Internet subscribers, and 150 gigabytes on DSL subscribers, with a charge of approximately 20¢ per gigabyte over those limits. We'll study how this affects consumer attitudes about Internet video, and take appropriate steps if needed. Comcast has had 250 gigabytes caps for years without overage charges and that hasn't been a problem for Comcast customers or for us").

86. Tony Werner, *The Facts about Xfinity TV and Xbox 360: Comcast Is Not Prioritizing*, Comcast Voices (May 15, 2012), http://corporate.comcast.com/comcast-voices/the-facts-about-xfinity-tv-and-xbox-360-comcast-is-not-prioritizing.html. Comcast asserted that although it was using Differentiated Services Code Point ("DSCP") markings to distinguish Xfinity Xbox packets from Internet packets, it did so to route the Xfinity Xbox packets over a logically separate "service flow" within the same broadband pipe while leaving the Internet packets unaffected.

87. *See, e.g.*, Harold Feld, *Michael Powell Works the Ref on the XBox360 Play*, Public Knowledge Blog (Mar. 29, 2012), http://www.publicknowledge.org/blog/michael-powell-works-ref-xbox360-play; Daniel Ionescu, *Netflix Boss Blasts Comcast over Bandwidth Caps, Net Neutrality*, PCWorld (Apr. 16, 2012), http://www.pcworld.com/article/253850/netflix_boss_blasts_comcast_over_bandwidth_caps_net_neutrality.html.

88. Reed Hastings, Facebook post, Apr. 15, 2012, http://www.facebook.com/reed1960/posts/10150706947044584.

89. *Quoted in* Todd Shields, *Netflix Wants Help from U.S. against Cable Data Caps*, Bloomberg (June 27, 2012), http://www.bloomberg.com/news/2012-06-27/netflix-wants-help-from-u-s-against-cable-data-caps.html.

90. Werner, *The Facts about Xfinity TV*; *see also* Michael Powell, *No Good Deed Goes Unpunished: Washington Advocacy Run Amok*, CableTechTalk (Mar. 2012), http://www.cabletechtalk.com/tech-discussions/2012/03/28/no-good-deed-goes-unpunished-%e2%80%93-washington-advocacy-run-amok/.

91. *Open Internet Order* ¶ 76.

92. *Id.* ¶ 67 n.209 (emphasis added).

93. *Announcing the Netflix Open Connect Network*, Netflix U.S. and Canada Blog (June 4, 2012), http://blog.netflix.com/2012/06/announcing-netflix-open-connect-network.html.

94. 47 C.F.R. § 8.7.

95. *See generally* United States v. Microsoft Corp., 253 F.3d 34 (D.C. Cir. 2001) (setting forth the standard for antitrust liability); Howard A. Shelanski & J. Gregory Sidak, *Antitrust Divestiture in Network Industries*, 68 U. Chi. L. Rev. 1, 19–20 (2001) (discussing concerns about remedies).

96. The two of us have written articles taking opposite positions on the latter institutional issue. *See* Weiser, *Future of Internet Regulation*, at 550–51 (rejecting reliance on antitrust process for net neutrality disputes); Nuechterlein, *Antitrust Oversight* (endorsing such reliance).

97. For example, Jon Leibowitz, then a commissioner (and later chairman) of the FTC, embraced this distinction when he opined that "while antitrust may be a good way of thinking about [consumers' 'Internet Freedoms'], it is not necessarily well-suited to *protecting* them." Commissioner Jon Leibowitz, *Concurring Statement Regarding the Staff Report:"Broadband Connectivity Competition Policy,"* at 1 (FTC 2007), http://www.ftc.gov/speeches/leibowitz/V070000statement.pdf; *see also* Commissioner J. Thomas Rosch, *Address at the Broadband Policy Summit IV: Broadband Access Policy: The Role of Antitrust*, at 6 (June 13, 2008), http://www.ftc.gov/speeches/rosch/080613broadbandaccess.pdf (noting that an antitrust court might well have refused to order a remedy against Madison River).

98. *See, e.g.*, Verizon Commc'ns Inc. v. Law Offices of Curtis V. Trinko, LLP, 540 U.S. 398 (2004); Brooke Group Ltd. v. Brown & Williamson Tobacco Corp., 509 U.S. 209 (1993).

99. For example, the FCC requires incumbent LECs to share local loops and some transport links with their rivals (see chapter 2) and requires cable companies to share certain "must see" programming with their subscription video rivals (see chapter 9). After the *Trinko* decision, it seems unlikely that antitrust courts would recognize any similar duties to deal (see chapter 10). In each case, however, the FCC applied antitrust-oriented concepts to reach its conclusions, including an analysis of the market, market power, and the costs and benefits of regulatory intervention.

100. *See, e.g.*, Brett M. Frischmann, Infrastructure: The Social Value of Shared Resources 330–32 (Oxford U. Press, 2012). *See generally* Barbara van Schewick, Internet Architecture and Innovation (MIT Press, 2010).

101. Lessig, Future of Ideas, at 176 (internal quotation marks omitted).

102. *See, e.g.*, SavetheInternet.com Coalition, *Frequently Asked Questions*, http://www.savetheinternet.com/faq (last visited August 19, 2012) ("If we allow telecom corporations to take control of the Internet, everyone who goes online will be affected. . . . Independent voices and political groups are especially vulnerable. The cost of sharing video and images could skyrocket, silencing independent media and amplifying the voices of big media companies"); *see also Open Internet Order* ¶ 68 (regulation is necessary because "fixed broadband providers have incentives and the ability to discriminate in their handling of network traffic in ways that can harm innovation, investment, competition, end users, and free expression").

103. *Open Internet Order* ¶ 15.

104. Wu, *Network Neutrality, Broadband Discrimination*, at 142.

105. News Release, *Wyden Moves to Ensure Fairness of Internet Usage with New Net Neutrality Bill* (Mar. 2, 2006), http://www.wyden.senate.gov/news/press-releases/wyden-moves-to-ensure-fairness-of-internet-usage-with-new-net-neutrality-bill.

106. Brett Frischman disputes this point, reasoning that the "social value of the Internet greatly exceeds [the] market value" that would be reflected in consumers' decisions even in fully competitive broadband markets. FRISCHMANN, INFRASTRUCTURE, at 331. That objection appears to assume that the disputed forms of net neutrality regulation—such as the presumptive ban on differential packet treatment—are necessary to keep the Internet from fragmenting and thus losing much of its social value as a shared public good. But proponents of such regulation have not substantiated that concern or explained why such regulation is currently necessary, given that policymakers would presumably be able to take stronger measures in the future if a genuine threat to the Internet's integrity later arose.

107. *See, e.g.*, Pet. for Declaratory Ruling of Free Press, et al., WC Dkt. 07-52, at ii (FCC filed Nov. 1, 2007) (identifying "[t]he paradigmatic fear of network neutrality defenders" as the possibility "that network providers who compete[] (or [seek] to compete) with independent applications [will] secretly degrade those applications in ways prompting consumers to abandon those degraded applications, undermining consumer choice, innovation, and a competitive market").

108. *See generally* Timothy J. Tardiff, *Changes in Industry Structure and Technological Convergence: Implications for Competition Policy and Regulation in Telecommunications*, 4 INT'L ECON. & ECON. POL. 109 (2006); Dennis L. Weisman, *When Can Regulation Defer to Competition for Constraining Market Power? Complements and Critical Elasticities*, 2 J. COMPETITION L. & ECON. 101, 102 (2006) ("[P]rice increases that produce even small reductions in demand can generate large losses in contribution to joint and common costs because the firm's revenues decline much more than the costs it can avoid. It is in this manner that high margins can serve to discipline the [de]regulated firm's pricing behavior.").

109. *Open Internet Order* ¶ 32.

110. *See, e.g.*, RICHARD A. POSNER, ANTITRUST LAW 223–29 (U. Chi. Press, 2d ed. 2001); Christopher S. Yoo, *Network Neutrality and the Economics of Congestion*, 94 GEO. L.J. 1847, 1885–87 (2006); Christopher S. Yoo, *Vertical Integration and Media Regulation in the New Economy*, 19 YALE J. ON REG. 171 (2002); *see generally* Herbert Hovenkamp, *Antitrust Policy after Chicago*, 84 MICH. L. REV. 213, 255–83 (1985).

111. *See, e.g.*, Farrell & Weiser, *Modularity, Vertical Integration, and Open Access Policies*, at 104.

112. For classic expositions of the efficiencies of vertical integration, see OLIVER E. WILLIAMSON, THE MECHANISMS OF GOVERNANCE (Oxford Univ. Press, 1996), and R. H. COASE, THE FIRM, THE MARKET, AND THE LAW (U. Chi. Press, 1990).

113. *See* WILLIAM J. BAUMOL & ALAN S. BLINDER, ECONOMICS: PRINCIPLES AND POLICY, 248–52 (Dryden Press, 8th ed. 2000) (discussing consumer benefits of product differentiation and "monopolistic competition"); Hermalin & Katz, *The Economics of Product-Line Restrictions*.

114. *See generally* Brunswick Corp. v. Pueblo Bowl-O-Mat, Inc., 429 U.S. 477, 488 (1977) (antitrust laws are enforced "for the protection of competition not competitors") (quotation marks omitted).

115. *See* Farrell & Weiser, *Modularity, Vertical Integration, and Open Access Policies*, at 105–19; *see also* Barbara van Schewick, *Toward an Economic Framework for Network Neutrality Regulation*, 5 J. TELECOMM. & HIGH TECH. L. 329 (2007) (arguing for recognition of additional exceptions beyond those acknowledged in existing economic literature).

116. *See* Farrell & Weiser, *Modularity, Vertical Integration, and Open Access Policies*, at 105–07.

117. *See* United States v. Microsoft Corp., 253 F.3d 34 (D.C. Cir. 2001); Farrell & Weiser, *Modularity, Vertical Integration, and Open Access Policies*, at 110–11. The precise empirical basis for the government's antitrust suit against Microsoft is subject to debate. *See, e.g.*, Thomas Hazlett, *US v Microsoft: Who Really Won*, FT.COM (Jan. 28, 2008).

118. Farrell & Weiser, *Modularity, Vertical Integration, and Open Access Policies*, at 108–09.

119. *See id.* at 119.

120. *See id.* at 114–17.

121. JAMES F. KUROSE & KEITH W. ROSS, COMPUTER NETWORKING: A TOP-DOWN APPROACH 598 (Addison Wesley, 5th ed. 2010).

122. Lawrence Lessig & Robert W. McChesney, *No Tolls on the Internet*, WASH. POST (June 8, 2006), http://www.washingtonpost.com/wp-dyn/content/article/2006/06/07/AR2006060702108.html.

123. *Open Internet Order* ¶ 29.

124. *See* Atkinson & Weiser, *A "Third Way."*

125. *Comcast–NBCU Final Judgment*, § V.G.5; *Comcast–NBCU Order*, Appx. A § IV.E.3.

126. *See* Gerald R. Faulhaber & David J. Farber, *The Open Internet: A Customer-Centric Framework*, 4 INT'L J. OF COMMUNICATION 302, 324 (2010) ("'Just add capacity' is a recipe for a very expensive Internet, primarily because of the bursty nature of Internet traffic.").

127. Hemphill, *False Promise*, at 139.

128. *See, e.g.*, Christopher S. Yoo, *Would Mandating Broadband Network Neutrality Help or Hurt Competition? A Comment on the End-to-End Debate*, 3 J. TELECOMM. & HIGH TECH. L. 23 (2004); Wu & Yoo, *Keeping the Internet Neutral?* at 587–90; *see also* Hermalin & Katz, *The Economics of Product-Line Restrictions*; Thomas Hazlett & Anil Caliskan, *Natural Experiments in U.S. Broadband Regulation*, GEORGE MASON UNIV. LAW AND ECONOMICS

RESEARCH PAPER SERIES, No. 08-04 (2007), http://www.law.gmu.edu/assets/files/publications/working_papers/08-04%20Natural%20Experiments.pdf.

129. Wu & Yoo, *Keeping the Internet Neutral?* at 590–92.

130. *See, e.g.*, LESSIG, FUTURE OF IDEAS, at 147–76.

131. *See, e.g.*, *Open Internet Order* ¶ 126 ("The Commission has not determined whether . . . VoIP providers are telecommunications carriers").

132. *See, e.g.*, *id.* ¶ 129 n.407 (noting that the Commission has never resolved disputes about "whether online-only video programming aggregators are themselves MVPDs"); *see generally* chapter 9 in this volume.

133. *Brand X*, 545 U.S. at 990.

134. *Id.* at 996.

135. 47 U.S.C. §§ 151, 152(a).

136. *See, e.g.*, Motion Picture Ass'n of America, Inc. v. FCC, 309 F.3d 796, 806–07 (D.C. Cir. 2002).

137. 47 U.S.C. § 154(i); *cf.* U.S. CONST. art. I, § 8.

138. In justifying one of the conditions it placed on its approval of the AOL–Time Warner merger, the Commission asserted Title I jurisdiction over instant-messaging services as a form of interstate communications. *See* Memorandum Opinion & Order, *Applications for Consent to the Transfer of Control of Licenses and Section 214 Authorizations by Time Warner Inc. and America Online, Inc., Transferors, to AOL Time Warner Inc., Transferee*, 16 FCC Rcd 6547, ¶ 148 (2001).

139. United States v. Southwestern Cable, 392 U.S. 157, 178 (1968).

140. United States v. Midwest Video Corp., 406 U.S. 649, 665 n.23 (1972) ("*Midwest Video I*") (quoting Gen. Tel. Co. of Cal. v. FCC, 413 F.2d 390, 398 (D.C. Cir. 1969)); *see also* Glen O. Robinson, *The Federal Communications Act: An Essay on Origins and Regulatory Purposes, in* A LEGISLATIVE HISTORY OF THE COMMUNICATIONS ACT OF 1934, at 24 (Oxford Univ. Press, Max D. Paglin, ed. 1989) (commenting, with respect to the FCC's role, that "with each passing era [the FCC's statutory charter] is beginning to look more like a 'living constitution' than a fixed statutory mandate"); Philip J. Weiser, *Federal Common Law, Cooperative Federalism, and the Enforcement of the Telecom Act*, 76 N.Y.U. L. REV. 1692, 1753–57 (2001) (explaining the rationale for agency leeway in implementing regulatory statutes). For a seminal early discussion of the FCC's ancillary jurisdiction, see Thomas G. Krattenmaker & A. Richard Metzger Jr., *FCC Regulatory Authority over Commercial Television Networks: The Role of Ancillary Jurisdiction*, 77 NW. U.L. REV. 403 (1982).

141. *Southwestern Cable*, 392 U.S. at 178.

142. FCC v. Midwest Video Corp., 440 U.S. 689, 706 (1979) ("*Midwest Video II*").

143. *Midwest Video I.*

144. *Id.* at 676 (Burger, C.J., concurring).

145. *Midwest Video II*, 440 U.S. 689.

146. *Id.* at 702–08. Congress perceived no such tension as a policy matter: it responded to the Supreme Court's decision by subsequently including similar public access requirements in the Cable Act of 1984, which added Title VI. *See* 47 U.S.C. § 531.

147. Motion Picture Ass'n. of Am., Inc. v. FCC, 309 F.3d 796, 805 (D.C. Cir. 2002) ("The FCC's position seems to be that the adoption of rules mandating video description is permissible because Congress did not expressly foreclose the possibility. This is an entirely untenable position").

148. *Id.* at 805–06.

149. American Library Ass'n v. FCC, 406 F.3d 689, 704 (D.C. Cir. 2005).

150. Computer & Communications Indus. Ass'n v. FCC, 693 F.2d 198 (D.C. Cir. 1982). *But cf.* GTE Serv. Corp. v. FCC, 474 F.2d 724, 735–36 (2d Cir. 1973) (invalidating the FCC's rule regarding the use of a common carrier's name on its information services affiliate because this regulation did not relate to communications and thus constituted an impermissible extension of the FCC's authority).

151. *See* 47 U.S.C. § 151.

152. *Comcast*, 600 F.3d at 654–55.

153. *Id.* at 655.

154. *Id.* at 659 (emphasis added).

155. *Id.* at 658–61.

156. *Id.* at 645.

157. Austin Schlick, *A Third-Way Legal Framework for Addressing the Comcast Dilemma*, at 2 (May 6, 2010), http://hraunfoss.fcc.gov/edocs_public/attachmatch/DOC-297945A1.pdf.

158. *Id.* at 7 (quoting *Brand X*, 545 U.S. at 1001).

159. Julius Genachowski, *The Third Way: A Narrowly Tailored Broadband Framework*, Broadband.gov (May 6, 2010), http://www.broadband.gov/the-third-way-narrowly-tailored-broadband-framework-chairman-julius-genachowski.html.

160. Notice of Inquiry, *Framework for Broadband Industry Practices*, 25 FCC Rcd 7866 (2010).

161. Schlick, *A Third-Way Legal Framework*, at 2.

162. For a highly readable account from the pro-reclassification camp, see Harold Feld, *Want to Play FCC Fantasy Baseball? Follow the Title II Debate*, Wetmachine (May 16, 2010), http://tales-of-the-sausage-factory.wetmachine.com/want-to-play-fcc-fantasy-baseball-follow-the-title-ii-debate/.

163. 47 U.S.C. § 153(24).

164. *Id.*

165. *See generally* First Report & Order, *Implementation of the Non-Accounting Safeguards of Sections 271 and 272 of the Communications Act of 1934*, 11 FCC Rcd 21905 ¶ 107 (1996).

166. 47 U.S.C. § 1302(a).

167. 600 F.3d at 658–59.

168. *Open Internet Order* ¶ 122.

169. *Id.* ¶ 125.

170. 47 U.S.C. § 153(51) (emphasis added). In a similar vein, opponents of regulation also relied on sections 332(c)(2) and 332(d) of the Communications Act, which, they said, collectively bar common carrier treatment for mobile broadband services. That issue is somewhat related to the main legal dispute presented in Verizon's separate appeal of the FCC's *Data Roaming Order*, discussed in chapter 4.

171. *Open Internet Order* ¶ 79. The FCC alternatively contended that even if it is appropriate to consider Internet content providers the "customers" of broadband ISPs for these purposes, its ban on "unreasonable discrimination" among such content providers stops short of common carrier regulation because the Commission had preserved the ISPs' discretion to engage in reasonable network management. *Id.* ¶ 79 n.251. The FCC's opponents responded that a ban on "unreasonable discrimination" is classic common carrier regulation with or without an exception for "reasonable" network management.

172. *See* Verizon v. FCC, No. 11-1355 and consolidated cases (D.C. Cir.).

173. *See* Br. for Verizon, No. 11-1355 (D.C. Cir. filed July 2, 2012).

Chapter 7

1. Report & Order and Further Notice of Proposed Rulemaking, *Connect America Fund*, 26 FCC Rcd 17663 (2011) ("*USF–ICC Reform Order*").

2. *See* 47 U.S.C. § 153(51) (defining "telecommunications carrier"); *Virgin Is. Tel. Corp. v. FCC*, 198 F.3d 921 (D.C. 1999) (noting equivalence of "telecommunications service" and "common carrier service").

3. See chapter 2 for a brief summary of the FCC's physical collocation rules, both before and after enactment of 47 U.S.C. § 251(c)(6).

4. 47 U.S.C. § 251(a)(1) (emphasis added). As noted in chapter 2 and later in this chapter, all LECs, including CLECs, have additional obligations under section 251(b), including the "reciprocal compensation" requirement of section 251(b)(5), but these obligations are distinct from the interconnection obligations specified in sections 251(a) and 251(c)(2).

5. 47 U.S.C. § 251(c)(2). The terms *exchange access* and *telephone exchange service* are separately defined in 47 U.S.C. §§ 153(20) and 153(54); in a nutshell, they are the main local services that LECs provide (whether to end users or to interconnecting long-distance carriers). *See* 47 U.S.C. § 153(32) (definition of "local exchange carrier"). The FCC has found that this clause—"for the transmission and routing of telephone exchange service and exchange access"—limits section 251(c)(2) interconnection rights to carriers that *provide* these LEC functions to their own customers and excludes carriers, such as pure long-distance providers, that merely wish to *obtain* these functions from the ILEC. First Report & Order, *Implementation of the Local Competition Provisions of the*

Telecommunications Act of 1996, 11 FCC Rcd 15499, ¶ 191 (1996) ("*Local Competition Order*").

6. See chapter 6 for an analysis of the indistinct line between "information service" providers and "telecommunications service" providers. As noted in chapter 2, a "private carrier" is a provider of transmission services that, unlike a common carrier, can and does offer its services on non-standardized terms and does not hold itself out as serving all comers. *See* Declaratory Ruling, *NORLIGHT Request for a Declaratory Ruling*, 2 FCC Rcd 132, ¶¶19–21 (1987); *see generally* National Ass'n of Reg. Util. Comm'rs v. FCC, 525 F.2d 630 (D.C. Cir. 1976).

7. *See, e.g.*, Fourth Report & Order, *Deployment of Wireline Services Offering Advanced Telecommunications Capability*, 16 FCC Rcd 15435 (2001) (specifying rules for physical collocation; see chapter 2 in this volume), *aff'd*, Verizon Tel. Cos. v. FCC, 292 F.3d 903, 906–07 (D.C. Cir. 2002); Report & Order, *Cellular Communications Systems*, 86 F.C.C.2d 469, ¶ 56 (1981) (ruling that an interconnecting "cellular system operator is a common carrier and not merely a customer" of an ILEC and requiring that interconnection arrangements be designed to minimize "unnecessary duplication of switching facilities").

8. *USF–ICC Reform Order* ¶ 682; Memorandum Opinion & Order, *Application by Verizon Maryland Inc. et al. for Authorization to Provide In-Region, InterLATA Services in Maryland, Washington, D.C., and West Virginia*, 18 FCC Rcd 5212, ¶ 103 (2003).

9. 131 S.Ct. 2254.

10. *See id.* The FCC itself had never ruled on how to apply section 251(c)(2) in these circumstances, and the Supreme Court thus relied on the analysis in an amicus brief filed by the FCC's general counsel. In theory, a future FCC may adopt the opposite position upon a vote of its commissioners.

11. Stephen Labaton, *MCI Faces Inquiry on Fees for Long Distance*, N.Y. TIMES, July 27, 2003, at A1.

12. *Id.*

13. The counterparts to terminating access charges in international calling are known as *settlement rates*. Carriers in many countries (or their governments, which sometimes have ownership interests in those carriers) have viewed the opportunity to "tax" incoming international calls with high settlement rates as an easy form of revenue, so they historically set these charges very high, sometimes at levels exceeding one U.S. dollar per minute. Consequently, as late as 1997, U.S. consumers paid on average 88 cents per minute for an international call, with much of that price reflecting a subsidy from U.S. consumers to foreign providers or governments. To address this issue, the FCC developed a set of benchmark rates to bring settlement rates down to 15–23 cents per minute, depending on the country at issue. Report & Order, *International Settlement Rates*, 12 FCC Rcd 19806, ¶ 19 (1997); *see also* First Report & Order, *International Settlements Policy Reform*, 19 FCC Rcd 5709, ¶ 82 (2004) (reiterating the importance of benchmarks policy, declining to eliminate it, and declining to initiate proceeding to revise benchmarks downward). In so doing, the FCC recog-

nized that, although inefficient, high settlement rates constitute an implicit social policy to subsidize less well-off nations through telecommunications regulation. As with inflated access charges, these rates face pressures not simply from FCC regulatory reforms, but also from various arbitrage mechanisms, including VoIP. Indeed, by some indications, the Internet-based VoIP services of Skype alone, which are exempt from the settlements regime because they avoid the PSTN, are rapidly supplanting traditional international telecommunications services subject to that regime. *See International Call Traffic Growth Slows as Skype's Volumes Soar*, TeleGeography (Jan. 9, 2012), http://www.telegeography.com/press/press-releases/2012/01/09/international-call-traffic-growth-slows-as-skypes-volumes-soar/index.html ("In contrast to international phone traffic, Skype's cross-border traffic has continued to soar. TeleGeography estimates that cross-border Skype-to-Skype calls (including video calls) grew 48 percent in 2011, to 145 billion minutes. . . . Skype added 47 billion minutes of international traffic in 2011—more than twice as much as all the telephone companies in the world, combined").

14. 47 U.S.C. § 251(g).

15. 47 U.S.C. §§ 251(b)(5), 252(d)(2)(A)(ii). By its terms, the provision containing the "additional costs" language applies only to ILECs, but the FCC has extended the requirement to all carriers terminating local calls.

16. 47 U.S.C. § 252(d)(2)(B)(i).

17. *See Local Competition Order* ¶¶ 1111–12. The FCC recognized an exception in the special case where the call volume from carrier A to carrier B is so similar to the call volume from carrier B to carrier A that the carriers' mutual obligations all but cancel out—in which event the FCC permitted bill-and-keep as a simplifying mechanism.

18. *USF–ICC Reform Order* ¶ 703. In the *USF–ICC Reform Order*, the FCC took various steps to combat the problem of phantom traffic during the transitional period in which termination rates will continue to vary with the origin of a call. *Id.* ¶¶ 702–35.

19. Similar calls date back to the 1980s, when providers offered information services ranging from Lexis-Nexis to email (MCIMail, for example) to online services such as Prodigy. But these services were fairly small compared to those later offered by AOL and EarthLink, which benefited from the World Wide Web's emergence in the mid-1990s and signed up millions of subscribers.

20. *See*, *e.g.*, Sam Biddle, *3.5 Million People Are Still Using AOL Dialup (!!!)*, Gizmodo (Nov. 3, 2011), http://gizmodo.com/5856113/35-million-people-are-still-using-aol-dialup-.

21. *See* Memorandum Opinion & Order, *MTS and WATS Market Structure*, 97 F.C.C.2d 682, ¶¶ 76–83 (1983); Notice of Proposed Rulemaking, *Amendments of Part 69 of the Commission's Rules Relating to Enhanced Service Providers*, 2 FCC Rcd 4305, ¶ 2 (1987); *see generally* Notice of Proposed Rulemaking, *IP-Enabled Services*, 19 FCC Rcd 4863 (2004) ("*IP-Enabled Servs. NPRM*").

22. *See Local Competition Order* ¶ 1054.

23. From an ILEC's perspective, the large sums paid to such CLECs as reciprocal compensation might still have been tolerable if the ILEC could have recouped those sums from its own individual residential customers to the extent that the customers were placing these ISP-bound calls to begin with. But that approach was politically untenable. First, many state public utility commissions required ILECs to charge flat-rated monthly fees for local service, not usage-sensitive rates, and most Americans had come to think of "local" calls as "free." Second, no state commission would have entitled ILECs to charge more to households placing high volumes of calls to ISPs, much less to CLEC-served ISPs. That would have been viewed—by political actors with only a partial understanding of the regulatory picture—as "taxing" the Internet.

The only other remedy available to the ILECs—short of a change in the reciprocal compensation rules themselves—would have been an increase in the monthly local service rates for everyone, irrespective of a given end user's actual use of the Internet. In the late 1990s, some ILECs argued, without success, that the skyrocketing use of the public telephone network for Internet access had placed unprecedented strains on the capacity of existing facilities, had required unforeseen capital expenditures on facility upgrades, and thus justified higher retail rates. Even apart from the political unsustainability of this argument, however, raising everyone's monthly rates would have been problematic as a theoretical matter because it would have exacerbated the existing cross-subsidy flowing from nonusers of the Internet to heavy users of dial-up Internet services.

24. Order on Remand and Report & Order, *Implementation of the Local Competition Provisions in the Telecommunications Act of 1996; Intercarrier Compensation for ISP-Bound Traffic*, 16 FCC Rcd 9151, ¶ 7 (2001) ("*ISP Reciprocal Compensation Remand Order*"), *remanded*, WorldCom, Inc. v. FCC, 288 F.3d 429 (D.C. Cir. 2002). In reality, the rates chosen by the FCC for these purposes did not quite fall to zero; instead, they fell to $0.0007 per minute, where they stayed for many years. The FCC did not follow through on its plans to implement a full-blown bill-and-keep transition until a decade later, when, as discussed later in this chapter, it finally issued a transitional schedule for reducing all PSTN-based termination charges to zero (i.e., to bill-and-keep).

25. In particular, if the ESP exemption were eliminated, then under the doctrine of "jointly provided access," the dial-up ISP would have to pay the ILEC for the functions it performs in connecting a dial-up Internet session to the ILEC's end users, even though the ISP itself is interconnecting directly only with a CLEC. *See* Declaratory Ruling, *Implementation of the Local Competition Provisions in the Telecommunications Act of 1996*, 14 FCC Rcd 3689, ¶ 9 (1999) ("When two carriers jointly provide interstate access (e.g., by delivering a call to an interexchange carrier (IXC)), the carriers will share access revenues received from the interstate service provider"), *vacated and remanded on other grounds*, Bell Atl. Tel. Cos. v. FCC, 206 F.3d 1 (D.C. Cir. 2000); *see also* Memorandum Opinion & Order, *Waiver of Access Billing Requirements and Investigation of Permanent Modifications*, 2 FCC Rcd 4518, ¶ 2 (1987).

26. *ISP Reciprocal Compensation Remand Order* ¶ 89. More specifically, the mirroring rule was confined to calls that were "subject to section 251(b)(5),"

which the FCC at the time essentially equated with local calls. The main beneficiaries of this rule were mobile wireless providers, which tended to place more calls than they received, in part because landline telephone numbers were better known then than mobile phone numbers and in part because some wireless subscribers kept their cellphones turned off when not in use.

27. *See* WorldCom, Inc. v. FCC, 288 F.3d 429 (D.C. Cir. 2002). For jurisdictional purposes, the FCC has long viewed a dial-up "call" to a Web site via a dial-up ISP as a single long-distance call, not as two separate calls—a local call from the end user to the ISP and then a long-distance transmission from the ISP to the website. *See generally* Bell Atl. Tel. Cos. v. FCC, 206 F.3d 1, 5 (D.C. Cir. 2000). Through various legal theories, the FCC invoked the "one long-distance call" construct to justify exercising exclusive jurisdiction over dial-up ISP-bound traffic. Indeed, in the early years of this dispute, the FCC focused almost single-mindedly on that construct because it wished to eliminate the inflated termination rates that the states had prescribed for these calls under section 251(b)(5) and because, as discussed later in this chapter, it had long interpreted that provision to grant states pricing jurisdiction only over "local" calls, not interstate "access" (long-distance) calls.

28. In re Core Communications, Inc., 531 F.3d 849 (D.C. Cir. 2008).

29. Core Communications, Inc. v. FCC, 592 F.3d 139 (D.C. Cir. 2010), *aff'g* Order on Remand and Report & Order, *Implementation of the Local Competition Provisions in the Telecommunications Act of 1996*, 24 FCC Rcd 6475 (2008). The FCC reiterated that these "calls" to distant websites via dial-up ISPs are jurisdictionally interstate and now further found that the "savings clause" of section 251(i) preserved its traditional authority under section 201 to set "just and reasonable rates" for interstate calls despite any more specific rate methodology prescribed by section 251.

30. *See generally IP-Enabled Servs. NPRM* ¶ 61 & n.179 (posing this question).

31. *See id.* (citing FCC precedent).

32. *USF–ICC Reform Order* ¶ 944.

33. *See* Eighth Report & Order and Fifth Order on Reconsideration, *Access Charge Reform; Reform of Access Charges Imposed by Competitive Local Exchange Carriers*, 19 FCC Rcd 9108, ¶¶ 59–61 (2004) ("*Eighth Report*"). *But cf.* AT&T Corp. v. FCC, 292 F.3d 808 (D.C. Cir. 2002).

34. *See* Seventh Report & Order and Further Notice of Proposed Rulemaking, *Access Charge Reform*, 16 FCC Rcd 9923 (2001) ("*CLEC Access Charge Order*"); *see also Eighth Report*.

35. *See, e.g.*, *USF–ICC Reform Order* ¶ 524 & n.774; Notice of Proposed Rulemaking, *Developing a Unified Intercarrier Compensation Regime*, 16 FCC Rcd 9610, ¶ 133 (2001) ("*2001 Intercarrier Comp. NPRM*").

36. *See CLEC Access Charge Order* ¶ 2 (acting "to prevent use *of the regulatory process* to impose excessive access charges") (emphasis added).

37. *USF–ICC Reform Order* ¶ 656 (footnotes and paragraph break omitted).

38. *Id.* at ¶ 657.

39. *Id.* at ¶ 656.

40. *Id.* ¶ 734.

41. *E.g.*, Declaratory Ruling & Order, *Establishing Just and Reasonable Rates for Local Exchange Carriers; Call Blocking by Carriers*, 22 FCC Rcd 11629 (2007).

42. *USF–ICC Reform Order* ¶ 658.

43. Notice of Proposed Rulemaking, *Interconnection between Local Exchange Carriers and Commercial Mobile Radio Service Providers*, 11 FCC Rcd 5020, ¶¶ 115–16 (1996).

44. *See* Declaratory Ruling, *Petitions of Sprint PCS and AT&T Corp. for Declaratory Ruling Regarding CMRS Access Charges*, 17 FCC Rcd 13192 (2002), *appeal dismissed sub nom.* AT&T Corp. v. FCC, 349 F.3d 692 (D.C. Cir. 2003); *see also* Declaratory Ruling and Report & Order, *T-Mobile et al. Petition for Declaratory Ruling Regarding Incumbent LEC Wireless Termination Tariffs*, 20 FCC Rcd 4855 (2005) ("*T-Mobile Order*") (prohibiting the filing of state tariffs for non-access mobile wireless traffic), *pet'n for review pending*, Ronan Tel. Co. et al. v. FCC, No. 05-71999 (9th Cir. filed Apr. 8, 2005). A wireless carrier's inability to impose terminating access charges by tariff can present significant practical obstacles to collecting such charges. Those obstacles are particularly severe in the tandem transit context, where the wireless carrier may not even know which carrier originated a given call, let alone have a contractual arrangement with that carrier.

45. The FCC ultimately determined that "the rate that a [CLEC] charges for access components" when positioning itself between the long-distance carrier and the called party's wireless carrier "should be no higher than the rate charged by the competing incumbent LEC for the same functions," and in particular a CLEC "has no right to collect access charges for the portion of the service provided by the [wireless] carrier." *Eighth Report* ¶¶ 16-17; *see also id.* ¶ 21.

46. *See generally USF–ICC Reform Order* ¶ 1311 ("The Commission has not addressed whether transit services must be provided pursuant to section 251 of the Act; however, some state commissions and courts have addressed this issue"); Qwest Corp. v. Cox Nebraska Telecom, LLC, 2008 WL 5273687 (D. Neb. 2008) (addressing ILEC transit obligations under section 251); *T-Mobile Order* (addressing role of tariffs and private contracts in wireless-related transit contexts); Memorandum Opinion & Order, *Texcom, Inc. v. Bell Atl. Corp.*, 16 FCC Rcd 21493, ¶¶ 4, 6, *recon. denied*, 17 FCC Rcd 6275 (2002) (ruling that a transiting carrier may charge the terminating carrier for handing off the calls originated by some other carrier); *see also* Iowa Network Servs., Inc. v. Qwest Corp., 363 F.3d 683 (8th Cir. 2004).

47. *2001 Intercarrier Comp. NPRM.*

48. *See, e.g.*, Order, *Cost-Based Terminating Comp. for CMRS Providers*, 18 FCC Rcd 18441 (2003) (opening door for wireless carriers to argue for higher termination rates than wireline ILECs), *pet. for review denied*, SBC Communica-

tions Inc. v. FCC, 414 F.3d 486 (3d Cir. 2005). In its 1996 *Local Competition Order*, the FCC established a rebuttable presumption that, with the limited exception of paging companies, any given carrier has termination costs equivalent to the ILEC's termination costs. *Local Competition Order* ¶¶ 1085–93; *see also Intercarrier Comp. NPRM* ¶¶ 102–04.

49. For a classic treatment of rate-setting challenges in the regulated utility context, see ALFRED E. KAHN, THE ECONOMICS OF REGULATION: PRINCIPLES AND INSTITUTIONS (MIT Press, reissue ed. 1988).

50. *See Local Competition Order* ¶ 1064; *ISP Reciprocal Compensation Remand Order* ¶ 76.

51. *See* Texas Office of Public Util. Counsel v. FCC, 265 F.3d 313, 328–29 (5th Cir. 2001); United States Tel. Ass'n v. FCC, 188 F.3d 521 (D.C. Cir. 1999).

52. Here we are using the terms "transport" and "termination" in their distinct, more technical senses.

53. *See USF–ICC Reform Order* ¶ 14 (extolling and purporting to quantify the ultimate consumer benefits of bill-and-keep).

54. *See id.* ¶ 744 (citing economic analyses for the propositions that "both parties generally benefit from participating in a call, and therefore, that both parties should split the cost of the call").

55. To be sure, the calling party can be said to cause more of the costs than the called party for undesired calls that end quickly because the "call set-up costs" associated with the first second of a call on a circuit-switched network occupy a disproportionate percentage of that call's total costs. But the proponents of bill-and-keep do not argue that the calling party and the called party each causes exactly 50% of the costs of a call. Instead, they argue that there can be no theoretically satisfying account of who "causes" what percentage of any call's costs—and thus no theoretically compelling reason to impose 100% of the costs of a call on the calling party's network. *See generally* Patrick DeGraba, *Bill and Keep at the Central Office as the Efficient Interconnection Regime*, OPP Working Paper No. 33, at 17–19 (FCC 2000), http://transition.fcc.gov/Bureaus/OPP/working_papers/oppwp33.pdf.

56. *See Local Competition Order* ¶ 1112.

57. *USF–ICC Reform Order* ¶ 755–58; *see also ISP Reciprocal Compensation Remand Order* ¶¶ 72–73 (adopting contrary position).

58. *See* DeGraba, *Bill and Keep at the Central Office*.

59. *USF–ICC Reform Order* ¶¶ 801, 1306 & n.2358.

60. As noted, the FCC's traditional approach permits each non-ILEC carrier to avail itself of a single point of interconnection per LATA.

61. Alternatively, regulators might prescribe actual physical points of interconnection between carriers across a range of scenarios, at least as default arrangements subject to negotiated alternatives. Depending on how the requirement is framed, however, this approach might be subject to challenge on the ground that section 251(c)(2) entitles CLECs to interconnect at "any technically feasible

point" on the ILEC's network. *See generally* US West Commun., Inc. v. Jennings, 304 F.3d 950, 961 (9th Cir. 2002); MCI Telecomm. Corp. v. Bell Atl.-Pa., 271 F.3d 491, 517–18 (3d Cir. 2001).

62. *USF–ICC Reform Order* ¶ 801 & fig. 9.

63. *See id.* ¶¶ 905–32.

64. *Id.* ¶ 14.

65. In re: FCC 11-161, Case No. 11-9900, et al. (10th Cir.).

66. Under the preemption provision of section 332, which bars state regulation of CMRS rates (see chapter 4), the Commission has much more straightforward jurisdiction over the intercarrier compensation paid to *mobile wireless carriers*.

67. AT&T Corp. v. Iowa Utils. Bd., 525 U.S. 366 (1999) (discussed in chapter 2).

68. 47 U.S.C. § 251(b)(5).

69. *Local Competition Order* ¶ 1034 (construing section 251(b)(5) to "apply only to traffic that originates and terminates within a local area").

70. *See* 47 U.S.C. § 152(b); Louisiana Pub. Serv. Comm'n v. FCC, 476 U.S. 355 (1986). The traditional jurisdictional divide between the FCC and the states over "interstate" and "intrastate" communications, respectively, is discussed in chapter 2.

71. *USF–ICC Reform Order* ¶¶ 761–68; *cf. ISP Reciprocal Compensation Remand Order* ¶ 37 & n.66.

72. *USF–ICC Reform Order* ¶¶ 761–62.

73. Advocates of a narrow construction have similarly noted that section 251(b)(5) imposes obligations only on "local exchange carriers" and have cited that as additional evidence that the provision covers only direct traffic exchanges between two LECs, which are almost always local. But the FCC rejected this argument in 1996. It reasoned that although section 251(b)(5) imposes obligations only on LECs, it "does not explicitly state *to whom [each] LEC's obligation runs*," and it concluded that all LECs "have a duty to establish reciprocal compensation arrangements with respect to local traffic originated by or terminating to *any* telecommunications carriers." *Local Competition Order* ¶ 1041 (emphasis added). At the time, the main beneficiaries of that determination were mobile wireless providers, which Congress excluded from the default statutory definition of "local exchange carrier." (Although the FCC has broad discretion to treat mobile providers as LECs, it found no need to exercise that discretion here.) In essence, the *USF–ICC Reform Order* followed the same approach but eliminated the qualifier "local traffic" in the quoted sentence.

74. 47 U.S.C. § 252(d)(2)(A).

75. *See USF–ICC Reform Order* ¶ 763.

76. *See, e.g.*, AT&T v. Iowa Utils. Bd., 525 U.S. at 397.

77. *Id.*

78. *Id.*

79. Even viewed in isolation, this language is ambiguous: What exactly is "a reasonable approximation of the additional costs of terminating such calls"? Remember that the "costs" of termination are composed almost entirely of the fixed costs of building in enough network capacity to handle peak traffic loads, and the short-term marginal cost of terminating any given call is therefore negligible. The term "additional costs" is not defined in the statute, and it might plausibly be defined in a number of different ways, including as the relevant TELRIC value for the switching and transport elements as a whole (the FCC's interpretation under the original calling-network-pays rule) or as short-term marginal cost, which arguably approaches zero (bill-and-keep).

80. *Local Competition Order* ¶¶ 1111–12.

81. *USF–ICC Reform Order* ¶ 775. The D.C. Circuit seemed to endorse that interpretation when, in one of its decisions addressing the ISP reciprocal compensation controversy, it cited the bill-and-keep savings clause for the proposition that "there is plainly a non-trivial likelihood that the Commission has authority to elect" bill-and-keep for unbalanced ISP-bound traffic. WorldCom, 288 F.3d at 434. Under this interpretation, section 252(d)(2) permits the FCC to choose either bill-and-keep or a truly cost-based calling-network-pays scheme. For this interpretation to be plausible, however, it must prohibit some other regulatory option; otherwise, all these convoluted statutory provisions would be so many wasted words. That excluded option would be any scheme in which money changes hands, but for reasons unrelated to the actual costs of transport and termination. Thus, intercarrier compensation for traffic falling within the scope of sections 251(b)(5) and 252(d)(2) may not be inflated for the purpose of subsidizing universal service needs. And ILECs may not charge terminating carriers for the right to receive calls originated by the ILECs themselves, as they sometimes did in their dealings with wireless carriers before 1996—at least if they enter into interconnection agreements with them. *See Local Competition Order* ¶¶ 1042, 1087.

82. *See USF–ICC Reform Order* ¶ 776.

83. *See* FCC Technology Advisory Council, *Status of Recommendations*, at 11, 15–16 (June 29, 2011), http://transition.fcc.gov/oet/tac/TACJune2011mtgfullpresentation.pdf.

84. *See* Fred Kemmerer, *Debunking the Myths of PSTN Sunset*, TMCnet.com (Apr. 19, 2012), http://www.tmcnet.com/topics/articles/2012/04/19/286901-debunking-myths-pstn-sunset-part.htm.

85. *See generally ENUM*, Int'l Telecomm'ns Union (Apr. 4, 2011), http://www.itu.int/osg/spu/enum/; *Telephone Number Mapping*, Wikipedia, http://en.wikipedia.org/wiki/Telephone_number_mapping. For security reasons, VoIP providers and customers may be loath to "advertise" individual users' IP addresses to the Internet as a whole. But VoIP providers could advertise an IP address for a session border controller that it uses for a large group of VoIP customers, and it could then use non–publicly advertised information to route incoming VoIP traffic to specific customers.

86. *See* 47 U.S.C. § 251(e)(1).

87. Memorandum Opinion & Order, *Applications for Consent to the Transfer of Control of Licenses and Section 214 Authorizations by Time Warner Inc. and America Online, Inc., Transferors, to AOL Time Warner Inc., Transferee*, 16 FCC Rcd 6547, ¶¶ 191–200 (2001). For a critical evaluation of the decision, see Philip J. Weiser, *Internet Governance, Standard Setting, and Self-Regulation*, 28 N. Ky. L. Rev. 822, 844 (2001).

88. Memorandum Opinion & Order, *Applications for Consent to the Transfer of Control of Licenses and Section 214 Authorizations by Time Warner Inc. and America Online, Inc., Transferors, to AOL Time Warner Inc., Transferee*, 18 FCC Rcd 16835 (2003) (lifting condition).

89. Indirect interconnection is common on the PSTN, too, in the form of tandem transit arrangements, but it is somewhat less efficient there. Traditional PSTN networks are hierarchical rather than distributed, which reduces the number of intermediate routes available for traffic to flow. And they are circuit switched rather than packet switched, which means that an entire circuit must be set up between the calling and called parties for the duration of a call, wasting capacity on every intermediate network involved in the call.

90. *See, e.g.*, DrPeering International, *Internet Transit Prices—Historical and Projected* (Aug. 2010), http://drpeering.net/white-papers/Internet-Transit-Pricing-Historical-And-Projected.php; *Global Internet Geography (Executive Summary)*, Telegeography, at 4–5 (2012), http://www.telegeography.com/page_attachments/products/website/research-services/global-internet-geography/0003/1871/GIG_Executive_Summary.pdf.

91. Peyman Faratin, David Clark et al., *The Growing Complexity of Internet Interconnection*, 72 Communications & Strategies 51, at 63 (4Q 2008) (if one network denies settlement-free peering privileges to others, those other networks, "if they can control the routing of their traffic," can "cause their traffic to/from the prospective peer to route over the peer's transit connection to raise the peer's transit costs in order to induce it to peer"); Rudolph van der Berg, *How the 'Net Works: An Introduction to Peering and Transit*, Ars Technica (Sept. 1, 2008) ("Allegedly, a big American software company was refused peering by one of the incumbent telco networks in the north of Europe. The American firm reacted by finding the most expensive transit route for that telco and then routing its own traffic to Europe over that link. Within a couple of months, the European CFO was asking why the company was paying out so much for transit. Soon afterward, there was a peering arrangement between the two networks"); Mark A. Israel & Stanley M. Besen, *The Evolution of Internet Interconnection from Hierarchy to "Mesh": Implications for Government Regulation*, at 27 (July 11, 2012), http://papers.ssrn.com/sol3/papers.cfm?abstract_id=2104323 ("[I]n negotiations with an ISP about the terms of paid peering, a CDN can threaten to exploit transit alternatives that would leave the ISP worse off than if it had entered into a reasonably priced paid peering relationship with the CDN").

92. *See* Israel & Besen, *Evolution of Internet Interconnection*, at 26 ("CDNs and ISPs have alternatives to direct peering, and those alternatives limit whatever negotiating leverage an ISP would otherwise have"); Michael Kende, *The Digital Handshake: Connecting Internet Backbones*, FCC OPP Working Paper No. 32,

at 7 (2000), http://transition.fcc.gov/Bureaus/OPP/working_papers/oppwp32.pdf ("In negotiating peering, one important bargaining chip is the number of customers to which a backbone provides access; this includes the number of transit customers. . . . [In addition,] large backbones will compete for the transit business of smaller backbones in order to increase their revenues, which will keep transit prices down").

93. *See* Complaint, United States v. WorldCom, No. 00-CV-1526 (D.D.C. filed June 26, 2000), http://www.usdoj.gov/atr/cases/f5000/5051.pdf; Constance K. Robinson, *Network Effects in Telecommunications Mergers; MCI WorldCom Merger: Protecting the Future of the Internet* (August 23, 1999), http://www.justice.gov/atr/public/speeches/3889.htm.

94. *Level 3 and Cogent Reach Agreement on Equitable Peering Terms*, PR Newswire (Oct. 28, 2005), http://www.prnewswire.com/news-releases/level-3-and-cogent-reach-agreement-on-equitable-peering-terms-55637437.html ("The modified peering arrangement allows for the continued exchange of traffic between the two companies' networks, and includes commitments from each party with respect to the characteristics and volume of traffic to be exchanged. Under the terms of the agreement, the companies have agreed to the settlement-free exchange of traffic subject to specific payments if certain obligations are not met").

95. For a lively exchange between the parties to this dispute, see Press Release, *Level 3 Releases Statement to Clarify Issues in Comcast/Level 3 Interconnection Dispute* (Dec. 3, 2010), http://lvlt.client.shareholder.com/releasedetail.cfm?ReleaseID=534802; Joe Waz, *20 Q's—with Accurate A's—about Level 3's Peering Dispute*, Comcast Voices (Dec. 7, 2010), http://blog.comcast.com/2010/12/20-qs---with-accurate-as---about-level-3s-peering-dispute.html; *see also* Daniel Golding, *The Real Story behind the Comcast–Level 3 Battle*, GigaOm (Dec. 1, 2010), http://gigaom.com/2010/12/01/comcast-level-3-battle/.

96. *See* Israel & Besen, *Evolution of Internet Interconnection.*

97. Karl Bode, *Claims Resurface Concerning Congested Comcast TATA Links: Level 3 Dispute Continues . . .*, DSL Reports (Dec. 14, 2010), http://www.dslreports.com/shownews/Claims-Resurface-Concerning-Congested-Comcast-TATA-Links-111818 ("TATA [Comcast's transit provider] cannot force Comcast to upgrade its links[.] Comcast elects to simply not purchase enough capacity and lets them run full. When Comcast demanded that Level 3 pay them, the only choice Level 3 had was to give in or have its traffic (such as Netflix) routed via the congested TATA links. If Level 3 didn't agree to pay, that means Netflix and large portions of the Internet . . . would be simply unusable for the majority of the day for Comcast subscribers. That falls in line with Level 3's claim that this is not simply a peering dispute, but a new effort that involves Comcast using their massive last mile customer base as leverage in order to milk tier 1 operators out of additional revenue") (internal parentheses omitted).

98. Joe Waz, *Comcast Answers Today's Traffic Question*, Comcast Voices (Dec. 14, 2010), http://blog.comcast.com/2010/12/comcast-answers-todays-traffic-question.html ("[T]he allegation that these graphs demonstrate that Comcast is

engaging in what the blogger calls 'congestion by choice' are patently false. Internet traffic shifts and growth happen all the time and many of these are beyond any one network's control, including ours. When they do occur, or when we can reasonably anticipate them, we adjust capacity to optimize the traffic flow between our network and our fellow networks").

99. Report & Order, *Preserving the Open Internet*, 25 FCC Rcd 17905, ¶ 76 (2010).

100. *Id.* ¶ 29.

101. *Id.* ¶ 67 n.209.

102. *See, e.g.*, Bode, *Claims Resurface.*

103. *Announcing the Netflix Open Connect Network*, Netflix U.S. and Canada Blog (June 4, 2012), http://blog.netflix.com/2012/06/announcing-netflix-open-connect-network.html.

104. Scott Woolley, *The Day the Web Went Dead*, Forbes.com (Dec. 2, 2008), http://www.forbes.com/2008/12/01/cogent-sprint-regulation-tech-enter-cz_sw_1202cogent.html?feed=rss_news.

105. *Id.*

106. *See* Philip J. Weiser, *The Future of Internet Regulation*, 43 U.C. Davis L. Rev. 529 (2009). Similar concerns arise in the "retransmission consent" context discussed in chapter 9, in which cable operators and other MVPDs negotiate for the right to carry the signals of broadcast television stations. The FCC has narrow authority to force broadcasters and MVPDs to negotiate in good faith, but no authority to prescribe substantive outcomes for those negotiations. Against that backdrop, some have urged the FCC to impose "standstill" remedies that basically freeze existing retransmission-consent arrangements in place for the duration of negotiations. Others have proposed that the FCC institute "cooling off" periods that would automatically extend contracts by some short period—say, 30 days—whenever an existing contract would otherwise expire with no replacement contract in sight. Apart from their legal qualms, opponents of such regulation argue that standstill remedies would undermine incentives to negotiate because one side will generally view the existing contract as more favorable than any new agreement it might obtain. And opponents argue that "cooling off" periods would accomplish little over the long term because once they become a known component of the legal regime, they will simply push back all contractual impasses by 30 days but not make them less frequent or severe.

107. *See, e.g.*, Remarks of Assistant Secretary Lawrence E. Strickling at the Brookings Inst., *Principles of Internet Governance: An Agenda for Economic Growth and Innovation* (Jan. 11, 2012), http://www.ntia.doc.gov/speechtestimony/2012/remarks-assistant-secretary-strickling-brookings-institutions-center-technology ("Many governments have called for the ITU to play a greater role in regulating peering and termination charges in order to compensate for lost telecommunication fees . . . . These governments fail to acknowledge how fundamentally different the Internet is to the forms of communication which preceded it. The Internet . . . is a diverse, multi-layered system that thrives only through the cooperation of many different parties. All of these

parties together form the 'network of networks' that we call the Internet, and to disrupt even one would jeopardize the entire system").

Chapter 8

1. *Quoted in* Paul Starr, The Creation of the Media: Political Origins of Modern Communications 207 (Basic Books, 2004).

2. *Id.* (quoting 1910 AT&T annual report).

3. *Id.* at 446 (quoting 1911 AT&T annual report).

4. There is some dispute about exactly what Vail meant by the term "universal service." Paul Starr suggests that Vail intended it to encompass this system of cross-subsidies (*see id.*), whereas Milton Mueller argues that Vail meant the term to denote "the unification of telephone service under regulated local exchange monopolies" (Milton L. Mueller Jr., Universal Service: Monopoly in the Making of American Telephone System 92 (MIT Press, 1997)).

5. 47 U.S.C. § 254(b)(1), (3). As elsewhere in this book, references to "section 254" and other sections of the Communications Act of 1934 (as amended by the Telecommunications Act of 1996) correspond to sections of Title 47 of the U.S. Code, unless otherwise indicated.

6. FCC, *Connecting America: The National Broadband Plan*, at 140 (2010), http://download.broadband.gov/plan/national-broadband-plan.pdf ("*National Broadband Plan*").

7. In addition to these universal service programs, there are other federal initiatives, administered by separate agencies, that also seek to ensure widespread access to telecommunications services. For example, the Rural Utilities Service (RUS), a subagency of the Department of Agriculture, has traditionally provided loans and other benefits to rural providers in order to increase access to telecommunications services. And Congress authorized both RUS and NTIA (a subagency of the Department of Commerce) to make broadband-deployment grants as part of the 2009 economic stimulus legislation.

8. *See National Broadband Plan* at 140.

9. Report & Order and Further Notice of Proposed Rulemaking, *Connect America Fund*, 26 FCC Rcd 17663 (2011) ("*USF–ICC Reform Order*").

10. *See generally* Robert Crandall & Leonard Waverman, Who Pays for Universal Service? When Telephone Subsidies Become Transparent 21 (Brookings Inst. Press, 2000) ("Most studies of universal service subsidies conclude that they have minimal effect on telephone subscriptions . . . they do not address the real causes of nonsubscription—installation fees and excessive past fees for long distance calling").

11. *See* W. Kip Viscusi et al., Economics of Regulation and Antitrust 352 (MIT Press, 3d ed. 2000) ("The Ramsey pricing 'rule' that gives the prices that minimize the deadweight losses is to raise prices in inverse proportion to demand elasticities"); *see also* AT&T Corp. v. Iowa Utils. Bd., 525 U.S. 366, 426 (1999) (Breyer, J., concurring in part and dissenting in part).

12. *See* H.R. Rep. No. 204, 104th Cong., 2d Sess. 80 (1995) (recognizing the need to reform universal service support "in the context of a local market changing from one characterized by monopoly to one of competition").

13. Richard Posner, *Taxation by Regulation*, 2 BELL J. ECON. & MGMT. SCI. 22, 29 (1971).

14. See chapter 2 for a discussion of the difference between price-cap and rate-of-return regulation.

15. *See generally* Smith v. Ill. Bell Tel. Co., 282 U.S. 133, 148 (1930).

16. Under a 1970 federal–state compact known as the "Ozark Plan," a disproportionate amount of the costs of telephone service were placed in the federal jurisdiction. *See* Report & Order, *Prescription of Procedures for Separating and Allocating Plant Investment, Operating Expenses, Taxes, and Reserves between the Intrastate and Interstate Operations of Telephone Companies*, 26 F.C.C.2d 247 (1970). At the time, the proportion allocated to the respective jurisdictions worked out to a 66/33 state/federal split, but the FCC ultimately moved to a fixed 75/25 ratio. *See* Decision & Order, *Amendment of Part 67 of the Commission's Rules and Establishment of a Joint Board,* 96 F.C.C.2d 781 (1984). The FCC has frozen key separations rules since 2001 and appears unlikely to alter them anytime soon, even though state and federal regulators are notionally engaged in developing reform proposals. *See* Report & Order, *Jurisdictional Separations and Referral to the Federal–State Joint Board,* CC Docket No. 80-286, FCC No. 12-49, ¶ 2 (2012); *cf.* Notice of Proposed Rulemaking, *Jurisdictional Separations Reform and Referral to the Federal–State Joint Board*, 12 FCC Rcd 22120, ¶ 3 (1997) (seeking comment "on whether some form of separations must exist under the 1930 *Smith v. Illinois* decision, or whether statutory, regulatory and market changes since that decision have been so pronounced and persuasive as to make its holding inapplicable in our new deregulatory environment").

17. *See* 47 U.S.C. § 221(c) (empowering the FCC to "determine what property of said carrier shall be considered as used in interstate or foreign telephone toll service"); *id.* § 410 (requiring use of a Joint Board to address separations); *see also* Hawaiian Tel. Co. v. Pub. Util. Comm., 827 F.2d 1264, 1277–78 (9th Cir. 1987) (making clear that the FCC may oversee separations rules).

18. To some extent, this regime followed the framework adopted in the wake of the AT&T breakup, although the amounts of the different charges have changed dramatically. *See* National Ass'n of Regulatory Utils. Comm'rs v. FCC, 737 F.2d 1095 (D.C. Cir. 1984).

19. *See* Qwest Communications Int'l, Inc. v. FCC, 398 F.3d 1222 (10th Cir. 2005) ("*Qwest II*"); Qwest Corp. v. FCC, 258 F.3d 1191, 1203–04 (10th Cir. 2001) ("*Qwest I*").

20. *See* 47 U.S.C. §§ 254(a), 410(c). The Joint Board is composed of several FCC commissioners, several members of state public utility commissions, and a designated consumer advocate. Its recommendations cannot bind the FCC in any legal sense, but they do carry political significance.

21. *See* 47 U.S.C. § 254(b)(4).

22. Sixth Report & Order, *Access Charge Reform*, 15 FCC Rcd 12962 (2000) ("*CALLS Order*"). The exact amount of the needed increase in the size of the federal fund was subject to some dispute. Although the FCC originally increased the fund size by $650 million, the Fifth Circuit reversed that portion of the *CALLS Order* on the ground that the Commission had essentially picked the number out of a hat, and it remanded for a recalculation. *See* Texas Office of Pub. Util. Counsel v. FCC, 265 F.3d 313, 327–28 (5th Cir. 2001). On remand, the FCC conducted further cost proceedings and concluded that the right number was $650 million after all, though this time it purported to offer greater justification. *See* Order on Remand, *Access Charge Reform*, 18 FCC Rcd 14976 (2003).

23. As in the unbundling context, the FCC defined "cost" for these purposes on the basis of a "forward-looking" methodology that in essence looks to replacement costs rather than to book costs (*see* chapter 2 in this volume). That methodological choice has less significance than might appear at first blush because the FCC used the same forward-looking cost methodology in both the numerator (average state costs) and the denominator (average national costs) of the cost calculus.

24. *See* Vermont Pub. Serv. Bd. v. FCC, 661 F.3d 54, 58 (D.C. Cir. 2011) ("*Vermont PSB*"); *Qwest I*. In *Qwest I* and *Qwest II*, the Tenth Circuit twice remanded the FCC's approach to these issues, but the FCC responded each time by tweaking the details while maintaining its basic regime until in 2011 the D.C. Circuit finally upheld that regime in *Vermont PSB*.

25. *See* Alenco Communications, Inc. v. FCC, 201 F.3d 608, 617 (5th Cir. 2000); Fourteenth Report & Order, *Federal–State Joint Board on Universal Service, Multi-Association Group (MAG) Plan for Regulation of Interstate Services of Non–Price Cap Incumbent Local Exchange Carriers and Interexchange Carriers*, 16 FCC Rcd 11244 (2001) ("*Rural Task Force Order*"); *see also* Second Report & Order, *Federal–State Joint Board on Universal Service, Multi-Association Group (MAG) Plan for Regulation of Interstate Services of Non–Price Cap Incumbent Local Exchange Carriers and Interexchange Carriers*, 16 FCC Rcd 19613 (2001).

26. *Rural Task Force Order* ¶¶ 8–10. See chapter 2 for a discussion of the difference between forward-looking and embedded "historical" costs.

27. *Id.* ¶¶ 5, 11.

28. *See USF–ICC Reform Order* ¶¶ 498–511 (discussing regulatory history); *see also* 47 U.S.C. §§ 214(e), 254(e).

29. *See* Report & Order, *Federal–State Joint Board on Universal Service*, 12 FCC Rcd 8776, ¶¶ 287–88 (1997) ("*1997 Universal Service Order*").

30. *Rural Task Force Order* ¶ 207.

31. *Id.*

32. *See id.* ¶ 286.

33. *See USF–ICC Reform Order* ¶ 501 (of the $1.2 billion in competitive ETC funding in 2010, only "$23 million was disbursed to wireline competitive ETCs"); *id.* ¶ 503 ("The Commission anticipated that universal service support would be

driven to the most efficient providers as they captured customers from the incumbent provider in a competitive marketplace. It originally expected that growth in subscribership to a competitive ETC's services would necessarily result in a reduction in subscribership to the incumbent's services").

34. *Id.* ¶ 296.

35. *Id.* ¶ 505.

36. Recommended Decision, *Federal–State Joint Board on Universal Service*, 19 FCC Rcd 4257, ¶¶ 56–87 (2004).

37. Order, *High-Cost Universal Service Support*, 23 FCC Rcd 8834 (2008).

38. Rural Cellular Ass'n v. FCC, 588 F.3d 1095 (D.C. Cir. 2009).

39. *See USF–ICC Reform Order* ¶ 520. In chapter 10, we address the broader policy concerns raised by the Commission's frequent use of the merger-review process to impose regulatory initiatives that could be undertaken in rulemaking proceedings of industry-wide scope.

40. *Id.* ¶ 297. Like virtually every other major aspect of the *USF–ICC Reform Order*, the elimination of the identical-support rule was appealed and sent to the Tenth Circuit for review (see chapter 7), where it remains pending as this edition goes to press.

41. *See id.* ¶ 298 n.493.

42. *Id.* ¶ 28.

43. *Id.* ¶¶ 27, 301–478.

44. *See* Recommended Decision, *Federal State Joint Board on Universal Service*, 17 FCC Rcd 14095, ¶¶ 9–19 (2002) (recommending that broadband not be included within universal service); Order and Order on Reconsideration, *Federal State Joint Board on Universal Service*, 18 FCC Rcd 15090, ¶¶ 8–13 (2003) (adopting recommendation). Indeed, a 2004 poll showed that nearly 70% of Americans opposed paying additional fees to underwrite residential broadband subsidies. *See* Charles Cooper, *Poll Shows Tough Road for Broadband*, CNET News (July 26, 2004), http://news.cnet.com/Poll-shows-tough-road-for-broadband/2100-1034_3-5273082.html.

45. *Rural Task Force Order* ¶ 200.

46. *See, e.g., USF–ICC Reform Order* ¶ 128 n.201.

47. *Id.* ¶ 21.

48. American Recovery and Reinvestment Act of 2009, Pub. L. No. 111-5, 123 Stat. 115, 516, § 6001(k)(2). In the same stimulus legislation, Congress separately directed NTIA and RUS to make one-time disbursements of more than $7 billion to fund broadband infrastructure deployment. *See National Broadband Plan* at 139 (discussing Recovery Act subsidy programs). The sequencing of the stimulus legislation and the *Broadband Plan* was somewhat unfortunate. To jump-start the economy, Congress directed NTIA and RUS to disburse broadband stimulus funds as quickly as possible, *before* the federal government had developed a considered plan for the most effective way to stimulate greater broadband deployment.

49. *National Broadband Plan* at xi.

50. *Id.* at 147.

51. *USF–ICC Reform Order* ¶ 18.

52. The FCC also announced that if Phase II is delayed beyond January 1, 2013, a provider would be required to begin using even its frozen high-cost funding "to build and operate broadband-capable networks used to offer the provider's own retail broadband service in areas substantially unserved by an unsubsidized competitor." *Id.* ¶ 150.

53. *Id.* ¶¶ 93–94, 137–38.

54. *Id.* ¶¶ 134, 137 n.220.

55. *Id.* ¶ 180 (discussing transition from legacy to CAF funding for price-cap carriers).

56. *Id.* ¶ 160. To ensure progress toward this goal, the FCC will require recipients to demonstrate that by the end of the third year they have offered broadband "to at least 85 percent of their high-cost locations . . . covered by the state-level commitment." *Id.*

57. *Id.* ¶¶ 167, 170.

58. *Id.* ¶ 171.

59. For theoretical and historical background on the use of procurement auctions in the universal service context, *see, e.g.*, William J. Baumol et al., *Comments of 71 Concerned Economists: Using Procurement Auctions to Allocate Broadband Stimulus Grants* (Apr. 13, 2009), http://papers.ssrn.com/sol3/papers.cfm?abstract_id=1377523; Irene S. Su, *Maximum Impact for Minimum Subsidy: Reverse Auctions for Universal Access in Chile and India*, FCC Staff Working Paper No. 2 (Oct. 2010), http://www.fcc.gov/working-papers/maximum-impact-minimum-subsidy-reverse-auctions-universal-access-chile-and-india; Paul Milgrom, *Procuring Universal Service: Putting Universal Service to Work*, Lecture at the Royal Swedish Academy of Sciences (Dec. 9, 1996).

60. *See generally* Fresno Mobile Radio, Inc. v. FCC, 165 F.3d 965, 969 (D.C. Cir. 1999) (noting that licensees have incentives to put spectrum to productive use whether they paid for it at auction or not). Despite these market incentives, the FCC has imposed build-out requirements on auction winners to ensure that insolvency or other unforeseen problems will not delay the process of putting spectrum to productive uses. The NextWave debacle discussed in chapter 3 illustrates the significance of that concern.

61. *USF–ICC Reform Order* ¶¶ 1189–212.

62. *Id.* ¶ 175.

63. *Id.* ¶¶ 15, 82. The FCC deferred until further proceedings the parallel question of whether to eliminate or modify section 214 ETC obligations (the federal law counterparts to state-level carrier-of-last-resort obligations) for ILECs that refuse Phase II funding. *See id.* ¶¶ 1089–102.

64. *Id.* ¶ 64 (quoting 47 U.S.C. § 254(e)) (emphasis added).

65. *Id.*

66. *Id.* ¶ 66 (quoting 47 U.S.C. § 1302(b)).

67. *Id.* ¶ 67 (internal quotation marks and brackets omitted).

68. See chapter 6 for a discussion of section 706's role in the net neutrality context.

69. *See, e.g.*, Jerry Hausman & Howard Shelanski, *Economic Welfare and Telecommunications Regulation: The E-Rate Policy for Universal-Service Subsidies*, 16 YALE J. REG. 19, 30 (1999) ("[t]he alternative of subsidizing universal services through general tax revenues" is "a good option from the standpoint of efficient public finance").

70. 47 U.S.C. § 254(d).

71. *1997 Universal Service Order* ¶ 808.

72. *See, e.g.*, Texas Office of Pub. Util. Counsel v. FCC, 183 F.3d 393, 447–48 (5th Cir. 1999).

73. Further Notice of Proposed Rulemaking, *Universal Service Contribution Methodology*, WC Docket No. 06-122, FCC No. 12-46, ¶ 9 (Apr. 30, 2012) ("*2012 Contribution FNPRM*").

74. *See* Further Notice of Proposed Rulemaking, *Federal–State Joint Board on Universal Service*, 17 FCC Rcd 3752, ¶ 5–13 (2002); Ken Belson, *Trying to Revive Struggling AT&T: A Job Made, It Seems, for Sisyphus*, N.Y. TIMES, June 1, 2004, at C1 (reporting on the decline in revenue from the long-distance market from $100 billion to $80 billion from 2000 to 2002).

75. *See 2012 Contribution FNPRM* at 149 (Appx. D); *id.* at 181 (Statement of Commissioner McDowell); Public Notice, *Proposed Third Quarter 2004 Universal Service Contribution Factor*, 19 FCC Rcd 10194 (2004).

76. *See* Report & Order and Further Notice of Proposed Rulemaking, *Universal Service Contribution Methodology*, 21 FCC Rcd 7518 ¶ 16 (2006) ("*2006 Contribution Order*") (raising contribution obligation for mobile wireless services from 28.5% to 37.1%).

77. The definition of "interconnected" VoIP services is itself controversial and potentially fluid. In 2011, the Commission sought public comment on whether it should expand the definition to include not only VoIP services (such as Vonage's) that allow subscribers to call *and* be called by PSTN subscribers, but also services (such as SkypeOut) that allow subscribers to make outbound calls to the PSTN but *not* receive inbound calls from it. *See* Notice of Proposed Rulemaking and Third Report & Order, *Amending the Definition of Interconnected VoIP Service in Section 9.3 of the Commission's Rules*, 26 FCC Rcd 10074 (2011).

78. Vonage Holdings Corp. v. FCC, 489 F.3d 1232, 1239–41 (D.C. Cir. 2007).

79. The Commission first formally posed the question in 2004. *See* Notice of Proposed Rulemaking, *IP-Enabled Services*, 19 FCC Rcd 4863, ¶¶ 43–44 (2004). Significantly, this statutory classification issue is distinct from whether, as the FCC did determine in 2004, over-the-top VoIP services are indivisibly *interstate* and thus beyond the scope of *state*-level common carrier regulation. *See* Memorandum Opinion & Order, *Vonage Holdings Corp. Petition for Declaratory*

Ruling Concerning Order of the Minnesota Public Utilities Commission, 19 FCC Rcd 22404 (2004), *aff'd* Minnesota Pub. Utils. Comm'n v. FCC, 483 F.3d 570 (8th Cir. 2007). *But cf.* p. 428 n. 16 in this volume.

80. *2006 Contribution Order* ¶¶ 39–42.

81. *Vonage Holdings*, 489 F.3d at 1239–41. Some clever lawyering was required to square the FCC's conclusion in this context, which turned on the word "provide," with its earlier conclusion that a broadband ISP does not "offer" telecommunications to end users and thus does not sell them a "telecommunications service." *See id.*; *see generally* chapter 6 in this volume.

82. In 1997, the Commission had exercised its discretionary authority to impose contribution obligations on "private carriers," which by definition do not provide "telecommunications services" (defined as synonymous with "common carrier" services). *1997 Universal Service Order* ¶ 796; *see also* 47 U.S.C. § 153(51); Virgin Is. Tel. Corp. v. FCC, 198 F.3d 921 (D.C. Cir. 1999). It is unclear whether the FCC meant to include Internet backbone providers within the scope of that requirement, given that the FCC did not even mention them in this context.

83. *See* Report to Congress, *Federal–State Joint Board on Universal Service*, 13 FCC Rcd 11501, ¶¶ 46–47, 82 (1998).

84. This policy has always applied to cable modem service, and the FCC likewise applied it to telco-provided broadband Internet access services in 2005, when it clarified the jurisdictional classification of those services. *See 2012 Contribution FNPRM* ¶ 10 n.33. To the extent that some telcos provide DSL transmission services on a common carrier basis separately from Internet access, they provide a telecommunications service and thus retain contribution obligations.

85. *2006 Contribution Order* ¶ 43.

86. *Id.* ¶ 44.

87. *See id.* ¶¶ 52–53.

88. *Id.* ¶ 55. On appeal, the D.C. Circuit rejected Vonage's challenge to this justification. *Vonage Holdings*, 489 F.3d at 1242–43.

89. *2012 Contribution FNPRM* ¶ 124–25.

90. *Id.* ¶ 42 (quoting BT Americas).

91. *Id.* ¶ 61.

92. *Id.* ¶ 101.

93. *Id.* ¶ 102. For a discussion of that economic challenge in the antitrust context, where it typically arises in cases alleging predatory pricing or unlawful tying, see Cascade Health Solutions v. PeaceHealth, 515 F.3d 883 (9th Cir. 2008); Ortho Diagnostic Sys., Inc. v. Abbott Labs., Inc., 920 F.Supp. 455 (S.D.N.Y. 1996).

94. *See 2012 Contribution FNPRM* ¶ 105.

95. *Id.* ¶ 11.

96. *Id.* ¶ 147 (emphasis added).

97. *Id.* ¶ 148 (internal quotation marks omitted).

98. *Id.*

99. Order on Remand and Report & Order and Further Notice of Proposed Rulemaking, *High Cost Universal Service Support*, 24 FCC Rcd 6475 (2008); Report & Order and Notice of Proposed Rulemaking, *Universal Service Contribution Methodology*, 21 FCC Rcd 7518 (2006); Further Notice of Proposed Rulemaking and Report & Order, *Federal–State Joint Board on Universal Service*, 17 FCC Rcd 3752 (2002).

100. *2012 Contribution FNPRM.*

101. *Id.* ¶ 20.

102. *Id.* ¶¶ 36–73.

103. *Id.* ¶¶ 74–75 (capitalization altered and emphasis omitted).

104. *Id.* ¶ 76.

105. *Id.* ¶ 80.

106. *Vonage Holdings*, 489 F.3d at 1239–41.

107. *2012 Contribution FNPRM* ¶ 148 (internal quotation marks omitted).

108. *Id.* ¶ 149.

109. *Id.* ¶ 153.

110. *Id.* ¶ 232.

111. *Id.* ¶ 222.

112. *Id.*

113. *Id.* ¶¶ 238, 249–63.

114. There are also a number of legal questions about how to configure a connections approach to make it comport with section 254(d), which requires contributions from all carriers that provide interstate telecommunications services. For that reason, the FCC has sought comment on whether to supplement any connections-based approach with "mandatory minimum contributions" from any provider of interstate telecommunications. *Id.* ¶ 224.

115. *See id.* ¶¶ 284–343.

116. *See id.* ¶ 287.

117. *Cf. id.* ¶ 307 & n.498.

Chapter 9

1. That said, technological advances such as *IP multicast* blur the lines between "broadcast" and "unicast" content delivery and challenge the assumption that traditional broadcast architecture is inevitably more efficient in transmitting widely viewed video-programming events. *See, e.g.*, Sprint, *Multicast Basics* (visited May 2012), https://www.sprint.net/index.php?p=faq_multicasting ("Multicast is an IP technology that allows for streams of data to be sent efficiently from one to many destinations. Instead of setting up separate unicast sessions for each destination, multicast will replicate packets at router hops where the path to different multicast group members diverges. This allows a source to send a single copy of a stream of data, while reaching any number

of possible receivers"); Cisco White Paper, *IP Multicast Technical Overview*, at 1 (Aug. 2007), http://www.cisco.com/en/US/prod/collateral/iosswrel/ps6537/ps6552/prod_white_paper0900aecd804d5fe6.html; see also Jon Hardwick, *Whitepapers, IP Multicast Explained*, Metaswitch Networks, at 2 (June 2004), http://network-technologies.metaswitch.com/download/multicast.pdf.

2. *See, e.g.*, Thomas W. Hazlett, *All Broadcast Regulation Politics Are Local: A Response to Christopher Yoo's Model of Broadcast Regulation*, 53 EMORY L.J. 233, 235 (2004) ("Licenses are awarded without competitive bidding, and when rival technologies (such as cable or satellite television) pose a competitive threat, regulators first attempt to thwart the new entrants and, at the point that they lose that battle, enforce rules mandating that the new rivals help broadcast licensees distribute their programs"). Indeed, at the D.C. Circuit's direction, the FCC expressly rejected competition for 30 years in favor of a protectionist policy toward local broadcasters as part of the "*Carroll* doctrine," which prevented the assignment of additional stations that could endanger the vitality of current ones. *See* Carroll Broadcasting Co. v. FCC, 258 F.2d 440 (D.C. Cir. 1958); Report & Order, *Regarding Detrimental Effects of Proposed New Broadcasting Stations on Existing Stations*, 3 FCC Rcd 638 (1988) (abolishing the *Carroll* doctrine).

3. As Cass Sunstein has put it, "There is a large difference between the public interest and what interests the public. This is so especially in light of the character and consequences of the communications market. One of the central goals of the system of broadcasting, private as well as public, should be to promote the American aspiration to deliberative democracy." Cass R. Sunstein, *Television and the Public Interest*, 88 CAL. L. REV. 499, 501 (2000); *see also* Mark Cooper, *Open Communications Platforms: The Physical Infrastructure as the Bedrock of Innovation and Democratic Discourse in the Internet Age*, 2 J. TELECOMM.& HIGH TECH. L. 177, 193 (2003) (arguing that speech is not just "an economic commodity").

4. *See, e.g.*, Prometheus Radio Project v. FCC, 373 F.3d 372, 414 (3rd Cir. 2004) ("*Prometheus I*") ("The Commission ensures that license transfers serve public goals of diversity, competition, and localism, while the antitrust authorities have a different purpose: ensuring that merging companies do not raise prices above competitive levels").

5. Fowler made this statement in an interview published in the November 1, 1981, issue of *Reason* magazine. The statement can be found at http://www.conservativeforum.org/quotesmast.asp. *See also* C. EDWIN BAKER, MEDIA, MARKETS, AND DEMOCRACY 3 (Cambridge Univ. Press, 2001).

6. Fourteenth Report, *Annual Assessment of the Status of Competition in the Market for the Delivery of Video Programming*, MB Docket No. 07-269, FCC No. 12-81, at ¶ 211 (July 20, 2012) ("*Fourteenth Video Report*") ("After a steady decline over the last few years, the percentage of television households relying exclusively on over-the-air broadcast service (as opposed to access to broadcast stations via an MVPD) has remained stable since 2010" at approximately "9.6 percent (10.97 million households) at the end of 2011"). Other surveys have placed the percentage somewhat higher, in the midteens. *See, e.g.*, Press Release, *Over-the-Air TV Homes Now Include 46 Million Consumers*,

Knowledge Networks (June 6, 2011), http://www.knowledgenetworks.com/news/releases/2011/060611_ota.html.

7. At the end of 2010, incumbent cable companies accounted for slightly less than 60% of MVPD subscribers nationwide (down from 65% in 2006); telcos accounted for about 7% (up from almost zero in 2006), and the satellite companies DirecTV and DISH together accounted for about 34% (a very slight increase over 2006). *Fourteenth Video Report* ¶¶ 30–32, 139.

8. For years, telephone companies were banned from providing cable television service. 47 U.S.C. § 533(b) (repealed); *see, e.g.*, Gen. Tel. v. United States, 449 F.2d 846 (5th Cir. 1971) (upholding ban); *but see* Chesapeake & Potomac Tel. Co. of Va. v. United States, 42 F.3d 181, 202 (4th Cir. 1996) (invalidating ban), *vacated as moot*, 516 U.S. 415 (1996). In the 1996 Act, Congress repealed this ban. *See* 1996 Act, Pub. L. 104-104, § 302(b)(1). Telcos that provide MVPD services via conventional distribution methods, as Verizon has done in its FiOS system, accept the "cable service" characterization. In contrast, AT&T has contended that its IP-based video services do not fit the definition of "cable service" and are therefore not subject to the Title VI obligations applicable to cable operators. The FCC has never resolved this issue, in part because AT&T complies with many of those obligations anyway. But the issue briefly assumed significance when, in 2005, cable competitors argued that AT&T was subject to a federal provision requiring any "cable operator providing a cable service" to obtain local franchises, 47 U.S.C. § 541, which AT&T was initially slow to obtain. That dispute, which turned on several abstruse statutory definitions (*see* 47 U.S.C. §§ 522(5), (6), (7), (20)), was placed on the back burner when many states in AT&T's region enacted franchise-reform legislation that cut red tape in the franchising process or eliminated it altogether.

9. As noted in chapter 3, the FCC and its predecessor (the Federal Radio Commission) have long held an expansive view of the statutory "public interest" standard for licensing broadcast applicants, premised on the presumed "scarcity" of the airwaves. As Justice Felix Frankfurter explained in upholding this view, "the Act does not restrict the Commission merely to supervision of the [radio] traffic. It puts upon the Commission the burden of determining the composition of that traffic. The facilities of radio are not large enough to accommodate all who wish to use them. Methods must be devised for choosing from among the many who apply." National Broad. Co. v. United States, 319 U.S. 190, 215–16 (1943) ("*NBC v. U.S.*"). The traditional view has always been that "programming is the essence" of the public interest standard. *See, e.g.*, Johnston Broad. Co. v. FCC, 175 F.2d 351, 359 (D.C. Cir. 1949); *see generally En Banc Programming Inquiry*, 44 F.C.C. 2303, 2311 (1960). The FCC has further attributed particular significance to each licensee's obligation to air programming on issues of *local* importance, given the Commission's obligation under section 307(b) of the Communications Act "to provide a fair, efficient, and equitable distribution" of service "among the several States and communities."

10. The relationships between networks and affiliates can be quite complex. The classic study is contained in Network Inquiry Special Staff, *An Analysis of the Network–Affiliate Relationship in Television*, reprinted in *New Television Net-*

works, vol. II, 106–292 (1980); *see also* BRUCE M. OWEN & STEVEN S. WILDMAN, VIDEO ECONOMICS 151–210 (Harvard Univ. Press, 1992); STANLEY M. BESEN ET AL., MISREGULATING TELEVISION: NETWORK DOMINANCE AND THE FCC (Univ. of Chicago Press, 1984). The FCC has often been asked to intervene in these relationships. *See, e.g.*,, Declaratory Ruling, *Network Affiliated Stations Alliance (NASA) Petition for Inquiry into Network Practices*, 23 FCC Rcd 13610 (2008) (addressing, inter alia, disputes about affiliates' regulatory "right to reject" network programming).

11. In the Cable Act of 1992, Congress barred municipalities from awarding *exclusive* monopoly franchises. *See* 47 U.S.C. § 541(a)(1).

12. *Fourteenth Video Report* ¶ 5.

13. *Id.* ¶ 139, table 5.

14. *See* Hearing Designation Order, *Application of EchoStar Communications Corp., General Motors Corp., and Hughes Electronics Corp. (Transferors), and EchoStar Communications Corp. (Transferee)*, 17 FCC Rcd 20559 (2002).

15. *See* Pub. L. No. 102-385, 106 Stat. 1460 (1992); *see also* Report & Order, *Implementation of Sections of the Cable Television Consumer Protection and Competition Act of 1992 Rate Regulation*, 8 FCC Rcd 5631 (1993).

16. *See* 47 U.S.C. § 543(b)(1), (c)(4); *see also* U.S. General Accounting Office (GAO), *Issues Related to Competition and Subscriber Rates in the Cable Television Industry*, GAO-04-8, at 8 (Oct. 2003) (relating history). Cable operators can avoid rate regulation even of their "basic service tiers" (essentially consisting of local broadcast stations plus public, educational, and governmental channels and perhaps a few additional advertiser-supported channels) if they demonstrate to the FCC that they are subject to "effective competition." 47 U.S.C. § 543(a)(2); *see id.*, § 543(*l*)(1) (defining "effective competition").

17. Rebecca Greenfield, *Why Is Your Cable Bill so High? Ask ESPN*, Atlantic Wire (Dec. 6, 2011), http://www.theatlanticwire.com/technology/2011/12/why-is-your-cable-bill-so-high/45791/# ("as of a year ago, 27 percent . . . said that they would not cut the cord without ESPN's offerings online"). Of course, those large fees are not all profit to ESPN, which must pay immense sums to secure the rights to its programming from (for example) the NFL and other sports leagues. *See id.*; *see also* Brian Stelter & Amy Chozick, *Paying a "Sports Tax," Even If You Don't Watch*, N.Y. TIMES, Dec. 15, 2011, http://www.nytimes.com/2011/12/16/business/media/for-pay-tv-clients-a-steady-diet-of-sports.html ("Although 'sports' never shows up as a line item on a cable or satellite bill, American television subscribers pay, on average, about $100 a year for sports programming—no matter how many games they watch. A sizable portion goes to the National Football League, which dominates sports on television and which struck an extraordinary deal this week with the major networks—$27 billion over nine years—that most likely means the average cable bill will rise again soon").

18. Greenfield, *Why Is Your Cable Bill So High?*

19. *See* Media Bureau, *Report on the Packaging and Sale of Video Programming Services to the Public* (Nov. 18, 2004), http://hraunfoss.fcc.gov/edocs_public/

attachmatch/DOC-254432A1.pdf. As one congressionally sponsored analysis noted, it is generally undisputed that "[l]arge tiers generally benefit those households that prefer a wide variety of programming and/or niche programming, while a la carte pricing generally benefits those households that watch only a small number of networks and prefer general interest programing. Requiring operators to offer all options might not meet the needs of all households, however, because the migration of some threshold number of households to a la carte pricing could undermine the economic feasibility of large tiers." Charles B. Goldfarb, *The FCC's "à la Carte" Reports*, CRS Report for Congress, at 15 (Mar. 30, 2006), www.ncta.com/DocumentBinary.aspx?id=294.

20. Media Bureau, *Further Report on the Packaging and Sale of Video Programming Services to the Public* (Feb. 9, 2006), http://www.creativevoices.us/cgi-upload/news/news_article/ALaCarteFCCReport020906.pdf.

21. This is not to say that wholesale channel bundling is universally popular as a business matter within the cable industry; indeed, it has drawn increasing criticism from cable companies that are unaffiliated with major programmers. *See, e.g.*, George Simpson, *Why Your Cable Bill Is Just Ridiculous*, Online Media Daily (May 25, 2012), http://www.mediapost.com/publications/article/175562/why-your-cable-bill-is-just-ridiculous.html#ixzz21GQAYEzf ("Time Warner Cable CEO Glenn Britt apparently said at the National Cable & Telecommunications Association annual cable show: 'There are too many networks.' 'There are a lot of general-interest networks that have lower viewership, and the industry would take cost out of the system if they shut those networks down and offered lower prices to consumers,' he said. 'The companies involved would make just as much money as they do now because of the costs'"); Sarah Barry James, *Cable Group CEO: Programming Costs Reaching "Breaking Point,"* SNL Kagan (May 24, 2012) (quoting American Cable Association President Matthew Polka: "There is no vehicle in Washington, D.C., yet that is driving this debate but this issue has really caught on in the media, with analysts and in the industry. Consumers really understand that, in their cable bill, there's a lot they're paying for that they wouldn't pay for if they had a choice. And that is a concept that's really settling in, and I'm sure that frightens the content owners to death").

22. Majority Staff Report, House Committee on Energy and Commerce, *Deception and Distrust: The Federal Communications Commission under Chairman Kevin J. Martin*, at 11 (Dec. 2008), http://www.natoa.org/policy-advocacy/Documents/Deception%26DistrustHouseRptFCC.pdf ("Chairman Martin's peremptory reversal of the First Report's conclusions without seeking further public comment or conducting further studies was handled neither openly nor fairly. The Chairman's manipulation of the Second A La Carte Report may have damaged the credibility of the Commission, and it certainly undermined the integrity of the staff").

23. Malrite T.V. v. FCC, 652 F.2d 1140, 1144 (2d Cir. 1981). Thomas Hazlett has similarly described this regime as "a textbook example of anticompetitive regulation." Thomas W. Hazlett, *The Wireless Craze, the Unlimited Bandwidth Myth, the Spectrum Auction Faux Pas, and the Punchline to Ronald Coase's "Big*

Joke": An Essay on Airwave Allocation Policy, 14 HARV. J.L. & TECH. 335, 420 (2001); *see also* Stanley W. Besen & Robert W. Crandall, *The Deregulation of Cable Television*, 44 LAW & CONTEMP. PROB. 77 (1981) (criticizing early regulation of cable television); National Ass'n of Broad. v. FCC, 740 F.2d 1190, 1195 (D.C. Cir. 1984) (acknowledging criticisms and upholding deregulated status of DBS). Over time, the FCC abandoned some of these rules, such as the ones mandating program origination, and the courts invalidated others as unconstitutional or beyond the scope of the FCC's statutory authority. *See* Report & Order, *Amendment of Part 76, Subpart G of Comm'n Rules and Regulations Relative to Program Origination by Cable Television Sys.*, 49 F.C.C.2d 1090 (1974); FCC v. Midwest Video Corp. 440 U.S. 689 (1979) (invalidating, as beyond FCC's Title I authority, pre–Cable Act requirements for "leased access" channels and channels dedicated to "public, educational, and governmental" programming); Home Box Office v. FCC, 567 F.2d 9 (D.C. Cir. 1977) (invalidating restrictions on pay television).

24. Fortnightly Corp. v. United Artists Television, Inc., 392 U.S. 390 (1968); Teleprompter Corp. v. CBS, Inc., 415 U.S. 394 (1974).

25. *See* 17 U.S.C. § 111. For thoughtful discussions of the rationales behind this regime, see Tim Wu, *Copyright's Communications Policy*, 103 MICH. L. REV. 278 (2004), and Stanley M. Besen et al., *Copyright Liability for Cable Television: Compulsory Licensing and the Coase Theorem*, 21 J. L. & ECON. 67 (1978).

26. Copyright licensing is notoriously complex and cannot be resolved through a few negotiations between cable systems and broadcasters. Indeed, broadcasters themselves rarely own or obtain all the rights to any given program they air. Instead, production studios and other rights holders generally license distribution rights in content to broadcasters, cable systems, and other distributors only in specifically defined contexts—for example, prime-time viewing but not Internet streaming. Adding to the complexity, a given TV program may encompass several different copyrighted works (for example, multiple songs), and any given performance of a work may sometimes involve several independent rights holders (the performer, the composer, the songwriter, etc.). Compulsory copyrights are designed to alleviate this complexity.

27. *See* 17 U.S.C. § 111(d).

28. Library of Congress, *Distribution of 1998 and 1999 Cable Royalty Funds*, 69 Fed. Reg. 3606, 3607 (2004).

29. *See, e.g.*, Testimony of Preston Padden before the Senate Comm. on Commerce, Science, and Transp. (July 24, 2012), http://siliconflatirons.com/documents/publications/policy/PaddenTestimony.pdf.

30. 47 U.S.C. § 325(b)(1)(A); *see* United Video Inc. v. FCC, 890 F.2d 1173, 1187 (D.C. Cir. 1989) (confessing surprise that Congress would institute compulsory license alongside rules akin to retransmission consent, but upholding FCC rules according broadcasters related rights).

31. *See generally* Notice of Proposed Rulemaking, *Amendment of the Commission's Rules Related to Retransmission Consent*, 26 FCC Rcd 2718 (2011) ("*2011 Retransmission Consent NPRM*"); FCC, *Retransmission Consent and*

Exclusivity Rules: Report to Congress Pursuant to Section 208 of the Satellite Home Viewer Extension and Reauthorization Act of 2004 (Sept. 8, 2005), http://hraunfoss.fcc.gov/edocs_public/attachmatch/DOC-260936A1.pdf.

32. *See 2011 Retransmission Consent NPRM* ¶ 22.

33. *See id.* ¶ 15 (discussing these two episodes).

34. Time Warner Cable Inc. et al., Petition for Rulemaking to Amend the Commission's Rules Governing Retransmission Consent, MB Docket No. 10-71 (FCC filed Mar. 9, 2010).

35. Disney Reply Comments, MB Docket No. 10-71, at 6 (FCC filed June 3, 2010).

36. *2011 Retransmission Consent NPRM* ¶ 18.

37. *Id.* ¶ 8 & n.20.

38. 47 C.F.R. § 76.92 et seq.

39. See *id.* § 76.101 et seq. For a history of these provisions, see Christopher S. Yoo, *Rethinking the Commitment to Free, Local Television*, 52 Emory L.J. 1579, 1646 n.189 (2003).

40. *See 2011 Retransmission Consent NPRM* ¶ 44 ("Do these rules provide stations and networks with any rights that cannot be secured through a combination of network–affiliate contracts and retransmission consent?"). Some advocates ask the FCC to go one step further, intervene in the network–affiliate contractual relationship itself, and ban contractual clauses that prohibit affiliates from selling network signals outside their broadcast areas.

41. *See generally id.* ¶ 5 n.13.

42. 47 U.S.C. §§ 534, 535. The FCC had laid the groundwork for this statutory "must-carry" requirement by developing a regulatory scheme of its own in the 1960s "to ameliorate the adverse impact of [cable] competition upon local stations, existing and potential." First Report & Order, *Rules Re Microwave-Service CATV*, 38 F.C.C. 683, ¶ 77 (1966). Although the FCC adhered to that scheme over the ensuing decades, the D.C. Circuit invalidated it in 1985 on the grounds that the agency's purported justification—the preservation of local over-the-air television—was "fanciful" and that the rules were overbroad in any event because they indiscriminately protected all broadcasters regardless of whether they aired local content. *See* Quincy Cable TV, Inc. v. FCC, 768 F.2d 1434, 1454–62 (D.C. Cir. 1985). The FCC then adopted a narrower form of "must carry" and a modified rationale, but the D.C. Circuit invalidated the new rules too. *See* Century Communications v. FCC, 835 F.2d 292 (D.C. Cir. 1987), *clarified*, 837 F.2d 517 (D.C. Cir. 1987).

43. *See* Turner Broad. Sys. v. FCC, 520 U.S. 180 (1997). Justice Breyer provided the crucial fifth vote, reasoning that even though the must-carry regime hurt cable subscribers by restricting their available programming choices, it helped the over-the-air viewers more by expanding their array of local programming options. *Id.* at 228–29 (Breyer, J., concurring).

44. *See generally* Thomas W. Hazlett, *Digitizing "Must Carry" under Turner Broadcasting v. FCC*, 8 Sup. Ct. Econ. Rev. 141, 172 (2000); Glen O. Robinson, *The Electronic First Amendment: An Essay for the New Age*, 47 Duke L.J. 899,

937 (1998); *see also* Turner Broadcasting v. FCC, 910 F. Supp. 734, 776 (D.D.C. 1995) (Williams, J., dissenting), *aff'd*, *Turner Broadcasting.*

45. Robinson, *Electronic First Amendment*, at 941.

46. 17 U.S.C. § 119(d)(10)(a).

47. *See*, *e.g.*, ABC, Inc. v. PrimeTime 24 Joint Venture, 17 F. Supp. 2d 478 (M.D.N.C.), *aff'd*, 184 F.3d 348 (4th Cir. 1999).

48. *See* Satellite Broad. and Communications Ass'n v. FCC, 275 F.3d 337, 348 (4th Cir. 2001) (quoting EchoStar executive).

49. *Id.* at 350 n.5.

50. 17 U.S.C. § 122(a), (f), (j)(2). Congress later qualified that general prohibition in the Satellite Home Viewer Extension and Reauthorization Act of 2004, Pub. L. No. 108-447, 118 Stat. 2809 (2004). SHVERA, as this law is called, entitles satellite carriers to offer subscribers certain "significantly viewed" signals of out-of-market broadcast stations. *See* 47 U.S.C. § 340.

51. *See* 47 U.S.C. § 338.

52. *See Satellite Broad. and Communications Ass'n*, 275 F.3d at 366.

53. *See* 47 U.S.C. § 548 (codifying section 628 of the Communications Act, added by the Cable Act of 1992). As with other provisions of Title VI of the Communications Act, the section number of this provision as it appears in Title 47 of the U.S. Code is different from its section number in the Act. Title VI is the most significant exception to the usual rule that the section numbers of the Communications Act are identical to the corresponding section numbers of Title 47.

54. Technically, aspects of these rules prevent a cable operator from withholding not only its *own* affiliated programming from its rivals, but also the affiliated programming of *other* cable operators as well. *See* 47 U.S.C. § 548(c)(2)(D) (directing the FCC to "prohibit exclusive contracts" for satellite-delivered programming "between a cable operator and [programmer] in which a cable operator has an attributable interest").

55. In general, antitrust law tolerates such self-dealing unless it effectively "forecloses" rival firms from the markets for either distribution or production. Antitrust courts evaluate a variety of factors to judge whether the practice suppresses competition, including the duration of the exclusivity arrangement and the availability of reasonable alternatives. *See*, *e.g.*, Twin City Sportservice, Inc. v. Charles O. Finley & Co., 676 F.2d 1291, 1302 (9th Cir. 1982). In the 1990s, before direct-to-home satellite providers had obtained significant market share, antitrust officials concluded on a couple of occasions that vertically integrated cable providers had both the incentive and opportunity to exclude entry by new video-distribution platforms. *See* Competitive Impact Statement, *United States v. Tele-Communications, Inc. and Liberty Media Corp.*, 59 Fed. Reg. 24723 (1994); Competitive Impact Statement, *United States v. Primestar Partners, L.P.*, 58 Fed. Reg. 33944 (1993).

56. 47 U.S.C. § 548(b); Report & Order, *Development of Competition and Diversity in Video Programming Distribution and Carriage*, 8 FCC Rcd 3359, ¶¶ 36–41 (1993).

57. 47 U.S.C. § 548(c)(2).

58. *Id.* § 548(c)(5).

59. *See* Cablevision Sys. Corp. v. FCC, 597 F.3d 1306 (D.C. Cir. 2010) ("*Cablevision I*"), *aff'g* Report & Order, *Implementation of the Cable Television Consumer Protection and Competition Act of 1992*, 22 FCC Rcd 17791 (2007); Report & Order, *Implementation of the Cable Television Consumer Protection and Competition Act of 1992*, 17 FCC Rcd 12124, ¶ 4 (2002) ("*2002 Program Access Order*").

60. Report & Order and Further Notice of Proposed Rulemaking, *Revision of the Commission's Program Access Rules*, MB Docket No. 12-68, FCC No. 12-123 (Oct. 5, 2012) ("*2012 Program Access Order*"). The FCC emphasized that, "in addition to claims under Section 628(b) of the Act, additional causes of action under Section 628 will continue to apply after expiration of the exclusive contract prohibition, including claims . . . alleging discrimination under Section 628(c)(2)(B)," including "selective refusals to license . . . content to a particular MVPD (such as a new entrant or satellite provider) while simultaneously licensing . . . content to other MVPDs competing in the same geographic area. Even after the expiration of the exclusive contract prohibition, such conduct will remain a violation of the discrimination provision . . . unless the cable-affiliated programmer can establish a legitimate business reason for the conduct." *Id.* ¶ 4 (footnote omitted).

61. *2002 Program Access Order*, 17 FCC Rcd at 12177 (dissenting statement of Commissioner Abernathy) (emphasis added).

62. *Cablevision I*, 597 F.3d at 1314.

63. *Id.* at 1316 (Kavanaugh, J., dissenting) (internal quotation marks omitted).

64. For example, even Judge Kavanaugh announced in his *Cablevision I* dissent that he "would leave open the possibility that the Government might still impose a prospective ban on some exclusive agreements between video programming distributors and affiliated *regional* video programming networks," particularly those carrying "highly desirable 'must have' regional sports" programming. *Id.* at 1326 n.6 (Kavanaugh, J, dissenting).

65. *See generally* EchoStar Communications Corp. v. FCC, 292 F.3d 749 (D.C. Cir. 2002).

66. *See* First Report & Order, *Review of the Commission's Program Access Rules and Examination of Programming Tying Arrangements*, 25 FCC Rcd 746, ¶ 52 (2010).

67. *Id.* The Commission later made clear that the same case-by-case regime would apply to *satellite*-delivered regional sports programming when the section 628(c)(2) exclusive-contract ban expired. *2012 Program Access Order* ¶¶ 2–3.

68. 47 U.S.C. § 548(b) (barring certain acts that "hinder significantly or prevent" any MVPD "from providing *satellite cable programming or satellite broadcast programming*") (emphasis added).

69. Cablevision Sys. Corp. v. FCC, 649 F.3d 695 (D.C. Cir. 2011). The court remanded one issue to the FCC for better-reasoned decisionmaking, but the

remand did not threaten the basics of the new regime, and the FCC subsequently ruled for Verizon and AT&T in their complaints against Cablevision.

70. *See, e.g., Fourteenth Video Report* ¶ 341 (citing Ryan Lawler, *Deloitte: 9% Have Cut Cable, Another 11% Are Considering It*, GigaOm (Jan. 4, 2012), http://gigaom.com/video/deloitte-cord-cutters/). Apart from outright cord *cutting*, some viewers have begun engaging in cord *shaving*, which "generally refers to a downgrading of pay video services from the subscriber's MVPD." *Id.* ¶ 341 n.1089.

71. Memorandum Opinion & Order, *Applications of Comcast Corp., General Elec. Co., and NBC Universal, Inc. for Consent to Assign Licenses and Transfer Control of Licenses*, 26 FCC Rcd 4238, ¶ 79 (2011) ("*Comcast–NBCU Order*") (emphasis omitted); *accord* Department of Justice, Competitive Impact Statement, United States v. Comcast Corp., No. 1:11-cv-00106, at 18 (D.D.C. filed Jan. 18, 2011) ("When measured by the number of customers who are cord-shaving or cord-cutting, OVDs currently have a *de minimis* share of the video programming distribution market. Their current market share, however, greatly understates their potential competitive significance in this market").

72. *Fourteenth Video Report* ¶¶ 341–42.

73. 17 U.S.C. § 111(f)(3). As noted in this chapter, a separate regime is applicable to satellite providers.

74. WPIX, Inc. v. ivi, Inc., 691 F.3d 275 (2d Cir. 2012).

75. *FCC Fourteenth Video Report* ¶ 269.

76. *Id.*

77. *Id.*

78. *Id.*

79. *See* 47 U.S.C. § 325(b)(3)(C)(ii) (retransmission consent), *id.* § 548 (program access).

80. 47 U.S.C. § 522(13) (emphasis added).

81. 47 U.S.C. § 522(4). "Video programming" is defined as "programming provided by, or generally considered comparable to programming provided by, a television broadcast station." *Id.* § 522(20).

82. *See* Public Notice, *Media Bureau Seeks Comment on Interpretation of the Terms "Multichannel Video Programming Distributor" and "Channel" as Raised in Pending Program Access Complaint Proceeding*, MB Docket No. 12-83, DA 12-507, ¶ 2 (Mar. 30, 2012) ("*Media Bureau MVPD Inquiry*") (noting that regulatory obligations of MVPDs include "statutory and regulatory requirements relating to program carriage, the competitive availability of navigation devices . . . , the requirement to negotiate in good faith with broadcasters for retransmission consent, Equal Employment Opportunity ('EEO') requirements, closed captioning and emergency information requirements, various technical requirements (such as signal leakage restrictions), and cable inside wiring requirements") (footnotes omitted). In some cases, it may be meaningless or impossible to apply these obligations to over-the-top providers of streaming video services.

83. *See* Order, *Sky Angel U.S., LLC*, 25 FCC Rcd 3879 ¶ 2 (Media Bureau 2010) (denying interim relief after tentatively concluding that an OVD was not an "MVPD" entitled to program-access rights).

84. *See Media Bureau MVPD Inquiry*.

85. *Comcast–NBCU Order* ¶ 3.

86. *Id.* ¶ 4; *see* United States v. Comcast Corp., No. 1:11-cv-00106, 2011 WL 5402137 (D.D.C. Sept. 1, 2011) ("*Comcast–NBCU Final Judgment*").

87. *Comcast–NBCU Final Judgment* § V.G.2.

88. *Id.*

89. *Comcast–NBCU Order* ¶ 94.

90. *Id.*, Appx. A, § IV.D–E; *see also Comcast–NBCU Final Judgment*, § V.G.5. In 2012, the FCC penalized Comcast and extended the $49.95 6 Mbps condition for an additional year because it found that Comcast had inadequately marketed that service option. *See* Order, *Comcast Corp.*, File No. EB-11-IH-0163 (June 27, 2012).

91. *See* Petition of Public Knowledge to Enforce Merger Conditions, MB Docket No. 10-56 (FCC filed Aug. 1, 2012), http://www.publicknowledge.org/files/Comcast-Xbox%20FINAL.pdf; Harold Feld, *Michael Powell Works the Ref on the XBox360 Play*, Public Knowledge Blog (Mar. 29, 2012), http://www.publicknowledge.org/blog/michael-powell-works-ref-xbox360-play; Daniel Ionescu, *Netflix Boss Blasts Comcast over Bandwidth Caps, Net Neutrality*, PCWorld (Apr. 16, 2012), http://www.pcworld.com/article/253850/netflix_boss_blasts_comcast_over_bandwidth_caps_net_neutrality.html.

92. *See, e.g.*, Michael Powell, *No Good Deed Goes Unpunished—Washington Advocacy Run Amok*, CableTechTalk (Mar. 2012), http://www.cabletechtalk.com/tech-discussions/2012/03/28/no-good-deed-goes-unpunished-%e2%80%93-washington-advocacy-run-amok/. See chapter 6 in this volume for a fuller treatment of this dispute.

93. In the same vein, the FCC also "impose[d] conditions to foster the continued viability of Hulu, an emerging OVD in which NBCU was an original participant," providing that "neither Comcast nor Comcast–NBCU shall exercise any right to influence the conduct or operation of Hulu, including that arising from agreements, arrangements or operation of its equity interests." *Comcast–NBCU Order* ¶ 90.

94. *Id.*, Appx. A, § IV.A.

95. *Id.* Analogous provisions appear in the Department of Justice consent decree as well. *Comcast–NBCU Final Judgment* § IV.

96. For a compelling critique of the traditional restrictions on vertical integration in the media industry, see Christopher Yoo, *Vertical Integration and Media Regulation in the New Economy*, 19 Yale J. Reg. 171 (2002). For an explanation of when vertical integration questions pose valid antitrust concerns, see Michael H. Riordan & Steven C. Salop, *Evaluating Vertical Mergers: A Post-Chicago Approach*, 63 Antitrust L.J. 513 (1995).

97. *See* United States v. Paramount Pictures, 334 U.S. 131 (1948).

98. *See generally* Oliver E. Williamson, the Mechanisms of Governance (Oxford Univ. Press, 1996).

99. *See* Ronald Coase, The Firm, The Market, and the Law (Univ. of Chicago Press, 1990).

100. *See, e.g.*, Bill Carter, *Ailing ABC Turns to HBO in Search of TV Hits*, N.Y. Times, Aug. 5, 2002, at C1.

101. *See, e.g.*, Time Warner Entertainment, Co. v. FCC, 240 F.3d 1126, 1138 (D.C. Cir. 2001) ("*Time Warner*") ("even where an unaffiliated supplier offered a better cost-quality trade-off, a company might be reluctant to ditch or curtail an inefficient in-house operation because of the impact on firm executives or other employees, or the resulting spotlight on management's earlier judgment").

102. Associated Press v. United States, 326 U.S. 1, 20 (1945).

103. Schurz Communications, Inc. v. FCC, 982 F.2d 1043, 1051 (7th Cir. 1992).

104. *Id.*; *see generally* Besen et al., Misregulating Television, at 127–46.

105. *Schurz*, 982 F.2d at 1051.

106. *See, e.g.*, FCC v. National Citizens Comm. for Broad., 436 U.S. 775 (1978) ("*NCCB*").

107. *Schurz*, 982 F.2d at 1054–55.

108. *Id.* at 1050.

109. *Viacom Makes Split Official*, CBSNews.com (2005), http://www.cbsnews.com/2100-201_162-701875.html ("The move to break up the company essentially undoes Viacom's acquisition of CBS Corp., which was announced in 1999. Viacom's split-up is also the biggest example of a recent trend among media companies to trim down their holdings as they try to regain favor on Wall Street. Large diversified media conglomerates have generally lost favor with investors in recent years amid growing skepticism about the 'synergies' to be gained from owning such diverse assets.").

110. Ted Turner, *Monopoly or Democracy*, Wash. Post, May 30, 2003, at A23. Lawrence Lessig also argues that finsyn's repeal has had a "narrowing effect" on what programs are produced. He acknowledges, however, that "the efficiencies [from vertical integration] are important, and the effect on culture is hard to measure." Lawrence Lessig, Free Culture: How Big Media Uses Technology and the Law to Lock Down Culture and Control Creativity 166 (Penguin, 2004). For a recent critique of the risks of vertical integration in general, see Tim Wu, The Master Switch: The Rise and Fall of Information Empires (Knopf, 2010).

111. Remarks of FCC Commissioner Michael J. Copps at the Future of Music Coalition Policy Summit at 3 (May 3, 2004), http://hraunfoss.fcc.gov/edocs_public/attachmatch/DOC-246862A1.doc.

112. *See* Report & Order, *2002 Biennial Regulatory Review—Review of the Commission's Broadcast Ownership Rules and Other Rules Adopted Pursuant to Section 202 of the Telecommunications Act of 1996*, 18 FCC Rcd 13620,

¶¶ 45, 640–56 (2003) ("*2003 Media Ownership Order*") (rejecting "source diversity" as a valid goal for media policy and declining to reinstitute finsyn rules).

113. 47 U.S.C. § 533(f)(1).

114. *See* Third Report & Order, *Implementation of Section 11(c) of the Cable Television Consumer Protection and Competition Act of 1992*, 14 FCC Rcd 19098 (1999) ("*Third R&O*"); Fourth Report & Order, *The Commission's Cable Horizontal and Vertical Ownership Limits*, 23 FCC Rcd 2134 (2008) ("*Fourth R&O*").

115. *Time Warner*, 240 F.3d at 1137–40.

116. *See Fourth R&O* ¶¶ 135–45.

117. *See Third R&O* ¶ 1.

118. *Time Warner*, 240 F.3d at 1130–36.

119. *See Fourth R&O*.

120. *Comcast Corp. v. FCC*, 579 F.3d 1, 9 (D.C. Cir. 2009).

121. *Time Warner*, 240 F.3d at 1134–36; *see also Comcast*, 579 F.3d at 9.

122. *Prometheus I*, 373 F.3d at 414; *see also NBC v. U.S.*

123. *Compare, e.g.*, C. Edwin Baker, *Media Concentration: Giving Up on Democracy*, 54 Fla. L. Rev. 839 (2002) (arguing for greater role for FCC regulation), *with* Bruce M. Owen, *Regulatory Reform: The Telecommunications Act of 1996 and the FCC Media Ownership Rules*, 2003 Mich. St.-DCL L. Rev. 671 (arguing for antitrust review alone).

124. 47 U.S.C. § 536(a)(3).

125. Second Report & Order and Notice of Proposed Rulemaking, *Revision of the Commission's Program Carriage Rules*, 26 FCC Rcd 11494, ¶ 79 (2011) ("*2011 Program Carriage Order*") (noting that "[o]nly two program carriage cases have been decided on the merits to date" and that in each the FCC had determined "that the defendant would prevail even assuming that the burdens shifted to the defendant").

126. *See* Memorandum Opinion & Order, *Tennis Channel, Inc. v. Comcast Cable Communications*, MB Docket No. 10-204, FCC No. 12-78 (July 24, 2012), *stayed pending appeal*, Comcast Cable Communications, LLC v. FCC, No. 12-1337 (D.C. Cir. August 24, 2012).

127. *See, e.g., 2011 Program Carriage Order* ¶ 25 (creating procedures for aggrieved programmers to seek "standstill" orders freezing existing carriage arrangements in place after they file complaints alleging program carriage violations).

128. *See* Time Warner Cable v. FCC, No. 11-4138 (2d Cir. pet. for review filed Oct. 11, 2011).

129. Notice, *Time Warner, Inc., et al.: Proposed Consent Agreement with Analysis to Aid Public Comment*, 61 Fed. Reg. 50301 (Sept. 25, 1996).

130. *Comcast–NBCU Order* ¶¶ 120–22. In 2012, the FCC's Media Bureau ruled that Comcast had violated this "neighborhooding" requirement in its assignment

of channel locations to Bloomberg Television, a 24-hour business news channel. *See* Bloomberg L.P. v. Comcast Cable Communications, LLC, MB Docket No. 11-104, DA No. 12-694 (May 2, 2012).

131. *See Rules Governing Standard and High Frequency Broadcast Stations*, 6 Fed. Reg. 2282, 2284–85 (1941) (regarding TV); *see also Rules Governing Standard and High Frequency Broadcast Stations*, 8 Fed. Reg. 16065 (1943) (regarding AM radio); *Rules Governing Standard and High Frequency Broadcast Stations*, 5 Fed. Reg. 2382, 2384 (1940) (regarding FM radio). The Supreme Court upheld the FCC's authority to impose such restrictions in United States v. Storer Broadcasting, 351 U.S. 192 (1956).

132. *See, e.g.*, Second Report & Order, *Rules Relating to Multiple Ownership of Standard, FM, and Television Broadcast Stations*, 50 F.C.C.2d 1046 (1975).

133. *NCCB*, 436 U.S. at 796.

134. *See* Report & Order, *Multiple Ownership of AM, FM, and Television Broadcast Stations*, 100 F.C.C.2d 17 (1984); *see also* Mark S. Fowler & Daniel L. Brenner, *A Marketplace Approach to Broadcast Regulation*, 60 Tex. L. Rev. 207 (1982).

135. *See* Second Supplemental Appropriations Act, Pub. L. No. 98-396, Sec. 304, 98 Stat. 1369 (1984); *Multiple Ownership of AM, FM, and Television Broadcast Stations*, 100 F.C.C.2d 74 (1984).

136. *See, e.g.*, Second Report & Order, *Amendment of Section 73.3555 of the Commission's Rules, the Broadcast Multiple Ownership Rules*, 4 FCC Rcd 1741, ¶ 76 (1988).

137. Pub. L. No. 104-104, § 202(c)(1)(B), 110 Stat. 56, 110 (1996).

138. *Id.*, § 202(c)(2), (h).

139. Report & Order, *Review of the Commission's Regulations Governing Television Broadcasting*, 14 FCC Rcd 12903, ¶ 64 (1999) (regarding TV); *see also id.* ¶ 100 (regarding TV and radio). In Sinclair Broadcasting Group, Inc. v. FCC, 284 F.3d 148, 162 (D.C. Cir. 2002), the D.C. Circuit rejected the rationale for portions of these rules and remanded the matter back to the FCC for further proceedings (without, however, vacating the rules).

140. Biennial Review Report, *1998 Biennial Regulatory Review—Review of the Commission's Broadcast Ownership Rules and Other Rules Adopted Pursuant to Section 202 of the Telecommunications Act of 1996*, 15 FCC Rcd 11058, ¶ 25 (2000).

141. 280 F.3d 1027, 1044 (D.C. Cir. 2002).

142. *Id.*

143. *See* Benjamin M. Compaine & Douglas Gomery, Who Owns the Media? Competition and Concentration in the Mass Media Industry 135, 136 (Dimension, 3d ed. 2000).

144. *See* Baker, *Media Concentration*; *see also* Mark Cooper, Media Ownership and Democracy in the Digital Information Age 31 (Gillis Pub. Group,

2003), http://cyberlaw.stanford.edu/attachments/mediabooke.pdf (discussing public disquiet with media mergers).

145. *2003 Media Ownership Order*. Our discussion does not address the Commission's efforts in the same order to amend its restrictions on the ownership of multiple broadcast radio stations in the same community. *See Prometheus I*, 373 F.3d at 421–35 (upholding some aspects of the FCC's analysis but invalidating others).

146. *See 2003 Media Ownership Order* ¶¶ 499–500.

147. *See* 2004 Consolidated Appropriations Act, Pub. L. No. 108-199, § 629, 118 Stat. 3, 99 (2004); *see also* Ben Scott, *The Politics and Policy of Media Ownership*, 53 AM. U. L. REV. 645 (2004).

148. Scott, *The Politics and Policy of Media Ownership*, at 674 (noting that the compromise simply "legalized the status quo," which stemmed from earlier waivers of the 35% limit).

149. *2003 Media Ownership Order* ¶¶ 368–69 (abolishing newspaper–TV cross-ownership rule); *id.* ¶ 390 (abolishing radio–TV cross ownership rule); *id.* ¶¶ 432–98 (instituting cross-media limits).

150. *Id.* ¶¶ 185–87.

151. *Id.* ¶ 327.

152. *See* Cooper, *Open Communications Platforms*, at 215.

153. 373 F.3d 372. The Third Circuit also dealt at length with threshold questions about whether the statute required the FCC to apply a presumption of deregulation when considering proposals to relax its media ownership rules. The Third Circuit concluded that the answer is no, despite mixed signals on the subject from the D.C. Circuit. *See id.* at 393–94 (concluding that discussion of issue in Cellco Partnership v. FCC, 357 F.3d 88, 96–98 (D.C. Cir. 2004), trumped earlier suggestions in *Fox* and *Sinclair* that a deregulatory presumption should apply).

154. 373 F.3d at 397–412.

155. *Id.* at 415.

156. *Id.* at 418–20. Chief Judge Anthony Scirica dissented, arguing that "[p]reserving the 'marketplace of ideas' does not easily lend itself to mathematical certitude" of the type the majority was demanding" (*id.* at 436) and that "[i]t is not the role of the judiciary to second-guess the reasoned policy judgments of an administrative agency acting within the scope of its delegated authority" (*id.* at 435).

157. Report & Order and Order on Reconsideration, *2006 Quadrennial Regulatory Review—Review of the Commission's Broadcast Ownership Rules*, 23 FCC Rcd 2010 (2008).

158. *Id.* ¶ 20.

159. *Id.* ¶ 87. Under the pre-2003 rules, "an entity may own two television stations in the same DMA if: (1) the Grade B [signal] contours of the stations do not overlap; or (2) at least one of the stations in the combination is not ranked

among the top four stations in terms of audience share, and at least eight independently owned and operating commercial or non-commercial full-power broadcast television stations would remain in the DMA after the combination." *Id.* The FCC will "reverse the negative presumption" in markets with fewer than eight "voices" if one of the stations is "failing," but only in narrowly defined circumstances. *Id.* ¶ 65.

160. Prometheus Radio Project v. FCC, 652 F.3d 431 (3d Cir. 2011) ("*Prometheus II*").

161. *Id.* at 445–54.

162. *Id.* at 453 n.25.

163. The grandfather of such precedents is Red Lion Broad. Co. v. FCC, 395 U.S. 367 (1969), which upheld the FCC's "fairness doctrine." The central Supreme Court case applying lenient review in the broadcast ownership context is *NCCB*.

164. *See, e.g.*, Turner Broad. Sys. v. FCC, 512 U.S. 622, 638 n.5 (1994); FCC v. Fox Television Stations, Inc., 556 U.S. 502, 530–35 (2009) (Thomas, J., concurring).

Chapter 10

1. For a discussion of airline deregulation and its parallels to the Telecommunications Act of 1996, see James B. Speta, *Deregulating Telecommunications in Internet Time*, 61 WASH. & LEE L. REV. 1063, 1071–75 (2004).

2. *See* Richard D. Cudahy, *Whither Deregulation: A Look at the Portents*, 58 N.Y.U. ANN. SURV. AM. L. 155, 166 (2001); Michael E. Levine, *Airline Competition in Deregulated Markets: Theory, Firm Strategy, and Public Policy*, 4 YALE J. ON REG. 393, 394 (1987) ("[B]y the mid-1970's it was probably fair to say that no impartial academic observer of any standing doubted that the airline business, if unregulated, would reach something that more or less resembled a competitive equilibrium").

3. *See, e.g.*, Alfred E. Kahn, *The Theory and Application of Regulation*, 55 ANTITRUST L.J. 177, 178 (1986).

4. For a discussion of how Fred Kahn operated effectively as a political entrepreneur, see Philip J. Weiser, *Alfred Kahn as a Case Study of a Political Entrepreneur*, 7 REV. OF NETWORK ECON. 601 (2008).

5. *See* Airline Deregulation Act of 1978, Pub. L. No. 95-504, 92 Stat. 1705 (codified in scattered sections of 49 U.S.C.).

6. H.R. Rep. No. 104-204, at 48, *reprinted in* 1996 U.S.C.C.A.N. 10, 11.

7. For an explanation of this new model and its parallels in other regulated industries, see Joseph D. Kearney & Thomas W. Merrill, *The Great Transformation of Regulated Industries Law*, 98 COLUM. L. REV. 1323 (1998); *see also* Joseph Farrell, *Creating Local Competition*, 49 FED. COMM. L.J. 201, 211–12 (1996).

8. Alfred Kahn, Letting Go: Deregulating the Process of Deregulation 70 (Inst. of Pub. Utils. and Network Indus., 1998).

9. Peter W. Huber, Law and Disorder in Cyberspace: Abolish the FCC and Let Common Law Rule the Telecosm (Oxford Univ. Press, 1997).

10. Lawrence Lessig, *Reboot the FCC*, Newsweek (Dec. 22, 2008), http://www.thedailybeast.com/newsweek/2008/12/22/reboot-the-fcc.html.

11. 444 Parl. Deb., H.C. (5th ser.) (1947) 206–07, http://hansard.millbanksystems.com/commons/1947/nov/11/parliament-bill#column_206 (remarks of Winston Churchill in the House of Commons).

12. Burnet v. Coronado Oil & Gas Co., 285 U.S. 393, 406 (1932) (Brandeis, J., dissenting).

13. AT&T Corp. v. Iowa Utils. Bd., 525 U.S. 366, 397 (1999) ("It would be gross understatement to say that the 1996 Act is not a model of clarity. It is in many important respects a model of ambiguity or indeed even self-contradiction. That is most unfortunate for a piece of legislation that profoundly affects a crucial segment of the economy worth tens of billions of dollars"). This problem seems to have eluded the legislation's congressional sponsors. Senator Larry Pressler, for example, maintained that the FCC's rulemaking tasks were straightforward because, he said, Congress had already done "the heavy lifting" on telecommunications policy. *Pressler Attacks FCC Request for More Money*, Communications Daily, 1996 WLNR 3291872 (Feb. 22, 1996) (quoting Sen. Larry Pressler).

14. Philip J. Weiser, *Law and Information Platforms*, 1 J. Telecomm. & High Tech. L. 1, 12 n.51 (2002).

15. We briefly address this Title VI theory in connection with our analysis of the *Portland* litigation in chapter 6.

16. *See, e.g.*, Richard S. Whitt, *A Horizontal Leap Forward: Formulating a New Communications Public Policy Framework Based on the Network Layers Model*, 56 Fed. Comm. L.J. 587 (2004); Rob Frieden, *Adjusting the Horizontal and Vertical in Telecommunications Regulation: A Comparison of the Traditional and a New Layered Approach*, 55 Fed. Comm. L.J. 207, 215 (2003) ("The horizontal orientation . . . makes better sense in a convergent, increasingly Internet-dominated marketplace and also provides a more intelligent model than the existing vertical orientation that creates unsustainable service and regulatory distinctions"); Craig McTaggart, *A Layered Approach to Internet Legal Analysis*, 48 McGill L.J. 571 (2003); Philip J. Weiser, *Toward a Next Generation Regulatory Strategy*, 35 Loy. U. Chi. L.J. 41 (2003); Kevin Werbach, *A Layered Model for Internet Policy*, 1 J. Telecomm. & High Tech. L. 37, 38 (2002) (arguing that communications regulation should be based on "the technical architecture of the Internet itself"); Douglas C. Sicker & Joshua L. Mindel, *Refinements of a Layered Model for Telecommunications Policy*, 1 J. Telecomm. & High Tech. L. 69, 71 (2002); John T. Nakahata, *Regulating Information Platforms: The Challenges of Rewriting Communications Regulation from the Bottom Up*, 1 J. Telecomm. & High Tech. L. 95, 98 (2002).

17. Peter Huber et al., Federal Telecommunications Law 402–03 (Aspen, 2d ed. 1999); *see also* Huber, Law and Disorder in Cyberspace.

18. *See* Joseph D. Kearney, *From the Fall of the Bell System to the Telecommunications Act: Regulation of Telecommunications under Judge Greene*, 50 Hastings L.J. 1395 (1999).

19. SBC Communications, Inc. v. FCC, 981 F. Supp. 996 (N.D. Tex. 1997) (invalidating Bell-specific provisions as unconstitutional "bills of attainder"), *rev'd*, 154 F.3d 226 (5th Cir.1998).

20. Frank H. Easterbrook, *When Does Competition Improve Regulation*, 52 Emory L. Rev. 1297, 1297 (2003); *see also* Richard A. Posner, *Antitrust in the New Economy*, 68 Antitrust L.J. 925, 937 (2001) (noting challenge in antitrust cases).

21. *See, e.g.*, Stephen Breyer, Economic Reasoning and Judicial Review 11–13 (American Enterprise Institute Press, 2004) (discussing use of specialist courts and experts and embracing the latter option).

22. United States v. Microsoft Corp., 147 F.3d 935, 954–56 (D.C. Cir. 1998).

23. Reviews of Huber's argument have made this same point, observing that his failure to explain how antitrust courts would oversee "non-discriminatory and reasonably priced interconnection . . . is a notable shortcoming." Joseph D. Kearney, *Twilight of the FCC?*, 1 Green Bag 2d 327, 329 (1998). The government of New Zealand reached a similar conclusion and adopted a sector-specific regulatory regime after experimenting with an antitrust-like approach for telecommunications deregulation. *See* Mary Newcomer Williams, *Comparative Analysis of Telecommunications Regulation: Pitfalls and Opportunities*, 56 Fed. Comm. L.J. 269, 277 (2003); *see also* Jean-Jacques Laffont & Jean Tirole, Competition in Telecommunications 34 (MIT Press, 2000) (examining the New Zealand case and concluding that it demonstrates the "difficulty of ensuring competition in the absence of regulation").

24. For a discussion of the rationale for delegation of lawmaking authority to agencies as opposed to courts, see Philip J. Weiser, *Federal Common Law, Cooperative Federalism, and the Enforcement of the Telecom Act*, 76 N.Y.U. L. Rev. 1692, 1718–20 (2001).

25. 540 U.S. 398 (2004).

26. United States v. Am. Tel. & Tel. Co., 552 F. Supp. 131, 168 (D.D.C. 1982). In 1975, AT&T's argument had earlier persuaded Greene's predecessor on the case, Judge Joseph Waddy, to halt further discovery until this jurisdictional point was resolved, thereby winning the company several years of delay. *See* Steve Coll, The Deal of the Century 79–82 (Atheneum, 1986).

27. Pub. L. No. 104-104, § 601(b)(1), 110 Stat. 56 (1996); *see also* H.R. Rep. No. 104-458, at 201 (1996) (Conf. Rep.) (explaining that the clause "prevents affected parties from asserting that the [Act] impliedly pre-empts other laws").

28. *See, e.g.*, Goldwasser v. Ameritech Corp., 222 F.3d 390, 401 (7th Cir. 2000); James B. Speta, *Antitrust and Local Competition under the Telecommunications Act*, 71 Antitrust L.J. 99 (2003).

29. *See, e.g.*, Law Offices of Curtis V. Trinko, L.L.P. v. Bell Atl., 305 F.3d 89 (2d Cir. 2002), *rev'd*, 540 U.S. 398 (2004); Philip J. Weiser, *Goldwasser, the Telecom Act, and Reflections on Antitrust Remedies*, 55 Admin. L. Rev. 1 (2003); Steven Semeraro, *Speta on Antitrust and Local Competition under the Telecommunications Act: A Comment Respecting the Accommodation of Antitrust and Telecom Regulation*, 71 Antitrust L.J. 147 (2003).

30. The essential facilities doctrine was most famously applied to the telecommunications industry in MCI's own private antitrust case against AT&T, which proceeded parallel to the litigation handled by Judge Greene. In that case, the Seventh Circuit ruled that a competitor states a claim under the essential facilities doctrine where (1) a monopolist exercises control of an essential facility; (2) a competitor cannot practically or reasonably duplicate the facility; (3) the monopolist has denied the competitor access to the facility; and (4) the monopolist can feasibly provide access to the facility. MCI Communications Corp. v. Am. Tel. & Tel. Co., 708 F.2d 1081, 1132–33 (7th Cir. 1983). Many have criticized this doctrine on the ground that it requires antitrust courts to manage access to monopoly facilities and invites overbroad applications. *See, e.g.*, Phillip Areeda, *Essential Facilities: An Epithet in Need of Limiting Principles*, 58 Antitrust L.J. 841 (1989). Some antitrust courts have also relied on the related claim that incumbent providers owe a duty to deal with new entrants "where some cooperation is indispensable to effective competition." Olympia Equip. Leasing Co. v. W. Union Tel. Co., 797 F.2d 370, 379 (7th Cir. 1986), *cert. denied*, 480 U.S. 934 (1987); *see, e.g.*, Covad Communications Co. v. BellSouth Corp., 299 F.3d 1272 (11th Cir. 2002), *vacated*, 124 S. Ct. 1143 (2004); *see also* Aspen Skiing Co. v. Aspen Highlands Skiing Corp., 472 U.S. 585, 605 (1985) (noting that antitrust laws prevent firms from "exclud[ing] rivals on some basis other than efficiency") (internal quotations omitted); Otter Tail v. United States, 410 U.S. 366, 380 (1973) (mandating cooperation from an incumbent electric utility to facilitate entry). *But see* Dennis W. Carlton, *A General Analysis of Exclusionary Conduct and Refusal to Deal—Why Aspen and Kodak Are Misguided*, 68 Antitrust L.J. 659, 659 (2001) (criticizing such theories).

31. *Trinko*, 540 U.S. at 407.

32. *See also* Pacific Bell Tel. Co. v. LinkLine Commc'ns, Inc., 129 S. Ct. 1109 (2009) (rejecting price-squeeze theory of liability).

33. *Trinko*, 540 U.S. at 408, 414 (internal quotation marks omitted).

34. *Id.* at 408.

35. *Id.* at 411.

36. Credit Suisse Sec. (USA) LLC v. Billing, 551 U.S. 264 (2007).

37. Jonathan E. Nuechterlein, *Antitrust Oversight of an Antitrust Dispute: An Institutional Perspective on the Net Neutrality Debate*, 7 J. Telecomm. & High Tech. L. 19 (2009). The FTC has historically played little role in the development of the telecommunications industry because in 1914 Congress fenced off from the FTC's jurisdiction the substantive subject areas assigned to other regulatory agencies. *See generally* FTC Staff Report, *Broadband Connectivity Competition Policy*, at 38–42 (June 27, 2007), http://www.ftc.gov/reports/broadband/

v070000report.pdf ("*FTC Net Neutrality Report*"). Here, section 5 of the Federal Trade Commission Act prohibits the FTC from exercising authority over "common carriers subject to the Acts to regulate commerce," 15 U.S.C. § 45(a)(2), a category that includes the later enacted Communications Act of 1934, *see* 15 U.S.C. § 44. In a 2007 report, however, the FTC contended that because the FCC does not treat broadband Internet access providers as "common carriers" (see chapter 6), this "common carrier exemption" no longer applies to the Internet access services those providers offer. *FTC Net Neutrality Report* at 38, 43–47. Significantly, the FTC has generalized authority to issue cease-and-desist orders against business practices that agency deems "unfair methods of competition" under section 5, 15 U.S.C. § 45(a)(1), even when a given practice "'does not infringe either the letter or spirit of the antitrust laws.'" *Negotiated Data Solutions LLC, Analysis of Proposed Consent Order to Aid Public Comment*, File No. 0510094, at 4 (FTC 2008) (quoting FTC v. Sperry & Hutchinson Co., 405 U.S. 233, 239 (1972)). The precise limits of that authority are unclear. *See generally* E.I. du Pont de Nemours & Co. v. FTC, 729 F.2d 128, 138–40 (2d Cir. 1984).

38. *See* Philip J. Weiser, *The Future of Internet Regulation*, 43 U.C. Davis L. Rev. 529 (2009).

39. *See id.*

40. Joe Waz & Phil Weiser, *Internet Governance: The Role of Multistakeholder Organizations*, Silicon Flatirons Roundtable Series on Entrepreneurship, Innovation, and Public Policy, at 8 (2012) (footnote omitted), http://siliconflatirons.com/documents/publications/report/InternetGovernanceRoleofMSHOrgs.pdf.

41. *See id.*; Philip J. Weiser, *Internet Governance, Standard Setting, and Self-Regulation*, 28 N. Ky. L. Rev. 822 (2001); *see also* Vint Cerf, *IETF and the Internet Society*, Internet Society (July 18, 1995), http://www.internetsociety.org/internet/what-internet/history-internet/ietf-and-internet-society; P. Hoffman & S. Harris, *The Tao of IETF: A Novice's Guide to the Internet Engineering Task Force*, RFC 4677 (Sept. 2006), http://www.rfc-editor.org/rfc/rfc4677.txt. See chapter 7 for a discussion of NANOG's potential role in Internet peering disputes.

42. American Broad. Co. v. FCC, 191 F.2d 492, 501 (D.C. Cir. 1951); *see also* In re Core Communications, Inc., 531 F.3d 849, 850, 861 (D.C. Cir. 2008) (issuing mandamus to remedy an "egregious" delay in "responding to our remand" and concluding: "Having repeatedly, and mistakenly, put our faith in the Commission, we will not do so again"); Radio–Television News Dirs. Ass'n v. FCC, 229 F.3d 269, 272 (D.C. Cir. 2000) (issuing writ of mandamus where the Commission "failed to act for nine months" after "acknowledg[ing] the need for a prompt decision," and "its response consists of an order that further postpones a final decision without any assurance of a final decision"); In re Monroe Communications Corp., 840 F.2d 942, 945 (D.C. Cir. 1988) (noting that "an undesirably large amount of time has passed during this [FCC] proceeding; the three years of administrative limbo following the *Initial Decision* have benefited neither the parties nor the public"); Sierra Club v. Thomas, 828 F.2d 783, 795 (D.C. Cir. 1987) (noting that "[t]he classic example of [delay depriving parties of rights granted by Congress] is the undue length of rate proceedings conducted

by the Federal Communications Commission," which "deprive[s] ratepayers of their statutory right to [just and reasonable] rates"); Southern Pac. Communications Co. v. Am. Tel. and Tel. Co., 740 F.2d 980, 1000 (D.C. Cir. 1984) ("At minimum, long regulatory delays often have preceded final FCC approval or disapproval of AT&T's allegedly predatory rates, refusals to interconnect, or unreasonable and discriminatory terms and conditions of access to local distribution facilities"); Telecommunications Research and Action Ctr. v. FCC, 750 F.2d 70, 80 (D.C. Cir. 1984) (noting "serious" delays and retaining jurisdiction over the case until final agency disposition "in light of the Commission's failure to meet its self-declared prior deadlines for these proceedings"); Nader v. FCC, 520 F.2d 182, 206–07 (D.C. Cir. 1975) (cautioning the Commission, again, "in the strongest terms" about its "dilatory pace" because court "foresee[s] the breakdown of the regulatory process if the public and the regulated carriers must wait as long as ten years to have important issues decided").

43. *See* 47 U.S.C. § 251(d)(1).

44. *Id.* § 271(d)(3). Beneath the surface, those proceedings lasted much longer than 90 days. A Bell company often spent months or years winning state commission support and then obtaining the FCC's own informal indication that a section 271 application would receive serious consideration. Then, if concerns remained that the FCC considered too serious to resolve by the deadline, the Bell company typically withdrew the application ("voluntarily") and refiled it with updated information, thereby restarting the 90-day clock.

45. 47 U.S.C. § 160(c). See Sprint-Nextel Corp. v. FCC, 508 F.3d 1129, 1132 (D.C. Cir. 2007) (concluding that forbearance was granted by operation of law when a short-handed FCC split two–two on a forbearance petition). We discuss this forbearance provision in chapter 2.

46. The D.C. Circuit has repeatedly rejected various rationales the FCC has devised for avoiding resolution of forbearance petitions on the merits. *See* AT&T Inc. v. FCC, 452 F.3d 830 (D.C. Cir. 2006) ("*AT&T II*"); Verizon Tel. Cos. v. FCC, 374 F.3d 1229 (D.C. Cir. 2004); AT&T Corp. v. FCC, 236 F.3d 729 (D.C. Cir. 2001). But that court has declined to require the FCC to resolve remanded forbearance petitions on any particular timetable. *See* Verizon Tel. Cos. v. FCC, 570 F.3d 294, 305 (D.C. Cir. 2009). And the FCC in fact lets those remand proceedings drag: for example, the Commission has still not acted on the Court's 2006 remand in *AT&T II*, which invalidated the FCC's procedural rationales for declining to rule on a petition seeking forbearance from Title II regulation of any IP platform service.

47. *See* Memorandum from the Majority Staff of the House Comm. on Energy and Commerce to the Members of the Subcomm. on Communications and Tech., Re: Subcomm. Hearing on "Reforming the FCC Process," at 3 (June 30, 2011), http://republicans.energycommerce.house.gov/Media/file/Hearings/Telecom/062211%20FCC%20Process%20Reform/Memo.pdf.

48. 47 U.S.C. § 154(b)(5), (c) (specifying rules on party affiliation and term of office).

49. *See, e.g.*, JAMES M. LANDIS, REPORT ON REGULATORY AGENCIES TO THE PRESIDENT-ELECT (1960).

50. Some categories of orders can be appealed only to the D.C. Circuit, whereas the remainder can be appealed either to that court or to any of the eleven regional circuits in which the appealing party resides or has its principal office. 47 U.S.C. § 402; 28 U.S.C. § 2343. Under 28 U.S.C. § 2342, known as the Hobbs Act, challenges to FCC orders *must* be filed in a court of appeals; they may not be filed in federal district court. That provision is generally construed to mean that when one private party sues another in district court, that court must assume the statutory validity of any FCC order that has not been vacated by a reviewing court of appeals. *See* US West Communications, Inc. v. Hamilton, 224 F.3d 1049, 1054–55 (9th Cir. 2000); Wilson v. A.H. Belo Corp., 87 F.3d 393, 400 (9th Cir. 1996); *see generally* FCC v. ITT World Communications, Inc., 466 U.S. 463, 468 (1984).

51. 467 U.S. 837, 866 (1984); *see also* Weiser, *Federal Common Law*, at 1715–18.

52. *See* 5 U.S.C. § 706(2) (Administrative Procedure Act; directing courts to invalidate, inter alia, agency actions that are "arbitrary, capricious, an abuse of discretion, or otherwise not in accordance with law").

53. Weiser, *Federal Common Law*, at 1725 n.177 (discussing, among other things, the Tenth Circuit's approach in *Qwest Corp. v. FCC*, 258 F.3d 1191, 1199–202 (10th Cir. 2001)).

54. *See* Cass R. Sunstein & Adrian Vermeule, *Interpretation and Institutions*, 101 MICH. L. REV. 885, 926 (2003) ("We think that the best defenses of *Chevron* attempt to read ambiguous congressional instructions in a way that is well-attuned to institutional considerations").

55. *See* chapter 2 (discussing the *Iowa Utilities Board* litigation).

56. Gulf Power Co. v. FCC, 208 F.3d 1263 (11th Cir. 2000), *rev'd sub nom.* National Cable & Telecommunications Ass'n, Inc. v. Gulf Power Co., 534 U.S. 327 (2002).

57. *See* chapter 6 (discussing the *Brand X* litigation).

58. The Act does not literally conscript the state agencies into service. Under section 252(e)(5), any state is free to opt out of this entire framework for implementing the local competition provisions, in which event the FCC stands in the state's shoes, sets the wholesale rates, and resolves any other disputes. So far, however, states have only rarely declined to participate in this regulatory scheme. Starpower Communications LLC v. FCC, 334 F.3d 1150 (D.C. Cir. 2003) (reviewing FCC decision rendered in the place of the Virginia State Corporation Commission).

59. Weiser, *Federal Common Law*, at 1731–33.

60. Jonathan E. Nuechterlein, *Incentives to Speak Honestly about Incentives: The Need for Structural Reform of the Local Competition Debate*, 2 J. TELECOMM. & HIGH TECH. L. 399, 402–05 (2003).

61. *See generally* Philip J. Weiser, *Reexamining the Legacy of Dual Regulation: Reforming Dual Merger Review by the DOJ and the FCC*, 61 FED. COMM. L.J. 167 (2008). By longstanding tradition, the Justice Department rather than the FTC reviews mergers between telecommunications carriers. But the FTC has taken the lead in reviewing several major mergers in the communications industry involving firms other than telecommunications carriers, such as Time Warner's merger with AOL in 2000–2001. Open questions about which of these two agencies should review a particular merger are generally decided after informal interagency consultation.

62. 15 U.S.C. § 18.

63. *See, e.g.*, United States v. Marine Bancorporation, Inc., 418 U.S. 602 (1974) (turning away potential competition argument).

64. Joel I. Klein, Assistant Attorney General, *Making The Transition from Regulation to Competition: Thinking about Merger Policy during the Process of Electric Power Restructuring* (Jan. 21, 1998), http://www.justice.gov/atr/public/speeches/1332.htm. The major post-1996 ILEC–ILEC combinations include SBC and Pacific Telesis (1996), Bell Atlantic and NYNEX (1997), SBC and Ameritech (1999), Bell Atlantic and GTE (2000, forming Verizon), AT&T (i.e., SBC) and BellSouth (2007), and CenturyLink and Qwest (2011).

65. 47 U.S.C. §§ 214(a), 310(d) (public interest authority to review license transfers); *see* 15 U.S.C. § 21(a) (FCC authorized to act under Clayton Act); James R. Weiss & Martin L. Stern, *Serving Two Masters: The Dual Jurisdiction of the FCC and the Justice Department over Telecommunications Transactions*, 6 COMM. L. CONSPECTUS 195, 198 (1998) (noting that the FCC rarely exercises its Clayton Act authority).

66. *See, e.g.*, Memorandum Opinion & Order, *Application of Nynex Corp., Transferor, and Bell Atlantic Corp., Transferee*, 12 FCC Rcd 19985, ¶ 7 (1997).

67. 15 U.S.C. § 25.

68. *See, e.g.*, Memorandum Opinion & Order, *AT&T Inc. and BellSouth Corporation Application for Transfer of Control*, 22 FCC Rcd 5662, ¶ 19 (2007).

69. *See, e.g., id.*, Appx. F, at 5814–15; Memorandum Opinion & Order, *Applications of Comcast Corp., General Elec. Corp., and NBC Universal, Inc. for Consent to Assign Licenses and Transfer Control of Licenses*, 26 FCC Rcd 4238, Appx. A (2011) ("*Comcast-NBCU Order*").

70. *See generally* Bryan N. Tramont, *Too Much Power, Too Little Restraint: How the FCC Expands Its Reach through Unenforceable and Unwieldy "Voluntary" Agreements*, 53 FED. COMM. L.J. 49 (2000). For a recent and particularly graphic example of this phenomenon, see *Comcast-NBCU Order*, Appx. A, §§ X–XIII (imposing various "diversity," "localism," "journalistic independence," and "children's programming" conditions in exchange for approval of Comcast–NBCU merger).

71. *See, e.g.*, Rachel E. Barkow & Peter W. Huber, *A Tale of Two Agencies: A Comparative Analysis of FCC and DOJ Review of Telecommunications Mergers*, 2000 U. CHI. LEGAL F. 29; Harold W. Furchtgott-Roth, Testimony before the

Antitrust Modernization Comm'n (Dec. 5, 2005), http://govinfo.library.unt.edu/amc/commission_hearings/pdf/Furchtgott_Roth_statement.pdf; Tramont, *Too Much Power, Too Little Restraint.*

72. Memorandum Opinion & Order, *Applications of Ameritech Corp., Transferor, and SBC Communications, Inc., Transferee*, 14 FCC Rcd 14712, 15197 (1999) (statement of Commissioner Michael K. Powell, concurring in part and dissenting in part).

73. *Id.* at 15197, 15201.

74. *See* Hearing Designation Order, *Application of EchoStar Communications Corp., General Motors Corp., and Hughes Elec. Corp. (Transferors) to EchoStar Communications Corp. (Transferee)*, 17 FCC Rcd 20559 (2002). In 2010, despite this commitment, the FCC's International Bureau unexpectedly conditioned Harbinger's acquisition of SkyTerra (later LightSquared) on the company's promise to avoid certain wholesale spectrum arrangements with the two largest wireless providers: AT&T and Verizon. *See* Memorandum Opinion & Order, *SkyTerra Communications, Inc., Transferor, and Harbinger Capital Partners Funds, Transferee, Applications for Consent to Transfer of Control of SkyTerra Subsidiary, LLC*, 25 FCC Rcd 3059, Appx. B, Attach. 2 (IB 2010). AT&T and Verizon cried foul, but the controversy diminished in importance when it became clear that interference concerns would likely keep LightSquared from providing major wholesale arrangements with any mobile wireless provider (see chapter 3).

75. *See* Donald J. Russell & Sherri Lynn Wolson, *Dual Antitrust Review of Telecommunications Mergers by the Department of Justice and the Federal Communications Commission*, 11 Geo. Mason L. Rev. 143 (2002); Barkow & Huber, *A Tale of Two Agencies.*

76. *See* Phil Weiser, *Paradigm Changes in Telecommunications Regulation*, 71 U. Colo. L. Rev. 819, 839–40 (2000) (contrasting approach of Illinois commission in the SBC–Ameritech merger with that of the New York commission in the Bell Atlantic–NYNEX merger); *see also* Ill. Bell Tel. Co. v. Ill. Commerce Comm'n, 816 N.E.2d 379 (Ill. App. 2004) (invalidating a condition of merger approval).

77. *See, e.g.*, Memorandum Opinion & Order, *General Motors Corp. and Hughes Electronic Corp., Transferors, and News Corp. Ltd., Transferee*, 19 FCC Rcd 473 (2004).

78. *See* 47 U.S.C. § 572 (d)(6)(A)(iii) (limiting cable–telephone company mergers to situations where the FCC concludes that "the anticompetitive effects of the proposed transaction are clearly outweighed in the public interest by the probable effect of the transaction in meeting the convenience and needs of the community to be served"); *see also* 141 Cong. Rec. S8464 (June 15, 1995) (Senator Leahy) (explaining rationale for the heightened standard for proposed combinations between cable and telephone companies on the ground that allowing "telephone companies to buy out cable companies—their most likely competitor—in the telephone companies' local service areas . . . would destroy the best hope of developing competition in both local telephone service and cable television markets").

79. *See* Weiser, *Reexamining the Legacy of Dual Regulation.*

80. *See* 47 C.F.R. § 1.1206.

81. *See* Report & Order and Further Notice of Proposed Rulemaking, *Amendment of the Commission's Ex Parte Rules and Other Procedural Rules*, 26 FCC Rcd 4517 (2011).

82. Those who favor greater disclosure argue that the FCC is unusual among federal agencies in that it allocates responsibility to lobbyists rather than to the lobbied agency to make public disclosure of their lobbying. *See*, *e.g.*, Comments of Marcus Spectrum Solutions LLC, GC Docket No. 10-43, at 2 (FCC Mar. 29, 2010) ("[O]nly the Commission has an ex parte system that depends entirely on statements filed by outside parties as the sole source of information in its docket files. As far as we can determine, every other federal agency involved in such rulemakings uses a notification prepared by its own staff, possibly with the inclusion of material provided by the outside party. . . . [This] is a root cause of the current problem. Outside parties do not always have the same transparency goals as the Commission"); Mike Marcus, *New FCC ex parte Rules: A Review*, Public Knowledge (Feb. 3, 2011), http://www.publicknowledge.org/blog/new-fcc-ex-parte-rules-review.

83. *See* 47 C.F.R. § 1.1208. Sometimes—as when no proceeding has formally begun—the FCC exempts communications from any disclosure rules at all. *See* 47 C.F.R. § 1.1204.

84. *See* Philip J. Weiser, *Institutional Design, FCC Reform, and the Hidden Side of the Administrative State*, 61 ADMIN. L. REV. 675 (2009); *see also* Philip J. Weiser, *Toward an International Dialogue on the Institutional Side of Antitrust*, 66 NYU ANN. SURVEY OF AM. L. 101 (2010).

85. 5 U.S.C. §552b(e)(1); *see also* 5 U.S.C. §552b(c) & (d)(1) (allowing closed meetings only when a majority of a commission votes to hold one); *see* Jim Rossi, *Participation Run Amok: The Costs of Mass Participation for Deliberative Agency Decisionmaking*, 92 NW. U. L. REV. 173 (1997).

86. Schurz Communications Inc. v. FCC, 982 F.2d 1043, 1050 (7th Cir. 1992).

87. Philip J. Weiser, *Regulatory Challenges and Models of Regulation*, 2 J. TELECOMM. & HIGH TECH. L. 1, 14–15 (2003); Weiser, *Paradigm Changes*, at 837–38.

88. KAHN, LETTING GO, at 70 (1998).

89. First Report & Order, *Rules Re Microwave-Served CATV*, 50 F.C.C. 683 (1966).

90. The change in the regulatory mentality over the past two generations has been nothing short of revolutionary. As Harold Demsetz summarized the old model of public utility regulation, "regulation has often been sought because of the inconvenience of competition." Harold Demsetz, *Why Regulate Utilities?* 11 J. L. & ECON. 55, 61 (1968).

Index

For sections of Title 47 of the U.S. Code, see Communications Act of 1934.